Springer Texts in Statistics

Advisors:
George Casella   Stephen Fienberg   Ingram Olkin

Springer
New York
Berlin
Heidelberg
Barcelona
Hong Kong
London
Milan
Paris
Singapore
Tokyo

Springer Texts in Statistics

Alfred: Elements of Statistics for the Life and Social Sciences
Berger: An Introduction to Probability and Stochastic Processes
Bilodeau and Brenner: Theory of Multivariate Statistics
Blom: Probability and Statistics: Theory and Applications
Brockwell and Davis: Introduction to Times Series and Forecasting, Second Edition
Chow and Teicher: Probability Theory: Independence, Interchangeability, Martingales, Third Edition
Christensen: Advanced Linear Modeling: Multivariate, Time Series, and Spatial Data—Nonparametric Regression and Response Surface Maximization, Second Edition
Christensen: Log-Linear Models and Logistic Regression, Second Edition
Christensen: Plane Answers to Complex Questions: The Theory of Linear Models, Second Edition
Creighton: A First Course in Probability Models and Statistical Inference
Davis: Statistical Methods for the Analysis of Repeated Measurements
Dean and Voss: Design and Analysis of Experiments
du Toit, Steyn, and Stumpf: Graphical Exploratory Data Analysis
Durrett: Essentials of Stochastic Processes
Edwards: Introduction to Graphical Modelling, Second Edition
Finkelstein and Levin: Statistics for Lawyers
Flury: A First Course in Multivariate Statistics
Jobson: Applied Multivariate Data Analysis, Volume I: Regression and Experimental Design
Jobson: Applied Multivariate Data Analysis, Volume II: Categorical and Multivariate Methods
Kalbfleisch: Probability and Statistical Inference, Volume I: Probability, Second Edition
Kalbfleisch: Probability and Statistical Inference, Volume II: Statistical Inference, Second Edition
Karr: Probability
Keyfitz: Applied Mathematical Demography, Second Edition
Kiefer: Introduction to Statistical Inference
Kokoska and Nevison: Statistical Tables and Formulae
Kulkarni: Modeling, Analysis, Design, and Control of Stochastic Systems
Lehmann: Elements of Large-Sample Theory
Lehmann: Testing Statistical Hypotheses, Second Edition
Lehmann and Casella: Theory of Point Estimation, Second Edition
Lindman: Analysis of Variance in Experimental Design
Lindsey: Applying Generalized Linear Models
Madansky: Prescriptions for Working Statisticians
McPherson: Applying and Interpreting Statistics: A Comprehensive Guide, Second Edition
Mueller: Basic Principles of Structural Equation Modeling: An Introduction to LISREL and EQS
Nguyen and Rogers: Fundamentals of Mathematical Statistics: Volume I: Probability for Statistics

(continued after index)

Colin Rose   Murray D. Smith

Mathematical Statistics with *Mathematica*®

With 134 Figures

 Springer

Colin Rose
Theoretical Research Institute
Sydney NSW 2023
Australia
colin@tri.org.au

Murray D. Smith
Discipline of Econometrics and Business Statistics
University of Sydney
Sydney NSW 2006
Australia
murray.smith@econ.usyd.edu.au

Editorial Board

George Casella
Department of Statistics
University of Florida
Gainesville, FL 32611-8545
USA

Stephen Fienberg
Department of Statistics
Carnegie Mellon University
Pittsburgh, PA 15213-3890
USA

Ingram Olkin
Department of Statistics
Stanford University
Stanford, CA 94305
USA

Library of Congress Cataloging-in-Publication Data
Rose, Colin.
  Mathematical statistics with Mathematica / Colin Rose, Murray D. Smith.
    p. cm. — (Springer texts in statistics)
  Includes bibliographical references and index.
  ISBN 0-387-95234-9 (alk. paper)
    1. Mathematical statistics—Data processing. 2. Mathematica (Computer file) I. Smith, Murray D. II. Title. III. Series.
  QA276.4 .R67   2001
  519.5'0285—dc21   00-067926

Printed on acid-free paper.

Mathematica is a registered trademark of Wolfram Research, Inc.

© 2002 Springer-Verlag New York, Inc.
This Work consists of a printed book and two CD-ROMs packaged with the book, all of which are protected by federal copyright law and international treaty. The book may not be translated or copied in whole or in part without the written permission of the publisher (Springer-Verlag New York, Inc., 175 Fifth Avenue, New York, NY 10010, USA), except for brief excerpts in connection with reviews or scholarly analysis. For copyright information regarding the CD-ROMs, please consult the printed information below the CD-ROM package in the back of this publication. Use of this Work in connection with any form of information storage and retrieval, electronic adaptation, computer software, or by similar or dissimilar methodology now known, or hereafter developed, other than as expressly granted in the CD-ROM copyright notice and disclaimer information, and as expressly limited by applicable law, is strictly forbidden.
The use in this publication of trade names, trademarks, service marks and similar terms, even if they are not identified as such, is not to be taken as an expression of opinion as to whether or not they are subject to proprietary rights. Where those designations appear in the book and Springer-Verlag was aware of a trademark claim, the designations follow the capitalization style used by the manufacturer.

Production managed by Steven Pisano; manufacturing supervised by Joe Quatela.
Photocomposed pages prepared from the authors' *Mathematica* files.
Printed and bound by Hamilton Printing Co., Rensselaer, NY.
Printed in the United States of America.

9 8 7 6 5 4 3 2 1

ISBN 0-387-95234-9   SPIN 10659720

Springer-Verlag   New York Berlin Heidelberg
A member of BertelsmannSpringer Science+Business Media GmbH

Contents

Preface   xi

Chapter 1   Introduction

1.1	Mathematical Statistics with *Mathematica*	1
	A A New Approach	1
	B Design Philosophy	1
	C If You Are New to *Mathematica*	2
1.2	Installation, Registration and Password	3
	A Installation, Registration and Password	3
	B Loading **mathStatica**	5
	C Help	5
1.3	Core Functions	6
	A Getting Started	6
	B Working with Parameters	8
	C Discrete Random Variables	9
	D Multivariate Random Variables	11
	E Piecewise Distributions	13
1.4	Some Specialised Functions	15
1.5	Notation and Conventions	24
	A Introduction	24
	B Statistics Notation	25
	C *Mathematica* Notation	27

Chapter 2   Continuous Random Variables

2.1	Introduction	31
2.2	Measures of Location	35
	A Mean	35
	B Mode	36
	C Median and Quantiles	37
2.3	Measures of Dispersion	40
2.4	Moments and Generating Functions	45
	A Moments	45
	B The Moment Generating Function	46
	C The Characteristic Function	50
	D Properties of Characteristic Functions (and mgf's)	52

	E	Stable Distributions	56
	F	Cumulants and Probability Generating Functions	60
	G	Moment Conversion Formulae	62
2.5		Conditioning, Truncation and Censoring	65
	A	Conditional/Truncated Distributions	65
	B	Conditional Expectations	66
	C	Censored Distributions	68
	D	Option Pricing	70
2.6		Pseudo-Random Number Generation	72
	A	*Mathematica*'s Statistics Package	72
	B	Inverse Method (Symbolic)	74
	C	Inverse Method (Numerical)	75
	D	Rejection Method	77
2.7		Exercises	80

Chapter 3   Discrete Random Variables

3.1		Introduction	81
3.2		Probability: 'Throwing' a Die	84
3.3		Common Discrete Distributions	89
	A	The Bernoulli Distribution	89
	B	The Binomial Distribution	91
	C	The Poisson Distribution	95
	D	The Geometric and Negative Binomial Distributions	98
	E	The Hypergeometric Distribution	100
3.4		Mixing Distributions	102
	A	Component-Mix Distributions	102
	B	Parameter-Mix Distributions	105
3.5		Pseudo-Random Number Generation	109
	A	Introducing `DiscreteRNG`	109
	B	Implementation Notes	113
3.6		Exercises	115

Chapter 4   Distributions of Functions of Random Variables

4.1		Introduction	117
4.2		The Transformation Method	118
	A	Univariate Cases	118
	B	Multivariate Cases	123
	C	Transformations That Are *Not* One-to-One; Manual Methods	127
4.3		The MGF Method	130
4.4		Products and Ratios of Random Variables	133
4.5		Sums and Differences of Random Variables	136
	A	Applying the Transformation Method	136
	B	Applying the MGF Method	141
4.6		Exercises	147

Chapter 5 Systems of Distributions

5.1	Introduction	149
5.2	The Pearson Family	149
	A Introduction	149
	B Fitting Pearson Densities	151
	C Pearson Types	157
	D Pearson Coefficients in Terms of Moments	159
	E Higher Order Pearson-Style Families	161
5.3	Johnson Transformations	164
	A Introduction	164
	B S_L System (Lognormal)	165
	C S_U System (Unbounded)	168
	D S_B System (Bounded)	173
5.4	Gram–Charlier Expansions	175
	A Definitions and Fitting	175
	B Hermite Polynomials; Gram–Charlier Coefficients	179
5.5	Non-Parametric Kernel Density Estimation	181
5.6	The Method of Moments	183
5.7	Exercises	185

Chapter 6 Multivariate Distributions

6.1	Introduction	187
	A Joint Density Functions	187
	B Non-Rectangular Domains	190
	C Probability and `Prob`	191
	D Marginal Distributions	195
	E Conditional Distributions	197
6.2	Expectations, Moments, Generating Functions	200
	A Expectations	200
	B Product Moments, Covariance and Correlation	200
	C Generating Functions	203
	D Moment Conversion Formulae	206
6.3	Independence and Dependence	210
	A Stochastic Independence	210
	B Copulae	211
6.4	The Multivariate Normal Distribution	216
	A The Bivariate Normal	216
	B The Trivariate Normal	226
	C CDF, Probability Calculations and Numerics	229
	D Random Number Generation for the Multivariate Normal	232
6.5	The Multivariate t and Multivariate Cauchy	236
6.6	Multinomial and Bivariate Poisson	238
	A The Multinomial Distribution	238
	B The Bivariate Poisson	243
6.7	Exercises	248

Chapter 7 Moments of Sampling Distributions

7.1	Introduction	251
	A Overview	251
	B Power Sums and Symmetric Functions	252
7.2	Unbiased Estimators of Population Moments	253
	A Unbiased Estimators of Raw Moments of the Population	253
	B h-statistics: Unbiased Estimators of Central Moments	253
	C k-statistics: Unbiased Estimators of Cumulants	256
	D Multivariate h- and k-statistics	259
7.3	Moments of Moments	261
	A Getting Started	261
	B Product Moments	266
	C Cumulants of k-statistics	267
7.4	Augmented Symmetrics and Power Sums	272
	A Definitions and a Fundamental Expectation Result	272
	B Application 1: Understanding Unbiased Estimation	275
	C Application 2: Understanding Moments of Moments	275
7.5	Exercises	276

Chapter 8 Asymptotic Theory

8.1	Introduction	277
8.2	Convergence in Distribution	278
8.3	Asymptotic Distribution	282
8.4	Central Limit Theorem	286
8.5	Convergence in Probability	292
	A Introduction	292
	B Markov and Chebyshev Inequalities	295
	C Weak Law of Large Numbers	296
8.6	Exercises	298

Chapter 9 Statistical Decision Theory

9.1	Introduction	301
9.2	Loss and Risk	301
9.3	Mean Square Error as Risk	306
9.4	Order Statistics	311
	A Definition and `OrderStat`	311
	B Applications	318
9.5	Exercises	322

Chapter 10 Unbiased Parameter Estimation

10.1	Introduction	325
	A Overview	325
	B `SuperD`	326

10.2	Fisher Information	326
	A Fisher Information	326
	B Alternate Form	329
	C Automating Computation: `FisherInformation`	330
	D Multiple Parameters	331
	E Sample Information	332
10.3	Best Unbiased Estimators	333
	A The Cramér–Rao Lower Bound	333
	B Best Unbiased Estimators	335
10.4	Sufficient Statistics	337
	A Introduction	337
	B The Factorisation Criterion	339
10.5	Minimum Variance Unbiased Estimation	341
	A Introduction	341
	B The Rao–Blackwell Theorem	342
	C Completeness and MVUE	343
	D Conclusion	346
10.6	Exercises	347

Chapter 11   Principles of Maximum Likelihood Estimation

11.1	Introduction	349
	A Review	349
	B `SuperLog`	350
11.2	The Likelihood Function	350
11.3	Maximum Likelihood Estimation	357
11.4	Properties of the ML Estimator	362
	A Introduction	362
	B Small Sample Properties	363
	C Asymptotic Properties	365
	D Regularity Conditions	367
	E Invariance Property	369
11.5	Asymptotic Properties: Extensions	371
	A More Than One Parameter	371
	B Non-identically Distributed Samples	374
11.6	Exercises	377

Chapter 12   Maximum Likelihood Estimation in Practice

12.1	Introduction	379
12.2	`FindMaximum`	380
12.3	A Journey with `FindMaximum`	384
12.4	Asymptotic Inference	392
	A Hypothesis Testing	392
	B Standard Errors and *t*-statistics	395

12.5	Optimisation Algorithms	399
	A Preliminaries	399
	B Gradient Method Algorithms	401
12.6	The BFGS Algorithm	405
12.7	The Newton–Raphson Algorithm	412
12.8	Exercises	418

Appendix

A.1	Is That the Right Answer, Dr Faustus?	421
A.2	Working with Packages	425
A.3	Working with =, →, == and :=	426
A.4	Working with Lists	428
A.5	Working with Subscripts	429
A.6	Working with Matrices	433
A.7	Working with Vectors	438
A.8	Changes to Default Behaviour	443
A.9	Building Your Own **mathStatica** Function	446

Notes

447

References

463

Index

469

Preface

Imagine computer software that can find expectations of *arbitrary* random variables, calculate variances, invert characteristic functions, solve transformations of random variables, calculate probabilities, derive order statistics, find Fisher's Information and Cramér–Rao Lower Bounds, derive symbolic (exact) maximum likelihood estimators, perform automated moment conversions, and so on. Imagine that this software was wonderfully easy to use, and yet so powerful that it can find corrections to mainstream reference texts and solve new problems in seconds. Then, imagine a book that uses that software to bring mathematical statistics to life …

Why *Mathematica*?

Why "Mathematical Statistics <u>with</u> *Mathematica*"? Why not Mathematical Statistics with Gauss, SPSS, Systat, SAS, JMP or S-Plus … ? The answer is four-fold:

(i) *Symbolic engine*

Packages like Gauss, SPSS, *etc.* provide a numerical/graphical toolset. They can illustrate, they can simulate, and they can find approximate numerical solutions to numerical problems, but they cannot solve the algebraic/symbolic problems that are of primary interest in mathematical statistics. Like all the other packages, *Mathematica* also provides a numerical engine and superb graphics. But, over and above this, *Mathematica* has a powerful symbolic/algebraic engine that is ideally suited to solving problems in mathematical statistics.

(ii) *Notebook interface*

Mathematica enables one to incorporate text, pictures, equations, animations and computer input into a single interactive live document that is known as a 'notebook'. Indeed, this entire book was written, typeset and published using *Mathematica*. Consequently, this book exists in two identical forms: (a) a printed book that has all the tactile advantages of printed copy, and (b) an electronic book on the **mathStatica** CD-ROM (included)—here, every input is live, every equation is at the reader's fingertips, every diagram can be generated on the fly, every example can be altered, and so on. Equations are hyperlinked, footnotes pop-up, cross-references are live, the index is hyperlinked, online HELP is available, and animations are a mouse-click away.

(iii) *Numerical accuracy*

Whereas most software packages provide only finite-precision numerics, *Mathematica* also provides an arbitrary-precision numerical engine: if accuracy is important, *Mathematica* excels. As McCullough (2000, p. 296) notes, "By virtue of its variable precision arithmetic and symbolic power, *Mathematica*'s performance on these reliability tests far exceeds any finite-precision statistical package".

(iv) *Cross-platform and large user base*
Mathematica runs on a wide variety of platforms, including Mac, OS X, Windows, Linux, SPARC, Solaris, SGI, IBM RISC, DEC Alpha and HP–UX. This is especially valuable in academia, where co-authorship is common.

What is mathStatica?

mathStatica is a computer software package—an add-on to *Mathematica*—that provides a sophisticated toolset specially designed for doing mathematical statistics. It automatically solves the types of problems that researchers and students encounter, over and over again, in mathematical statistics. The **mathStatica** software is bundled free with this book (Basic version). It is intended for use by researchers and lecturers, as well as postgraduate and undergraduate students of mathematical statistics, in any discipline in which the theory of statistics plays a part.

Assumed Knowledge

How much statistics knowledge is assumed? How much *Mathematica* knowledge?

Statistics: We assume the reader has taken one year of statistics. The level of the text is generally similar to Hogg and Craig (1995). The focus, of course, is different, with less emphasis on theorems and proofs, and more emphasis on problem solving.

Mathematica: No experience is required. We do assume the reader has *Mathematica* installed (this book comes bundled with a fully-functional trial copy of *Mathematica*) and that the user knows how to evaluate 2+2, but that's about it. Of course, there are important *Mathematica* conventions, such as a knowledge of bracket types (), [], {}, which are briefly discussed in Chapter 1. For the new user, the best approach is to try a few examples and the rest usually follows by osmosis ☺.

As a Course Textbook

This book can be used as a course text in mathematical statistics or as an accompaniment to a more traditional text. We have tried to pitch the material at the level of Hogg and Craig (1995). Having said that, when one is armed with **mathStatica**, the whole notion of what is difficult changes, and so we can often extend material to the level of, say, Stuart and Ord (1991, 1994) without any increase in 'difficulty'. We assume that the reader has taken preliminary courses in calculus, statistics and probability. Our emphasis is on problem solving, with less attention paid to the presentation of theorems and their associated proofs, since the latter are well-covered in more traditional texts. We make no assumption about the reader's knowledge of *Mathematica*, other than that it is installed on their computer.

In the lecture theatre, lecturers can use **mathStatica** to remove a lot of the annoying technical calculation often associated with mathematical statistics. For example, instead of spending time and energy laboriously deriving, step by step, a nasty expectation using integration by parts, the lecturer can use **mathStatica** to calculate the same expectation in a few seconds, in front of the class. This frees valuable lecture time to either explore the topic in more detail, or to tackle other topics. For students, this book serves three roles: first, as a text in mathematical statistics; second, as an interactive medium to explore; third, as a tool for tackling problems set by their professors — the book comes complete with 101 exercises (a solution set for instructors is available at www.mathstatica.com).

mathStatica has the potential to enliven the educational experience. At the same time, it is not a panacea for all problems. Nor should it be used as a substitute for thinking. Rather, it is a substitute for mechanical and dreary calculation, hopefully freeing the reader to solve higher-order problems. Armed with this new and powerful toolset, we hope that others go on to solve ever more challenging problems with consummate ease.

Acknowledgements

Work on **mathStatica** began in 1995 for an invited chapter published in Varian (1996). As such, our first thanks go to Hal Varian for providing us with the impetus to start this journey, which has taken almost five years to complete. Thanks go to Allan Wylde for his encouragement at the beginning of this project. Combining a book (print and electronic) with software creates many unforeseen possibilities and complexities. Fortunately, our publisher, John Kimmel, Springer's Executive Editor for Statistics, has guided the project with a professional savoir faire and friendly warmth, both of which are most appreciated.

Both the book and the software have gone through a lengthy and extensive beta testing programme. B. D. McCullough, in particular, subjected **mathStatica** to the most rigorous testing, over a period of almost two years. Marc Nerlove tested **mathStatica** out on his undergraduate classes at the University of Maryland. Special thanks are also due to flip phillips and Ron Mittelhammer, and to Luci Ellis, Maxine Nelson and Robert Kieschnick. Paul Abbott and Rolf Mertig have been wonderful sounding boards.

We are most grateful to Wolfram Research for their support of **mathStatica**, in so many ways, including a Visiting Scholar Grant. In particular, we would like to thank Alan Henigman, Roger Germundsson, Paul Wellin and Todd Stevenson for their interest and support. On a more technical level, we are especially grateful to Adam Strzebonski for making life Simple[] even when the leaf count suggests it is not, to PJ Hinton for helping to make **mathStatica**'s palette technology deliciously 'palatable', and Theo Gray for tips on 101 front-end options. We would also like to thank André Kuzniarek, John Fultz, Robby Villegas, John Novak, Ian Brooks, Dave Withoff, Neil Soiffer, and Victor Adamchik for helpful discussions, tips, tweaks, the odd game of lightning chess, and 'Champaign' dinners, which made it all so much fun. Thanks also to Jamie Peterson for keeping us up to date with the latest and greatest. Finally, our families deserve special thanks for their encouragement, advice and patience.

Sydney, November 2001

Chapter 1

Introduction

1.1 Mathematical Statistics with *Mathematica*

1.1 A A New Approach

The use of computer software in statistics is far from new. Indeed, hundreds of statistical computer programs exist. Yet, underlying existing programs is almost always a numerical/graphical view of the world. *Mathematica* can easily handle the numerical and graphical sides, but it offers in addition an extremely powerful and flexible symbolic computer algebra system. The **mathStatica** software package that accompanies this book builds upon that symbolic engine to create a sophisticated toolset specially designed for doing mathematical statistics.

While the subject matter of this text is similar to a traditional mathematical statistics text, this is not a traditional text. The reader will find few proofs and comparatively few theorems. After all, the theorem/proof text is already well served by many excellent volumes on mathematical statistics. Nor is this a cookbook of numerical recipes bundled into a computer package, for there is limited virtue in applying *Mathematica* as a mere numerical tool. Instead, this text strives to bring mathematical statistics to life. We hope it will make an exciting and substantial contribution to the way mathematical statistics is both practised and taught.

1.1 B Design Philosophy

mathStatica has been designed with two central goals: it sets out to be **general**, and it strives to be **delightfully simple**.

By **general**, we mean that it should *not* be limited to a set of special or well-known textbook distributions. It should *not* operate like a textbook appendix with prepared 'crib sheet' answers. Rather, it should know how to solve problems from first principles. It should seamlessly handle: univariate and multivariate distributions, continuous and discrete random variables, and smooth and kinked densities—all with and without parameters. It should be able to handle mixtures, truncated distributions, reflected distributions, folded distributions, and distributions of functions of random variables, as well as distributions no-one has ever thought of before.

By **delightfully simple**, we mean both (i) easy to use, and (ii) able to solve problems that seem difficult, but which are formally quite simple. Consider, for instance, playing a devilish game of chess against a strong chess computer: in the middle of the game, after a short pause, the computer announces, "Mate in 16 moves". The problem it has solved might seem fantastically difficult, but it is really just a 'delightfully simple' finite problem that is conceptually no different than looking just two moves ahead. The salient point is that as soon as one has a tool for solving such problems, the notion of what is difficult changes completely. A pocket calculator is certainly a delightfully simple device: it is easy to use, and it can solve tricky problems that were previously thought to be difficult. But today, few people bother to ponder at the marvel of a calculator any more, and we now generally spend our time either using such tools or trying to solve higher-order conceptual problems—and so, we are certain, it will be with mathematical statistics too.

In fact, while much of the material traditionally studied in mathematical statistics courses may appear difficult, such material is often really just delightfully simple. Normally, all we want is an expectation, or a probability, or a transformation. But once we are armed with say a computerised expectation operator, we can find any kind of expectation including the mean, variance, skewness, kurtosis, mean deviation, moment generating function, characteristic function, raw moments, central moments, cumulants, probability generating function, factorial moment generating function, entropy, and so on. Normally, many of these calculations are not attempted in undergraduate texts, because the mechanics are deemed too hard. And yet, underlying all of them is just the delightfully simple expectation operator.

1.1 C   If You Are New to *Mathematica*

For those readers who do not own a copy of *Mathematica*, this book comes bundled with a free trial copy of *Mathematica* Version 4. This will enable you to use **mathStatica**, and try out and evaluate all the examples in this book.

If you have never used *Mathematica* before, we recommend that you first read the opening pages of Wolfram (1999) and run through some examples. This will give you a good feel for *Mathematica*. Second, new users should learn how to enter formulae into *Mathematica*. This can be done via palettes, see

File Menu ▷ Palettes ▷ BasicInput,

or via the keyboard (see §1.5 below), or just by copy and pasting examples from this book. Third, both new and experienced readers may benefit from browsing Appendices A.1 to A.7 of this book, which cover a plethora of tips and tricks.

Before proceeding further, please ensure that *Mathematica* Version 4 (or later) is installed on your computer.

1.2 Installation, Registration and Password

1.2 A Installation, Registration and Password

Before starting, please make sure you have a working copy of *Mathematica* Version 4 (or later) installed on your computer.

Installing **mathStatica** is an easy 4-step process, irrespective of whether you use a Macintosh, Windows, or a flavour of UNIX.

Step 1: Insert the **mathStatica** CD-ROM into your computer.

Step 2: Copy the following files:

    (i)   `mathStatica.m`   (file)
    (ii) `mathStatica`   (folder/directory)

from the **mathStatica** CD-ROM *into* the

    `Mathematica ▷ AddOns ▷ Applications`

folder on your computer's hard drive. The installation should look something like Fig. 1.

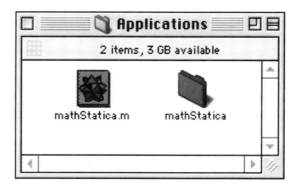

Fig. 1: Typical installation of **mathStatica**

Step 3: Get a password

To use **mathStatica**, you will need a password. To get a password, you will need to register your copy of **mathStatica** at the following web site:

    `www.mathstatica.com`

mathStatica is available in two versions: Basic and Gold. The differences are summarised in Table 1; for full details, see the web site.

class	description
Basic	• Fully functional **mathStatica** package code
	• Included on the CD-ROM
	• FREE to buyers of this book
	• Single-use license
Gold	• All the benefits of Basic, *plus* ...
	• *Continuous* and *Discrete* Distribution Palettes
	• Detailed interactive HELP system
	• Upgrades
	• Technical support
	• and more ...

Table 1: mathStatica—Basic and Gold

Once you have registered your copy, you will be sent a password file called: `pass.txt`. Put this file into the `Mathematica ▷ AddOns ▷ Applications ▷ mathStatica ▷ Password` directory, as shown in Fig. 2.

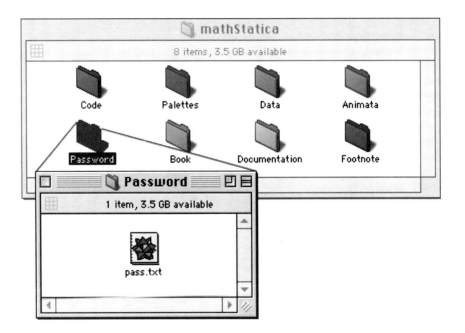

Fig. 2: Once you have received `"pass.txt"`, put it into the `Password` folder

Step 4: Run *Mathematica*, go to its HELP menu, and select: "Rebuild Help Index"

That's it—all done. Your installation is now complete.

1.2 B Loading mathStatica

If everything is installed correctly, first start up *Mathematica* Version 4 or later. Then **mathStatica** can be loaded by evaluating:

    `<< mathStatica.m`

or by clicking on a button such as this one: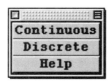

The **Book** palette should then appear, as shown in Fig. 3 (right panel). The **Book** palette provides a quick and easy way to access the electronic version of this book, including the live hyperlinked index. If you have purchased the Gold version of **mathStatica**, then the **mathStatica** palette will also appear, as shown in Fig. 3 (left panel). This provides the **Continuous** and **Discrete** distribution palettes (covering 37 distributions), as well as the detailed **mathStatica Help** system (complete with hundreds of extra examples).

Fig. 3: The **mathStatica** palette (left) and the **Book** palette (right)

WARNING: To avoid so-called 'context' problems, **mathStatica** should always be loaded from a fresh *Mathematica* kernel. If you have already done some calculations in *Mathematica*, you can get a fresh kernel by either typing `Quit` in an Input cell, or by selecting Kernel Menu ▷ Quit Kernel.

1.2 C Help

Both Basic Help and Detailed Help are available for any **mathStatica** function:

(i)   Basic Help is shown in Table 2.

function	description
? Name	show information on Name

Table 2: Basic Help on function names

For example, to get Basic Help on the **mathStatica** function `CentralToRaw`, enter:

    `? CentralToRaw`

    CentralToRaw[r] expresses the rth central
       moment μ_r in terms of raw moments μ_i'. To obtain a
       multivariate conversion, let r be a list of integers.

(ii)   Detailed Help (Gold version only) is available via the **mathStatica** palette (Fig. 3).

1.3 Core Functions

1.3 A Getting Started

mathStatica adds about 100 new functions to *Mathematica*. But most of the time, we can get by with just four of them:

function	description	
`PlotDensity[f]`	Plotting (automated)	
`Expect[x, f]`	Expectation operator	$E[X]$
`Prob[x, f]`	Probability	$P(X \le x)$
`Transform[eqn, f]`	Transformations	

Table 3: Core functions for a random variable X with density $f(x)$

This ability to handle plotting, expectations, probability, and transformations, with just four functions, makes the **mathStatica** system very easy to use, even for those not familiar with *Mathematica*.

To illustrate, let us suppose the continuous random variable X has probability density function (pdf)

$$f(x) = \frac{1}{\pi \sqrt{1-x} \sqrt{x}}, \quad \text{for } x \in (0, 1).$$

In *Mathematica*, we enter this as:

```
f = 1/(π √(1 - x) √x);     domain[f] = {x, 0, 1};
```

This is known as the Arc–Sine distribution. Here is a plot of $f(x)$:

```
PlotDensity[f];
```

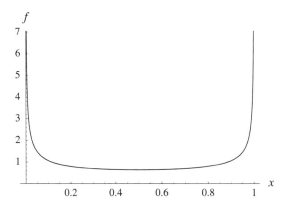

Fig. 4: The Arc–Sine pdf

Here is the cumulative distribution function (cdf), $P(X \leq x)$, which also provides the clue to the naming of this distribution:

Prob[x, f]

$$\frac{2 \, \text{ArcSin}\left[\sqrt{x}\,\right]}{\pi}$$

The mean, $E[X]$, is:

Expect[x, f]

$$\frac{1}{2}$$

while the variance of X is:

Var[x, f]

$$\frac{1}{8}$$

The rth moment of X is $E[X^r]$:

Expect[x$^\text{r}$, f]

— This further assumes that: $\left\{r > -\frac{1}{2}\right\}$

$$\frac{\Gamma\left[\frac{1}{2} + r\right]}{\sqrt{\pi} \; \Gamma[1 + r]}$$

Now consider the transformation to a new random variable Y such that $Y = \sqrt{X}$. By using the Transform and TransformExtremum functions, the pdf of Y, say $g(y)$, and the domain of its support can be found:

g = Transform$\left[\text{y} == \sqrt{\text{x}}, \text{f}\right]$

$$\frac{2y}{\pi \sqrt{y^2 - y^4}}$$

domain[g] = TransformExtremum$\left[\text{y} == \sqrt{\text{x}}, \text{f}\right]$

$\{y, 0, 1\}$

So, we have started out with a quite arbitrary pdf $f(x)$, transformed it to a new one $g(y)$, and since both density g and its domain have been entered into *Mathematica*, we can also apply the **mathStatica** tool set to density g. For example, use PlotDensity[g] to plot the pdf of $Y = \sqrt{X}$.

1.3 B Working with Parameters (Assumptions technology 💡)

mathStatica has been designed to seamlessly support parameters. It does so by taking full advantage of the new *Assumptions technology* introduced in Version 4 of *Mathematica*, which enables us to make assumptions about parameters. To illustrate, let us consider the familiar Normal distribution with mean μ and variance σ^2. That is, let $X \sim N(\mu, \sigma^2)$, where $\mu \in \mathbb{R}$ and $\sigma > 0$. We enter the pdf $f(x)$ in the standard way, but this time we have some extra information about the parameters μ and σ. We use the And function, &&, to add these assumptions to the end of the domain[f] statement:

$$f = \frac{1}{\sigma \sqrt{2\pi}} \text{Exp}\left[-\frac{(x-\mu)^2}{2\sigma^2}\right];$$

$$\text{domain}[f] = \{x, -\infty, \infty\} \;\&\&\; \{\mu \in \text{Reals}, \sigma > 0\};$$

From now on, the assumptions about μ and σ will be 'attached' to density f, so that whenever we operate on density f with a **mathStatica** function, these assumptions will be applied automatically in the background. With this new technology, **mathStatica** can usually produce remarkably crisp textbook-style answers, even when working with very complicated distributions.

The **mathStatica** function, PlotDensity, makes it easy to examine the effect of changing parameter values. The following input reads: "Plot density $f(x)$ when μ is 0, and σ is 1, 2 and 3". For more detail on using the /. operator, see Wolfram (1999, Section 2.4.1).

$$\text{PlotDensity}[f \;/.\; \{\mu \to 0, \sigma \to \{1, 2, 3\}\}];$$

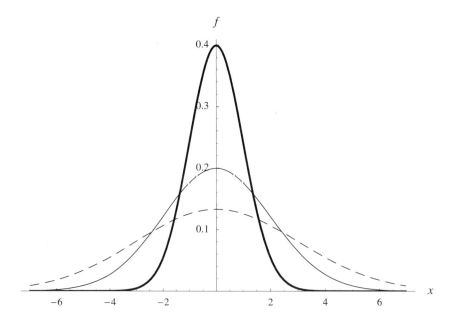

Fig. 5: The pdf of a Normal random variable, when $\mu = 0$ and $\sigma = 1(\text{———})$, $2(\text{—})$, $3(\text{- - -})$

It is well known that $E[X] = \mu$ and $\text{Var}(X) = \sigma^2$, as we can easily verify:

Expect[x, f]

μ

Var[x, f]

σ^2

Because **mathStatica** is general in its design, we can just as easily solve problems that are both less well-known and more 'difficult', such as finding $\text{Var}(X^2)$:

Var[x², f]

$2 \, (2 \, \mu^2 \, \sigma^2 + \sigma^4)$

Assumptions technology is a very important addition to *Mathematica*. In order for it to work, one should enter as much information about parameters as possible. The resulting answer will be much neater, it may also be obtained faster, and it may make it possible to solve problems that could not otherwise be solved. Here is an example of some Assumptions statements:

$$\{\alpha > 1, \quad \beta \in \text{Integers}, \quad -\infty < \gamma < \pi, \quad \delta \in \text{Reals}, \quad \theta > 0\}$$

mathStatica implements Assumptions technology in a *distribution*-specific manner. This means the assumptions are attached to the density $f(x; \theta)$ and not to the parameter θ. What if we have two distributions, both using the same parameter? No problem ... suppose the two pdf's are

(i) $f(x; \theta)$ $\quad \theta > 0$
(ii) $g(x; \theta)$ $\quad \theta < 0$

Then, when we work with density f, **mathStatica** will assume $\theta > 0$; when we work with density g, it will assume $\theta < 0$. For example,

(i) $\text{Expect}[x, f]$ will assume $\theta > 0$
(ii) $\text{Prob}[x, g]$ will assume $\theta < 0$

It is important to realise that the assumptions will only be automatically invoked when using the suite of **mathStatica** functions. By contrast, *Mathematica*'s built-in functions, such as the derivative function, $D[f, x]$, will not automatically assume that $\theta > 0$.

1.3 C Discrete Random Variables

mathStatica automatically handles discrete random variables in the same way. The only difference is that, when we define the density, we add a flag to tell *Mathematica* that the random variable is {Discrete}. To illustrate, let the discrete random variable X have probability mass function (pmf)

$$f(x) = P(X = x) = \binom{r + x - 1}{x} p^r (1 - p)^x, \quad \text{for } x \in \{0, 1, 2, \ldots\}.$$

Here, parameter *p* is the probability of success, while parameter *r* is a positive integer. In *Mathematica*, we enter this as:

```
f = Binomial[r + x - 1, x] p^r (1 - p)^x;
domain[f] = {x, 0, ∞} && {Discrete} &&
            {0 < p < 1, r > 0, r ∈ Integers};
```

This is known as the Pascal distribution. Here is a plot of $f(x)$:

```
PlotDensity[f /. {p → 1/2, r → 10}];
```

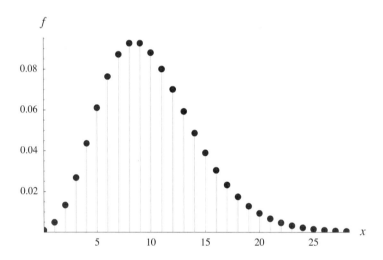

Fig. 6: The pmf of a Pascal discrete random variable

Here is the cdf, equal to $P(X \leq x)$:

```
Prob[x, f]
```

$$1 - \frac{1}{\Gamma[r]\,\Gamma[2 + \text{Floor}[x]]}\left((1-p)^{1+\text{Floor}[x]}\, p^r\, \Gamma[1 + r + \text{Floor}[x]]\, \text{Hypergeometric2F1}[1,\, 1 + r + \text{Floor}[x],\, 2 + \text{Floor}[x],\, 1 - p]\right)$$

The mean $E[X]$ and variance of X are given by:

```
Expect[x, f]
```

$$\left(-1 + \frac{1}{p}\right) r$$

```
Var[x, f]
```

$$\frac{r - p\,r}{p^2}$$

The probability generating function (pgf) is $E[t^X]$:

Expect[t^x, f]

$p^r \, (1 + (-1 + p) \, t)^{-r}$

For more detail on discrete random variables, see Chapter 3.

1.3 D Multivariate Random Variables

mathStatica extends naturally to a multivariate setting. To illustrate, let us suppose that X and Y have joint pdf $f(x, y)$ with support $x > 0, y > 0$:

f = e^{-2 (x+y)} (e^{x+y} + α (e^x - 2) (e^y - 2));

domain[f] = {{x, 0, ∞}, {y, 0, ∞}} && {-1 < α < 1};

where parameter α is such that $-1 < \alpha < 1$. This is known as a Gumbel bivariate Exponential distribution. Here is a plot of $f(x, y)$. To display the code that generates this plot, simply click on the ▷ adjacent to Fig. 7 in the electronic version of this chapter. Clicking the 'View Animation' button in the electronic notebook brings up an animation of $f(x, y)$, allowing parameter α to vary from -1 to 0 in step sizes of $1/20$. This provides a rather neat way to visualise how the shape of the joint pdf changes with α. In the printed text, the symbol 🎬 indicates that an animation is available.

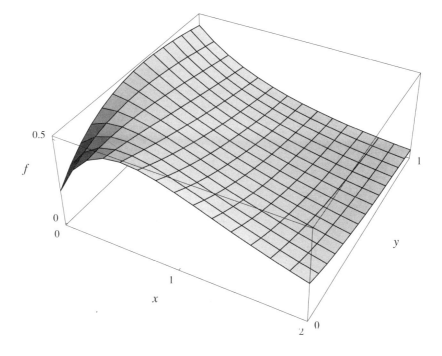

Fig. 7: A Gumbel bivariate Exponential pdf when $\alpha = -0.8$ 🎬

Here is the cdf, namely $P(X \le x, Y \le y)$:

Prob[{x, y}, f]

$e^{-2(x+y)} (-1 + e^x) (-1 + e^y) (e^{x+y} + \alpha)$

Here is Cov(X, Y), the covariance between X and Y:

Cov[{x, y}, f]

$\dfrac{\alpha}{4}$

More generally, here is the variance-covariance matrix:

Varcov[f]

$\begin{pmatrix} 1 & \frac{\alpha}{4} \\ \frac{\alpha}{4} & 1 \end{pmatrix}$

Here is the marginal pdf of X:

Marginal[x, f]

e^{-x}

Here is the conditional pdf of Y, given $X = x$:

Conditional[y, f]

− Here is the conditional pdf $f(y \mid x)$:

$e^{x-2(x+y)} (e^{x+y} + (-2 + e^x)(-2 + e^y)\alpha)$

Here is the bivariate mgf $E[e^{t_1 X + t_2 Y}]$:

mgf = Expect[$e^{t_1 x + t_2 y}$, f]

This further assumes that: $\{t_1 < 1, t_2 < 1\}$

$\dfrac{4 - 2 t_2 + t_1 (-2 + (1 + \alpha) t_2)}{(-2 + t_1)(-1 + t_1)(-2 + t_2)(-1 + t_2)}$

Differentiating the mgf is one way to derive moments. Here is the product moment $E[X^2 Y^2]$:

D[mgf, {t₁, 2}, {t₂, 2}] /. t_ → 0 // Simplify

$4 + \dfrac{9\alpha}{4}$

which we could otherwise have found directly with:

Expect[x² y², f]

$$4 + \frac{9\alpha}{4}$$

Multivariate transformations pose no problem to **mathStatica** either. For instance, let $U = \frac{Y}{1+X}$ and $V = \frac{1}{1+X}$ denote transformations of X and Y. Then our transformation equation is:

$$\text{eqn} = \left\{ u == \frac{y}{1+x},\ v == \frac{1}{1+x} \right\};$$

Using Transform, we can find the joint pdf of random variables U and V, denoted $g(u, v)$:

g = Transform[eqn, f]

$$\frac{e^{\frac{-2-2u+v}{v}} \left(4 e \alpha - 2 e^{\frac{1}{v}} \alpha - 2 e^{\frac{u+v}{v}} \alpha + e^{\frac{1+u}{v}} (1+\alpha) \right)}{v^3}$$

while the extremum of the domain of support of the new random variables are:

TransformExtremum[eqn, f]

$$\{\{u, 0, \infty\}, \{v, 0, 1\}\}$$

For more detail on multivariate random variables, see Chapter 6.

1.3 E Piecewise Distributions

Some density functions take a bipartite form. To illustrate, let us suppose X is a continuous random variable, $0 < x < 1$, with pdf

$$f(x) = \begin{cases} 2\left(\frac{c-x}{c}\right) & \text{if } x < c \\ 2\left(\frac{x-c}{1-c}\right) & \text{if } x \geq c \end{cases}$$

where $0 < c < 1$. We enter this as:

$$\mathbf{f} = \mathbf{If}\left[\mathbf{x < c}, \ 2\ \frac{\mathbf{c-x}}{\mathbf{c}}, \ 2\ \frac{\mathbf{x-c}}{\mathbf{1-c}} \right];$$

domain[f] = {x, 0, 1} && {0 < c < 1};

This is known as the Inverse Triangular distribution, as is clear from a plot of $f(x)$, as illustrated in Fig. 8.

PlotDensity$\left[\text{f} \; / . \; \text{c} \to \{\frac{1}{4}, \frac{1}{2}, \frac{3}{4}\}\right];$

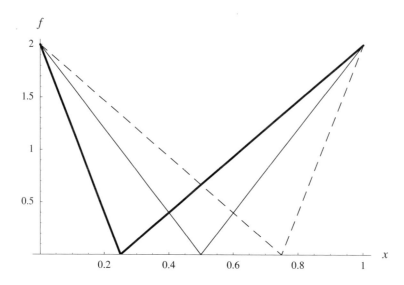

Fig. 8: The Inverse Triangular pdf, when $c = \frac{1}{4}$(——), $\frac{1}{2}$(—), $\frac{3}{4}$(– – –)

Here is the cdf, $P(X \le x)$:

Prob[x, f]

$\text{If}\left[x < c, \; x\left(2 - \frac{x}{c}\right), \; \frac{c - 2cx + x^2}{1 - c}\right]$

Note that the solution depends on whether $x < c$ or $x \ge c$. Figure 9 plots the cdf at the same three values of c used in Fig. 8.

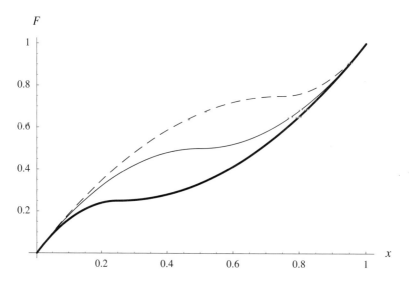

Fig. 9: The Inverse Triangular cdf, when $c = \frac{1}{4}$(——), $\frac{1}{2}$(—), $\frac{3}{4}$(– – –)

mathStatica operates on bipartite distributions in the standard way. For instance, the mean $E[X]$ is given by:

Expect[x, f]

$$\frac{2-c}{3}$$

while the entropy is given by $E[-\log(f(X))]$:

Expect[-Log[f], f]

$$\frac{1}{2} - \text{Log}[2]$$

1.4 Some Specialised Functions

⊕ *Example 1:* Moment Conversion Functions

mathStatica allows one to express any moment (raw μ', central μ, or cumulant κ) in terms of any other moment (μ', μ, or κ). For instance, to express the second central moment (the variance) $\mu_2 = E\big[(X - E[X])^2\big]$ in terms of raw moments, we enter:

CentralToRaw[2]

$$\mu_2 \to -\mu_1'^2 + \mu_2'$$

This is just the well-known result that $\mu_2 = E[X^2] - (E[X])^2$. As a further example, here is the sixth cumulant expressed in terms of raw moments:

CumulantToRaw[6]

$$\kappa_6 \to -120\,\mu_1'^6 + 360\,\mu_1'^4\,\mu_2' - 270\,\mu_1'^2\,\mu_2'^2 + 30\,\mu_2'^3 - 120\,\mu_1'^3\,\mu_3' + 120\,\mu_1'\,\mu_2'\,\mu_3' - 10\,\mu_3'^2 + 30\,\mu_1'^2\,\mu_4' - 15\,\mu_2'\,\mu_4' - 6\,\mu_1'\,\mu_5' + \mu_6'$$

The moment converter functions are completely general, and extend in the natural manner to a multivariate framework. Here is the bivariate central moment $\mu_{2,3}$ expressed in terms of bivariate cumulants:

CentralToCumulant[{2, 3}]

$$\mu_{2,3} \to 6\,\kappa_{1,1}\,\kappa_{1,2} + \kappa_{0,3}\,\kappa_{2,0} + 3\,\kappa_{0,2}\,\kappa_{2,1} + \kappa_{2,3}$$

For more detail, see Chapter 2 (univariate) and Chapter 6 (multivariate). ■

⊕ *Example 2:* Pseudo-Random Number Generation

Let X be *any* discrete random variable with probability mass function (pmf) $f(x)$. Then, the **mathStatica** function `DiscreteRNG[n, f]` generates n pseudo-random copies of X. To illustrate, let us suppose $X \sim \text{Poisson}(6)$:

$$f = \frac{e^{-\lambda} \lambda^x}{x!} \;/.\; \lambda \to 6; \quad \text{domain}[f] = \{x, 0, \infty\} \;\&\&\; \{\text{Discrete}\};$$

As usual, `domain[f]` must *always* be entered along with `f`, as it passes important information onto `DiscreteRNG`. Here are 30 copies of X:

DiscreteRNG[30, f]

{10, 4, 8, 3, 5, 6, 3, 2, 9, 6, 3, 5, 6, 5,
 5, 4, 3, 5, 3, 8, 2, 3, 6, 5, 3, 10, 8, 5, 8, 5}

Here, in a fraction of a second, are 50000 more copies of X:

data = DiscreteRNG[50000, f]; // Timing

{0.39 Second, Null}

`DiscreteRNG` is not only completely general, but it is also very efficient. We now contrast the empirical distribution of `data` with the true distribution of X:

FrequencyPlotDiscrete[data, f];

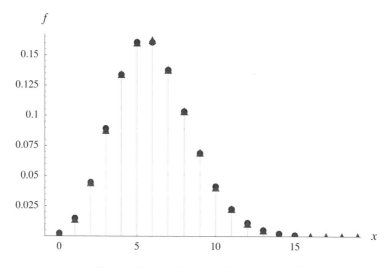

Fig. 10: The empirical pmf (▲) and true pmf (●)

The triangular dots denote the empirical pmf, while the round dots denote the true density $f(x)$. One obtains a superb fit because `DiscreteRNG` is an exact solution. This may make it difficult to distinguish the triangles from the round dots. For more detail, see Chapter 3. ∎

§1.4 INTRODUCTION

⊕ *Example 3:* Pearson Fitting

Karl Pearson showed that if we know the first four moments of a distribution, we can construct a density function that is consistent with those moments. This can provide a neat way to build density functions that approximate a given set of data. For instance, for a given data set, let us suppose that:

```
mean = 37.875;
μ̂₂₃₄ = {191.55, 1888.36, 107703.3};
```

denoting estimates of the mean, and of the second, third and fourth central moments. The Pearson family consists of 7 main *Types*, so our first task is to find out which type this data is consistent with. We do this with the `PearsonPlot` function:

```
PearsonPlot[μ̂₂₃₄];
```

$\{\beta_1 \to 0.507368,\ \beta_2 \to 2.93538\}$

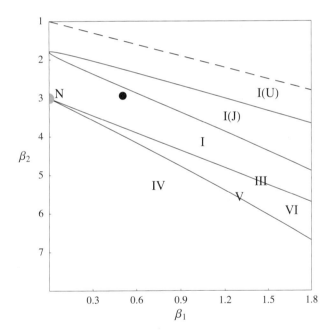

Fig. 11: The β_1, β_2 chart for the Pearson system

The big black dot in Fig. 11 is in the *Type I* zone. Then, the fitted Pearson density $f(x)$ and its domain are immediately given by:

```
{f, domain[f]} = PearsonI[mean, μ̂₂₃₄, x]
```

$\{9.62522 \times 10^{-8}\ (94.3127 - 1.\ x)^{2.7813}$
$(-16.8709 + 1.\ x)^{0.407265},\ \{x,\ 16.8709,\ 94.3127\}\}$

The actual data used to create this example is grouped data (see *Example 3* of Chapter 5) depicting the number of sick people (`freq`) at different ages (X):

```
X = {17, 22, 27, 32, 37, 42, 47, 52, 57, 62, 67, 72, 77, 82, 87};
freq = {34, 145, 156, 145, 123, 103, 86, 71, 55, 37, 21, 13, 7, 3, 1};
```

We can easily compare the histogram of the empirical data with our fitted Pearson pdf:

```
FrequencyGroupPlot[{X, freq}, f];
```

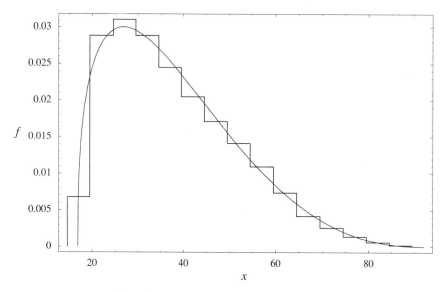

Fig. 12: The data histogram and the fitted Pearson pdf

Related topics include Gram–Charlier expansions, and the Johnson family of distributions. For more detail, see Chapter 5. ∎

⊕ **Example 4:** Fisher Information

The Fisher Information on a parameter can be constructed from first principles using the Expect function. Alternatively, we can use **mathStatica**'s FisherInformation function, which automates this calculation. To illustrate, let $X \sim$ InverseGaussian(μ, λ) with pdf $f(x)$:

$$f = \sqrt{\frac{\lambda}{2\pi x^3}} \; \text{Exp}\left[-\lambda \frac{(x-\mu)^2}{2\mu^2 x}\right];$$

```
domain[f] = {x, 0, ∞} && {μ > 0, λ > 0};
```

Then, Fisher's Information on (μ, λ) is the (2×2) matrix:

```
FisherInformation[{μ, λ}, f]
```

$$\begin{pmatrix} \frac{\lambda}{\mu^3} & 0 \\ 0 & \frac{1}{2\lambda^2} \end{pmatrix}$$

For more detail on Fisher Information, see Chapter 10. ∎

⊕ ***Example 5:*** Non-Parametric Kernel Density Estimation

Here is some raw data measuring the diagonal length of 100 forged Swiss bank notes and 100 real Swiss bank notes (Simonoff, 1996):

```
data = ReadList["sd.dat"];
```

Non-parametric kernel density estimation involves two components: (i) the choice of a kernel, and (ii) the selection of a bandwidth. Here we use a Gaussian kernel f:

$$f = \frac{e^{-\frac{x^2}{2}}}{\sqrt{2\pi}}; \qquad \text{domain}[f] = \{x, -\infty, \infty\};$$

Next, we select the bandwidth c. Small values for c produce a rough estimate while large values produce a very smooth estimate. A number of methods exist to automate bandwidth choice; **mathStatica** implements both the Silverman (1986) approach and the more sophisticated Sheather and Jones (1991) method. For the Swiss bank note data set, the Sheather–Jones optimal bandwidth (using the Gaussian kernel f) is:

```
c = Bandwidth[data, f, Method → SheatherJones]
```

0.200059

We can now plot the smoothed non-parametric kernel density estimate using the NPKDEPlot[*data*, *kernel*, *c*] function:

```
NPKDEPlot[data, f, c];
```

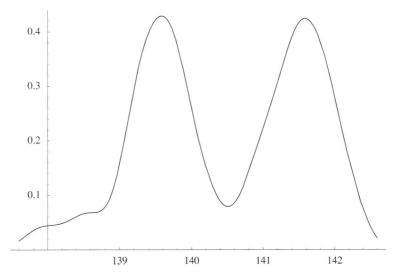

Fig. 13: The smoothed non-parametric kernel density estimate (Swiss bank notes)

For more detail, see Chapter 5. ∎

⊕ **Example 6:** Unbiased Estimation of Population Moments; Moments of Moments

mathStatica can find unbiased estimators of population moments. For instance, it offers h-statistics (unbiased estimators of population central moments), k-statistics (unbiased estimators of population cumulants), multivariate varieties of the same, polykays (unbiased estimators of products of cumulants) and more. Consider the k-statistic k_r which is an unbiased estimator of the rth cumulant κ_r; that is, $E[k_r] = \kappa_r$, for $r = 1, 2, \ldots$. Here are the 2nd are 3rd k-statistics:

```
k2 = KStatistic[2]
k3 = KStatistic[3]
```

$$k_2 \to \frac{-s_1^2 + n\, s_2}{(-1+n)\, n}$$

$$k_3 \to \frac{2\, s_1^3 - 3\, n\, s_1\, s_2 + n^2\, s_3}{(-2+n)\, (-1+n)\, n}$$

As per convention, the solution is expressed in terms of power sums $s_r = \sum_{i=1}^{n} X_i^r$.

Moments of moments: Because the above expressions (sample moments) are functions of random variables X_i, we might want to calculate population moments of them. With **mathStatica**, we can find any moment (raw, central, or cumulant) of the above expressions. For instance, k_3 is meant to have the property that $E[k_3] = \kappa_3$. We test this by calculating the first raw moment of k_3, and express the answer in terms of cumulants:

```
RawMomentToCumulant[1, k3〚2〛]
```

κ_3

In 1928, Fisher published the product cumulants of the k-statistics, which are now listed in reference bibles such as Stuart and Ord (1994). Here is the solution to $\kappa_{2,2}(k_3, k_2)$:

```
CumulantMomentToCumulant[{2, 2}, {k3〚2〛, k2〚2〛}]
```

$$\frac{288\, n\, \kappa_2^5}{(-2+n)\, (-1+n)^3} + \frac{288\, (-23+10\, n)\, \kappa_2^2\, \kappa_3^2}{(-2+n)\, (-1+n)^3} + \frac{360\, (-7+4\, n)\, \kappa_2^3\, \kappa_4}{(-2+n)\, (-1+n)^3} +$$
$$\frac{36\, (160 - 155\, n + 38\, n^2)\, \kappa_3^2\, \kappa_4}{(-2+n)\, (-1+n)^3\, n} + \frac{36\, (93 - 103\, n + 29\, n^2)\, \kappa_2\, \kappa_4^2}{(-2+n)\, (-1+n)^3\, n} +$$
$$\frac{24\, (202 - 246\, n + 71\, n^2)\, \kappa_2\, \kappa_3\, \kappa_5}{(-2+n)\, (-1+n)^3\, n} + \frac{2\, (113 - 154\, n + 59\, n^2)\, \kappa_5^2}{(-1+n)^3\, n^2} +$$
$$\frac{6\, (-131 + 67\, n)\, \kappa_2^2\, \kappa_6}{(-2+n)\, (-1+n)^2\, n} + \frac{3\, (117 - 166\, n + 61\, n^2)\, \kappa_4\, \kappa_6}{(-1+n)^3\, n^2} +$$
$$\frac{6\, (-27 + 17\, n)\, \kappa_3\, \kappa_7}{(-1+n)^2\, n^2} + \frac{37\, \kappa_2\, \kappa_8}{(-1+n)\, n^2} + \frac{\kappa_{10}}{n^3}$$

This is the correct solution. Unfortunately, the solutions given in Stuart and Ord (1994, equation (12.70)) and Fisher (1928) are actually incorrect (see *Example 14* of Chapter 7). ■

⊕ ***Example 7:*** Symbolic Maximum Likelihood Estimation

Although statistical software has long been used for maximum likelihood (ML) estimation, the focus of attention has almost always been on obtaining ML estimates (a *numerical* problem), rather than on deriving ML estimators (a *symbolic* problem). **mathStatica** makes it possible to derive *exact* symbolic ML estimators from first principles with a computer algebra system.

For instance, consider the following simple problem: let $(X_1, \ldots, X_n)$ denote a random sample of size n collected on $X \sim \text{Rayleigh}(\sigma)$, where parameter $\sigma > 0$ is unknown. We wish to find the ML estimator of σ. We begin in the usual way by inputting the likelihood function into *Mathematica*:

$$L = \prod_{i=1}^{n} \frac{x_i}{\sigma^2} \, \text{Exp}\left[-\frac{x_i^2}{2\sigma^2}\right];$$

If we try to evaluate the log-likelihood:

Log[L]

$$\text{Log}\left[\prod_{i=1}^{n} \frac{e^{-\frac{x_i^2}{2\sigma^2}} x_i}{\sigma^2}\right]$$

… nothing happens! (*Mathematica* assumes nothing about the symbols that have been entered, so its inaction is perfectly reasonable.) But we can enhance Log to do what is wanted here using the **mathStatica** function SuperLog. To activate this enhancement, we switch it on:

SuperLog[On]

— SuperLog is now On.

If we now evaluate Log[L] again, we obtain a much more useful result:

logL = Log[L]

$$-2n \, \text{Log}[\sigma] + \sum_{i=1}^{n} \text{Log}[x_i] - \frac{\sum_{i=1}^{n} x_i^2}{2\sigma^2}$$

To derive the first-order conditions for a maximum:

FOC = D[logL, σ]

$$-\frac{2n}{\sigma} + \frac{\sum_{i=1}^{n} x_i^2}{\sigma^3}$$

... we solve FOC==0 using *Mathematica*'s Solve function. The ML estimator $\hat{\sigma}$ is given as a replacement rule → for σ:

$\hat{\sigma}$ = Solve[FOC == 0, σ]〚2〛

$$\left\{\sigma \to \frac{\sqrt{\sum_{i=1}^{n} x_i^2}}{\sqrt{2}\,\sqrt{n}}\right\}$$

The second-order conditions (evaluated at the first-order conditions) are always negative, which confirms that $\hat{\sigma}$ is indeed the ML estimator:

SOC = D[logL, {σ, 2}] /. $\hat{\sigma}$

$$-\frac{8\,n^2}{\sum_{i=1}^{n} x_i^2}$$

Finally, let us suppose that an observed random sample is {1, 6, 3, 4}:

data = {1, 6, 3, 4};

Then the ML estimate of σ is obtained by substituting this data into the ML estimator $\hat{\sigma}$:

$\hat{\sigma}$ /. {n → 4, x$_{i\_}$:> data〚i〛}

$$\left\{\sigma \to \frac{\sqrt{31}}{2}\right\}$$

Figure 14 plots the observed likelihood (for the given data) against values of σ, noting the derived exact optimal solution $\hat{\sigma} = \frac{\sqrt{31}}{2}$.

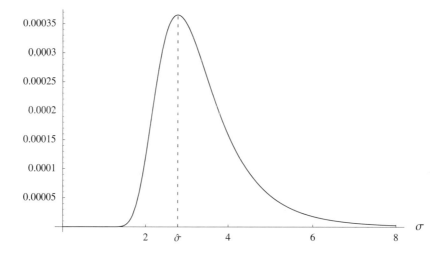

Fig. 14: The observed likelihood and $\hat{\sigma}$

Before continuing, we return Log to its default condition:

SuperLog[Off]

— SuperLog is now Off.

For more detail, see Chapter 11. ∎

⊕ ***Example 8:*** Order Statistics

Let random variable X have a Logistic distribution with pdf $f(x)$:

$$\mathtt{f} = \frac{e^{-x}}{(1+e^{-x})^2}; \qquad \mathtt{domain[f]} = \{\mathtt{x}, -\infty, \infty\};$$

Let $(X_1, X_2, \ldots, X_n)$ denote a sample of size n drawn on X, and let $(X_{(1)}, X_{(2)}, \ldots, X_{(n)})$ denote the ordered sample, so that $X_{(1)} < X_{(2)} < \cdots < X_{(n)}$. The pdf of the rth order statistic, $X_{(r)}$, is given by the **mathStatica** function:

OrderStat[r, f]

$$\frac{(1+e^{-x})^{-r} (1+e^x)^{-1-n+r} n!}{(n-r)! \, (-1+r)!}$$

The joint pdf of $X_{(r)}$ and $X_{(s)}$, for $r < s$, is given by:

OrderStat[{r, s}, f]

$$\frac{e^{x_s} (1+e^{-x_r})^{-r} (1+e^{x_s})^{-1-n+s} \left(\frac{1}{1+e^{x_r}} - \frac{1}{1+e^{x_s}}\right)^{-r+s} \Gamma[1+n]}{(-e^{x_r} + e^{x_s}) \, \Gamma[r] \, \Gamma[1+n-s] \, \Gamma[-r+s]}$$

The OrderStat function also supports piecewise pdf's. For example, let random variable $X \sim \text{Laplace}(\mu, \sigma)$ with pdf $f(x)$:

$$\mathtt{f} = \mathtt{If}\left[\mathtt{x} < \mu, \frac{e^{\frac{x-\mu}{\sigma}}}{2\sigma}, \frac{e^{-\frac{x-\mu}{\sigma}}}{2\sigma}\right];$$

$$\mathtt{domain[f]} = \{\mathtt{x}, -\infty, \infty\} \, \&\& \, \{\mu \in \mathtt{Reals}, \sigma > 0\};$$

Then, the pdf of the rth order statistic, $X_{(r)}$, is:

OrderStat[r, f]

$$\mathtt{If}\left[\mathtt{x} < \mu, \frac{2^{-r} e^{\frac{r(x-\mu)}{\sigma}} \left(1 - \frac{1}{2} e^{\frac{x-\mu}{\sigma}}\right)^{n-r} n!}{\sigma \, (n-r)! \, (-1+r)!}, \frac{2^{1-n-r} e^{\frac{(1+n-r)(-x+\mu)}{\sigma}} \left(1 - \frac{1}{2} e^{\frac{-x+\mu}{\sigma}}\right)^{-1+r} n!}{\sigma \, (n-r)! \, (-1+r)!}\right]$$

The textbook reference solution, given in Johnson *et al.* (1995, p. 168), is alas incorrect. For more detail on order statistics, see Chapter 9. ∎

1.5 Notation and Conventions

1.5 A Introduction

This book brings together two conceptually different worlds: on the one hand, the *statistics* literature has a set of norms and conventions, while on the other hand *Mathematica* has its own (and different) norms and conventions for symbol entry, typefaces and notation. For instance, Table 4 describes the different conventions for upper and lower case letters, say X and x:

Statistics	X denotes a random variable,
	x denotes a realisation of that random variable, such as $x = 3$.
Mathematica	Since *Mathematica* is case-specific, X and x are interpreted as completely different symbols, just as different as y is to Z.

Table 4: Upper and lower case conventions

While one could try to artificially fuse these disparate worlds together, the end solution would most likely be a forced, unnatural and ultimately irritating experience. Instead, the approach we have adopted is to keep the two worlds separate, in the obvious way:

- In Text cells: standard statistics notation is used.
- In Input cells: standard *Mathematica* notation is used.

Thus, the Text of this book reads exactly like a standard mathematical statistics text. For instance,

"Let X have pdf $f(x) = \dfrac{\text{sech}(x)}{\pi}$, $x \in \mathbb{R}$. Find $E[X^2]$."

By contrast, the computer Input for the same problem follows *Mathematica* conventions, so lower case x is used throughout (no capital X), functions use square brackets (not round ones), and the names of mathematical functions are capitalised so that sech(x) becomes Sech[x]:

f = $\dfrac{\text{Sech[x]}}{\pi}$;   domain[f] = {x, -∞, ∞};   Expect[x², f]

$\dfrac{\pi^2}{4}$

If it is necessary to use *Mathematica* notation in the main text, this is indicated by using Courier font. This section summarises these notational conventions in both statistics (Part B) and *Mathematica* (Part C). Related material includes Appendices A.3 to A.8.

1.5 B Statistics Notation

abbreviation	description
cdf	cumulative distribution function
cf	characteristic function
cgf	cumulant generating function
$Cov(X_i, X_j)$	covariance of X_i and X_j
$E[X]$	the expectation of X
iid	independent and identically distributed
mgf	moment generating function
mgfc	central mgf
$M(t)$	mgf: $M(t) = E[e^{tX}]$
MLE	maximum likelihood estimator
MSE	mean square error
$N(\mu, \sigma^2)$	Normal distribution with mean μ and variance σ^2
pdf	probability density function
pgf	probability generating function
pmf	probability mass function
$\Pi(t)$	pgf: $\Pi(t) = E[t^X]$
$P(X \leq x)$	probability
$Var(X)$	the variance of X
$Varcov()$	the variance-covariance matrix

Table 5: Abbreviations

symbol	description
$\mathbb{R}$	set of real numbers
$\mathbb{R}^2$	two-dimensional real plane
$\mathbb{R}_+$	set of positive real numbers
$\vec{X}$	$\vec{X} = (X_1, X_2, \ldots, X_m)$
Σ	summation operator
Π	product operator
d	total derivative
∂	partial derivative
$\log(x)$	natural logarithm of x
H^T	transpose of matrix H
$\binom{n}{r}$	Binomial coefficient

Table 6: Sets and operators

symbol	description	
μ	the population mean (same as μ'_1)	
μ'_r	r^{th} raw moment	$\mu'_r = E[X^r]$
μ_r	r^{th} central moment	$\mu_r = E[(X-\mu)^r]$
κ_r	r^{th} cumulant	
$\mu'_{r,s,\ldots}$	multivariate raw moment	$\mu'_{r,s} = E[X_1^r X_2^s]$
$\mu_{r,s,\ldots}$	multivariate central moment	$\mu_{r,s} = E\left[(X_1 - E[X_1])^r (X_2 - E[X_2])^s\right]$
$\kappa_{r,s,\ldots}$	multivariate cumulant	
$\mu'_{[r]}$	r^{th} factorial moment	
$\mu'_{[r,s]}$	multivariate factorial moments	
β_1	Pearson skewness measure is $\sqrt{\beta_1}$, where $\beta_1 = \mu_3^2/\mu_2^3$	
β_2	Pearson kurtosis measure	$\beta_2 = \mu_4/\mu_2^2$
p	success probability in Bernoulli trials	
ρ or ρ_{ij}	correlation between two random variables	
s_r	power sums	$s_r = \sum_{i=1}^n X_i^r$
m'_r	sample raw moments	$m'_r = \frac{1}{n} \sum_{i=1}^n X_i^r$
m_r	sample central moments	$m_r = \frac{1}{n} \sum_{i=1}^n (X_i - m'_1)^r$
S_n	sample sum, for a sample of size n (same as s_1)	
$\bar{X}_n$	the sample mean, for a sample of size n (same as m'_1)	
θ	population parameter	
$\hat{\theta}$	estimate or estimator of θ	
h_r	h-statistic: $E[h_r] = \mu_r$	
k_r	k-statistic: $E[k_r] = \kappa_r$	
i_θ	Fisher Information on parameter θ	
I_θ	Sample Information on parameter θ	
$\sim$	distributed as; e.g. $X \sim$ Chi-squared(n)	
$\overset{a}{\sim}$	asymptotically distributed	
$\overset{d}{\longrightarrow}$	convergence in distribution	
$\overset{p}{\longrightarrow}$	convergence in probability	

Table 7: Statistics notation

1.5 C *Mathematica* Notation

Common: Table 8 lists common *Mathematica* expressions.

- Note that ⁝ denotes the ESC key.
- *Mathematica* only understands that $\Gamma[x]$ is equal to Gamma[x] if **mathStatica** has been loaded (see Appendix A.8).

expression	description	short cut
π	Pi	⁝p⁝
∞	Infinity	⁝inf⁝
i	$\sqrt{-1}$	⁝ii⁝
e	e^x or Exp[x]	⁝ee⁝
Γ	$\Gamma[x]$ = Gamma[x]	⁝G⁝
$\in$	Element : {$x \in$ Reals}	⁝elem⁝
lis〚4〛	Part 4 of lis	⁝[[⁝ or [[
Binomial[n, r]	Binomial coefficient : $\binom{n}{r}$	

Table 8: *Mathematica* notation (common)

Brackets: In *Mathematica*, each kind of bracket has a very specific meaning. Table 9 lists the four main types.

bracket	description	example		
{ }	Lists	lis = {1, 2, 3, 4}		
[ ]	Functions	y = Exp[x]	not	Exp(x)
( )	Grouping	$(y(x+2)^3)^4$	not	$\{y(x+2)^3\}^4$
lis〚4〛	Part 4 of lis	⁝[[⁝ or [[		

Table 9: *Mathematica* bracketing

Replacements: Table 10 lists notation for making replacements; see also Wolfram (1999, Section 2.4.1). Note that :→ is entered as :> and *not* as :->. Example:

$3 x^2 /. x \to \theta$

$3 \theta^2$

operator	description	short cut		
/.	ReplaceAll			
→	Rule	⁝->⁝	or	->
:→	RuleDelayed	⁝:>⁝	or	:>

Table 10: *Mathematica* replacements

Greek alphabet (common):

letter		short cut	name
α		:a:	alpha
β		:b:	beta
γ,	Γ	:g: , :G:	gamma
δ,	Δ	:d: , :D:	delta
ε		:ce:	epsilon
θ,	Θ	:q: , :Q:	theta
κ		:k:	kappa
λ,	Λ	:l: , :L:	lambda
μ		:m:	mu
ξ		:x:	xi
π		:p:	pi
ρ		:r:	rho
σ,	Σ	:s: , :S:	sigma
ϕ,	Φ	:f: , :F:	phi
χ		:c:	chi
ψ,	Ψ	:y: , :Y:	psi
ω,	Ω	:w: , :W:	omega

Table 11: Greek alphabet (common)

Notation entry: Mathematica's sophisticated typesetting engine makes it possible to use standard statistical notation such as $\hat{x}$ instead of typing xHAT, and x_1 instead of x1 (see Appendix A.5). This makes the transition from paper to computer a much more natural, intuitive and aesthetically pleasing experience. The disadvantage is that we have to learn how to enter expressions like $\hat{x}$. One easy way is to use the BasicTypesetting palette, which is available via File Menu ▷ Palettes ▷ BasicTypesetting. Alternatively, Table 12 lists five essential notation short cuts which are well worth mastering.

notation	short cut
$\dfrac{x}{y}$	x [CTRL] / y
x^r	x [CTRL] 6 r
x_1	x [CTRL] – 1
$\overset{2}{x}$	x [CTRL] 7 2
$\underset{3}{x}$	x [CTRL] = 3

Table 12: Five essential notation short cuts

These five notation types

$$\left\{\frac{x}{y},\ x^r,\ x_1,\ \overset{2}{x},\ \underset{3}{x}\right\}$$

can generate almost any expression used in this book. For instance, the expression $\hat{x}$ has the same form as $\overset{2}{x}$ in Table 12, so we can enter $\hat{x}$ with x [CTRL]7 ^. If the expression is a well-known notational type, *Mathematica* will represent it internally as a 'special' function. For instance, the internal representation of $\hat{x}$ is actually:

    `x̂ // InputForm`

    `OverHat[x]`

Table 13 lists these special functions—they provide an alternative way to enter notation. For instance, to enter $\vec{x}$ we could type in OverVector[x], then select the latter with the mouse, and then choose **Cell Menu ▷ Convert to StandardForm**. This too yields $\vec{x}$.

notation	short cut			function name
x^+	x	[CTRL] 6	+	SuperPlus[x]
x^-	x	[CTRL] 6	−	SuperMinus[x]
x^*	x	[CTRL] 6	*	SuperStar[x]
$x^\dagger$	x	[CTRL] 6	†	SuperDagger[x]
x_+	x	[CTRL] −	+	SubPlus[x]
x_-	x	[CTRL] −	−	SubMinus[x]
x_*	x	[CTRL] −	*	SubStar[x]
$\bar{x}$	x	[CTRL] 7	_	OverBar[x]
$\vec{x}$	x	[CTRL] 7	:vec:	OverVector[x]
$\tilde{x}$	x	[CTRL] 7	~	OverTilde[x]
$\hat{x}$	x	[CTRL] 7	^	OverHat[x]
$\dot{x}$	x	[CTRL] 7	.	OverDot[x]
$\underline{x}$	x	[CTRL] =	_	UnderBar[x]

Table 13: Special forms

Even more sophisticated structures can be created with `Subsuperscript` and `Underoverscript`, as Table 14 shows.

notation	function name
x_1^r	Subsuperscript[x, 1, r]
$\overset{b}{\underset{a}{x}}$	Underoverscript[x, a, b]

Table 14: `Subsuperscript` and `Underoverscript`

Entering $\overset{\prime}{\mu}_r$: This text uses $\overset{\prime}{\mu}_r$ to denote the rth raw moment. The prime ′ above μ is entered by typing [ESC] ' [ESC]. This is because the keyboard ' is reserved for other purposes by *Mathematica*. Further, notation such as x' (where the prime comes *after* the x, rather than above it) should generally be avoided, as *Mathematica* may interpret the prime as a derivative. This problem does not occur with $\overset{\prime}{x}$ notation.

$\overset{\prime}{x}$ // **InputForm**

Overscript[x, ′]

x′ // **InputForm**

Derivative[1][x]

Animations: In the printed text, the symbol ▦ is used to indicate that an animation is available at the marked point in the electronic version of the chapter.

Magnification: If the on-screen notation seems too small, magnification can be used: Format Menu ▷ Magnification.

Notes: Here is an example of a note.[1] In the electronic version of the text, notes are live links that can be activated with the mouse. In the printed text, notes are listed near the end of the book in the Notes section.

Timings: All timings in this book are based on *Mathematica* Version 4 running on a PC with an 850 MHz Pentium III processor.

Finally, the Appendix provides several tips for both the new and advanced user on the accuracy of symbolic and numeric computation, on working with Lists, on using Subscript notation, on working with matrices and vectors, on changes to default behaviour, and on how to expand the **mathStatica** framework with your own functions.

Chapter 2

Continuous Random Variables

2.1 Introduction

Let the continuous random variable X be defined on a domain of support $\Lambda \subset \mathbb{R}$. Then a function $f : \Lambda \to \mathbb{R}_+$ is a *probability density function* (pdf) if it has the following properties:

$$f(x) > 0 \text{ for all } x \in \Lambda$$

$$\int_\Lambda f(x)\,dx = 1 \tag{2.1}$$

$$P(X \in S) = \int_S f(x)\,dx, \quad \text{for } S \subset \Lambda$$

The *cumulative distribution function* (cdf) of X, denoted $F(x)$, is defined by

$$F(x) = P(X \le x) = \int_{-\infty}^{x} f(w)\,dw, \quad -\infty < x < \infty. \tag{2.2}$$

The **mathStatica** function Prob[x, f] calculates $P(X \le x)$. Random variable X is said to be a *continuous random variable* if $F(x)$ is continuous. In fact, although our starting point in **mathStatica** is typically to enter a pdf, it should be noted that the fundamental statistical concept is really the cdf, not the pdf. Table 1 summarises some properties of the cdf for a continuous random variable (a and b are constants).

(i)	$0 \le F(x) \le 1$
(ii)	$F(x)$ is a non-decreasing function of x
(iii)	$F(-\infty) = 0, \ F(\infty) = 1$
(iv)	$P(a < X \le b) = F(b) - F(a), \ \text{for } a < b$
(v)	$P(X = x) = 0$
(vi)	$\dfrac{dF(x)}{dx} = f(x)$

Table 1: Properties of the cdf $F(x)$ for a continuous random variable

The *expectation* of a function $u(X)$ is defined to be:

$$E[u(X)] = \int_x u(x) f(x) \, dx \qquad (2.3)$$

The **mathStatica** function `Expect[u, f]` calculates $E[u]$, where $u = u(X)$. Table 2 summarises some properties of the expectation operator, where a and b are again constants.

(i)	$E[a]$	$= a$
(ii)	$E[a\,u(X)]$	$= a\,E[u(X)]$
(iii)	$E[u(X) + b]$	$= b + E[u(X)]$
(iv)	$E[\sum_{i=1}^{n} a_i X_i]$	$= \sum_{i=1}^{n} a_i\,E[X_i]$

Table 2: Basic properties of the expectation operator

⊕ **Example 1:** Maxwell–Boltzmann: The Distribution of Molecular Speed in a Gas

The Maxwell–Boltzmann speed distribution describes the distribution of the velocity X of a random molecule of gas in a closed container. The pdf can be entered directly from **mathStatica**'s *Continuous* palette:

```
f = √(2/π)/σ³ x² e^(-x²/(2σ²)) ;      domain[f] = {x, 0, ∞} && {σ > 0};
```

From a statistical point of view, the distribution depends on just a single parameter $\sigma > 0$. Formally though, in physics, $\sigma = \sqrt{T k_B / m}$ where k_B denotes Boltzmann's constant, T denotes temperature in Kelvin, and m is the mass of the molecule. The cdf $F(x)$ is $P(X \leq x)$:

```
F = Prob[x, f]
```

$$-\frac{e^{-\frac{x^2}{2\sigma^2}} \sqrt{\frac{2}{\pi}} \, x}{\sigma} + \text{Erf}\left[\frac{x}{\sqrt{2}\,\sigma}\right]$$

Figure 1 plots the pdf (left panel) and cdf (right panel) at three different values of σ.

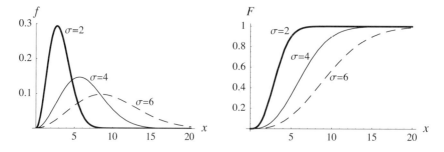

Fig. 1: The Maxwell–Boltzmann pdf (left) and cdf (right), when $\sigma = 2, 4, 6$

The average molecular speed is $E[X]$:

Expect[x, f]

$$2\sqrt{\frac{2}{\pi}}\,\sigma$$

The average kinetic energy per molecule is $E\left[\frac{1}{2}\,m\,X^2\right]$:

Expect$\left[\frac{1}{2}$ m x^2, f$\right]$ /. $\sigma \to \sqrt{T\,k_B\,/\,m}$

$$\frac{3\,T\,k_B}{2}$$

⊕ *Example 2:* The Reflected Gamma Distribution

Some density functions take a piecewise form, such as:

$$f(x) = \begin{cases} f_1(x) & \text{if } x < \alpha \\ f_2(x) & \text{if } x \geq \alpha \end{cases}$$

Such functions are often not smooth, with a kink at the point $x = \alpha$. In *Mathematica*, the natural way to enter such expressions is with the If[condition is true, then f_1, else f_2] function. That is,

f = If[x<α, f1, f2];   domain[f] = {x,-∞,∞}

where f1 and f2 must still be stated. **mathStatica** has been designed to seamlessly handle If statements, without the need for any extra thought or work. In fact, by using this structure, **mathStatica** can solve many integrals that *Mathematica* could not normally solve by itself. To illustrate, let us suppose X is a continuous random variable such that $X = x \in \mathbb{R}$ with pdf

$$f(x) = \begin{cases} \dfrac{(-x)^{\alpha-1}\,e^x}{2\,\Gamma[\alpha]} & \text{if } x < 0 \\ \dfrac{x^{\alpha-1}\,e^{-x}}{2\,\Gamma[\alpha]} & \text{if } x \geq 0 \end{cases}$$

where $0 < \alpha < 1$. This is known as a Reflected Gamma distribution, and it nests the standard Laplace distribution as a special case when $\alpha = 1$. We enter $f(x)$ as follows:

f = If$\left[$x < 0, $\dfrac{(-x)^{\alpha-1}\,e^x}{2\,\Gamma[\alpha]}$, $\dfrac{x^{\alpha-1}\,e^{-x}}{2\,\Gamma[\alpha]}\right]$;

domain[f] = {x, -∞, ∞} && {α > 0};

Here is a plot of $f(x)$ when $\alpha = 1$ and 3:

PlotDensity[f /. α → {1, 3}];

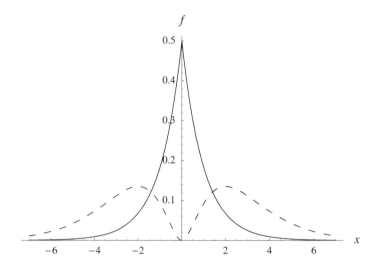

Fig. 2: The pdf of the Reflected Gamma Distribution, when $\alpha = 1$ (—) and 3 (– – –)

Here is the cdf, $P(X \leq x)$:

cdf = Prob[x, f]

$$\text{If}\left[x < 0, \frac{\text{Gamma}[\alpha, -x]}{2\,\Gamma[\alpha]},\ 1 - \frac{\text{Gamma}[\alpha, x]}{2\,\Gamma[\alpha]}\right]$$

Figure 3 plots the cdf when $\alpha = 1$ and 3.

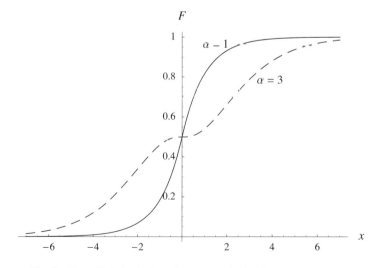

Fig. 3: The cdf of the Reflected Gamma Distribution ($\alpha = 1$ and 3)

2.2 Measures of Location

2.2 A Mean

Let the continuous random variable X have pdf $f(x)$. Then the population mean, or *mean* for short, notated by μ or μ_1', is defined by

$$\mu_1' = E[X] = \int_x x f(x)\, dx \qquad (2.4)$$

if the integral converges.

⊕ ***Example 3:*** The Mean for Sinc^2 and Cauchy Random Variables

Let random variable X have a Sinc^2 distribution with pdf $f(x)$, and let Y have a Cauchy distribution with pdf $g(y)$:

```
f = 1   Sin[x]²
    ─ ─────── ;      domain[f] = {x, -∞, ∞};
    π    x²

g =     1
    ─────────── ;    domain[g] = {y, -∞, ∞};
     π (1 + y²)
```

Figure 4 compares the pdf's of the two distributions.

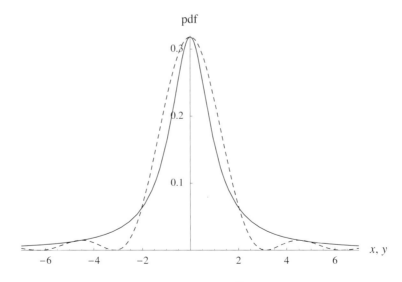

Fig. 4: Cauchy pdf (——) and Sinc^2 pdf (– – –)

The tails of the Sinc^2 pdf are snake-like, and they contact the axis repeatedly at non-zero integer multiples of π.

The mean of the Sinc^2 random variable is $E[X]$:

```
Expect[x, f]
```

0

By contrast, the mean of the Cauchy random variable, $E[Y]$, does not exist:

```
Expect[y, g]
```

— Integrate::idiv :
  Integral of $\frac{y}{1+y^2}$ does not converge on $\{-\infty, \infty\}$.
— Integrate::idiv :
  Integral of $\frac{y}{1+y^2}$ does not converge on $\{-\infty, \infty\}$.

$$\frac{\int_{-\infty}^{\infty} \frac{y}{1+y^2}\, dy}{\pi}$$

2.2 B Mode

Let random variable X have pdf $f(x)$. If $f(x)$ has a local maximum at value x_m, then we say there is a *mode* at x_m. If there is only one mode, then the distribution is said to be unimodal. If the pdf is everywhere continuous and twice differentiable, and there is no corner solution, then a mode is the solution to

$$\frac{df(x)}{dx} = 0, \qquad \frac{d^2 f(x)}{dx^2} < 0. \tag{2.5}$$

Care should always be taken to check for corner solutions.

⊕ *Example 4:* The Mode for a Chi-squared Distribution

Let random variable $X \sim$ Chi-squared(n) with pdf $f(x)$:

```
f = x^(n/2-1) e^(-x/2) / (2^(n/2) Γ[n/2]) ;     domain[f] = {x, 0, ∞} && {n > 0};
```

The first-order condition for a maximum is obtained via:

```
FOC = D[f, x] // Simplify;    Solve[FOC == 0, x]
```

— Solve::ifun : Inverse functions are being
  used by Solve, so some solutions may not be found.

$$\left\{\left\{x \to 0^{-\frac{2}{4-n}}\right\}, \{x \to -2 + n\}\right\}$$

Consider the interior solution, $x_m = n - 2$, for $n > 2$. The second-order condition for a maximum, at $x_m = n - 2$, is:

§2.2 B   CONTINUOUS RANDOM VARIABLES   37

```
SOC = D[f, {x, 2}] /. x → n - 2 // Simplify
```

$$-\frac{2^{-1-\frac{n}{2}} e^{1-\frac{n}{2}} (-2+n)^{\frac{1}{2}(-4+n)}}{\Gamma[\frac{n}{2}]}$$

which is negative for $n > 2$. Hence, we conclude that x_m is indeed a mode, when $n > 2$. If $n \leq 2$, the mode is the corner solution $x_m = 0$. Figure 5 illustrates the two scenarios by plotting the pdf when $n = 1.98$ and $n = 3$.

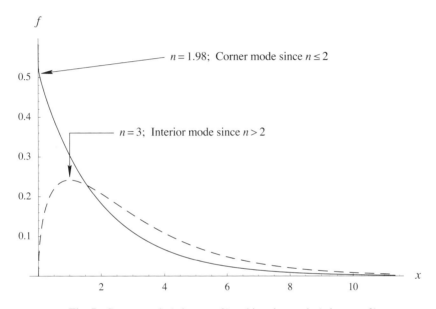

Fig. 5: Corner mode (when $n \leq 2$) and interior mode (when $n > 2$)

2.2 C   Median and Quantiles

Let the continuous random variable X have pdf $f(x)$ and cdf $F(x) = P(X \leq x)$. Then, the *median* is the value of X that divides the total probability into two equal halves; *i.e.* the value x at which $F(x) = \frac{1}{2}$. More generally, the pth quantile is the value of X, say x_p, at which $F(x_p) = p$, for $0 < p < 1$. Quantiles are calculated by deriving the inverse cdf, $x_p = F^{-1}(p)$. Ideally, inversion should be done symbolically (algebraically). Unfortunately, for many distributions, symbolic inversion can be difficult, either because the cdf can not be found symbolically and/or because the inverse cdf can not be found. In such cases, one can often resort to numerical methods. Symbolic and numerical inversion are also discussed in §2.6 B and §2.6 C, respectively.

⊕ **Example 5:** Symbolic Inversion: The Median for the Pareto Distribution

Let random variable $X \sim \text{Pareto}(a, b)$ with pdf $f(x)$:

```
f = a bᵃ x⁻⁽ᵃ⁺¹⁾;    domain[f] = {x, b, ∞} && {a > 0, b > 0};
```

and cdf $F(x)$:

```
F = Prob[x, f]
```

$$1 - \left(\frac{b}{x}\right)^a$$

The median is the value of X at which $F(x) = \frac{1}{2}$:

```
Solve[F == 1/2, x]
```

— Solve::ifun : Inverse functions are being
     used by Solve, so some solutions may not be found.

$$\{\{x \to 2^{\frac{1}{a}} b\}\}$$

More generally, if *Mathematica* can find the inverse cdf, the pth quantile is given by:

```
Solve[F == p, x]
```

— Solve::ifun : Inverse functions are being
     used by Solve, so some solutions may not be found.

$$\{\{x \to b(1-p)^{-1/a}\}\}$$

Figure 6 plots the cdf and inverse cdf, when $a = 4$ and $b = 2$.

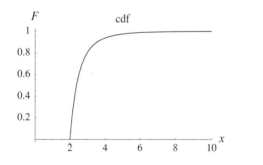
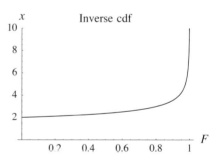

Fig. 6: cdf and inverse cdf

⊕ **Example 6:** Numerical Inversion: Quantiles for a Birnbaum–Saunders Distribution

Let $f(x)$ denote the pdf of a Birnbaum–Saunders distribution, with parameters $\alpha = \frac{1}{2}$ and $\beta = 4$:

$$f = \frac{e^{-\frac{(x-\beta)^2}{2\alpha^2 \beta x}} (x + \beta)}{2\alpha \sqrt{2\pi\beta} \; x^{3/2}} \; /. \; \{\alpha \to \frac{1}{2}, \beta \to 4\};$$

```
domain[f] = {x, 0, ∞} && {α > 0, β > 0};
```

Mathematica cannot find the cdf symbolically; that is, `Prob[x, f]` fails. Instead, we can construct a numerical cdf function `NProb`:

NProb[w_] := NIntegrate[f, {x, 0, w}]

For example, $F(8) = P(X \le 8)$ is given by:

NProb[8]

0.92135

which means that $X = 8$ is approximately the 0.92 quantile. Suppose we want to find the 0.7 quantile: one approach would be to manually try different values of X. As a first guess, how about $X = 6$?

NProb[6]

0.792892

Too big. So, try $X = 5$:

NProb[5]

0.67264

Too small. And so on. Instead of doing this iterative search manually, we can use *Mathematica*'s `FindRoot` function to automate the search for us. Here, we ask *Mathematica* to search for the value of X at which $F(x) = 0.7$, starting the search by trying $X = 1$ and $X = 10$:

sol = FindRoot[NProb[x] == 0.7, {x, {1, 10}}]

{x → 5.19527}

This tells us that $X = 5.19527 \ldots$ is the 0.7 quantile, as we can check by substituting it back into our numerical $F(x)$ function:

NProb[x /. sol]

0.7

Care is always required with numerical methods, in part because they are not exact, and in part because different starting points can sometimes lead to different 'solutions'. Finally, note that numerical methods can only be used if the pdf itself is numerical. Thus, numerical methods cannot be used to find quantiles as a function of parameters α and β—the method can only work given numerical values for α and β. ■

2.3 Measures of Dispersion

A number of methods exist to measure the dispersion of the distribution of a random variable X. The most well known is the *variance* of X, defined as the second central moment

$$\text{Var}(X) = \mu_2 = E[(X-\mu)^2] \tag{2.6}$$

where μ denotes the mean $E[X]$. The **mathStatica** function Var[*x, f*] calculates Var(X). The *standard deviation* is the (positive) square root of the variance, and is often denoted σ.[1] Another measure is the *mean deviation* of X, defined as the first absolute central moment

$$E[\,|X-\mu|\,]. \tag{2.7}$$

The above measures of dispersion are all expressed in terms of the units of X. This can make it difficult to compare the dispersion of one population with another. By contrast, the following statistics are independent of the variable's units of measurement. The *coefficient of variation* is defined by

$$\sigma/\mu. \tag{2.8}$$

Gini's coefficient lies within the unit interval; it is discussed in *Example 9*. Alternatively, one can often compare the dispersion of two distributions by standardising them. A *standardised* random variable Z has zero mean and unit variance:

$$Z = \frac{X-\mu}{\sigma}. \tag{2.9}$$

Related measures are $\sqrt{\beta_1}$ and β_2, where

$$\begin{aligned} \sqrt{\beta_1} &= \frac{\mu_3}{\mu_2^{3/2}} = E\Big[\Big(\frac{X-\mu}{\sigma}\Big)^3\Big] \\ \beta_2 &= \frac{\mu_4}{\mu_2^2} = E\Big[\Big(\frac{X-\mu}{\sigma}\Big)^4\Big] \end{aligned} \tag{2.10}$$

Here, the μ_i terms denote central moments, which are introduced in §2.4 A. If a density is not symmetric about μ, it is said to be skewed. A common measure of *skewness* is $\sqrt{\beta_1}$. If the distribution of X is symmetric about μ, then $\mu_3 = E[(X-\mu)^3] = 0$ (assuming μ_3 exists). However, $\mu_3 = 0$ does not guarantee symmetry; Ord (1968) provides examples. Densities with long tails to the right are called *skewed to the right* and they tend to have $\mu_3 > 0$, while densities with long tails to the left are called *skewed to the left* and tend to have $\mu_3 < 0$. *Kurtosis* is commonly said to measure the peakedness of a distribution. More correctly, kurtosis is a measure of both the peakedness (near the centre) and the tail weight of a distribution. Balanda and MacGillivray (1988, p. 116) define kurtosis as "the location- and scale-free movement of probability mass from the shoulders of a distribution into its centre and tails. In particular, this definition implies that peakedness and tail weight are best viewed as components of kurtosis, since any movement of mass from the shoulders into the tails must be accompanied by a movement of mass into the centre if the scale is to be left unchanged." The expression β_2 is Pearson's measure of the *kurtosis* of a distribution. For the Normal distribution, $\beta_2 = 3$, and so the value 3 is often used as a reference point.

⊕ *Example 7:* Mean Deviation for the Chi-squared(n) Distribution

Let $X \sim$ Chi-squared(n) with pdf $f(x)$:

$$f = \frac{x^{n/2-1} e^{-x/2}}{2^{n/2} \Gamma[\frac{n}{2}]}; \qquad \text{domain}[f] = \{x, 0, \infty\} \,\&\&\, \{n > 0\};$$

The mean μ is:

$$\mu = \text{Expect}[x, f]$$

$$n$$

The mean deviation is $E[|X - \mu|]$. Evaluating this directly using Abs[] fails to yield a solution:

$$\text{Expect}[\text{Abs}[x - \mu], f]$$

$$\frac{2^{-n/2} \int_0^\infty e^{-x/2} x^{-1+\frac{n}{2}} \text{Abs}[n - x] \, dx}{\Gamma[\frac{n}{2}]}$$

In fact, quite generally, *Mathematica* Version 4 is not very successful at integrating expressions containing absolute values. Fortunately, **mathStatica**'s support for If[*a*, *b*, *c*] statements provides a backdoor way of handling absolute values — to see this, express $y = |x - \mu|$ as:

$$y = \text{If}[x < \mu, \mu - x, x - \mu];$$

Then the mean deviation $E[|X - \mu|]$ is given by:[2]

$$\text{Expect}[y, f]$$

$$\frac{4 \, \text{Gamma}[1 + \frac{n}{2}, \frac{n}{2}] - 2 \, n \, \text{Gamma}[\frac{n}{2}, \frac{n}{2}]}{\Gamma[\frac{n}{2}]}$$

⊕ *Example 8:* β_1 and β_2 for the Weibull Distribution

Let $X \sim \text{Weibull}(a, b)$ with pdf $f(x)$:

```
       a x^(a-1)
f = ─────────── ;    domain[f] = {x, 0, ∞} && {a > 0, b > 0};
     b^a e^((x/b)^a)
```

Here, a is termed the shape parameter, and b is termed the scale parameter. The mean μ is:

```
μ = Expect[x, f]
```

$$b\, \Gamma\!\left[1 + \frac{1}{a}\right]$$

while the second, third and fourth central moments are:

```
{μ₂, μ₃, μ₄} = Expect[(x - μ)^{2, 3, 4}, f];
```

Then, β_1 and β_2 are given by:

$$\{\beta_1, \beta_2\} = \left\{\frac{\mu_3^2}{\mu_2^3},\ \frac{\mu_4}{\mu_2^2}\right\}$$

$$\left\{ \frac{\left(2\,\Gamma[1+\tfrac{1}{a}]^3 - \frac{6\,\Gamma[\tfrac{1}{a}]\,\Gamma[\tfrac{2}{a}]}{a^2} + \Gamma[\tfrac{3+a}{a}]\right)^2}{\left(-\Gamma[1+\tfrac{1}{a}]^2 + \Gamma[\tfrac{2+a}{a}]\right)^3},\right.$$

$$\left. \frac{-\frac{3\,\Gamma[\tfrac{1}{a}]\left(\Gamma[\tfrac{1}{a}]^3 - 4a\,\Gamma[\tfrac{1}{a}]\,\Gamma[\tfrac{2}{a}] + 4a^2\,\Gamma[\tfrac{3}{a}]\right)}{a^4} + \Gamma[\tfrac{4+a}{a}]}{\left(-\Gamma[1+\tfrac{1}{a}]^2 + \Gamma[\tfrac{2+a}{a}]\right)^2} \right\}$$

Note that both β_1 and β_2 only depend on the shape parameter a; the scale parameter b has disappeared, as per intuition. Figure 7 plots β_1 and β_2 for different values of parameter a.

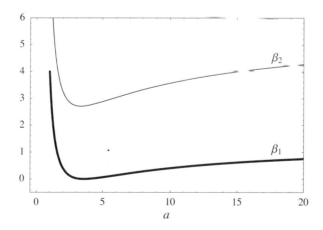

Fig. 7: β_1 and β_2 for the Weibull distribution (plotted as a function of parameter a)

§2.3   CONTINUOUS RANDOM VARIABLES

Note that the symbols μ_2, μ_3 and μ_4 are 'reserved' for use by **mathStatica**'s moment converter functions. To avoid any confusion, it is best to Unset them:

μ = .; μ_2 =.; μ_3 =.; μ_4 =.;

prior to leaving this example.   ∎

⊕ *Example 9:* The Lorenz Curve and the Gini Coefficient

ClearAll[a, b, p, x, u, f, F]

Let X be a positive random variable with pdf $f(x)$ and cdf $F(x)$, and let $p = F(x)$. The *Lorenz curve* is the graph of $L(p)$ against p, where

$$L(p) = \frac{1}{E[X]} \int_0^p F^{-1}(u)\,du \qquad (2.11)$$

and where $F^{-1}(\cdot)$ denotes the inverse cdf. In economics, the Lorenz curve is often used to measure the extent of inequality in the distribution of income. To illustrate, suppose income X is Pareto distributed with pdf $f(x)$:

f = a b^a x$^{-(a+1)}$;   domain[f] = {x, b, ∞} && {a > 0, b > 0};

and cdf $F(x)$:

F = Prob[x, f]

$1 - \left(\dfrac{b}{x}\right)^a$

The inverse cdf is found by solving the equation $p = F(x)$ in terms of x:

Solve[p == F, x]

− Solve::ifun : Inverse functions are being used by Solve, so some solutions may not be found.

{{x → b (1 − p)$^{-1/a}$}}

Equation (2.11) requires that the mean of X exists:

mean = Expect[x, f]

− This further assumes that: {a > 1}

$\dfrac{a\,b}{-1 + a}$

... so we shall impose the tighter restriction $a > 1$. We can now evaluate (2.11):

```
LC = ──────── Integrate[b (1 - u)^-1/a, {u, 0, p}]
      mean
```

$$\frac{(-1+a)\left(\frac{a}{-1+a} + \frac{a\,(1-p)^{-1/a}\,(-1+p)}{-1+a}\right)}{a}$$

Note that the solution does not depend on the location parameter *b*. The solution can be simplified further:

```
LC = FullSimplify[ LC, {0 < p < 1, a > 1}]
```

$$1 - (1-p)^{1-\frac{1}{a}}$$

The Lorenz curve is a plot of LC as a function of *p*, as illustrated in Fig. 8. The horizontal axis (*p*) measures quantiles of the population sorted by income; that is, $p = 0.25$ denotes the poorest 25% of the population. The vertical axis, $L(p)$, measures what proportion of society's total income accrues to the poorest *p* people. In the case of Fig. 8, where $a = 2$, the poorest 50% of the population earn only 29% of the total income:

```
LC /. {a → 2, p → .50}
```

0.292893

The 45° line, $L(p) = p$, represents a society with absolute income equality. By contrast, the line $L(p) = 0$ represents a society with absolute income inequality: here, all the income accrues to just one person.

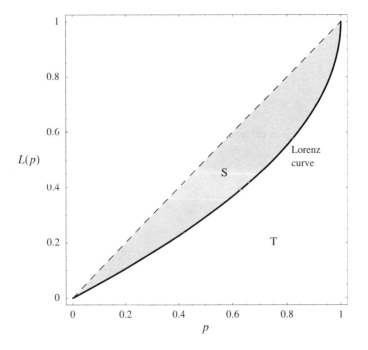

Fig. 8: The Lorenz Curve for a Pareto distribution ($a = 2$)

The *Gini coefficient* is often used in economics to quantify the extent of inequality in the distribution of income. The advantage of the Gini coefficient over the variance as a measure of dispersion is that the Gini coefficient is unitless and lies within the unit interval. Let S denote the shaded area in Fig. 8, and let T denote the area below the Lorenz curve. The Gini coefficient (GC) is defined by the ratio $GC = \frac{S}{S+T} = \frac{S}{1/2} = 2\,S$. That is, GC = twice the shaded area. Since it is easy to compute area T, and since $S = \frac{1}{2} - T$, we use GC = 2 S = 1 − 2 T. Then, for our Pareto example, the Gini coefficient is:

```
1 - 2 Integrate[ LC, {p, 0, 1}, Assumptions → a > 1] //
   Simplify
```

$$\frac{1}{-1 + 2\,a}$$

This corresponds to a Gini coefficient of $\frac{1}{3}$ for Fig. 8 where $a = 2$. If $a = 1$, then GC = 1 denoting absolute income inequality. As parameter a increases, the Lorenz curve shifts toward the 45° line, and the Gini coefficient tends to 0, denoting absolute income equality. ∎

2.4 Moments and Generating Functions

2.4 A Moments

The rth *raw moment* of the random variable X is denoted by $\mu'_r(X)$, or μ'_r for short, and is defined by

$$\mu'_r = E[X^r]. \qquad (2.12)$$

Note that $\mu'_0 = 1$, since $E[X^0] = E[1] = 1$. The first moment, $\mu'_1 = E[X]$, is the *mean of X*, and it is also denoted μ.

The rth *central moment* μ_r is defined by

$$\mu_r = E\!\left[(X - \mu)^r\right] \qquad (2.13)$$

where $\mu = E[X]$. This is also known as the rth *moment about the mean*. Note that $\mu_0 = 1$, since $E\!\left[(X - \mu)^0\right] = E[1]$. Similarly, $\mu_1 = 0$, since $E\!\left[(X - \mu)^1\right] = E[X] - \mu$. The second central moment, $\mu_2 = E\!\left[(X - \mu)^2\right]$, is known as the *variance* of X, and is denoted Var(X). The *standard deviation* of X is the (positive) square root of the variance, and is often denoted σ. Moments can also be obtained via generating functions; see §2.4 B. Further, the various types of moments can be expressed in terms of one another; this is discussed in §2.4 G.

⊕ *Example 10:* Raw Moments for a Standard Normal Random Variable

Let $X \sim N(0, 1)$ with pdf $f(x)$:

```
f = e^(-x²/2) / √(2π);   domain[f] = {x, -∞, ∞};
```

The rth raw moment $E[X^r]$ is given by:

```
sol = Expect[x^r, f]
```

— This further assumes that: {r > -1}

$$\frac{2^{\frac{1}{2}(-2+r)} (1 + (-1)^r) \Gamma[\frac{1+r}{2}]}{\sqrt{\pi}}$$

Then, the first 15 raw moments are given by:

```
sol /. r → Range[15]
```

{0, 1, 0, 3, 0, 15, 0, 105, 0, 945, 0, 10395, 0, 135135, 0}

The odd moments are all zero, because the standard Normal distribution is symmetric about zero. ■

2.4 B   The Moment Generating Function

The *moment generating function* (mgf) of a random variable X is a function that may be used to generate the moments of X. In particular, the mgf $M_X(t)$ is a function of a real-valued dummy variable t. When no confusion is possible, we denote $M_X(t)$ by $M(t)$. We first consider whether or not the mgf exists, and then show how moments may be derived from it, if it exists.

Existence: Let X be a random variable, and $t \in \mathbb{R}$ denote a dummy variable. Let $\underline{t}$ and $\bar{t}$ denote any two real-valued constants such that $\underline{t} < 0$ and $\bar{t} > 0$; thus, the open interval $(\underline{t}, \bar{t})$ includes zero in its interior. Then, the mgf is given by

$$M(t) = E[e^{tX}] \quad (2.14)$$

provided $E[e^{tX}] \in \mathbb{R}_+$ for all t in the chosen interval $\underline{t} < t < \bar{t}$. The condition that $M(t)$ be positive real for all $t \in (\underline{t}, \bar{t})$ ensures that $M(t)$ is differentiable with respect to t at zero. Note that when $t = 0$, $M(0)$ is always equal to 1. However, $M(t)$ may fail to exist for $t \neq 0$.

Generating moments: Let X be a random variable for which the mgf $M(t)$ exists. Then, the rth raw moment of X is obtained by differentiating the mgf r times with respect to t, followed by setting $t = 0$ in the resulting formula:

$$\mu'_r = \frac{d^r M(t)}{d t^r}\bigg|_{t=0}. \qquad (2.15)$$

Proof: If $M(t)$ exists, then $M(t)$ is 'r-times' differentiable at $t = 0$ (for integer $r > 0$) and $\frac{d E[e^{tX}]}{dt} = E\left[\frac{d e^{tX}}{dt}\right]$ for all $t \in (\underline{t}, \bar{t})$ (Mittelhammer (1996, p. 142)). Hence,

$$\frac{d^r E[e^{tX}]}{d t^r}\bigg|_{t=0} = E\left[\frac{d^r e^{tX}}{d t^r}\right]\bigg|_{t=0} = E[X^r e^{tX}]\bigg|_{t=0} = E[X^r] \quad \square$$

Using **mathStatica**, the expectation $E[e^{tX}]$ can be found in the usual way with Expect. However, before using the obtained solution as the mgf of X, one must check that the mgf definition (2.14) is satisfied; *i.e.* that $M(t)$ is positive real for all $t \in (\underline{t}, \bar{t})$.

⊕ *Example 11:* The mgf of the Normal Distribution

Let $X \sim \text{Normal}(\mu, \sigma^2)$. Derive the mgf of X, and derive the first 4 raw moments from it.

Solution: Input the pdf of X:

```
f = 1/(σ √(2 π)) Exp[-(x - μ)²/(2 σ²)];
domain[f] = {x, -∞, ∞} && {μ ∈ Reals, σ > 0};
```

Evaluating (2.14), we find:

```
M = Expect[e^(t x), f]
```

$e^{t \mu + \frac{t^2 \sigma^2}{2}}$

By inspection, $M \in \mathbb{R}_+$ for all $t \in \mathbb{R}$, and $M = 1$ when $t = 0$. Thus, M corresponds to the mgf of X. Then, to determine say μ'_2 from M, we apply (2.15) as follows:

```
D[M, {t, 2}] /. t → 0
```

$\mu^2 + \sigma^2$

More generally, to determine μ'_r, $r = 1, \ldots, 4$, from M:

```
Table[ D[M, {t, r}] /. t → 0, {r, 4}]
```

$\{\mu, \; \mu^2 + \sigma^2, \; \mu^3 + 3 \mu \sigma^2, \; \mu^4 + 6 \mu^2 \sigma^2 + 3 \sigma^4\}$

⊕ **Example 12:** The mgf of the Uniform Distribution

Let $X \sim \text{Uniform}(0, 1)$. Derive the mgf of X, and derive the first 4 raw moments from it.

Solution: Input the pdf of X, and derive M:

```
f = 1;   domain[f] = {x, 0, 1};   M = Expect[e^(t x), f]
```

$$\frac{-1 + e^t}{t}$$

Figure 9 plots M in the neighbourhood of $t = 0$.

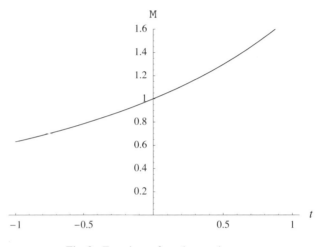

Fig. 9: Function M for $-1 < t < 1$

Clearly, $M \in \mathbb{R}_+$ in a neighbourhood of values about $t = 0$. At the particular value $t = 0$, the plot seems to indicate that $M = 1$. If we input M/.t→0, *Mathematica* replaces t with 0 to yield 0/0:

```
M /. t → 0
```

— Power::infy : Infinite expression $\frac{1}{0}$ encountered.

— ∞::indet :
   Indeterminate expression 0 ComplexInfinity encountered.

   Indeterminate

To correctly determine the value of M at $t = 0$, L'Hôpital's rule should be applied. This rule is incorporated into *Mathematica*'s Limit function:

```
Limit[M, t → 0]
```

1

Thus, $M = 1$ when $t = 0$, as required. Since all requirements of the mgf definition are now satisfied, M is the mgf of X.

To determine the first 4 raw moments of X, we again apply (2.15), but this time in tandem with the Limit function:

Table[Limit[D[M, {t, r}], t → 0], {r, 4}]

$$\left\{\frac{1}{2}, \frac{1}{3}, \frac{1}{4}, \frac{1}{5}\right\}$$

More generally, $E[X^r] = \frac{1}{1+r}$, as we can verify with Expect[x^r, f]. ∎

⊕ **Example 13:** The mgf of the Pareto Distribution?

Let X be Pareto distributed with shape parameter $a > 0$ and location parameter $b > 0$. Does the mgf of X exist?

Solution: Input the pdf of X via the **mathStatica** palette:

f = a b^a x^-(a+1);   domain[f] = {x, b, ∞} && {a > 0, b > 0};

The solution to $M(t) = E[e^{tX}]$ is given by **mathStatica** as:

M = Expect[e^{t x}, f]

a ExpIntegralE[1 + a, -b t]

If we consult *Mathematica*'s on-line help system on ExpIntegralE, we see that the ExpIntegralE function is complex-valued if the value of its second argument, $-bt$, is negative. Since $b > 0$, M will be complex-valued for any positive value assigned to t. To illustrate, suppose parameters a and b are given specific values, and M is evaluated for various values of $t > 0$:

M /. {a → 5, b → 1} /. t → {.2, .4, .6, .8}

{1.28704 - 0.0000418879 i, 1.66642 - 0.00134041 i,
 2.17384 - 0.0101788 i, 2.85641 - 0.0428932 i}

Hence, the requirement that M must be positive real in an open interval that includes the origin is not satisfied. Therefore, the mgf of X does not exist. The non-existence of the mgf does not necessarily mean that the moments do not exist. The Pareto is a case in point, for from:

Expect[x^r, f]

— This further assumes that: {a > r}

$$\frac{a\, b^r}{a - r}$$

... we see that the raw moment μ'_r exists, under the given conditions. ∎

2.4 C The Characteristic Function

As *Example 13* illustrated, the mgf of a random variable does not have to exist. This may occur if e^{tx} is unbounded (see (2.14)). However, the function e^{itx}, where i denotes the unit imaginary number, does not suffer from unboundedness. On an Argand diagram, for any $t \in \mathbb{R}$, e^{itx} takes values on the unit circle. This leads to the so-called *characteristic function* (cf) of random variable X, which is defined as

$$C(t) = E[e^{itX}]. \tag{2.16}$$

The cf of a random variable exists for any choice of $t \in \mathbb{R}$ that we may wish to make; note $C(0) = 1$. If the mgf of a random variable exists, the relationship between the cf and the mgf is simply $C(t) = M(it)$. Analogous to (2.15), raw moments can be obtained from the cf via

$$\mu'_r = i^{-r} \left. \frac{d^r C(t)}{d t^r} \right|_{t=0} \tag{2.17}$$

provided μ'_r exists.

⊕ *Example 14:* The cf of the Normal Distribution

Let $X \sim N(\mu, \sigma^2)$. Determine the cf of X.

Solution: Input the pdf of X:

$$\mathtt{f} \;=\; \frac{1}{\sigma\sqrt{2\pi}} \, \mathtt{Exp}\!\left[-\frac{(\mathbf{x}-\mu)^2}{2\sigma^2}\right];$$

$$\mathtt{domain[f]} \;=\; \{\mathtt{x}, -\infty, \infty\} \,\&\&\, \{\mu \in \mathtt{Reals}, \sigma > 0\};$$

Since we know from *Example 11* that the mgf exists, the cf of X can be obtained via $C(t) = M(it)$. This sometimes works better in *Mathematica* than trying to evaluate $\mathtt{Expect[e^{itx}, f]}$ directly:

$$\mathtt{cf = Expect[e^{tx}, f] \;/.\; t \to i\,t}$$

$$e^{i t \mu - \frac{t^2 \sigma^2}{2}}$$

Then, the first 4 moments are given by:

$$\mathtt{Table[i^{-r}\, D[cf, \{t, r\}] \;/.\; t \to 0, \{r, 4\}] \;//\; Simplify}$$

$$\{\mu,\; \mu^2 + \sigma^2,\; \mu^3 + 3\mu\sigma^2,\; \mu^4 + 6\mu^2\sigma^2 + 3\sigma^4\}$$

⊕ **Example 15:** The cf of the Lindley Distribution

Let the random variable X be Lindley distributed with parameter $\delta > 0$. Derive the cf, and derive the first 4 raw moments from it.

Solution: Input the pdf of X from the **mathStatica** palette:

$$f = \frac{\delta^2}{\delta + 1} (x + 1) \, e^{-\delta x}; \quad \text{domain}[f] = \{x, 0, \infty\} \,\&\&\, \{\delta > 0\};$$

The cf is given by

$$cf = \text{Expect}[e^{i t x}, f]$$

— This further assumes that: $\{\text{Im}[t] == 0\}$

$$\frac{\delta^2 \, (1 - i\,t + \delta)}{(1 + \delta) \, (-i\,t + \delta)^2}$$

The condition on t output by **mathStatica** is not relevant here, for we restrict the dummy variable t to the real number line. The first 4 raw moments of X are given by:

$$\text{Table}[i^{-r} \, D[cf, \{t, r\}]] \, /. \, t \to 0, \, \{r, 4\}] \, // \, \text{Simplify}$$

$$\left\{ \frac{2 + \delta}{\delta + \delta^2}, \; \frac{2 \, (3 + \delta)}{\delta^2 \, (1 + \delta)}, \; \frac{6 \, (4 + \delta)}{\delta^3 \, (1 + \delta)}, \; \frac{24 \, (5 + \delta)}{\delta^4 \, (1 + \delta)} \right\}$$

⊕ **Example 16:** The cf of the Pareto Distribution

Let X be Pareto distributed with shape parameter $a = 4$ and location parameter $b = 1$. Derive the cf, and from it, derive those raw moments which exist.

Solution: The Pareto pdf is:

$$f = a \, b^a \, x^{-(a+1)}; \quad \text{domain}[f] = \{x, b, \infty\} \,\&\&\, \{a > 0, b > 0\};$$

When $a = 4$ and $b = 1$, the solution to the cf of X is:

$$cf = \text{Expect}[e^{i t x}, f \, /. \, \{a \to 4, b \to 1\}]$$

$$\frac{1}{6} \, (e^{i t} \, (6 - i\,t \, (-2 + t \, (-i + t))) + t^4 \, \text{Gamma}[0, -i\,t])$$

From *Example 13*, we know that the mgf of X does not exist. However, the moments of X up to order $r < a = 4$ do exist, which we obtain from the cf by applying (2.17):

$$\text{Table}[\, \text{Limit}[i^{-r} \, D[cf, \{t, r\}], \, t \to 0], \, \{r, 4\}]$$

$$\left\{ \frac{4}{3}, \, 2, \, 4, \, \infty \right\}$$

Notice that we have utilised `Limit` to obtain the moments here, so as to avoid the 0/0 problem discussed in *Example 12*. ■

2.4 D Properties of Characteristic Functions (and mgf's)

§2.4 B and §2.4 C illustrated how the mgf and cf can be used to generate the moments of a random variable. A second (and more important) application of the mgf and cf is to prove that a random variable has a specific distribution. This methodology rests on the Uniqueness Theorem, which we present here using characteristic functions; of course, the theorem also applies to moment generating functions, provided the mgf exists, since then $C(t) = M(i\,t)$.

Uniqueness Theorem: There is a one-to-one correspondence between the cf and the pdf of a random variable.

Proof: The pdf determines the cf via (2.16). The cf determines the pdf via the Inversion Theorem below.

The Uniqueness Theorem means that if two random variables X and Y have the same distribution, then X and Y must have the same mgf. Conversely, if they have the same mgf, then they must have the same distribution. The following results can be especially useful when applying the Uniqueness Theorem. We present these results as the MGF Theorem, which holds provided the mgf exists. A similar result holds, of course, for any cf, with t replaced by $i\,t$.

MGF Theorem: Let random variable X have mgf $M_X(t)$, and let a and b denote constants. Then

$M_{X+a}(t) = e^{ta}\, M_X(t)$ *Proof:* $M_{X+a}(t) = E[e^{t(X+a)}] = e^{ta}\, M_X(t)$

$M_{bX}(t) = M_X(b\,t)$ *Proof:* $M_{bX}(t) = E[e^{t(bX)}] = E[e^{(tb)X}] = M_X(t\,b)$

$M_{a+bX}(t) = e^{ta}\, M_X(b\,t)$ *Proof:* via above.

Further, let $(X_1, \ldots, X_n)$ be independent random variables with mgf's $M_{X_i}(t)$, $i = 1, \ldots, n$, and let $Y = \sum_{i=1}^{n} X_i$. Then

$$M_Y(t) = \prod_{i=1}^{n} M_{X_i}(t)$$ *Proof:* via independence (see Table 3 of Chapter 6).

If we can match the functional form of $M_Y(t)$ with a well-known moment generating function, then we know the distribution of Y. This matching is usually done using a textbook that lists the mgf's for well-known distributions. Unfortunately, the matching process is often neither easy nor obvious. Moreover, if the pdf of Y is not well-known (or not listed in the textbook), the matching may not be possible. Instead of trying to match $M_Y(t)$ in a textbook appendix, we can (in theory) derive the pdf that is associated with it by means of the Inversion Theorem. This is particularly important if the derived cf is not of a standard (or common) form. Recall that the characteristic function (cf) is defined by

$$C(t) = \int_{-\infty}^{\infty} e^{itx} f(x)\,dt. \qquad (2.18)$$

Then, the Inversion Theorem is given by:

Inversion Theorem: The characteristic function $C(t)$ uniquely determines the pdf $f(x)$ via

$$f(x) = \frac{1}{2\pi} \int_{-\infty}^{\infty} e^{-itx} C(t)\,dt \qquad (2.19)$$

Proof: See Roussas (1997, p. 142) or Stuart and Ord (1994, p. 126).

If the mgf exists, one can replace $C(t)$ with $M(it)$ in (2.19). Inverting a characteristic function is often computationally difficult. With *Mathematica*, one can take two approaches: symbolic inversion and numerical inversion.

Symbolic inversion: If we think of (2.18) as the Fourier transform $f(x) \to C(t)$, then (2.19) is the inverse Fourier transform $C(t) \to f(x)$ which can be implemented in *Mathematica* via:

```
InverseFourierTransform[ cf, t, x, FourierParameters→{1,1}]
```

To further automate this mapping, we shall create a function `InvertCF[t → x, cf]`. Moreover, we shall allow this function to take an optional third argument, `InvertCF[t → x, cf, assume]`, which we can use to make assumptions about x, such as $x > 0$, or $x \in$ Reals. Here is the code for `InvertCF`:

```
InvertCF[t_ → x_, cf_, Assum_:{}] :=
  Module[{sol},
    sol = InverseFourierTransform[cf, t, x,
                        FourierParameters→{1,1}];
    If[Assum === {}, sol, FullSimplify[sol, Assum]]]
```

Numerical inversion: There are many characteristic functions that *Mathematica* cannot invert symbolically. In such cases, we can resort to numerical methods. We can automate the inversion (2.19) $C(t) \to f(x)$ using numerical integration, by constructing a function `NInvertCF[t → x, cf]`:

```
NInvertCF[t_ → x_, cf_] :=
  1
  ─── NIntegrate[ e^(-i t x) cf, {t, -∞, 0, ∞},
  2 π
                        Method → DoubleExponential]
```

The syntax `{t,-∞,0,∞}` tells *Mathematica* to check for singularities at 0.

⊕ *Example 17:* Linnik Distribution

The distribution whose characteristic function is

$$C(t) = \frac{1}{1 + |t|^\alpha}, \quad t \in \mathbb{R}, \ 0 < \alpha \le 2 \qquad (2.20)$$

is known as a Linnik distribution; this is also known as an α-Laplace distribution. The standard Laplace distribution is obtained when $\alpha = 2$. Consider the case $\alpha = \frac{3}{2}$:

```
cf =    1
     ─────────── ;
     1 + Abs[t]^(3/2)
```

Inverting the cf symbolically yields the pdf $f(x)$:

```
f = InvertCF[t → x, cf]
```

$$\frac{1}{4\sqrt{3}\,\pi^{7/2}} \text{MeijerG}\Big[\Big\{\Big\{\tfrac{1}{12}, \tfrac{1}{3}, \tfrac{7}{12}\Big\}, \{\}\Big\},$$

$$\Big\{\Big\{0, \tfrac{1}{12}, \tfrac{1}{3}, \tfrac{1}{3}, \tfrac{7}{12}, \tfrac{2}{3}, \tfrac{5}{6}\Big\}, \Big\{\tfrac{1}{6}, \tfrac{1}{2}\Big\}\Big\}, \tfrac{x^6}{46656}\Big]$$

where `domain[f] = {x, -∞, ∞}`. Figure 10 compares the $\alpha = \frac{3}{2}$ pdf to the $\alpha = 2$ pdf.

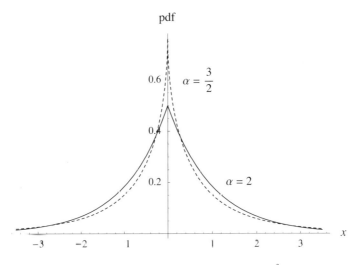

Fig. 10: The pdf of the Linnik distribution, when $\alpha = \frac{3}{2}$ and 2

⊕ *Example 18:* The Sum of Uniform Random Variables

Let $(X_1, \ldots, X_n)$ be independent Uniform(0, 1) random variables, each with characteristic function $C(t) = \frac{e^{it}-1}{it}$. It follows from the MGF Theorem that the cf of $Y = \sum_{i=1}^n X_i$ is:

```
cf = ( e^(it) - 1 )^n
     ( ─────────── )  ;
     (    it      )
```

§2.4 D   CONTINUOUS RANDOM VARIABLES   55

The pdf of Y is known as the Irwin–Hall distribution, and it can be obtained in *Mathematica*, for a given value of n, by inverting the characteristic function cf. For instance, when $n = 1, 2, 3$, the pdf's are, respectively, f1, f2, f3:

{f1, f2, f3} = InvertCF[t → y, cf /. n → {1, 2, 3}, y > 0]

$$\left\{\frac{1}{2}\left(1 + \text{Sign}[1-y]\right), \frac{1}{2}\left(y + \text{Abs}[-2+y] - 2\,\text{Abs}[-1+y]\right),\right.$$
$$\frac{1}{4}\left(y^2 + 3\,(-1+y)^2\,\text{Sign}[1-y] + \right.$$
$$\left.\left.(-3+y)^2\,\text{Sign}[3-y] + 3\,(-2+y)^2\,\text{Sign}[-2+y]\right)\right\}$$

Figure 11 plots the three pdf's. When $n = 1$, we obtain the Uniform(0, 1) distribution, $n = 2$ yields a Triangular distribution, while $n = 3$ already looks somewhat bell-shaped.

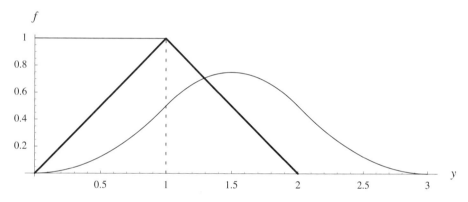

Fig. 11: The pdf of the sum of n Uniform(0, 1) random variables, when $n = 1, 2, 3$

⊕ *Example 19:* Numerical Inversion

Consider the distribution whose characteristic function is:

$$\text{cf} = e^{-\frac{t^2}{2}} + \sqrt{\frac{\pi}{2}}\,t\left(\text{Erf}\left[\frac{t}{\sqrt{2}}\right] - \text{Sign}[t]\right);$$

Alas, *Mathematica* Version 4 cannot invert this cf symbolically; that is, InvertCF[t→x, cf] fails. However, by using the NInvertCF function defined above, we can numerically invert the cf at a specific point such as $x = 2.9$, which yields the pdf evaluated at $x = 2.9$:

NInvertCF[t → 2.9, cf]

0.0467289 + 0. i

By doing this at many points, we can plot the pdf:

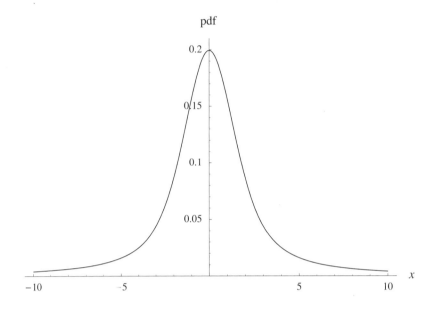

Fig. 12: The pdf, now obtained by numerically inverting the cf

2.4 E Stable Distributions

According to the Central Limit Theorem, the sum of a large number of iid random variables with finite variance converges to a Normal distribution (which is itself a special member of the stable family) when suitably standardised. If the finite variance assumption is dropped, one obtains a Generalised Central Limit Theorem, which states that the resulting limiting distribution must be a member of the stable class. The word 'stable' is used because, informally speaking, when iid members of a stable family are added together, the shape of the distribution does not change. Stable distributions are becoming increasingly important in empirical work. For example, in finance, financial returns are the sum of an enormous number of separate trades that arrive continuously in time. Yet, the distribution of financial returns often has fatter tails and more skewness than is consistent with Normality; by contrast, non-Gaussian stable distributions can often provide a better description of the data. For more detail on stable distributions, see Uchaikin and Zolotarev (1999), Nolan (2001), and McCulloch (1996).

Formally, a *stable distribution* $S(\alpha, \beta, c, a)$ is a 4-parameter distribution with characteristic function $C(t)$ given by

$$C(t) = \begin{cases} \exp\left(ait - c|t|^\alpha \{1 + i\beta\,\text{sign}(t)\,\tan(\tfrac{\pi}{2}\alpha)\}\right) & \text{if } \alpha \neq 1 \\ \exp\left(ait - c|t|\,\{1 + i\beta\,\text{sign}(t)\,\tfrac{2}{\pi}\log|t|\}\right) & \text{if } \alpha = 1 \end{cases} \qquad (2.21)$$

where $0 < \alpha \le 2$, $-1 \le \beta \le 1$, $c > 0$ and $a \in \mathbb{R}$. Parameter α is known as the 'characteristic exponent' and controls tail behaviour, β is a skewness parameter, c is a scale parameter, and a is a location parameter. Since the shape parameters α and β are of primary interest, we will let $S(\alpha, \beta)$ denote $S(\alpha, \beta, 1, 0)$. Then $C(t)$ reduces to

$$C(t) = \begin{cases} \exp\left(-|t|^\alpha \{1 + i\beta \operatorname{sign}(t) \tan(\tfrac{\pi}{2}\alpha)\}\right) & \text{if } \alpha \ne 1 \\ \exp\left(-|t| \{1 + i\beta \operatorname{sign}(t) \tfrac{2}{\pi} \log|t|\}\right) & \text{if } \alpha = 1 \end{cases} \quad (2.22)$$

with support,

$$\text{support } f(x) = \begin{cases} \mathbb{R}_+ & \text{if } \alpha < 1 \text{ and } \beta = -1 \\ \mathbb{R}_- & \text{if } \alpha < 1 \text{ and } \beta = 1 \\ \mathbb{R} & \text{otherwise} \end{cases} \quad (2.23)$$

If $\alpha \le 1$, the mean does not exist; if $1 < \alpha < 2$, the mean exists, but the variance does not; if $\alpha = 2$ (the Normal distribution), both the mean and the variance exist. A symmetry property is that $f(x; \alpha, \beta) = f(-x; \alpha, -\beta)$. Thus, if the skewness parameter $\beta = 0$, we have $f(x; \alpha, 0) = f(-x; \alpha, 0)$, so that the pdf is symmetrical about zero. In *Mathematica*, we shall stress the dependence of the cf on its parameters α and β by defining the cf (2.22) as a *Mathematica* function of α and β, namely cf[α,β]:

```
Clear[cf]
```

$$\texttt{cf}[\alpha\_, \beta\_] := \texttt{Exp}\left[-\texttt{Abs}[\texttt{t}]^\alpha \left(1 + \mathrm{i}\,\beta\,\texttt{Sign}[\texttt{t}] \,*\, \texttt{If}\left[\alpha == 1, \frac{2}{\pi}\texttt{Log}[\texttt{Abs}[\texttt{t}]], \texttt{Tan}\left[\frac{\pi}{2}\alpha\right]\right]\right)\right]$$

In the usual fashion, inverting the cf yields the pdf. Surprisingly, there are only three known stable pdf's that can be expressed in terms of *elementary* functions, and they are:

(i) *The Normal Distribution:* Let $\alpha = 2$; then the cf is:

```
cf[2, β]
```

$e^{-\texttt{Abs}[\texttt{t}]^2}$

which simplifies to e^{-t^2} for $t \in \mathbb{R}$. Inverting the cf yields a Normal pdf (the InvertCF function was defined in §2.4 D above):

```
f = InvertCF[t → x, cf[2, β]]
```

$$\frac{e^{-\frac{x^2}{4}}}{2\sqrt{\pi}}$$

(ii) *The Cauchy Distribution:* Let $\alpha = 1$ and $\beta = 0$; then the cf and pdf are:

cf[1, 0]

$e^{-\text{Abs}[t]}$

f = InvertCF[t → x, cf[1, 0]]

$$\frac{1}{\pi + \pi x^2}$$

(iii) *The Levy Distribution:* Let $\alpha = \frac{1}{2}$, $\beta = -1$; then the cf is:

cf[$\frac{1}{2}$, -1]

$e^{-\sqrt{\text{Abs}[t]}\,(1-i\,\text{Sign}[t])}$

which, when inverted, yields the Levy pdf:

f = InvertCF[t → x, cf[$\frac{1}{2}$, -1], x > 0]
domain[f] = {x, 0, ∞};

$$\frac{e^{-\frac{1}{2x}}}{\sqrt{2\pi}\, x^{3/2}}$$

Here is a plot of the Levy pdf:

PlotDensity[f, {x, 0, 6}];

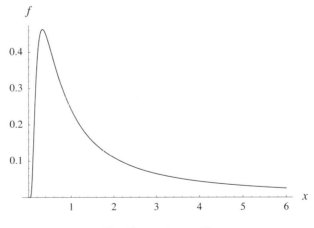

Fig. 13: The Levy pdf

The Levy distribution may also be obtained as a special case of the InverseGamma(γ, b) distribution with $\gamma = \frac{1}{2}$ and $b = 2$.

○ Only Three Known pdf's?

It is often claimed that, aside from the Normal, Cauchy and Levy, no other stable pdf can be expressed in terms of known functions. This is not quite true: it depends on which functions are *known*. Hoffman-Jørgenson (1993) showed that some stable densities can be expressed in terms of hypergeometric $_pF_q$ functions, while Zolotarev (1995) showed more generally that some stable pdf's can be expressed in terms of MeijerG functions. Quite remarkably, *Mathematica* can often derive symbolic stable pdf's in terms of $_pF_q$ functions, without any extra help! To illustrate, suppose we wish to find the pdf of $S(\frac{1}{2}, 0)$. Inverting the cf in the standard way yields:

```
ff = InvertCF[t → x, cf[1/2, 0], x ∈ Reals]
```

$$\frac{1}{4\pi \, \text{Abs}[x]^{7/2}}$$
$$\left(-2 \, \text{Abs}[x]^{3/2} \, \text{HypergeometricPFQ}\left[\{1\}, \left\{\frac{3}{4}, \frac{5}{4}\right\}, -\frac{1}{64 x^2}\right] + \sqrt{2\pi} \, x^2 \left(\text{Cos}\left[\frac{1}{4x}\right] + \text{Sign}[x] \, \text{Sin}\left[\frac{1}{4x}\right] \right) \right)$$

Since *Mathematica* does not handle densities containing Abs[x] very well, we shall eliminate the absolute value term by considering the $x < 0$ and $x > 0$ cases separately:

```
f_ = Simplify[ ff /. Abs[x] → -x, x < 0];
f_+ = Simplify[ ff, x > 0];
```

and then re-express the $S(\frac{1}{2}, 0)$ stable density as:

```
f = If[x < 0, f_, f_+];   domain[f] = {x, -∞, ∞};
```

Note that we are now working with a stable pdf in symbolic form that is *neither* Normal, Cauchy, nor Levy. Further, because it is a symbolic entity, we can apply standard **mathStatica** functions in the usual way. For instance, Expect[x, f] correctly finds that the integral does not converge, while the cdf $F(x) = P(X \le x)$ is obtained in the familiar way, as a symbolic entity!

```
F = Prob[x, f]
```

$$\text{If}\left[x < 0, \; i\left(\text{FresnelC}\left[\frac{1}{\sqrt{2\pi}\sqrt{x}}\right] - \text{FresnelS}\left[\frac{1}{\sqrt{2\pi}\sqrt{x}}\right]\right) + \frac{\text{HypergeometricPFQ}\left[\left\{\frac{1}{2}, 1\right\}, \left\{\frac{3}{4}, \frac{5}{4}, \frac{3}{2}\right\}, -\frac{1}{64 x^2}\right]}{2\pi x}, \right.$$
$$\left. 1 - \frac{\text{FresnelC}\left[\frac{1}{\sqrt{2\pi}\sqrt{x}}\right] - \text{FresnelS}\left[\frac{1}{\sqrt{2\pi}\sqrt{x}}\right] + \frac{\text{HypergeometricPFQ}\left[\left\{\frac{1}{2}, 1\right\}, \left\{\frac{3}{4}, \frac{5}{4}, \frac{3}{2}\right\}, -\frac{1}{64 x^2}\right]}{2\pi x} \right]$$

Figure 14 plots the pdf and cdf.

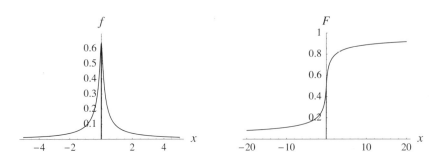

Fig. 14: The $S(\frac{1}{2}, 0)$ pdf and cdf

More generally, examples fall into two classes: those that can be inverted symbolically, and those that can only be inverted numerically. To illustrate, we shall consider $S(\frac{1}{2}, \beta)$ using symbolic methods; then $S(1, \beta)$ via numerical methods, and finally $S(\frac{3}{2}, \beta)$ with both numerical and symbolic methods, all plotted when $\beta = 0, \frac{1}{2}, 1$. Figures 15–17 illustrate these cases: as usual, the code to generate these diagrams is given in the electronic version of the text, along with some discussion.

2.4 F Cumulants and Probability Generating Functions

The *cumulant generating function* is the natural logarithm of the mgf. The rth *cumulant*, κ_r, is given by

$$\kappa_r = \left. \frac{d^r \log(M(t))}{d t^r} \right|_{t=0} \quad (2.24)$$

provided $M(t)$ exists. Unlike the raw and central moments, cumulants can not generally be obtained by direct integration. To find them, one must either derive them from the cumulant generating function, or use the moment conversion functions of §2.4 G.

The *probability generating function* (pgf) is

$$\Pi(t) = E[t^X] \quad (2.25)$$

and is mostly used when working with discrete random variables defined on the set of non-negative integers $\{0, 1, 2, \ldots\}$. The pgf provides a way to determine the probabilities. For instance:

$$P(X = r) = \frac{1}{r!} \left. \frac{d^r \Pi(t)}{d t^r} \right|_{t=0}, \quad r = 0, 1, 2, \ldots. \quad (2.26)$$

The pgf can also be used as a *factorial moment generating function*. For instance, the *factorial moment*

$$\mu[r] = E[X^{[r]}] = E[X(X-1) \cdots (X-r+1)]$$

may be obtained from $\Pi(t)$ as follows:

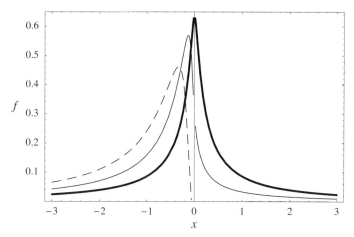

Fig. 15: $S(\frac{1}{2}, \beta)$ with $\beta = 0, \frac{1}{2}, 1$ (bold, plain, dashed)

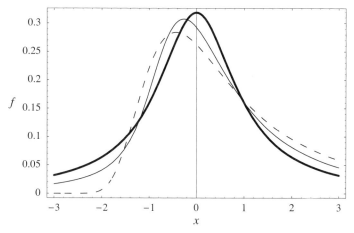

Fig. 16: $S(1, \beta)$ with $\beta = 0, \frac{1}{2}, 1$ (bold, plain, dashed)

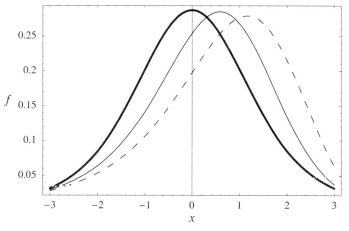

Fig. 17: $S(\frac{3}{2}, \beta)$ with $\beta = 0, \frac{1}{2}, 1$ (bold, plain, dashed)

$$\mu'[r] = E[X^{[r]}] = \left.\frac{d^r \Pi(t)}{d t^r}\right|_{t=1} \tag{2.27}$$

where we note that t is set to 1 and not 0. To convert from factorial moments to raw moments, see the `FactorialToRaw` function of §2.4 G.

2.4 G Moment Conversion Formulae

One can express any moment (μ', μ, or κ) in terms of any other moment (μ', μ, or κ). To this end, **mathStatica** provides a suite of functions to automate such conversions. The supported conversions are:

function	description
`RawToCentral[r]`	μ'_r in terms of μ_i
`RawToCumulant[r]`	μ'_r in terms of κ_i
`CentralToRaw[r]`	μ_r in terms of μ'_i
`CentralToCumulant[r]`	μ_r in terms of κ_i
`CumulantToRaw[r]`	κ_r in terms of μ'_i
`CumulantToCentral[r]`	κ_r in terms of μ_i
and	
`RawToFactorial[r]`	μ'_r in terms of $\mu'[i]$
`FactorialToRaw[r]`	$\mu'[r]$ in terms of μ'_i

Table 3: Univariate moment conversion functions

For instance, to express the 2nd central moment (the variance) $\mu_2 = E[(X - \mu)^2]$ in terms of raw moments μ'_i, we enter:

CentralToRaw[2]

$\mu_2 \to -{\mu'_1}^2 + \mu'_2$

This is just the well-known result that $\mu_2 = E[X^2] - (E[X])^2$. Here are the first 6 central moments in terms of raw moments:

Table[CentralToRaw[i], {i, 6}] // TableForm

$\mu_1 \to 0$

$\mu_2 \to -{\mu'_1}^2 + \mu'_2$

$\mu_3 \to 2{\mu'_1}^3 - 3\mu'_1 \mu'_2 + \mu'_3$

$\mu_4 \to -3{\mu'_1}^4 + 6{\mu'_1}^2 \mu'_2 - 4\mu'_1 \mu'_3 + \mu'_4$

$\mu_5 \to 4{\mu'_1}^5 - 10{\mu'_1}^3 \mu'_2 + 10{\mu'_1}^2 \mu'_3 - 5\mu'_1 \mu'_4 + \mu'_5$

$\mu_6 \to -5{\mu'_1}^6 + 15{\mu'_1}^4 \mu'_2 - 20{\mu'_1}^3 \mu'_3 + 15{\mu'_1}^2 \mu'_4 - 6\mu'_1 \mu'_5 + \mu'_6$

§2.4 G   CONTINUOUS RANDOM VARIABLES

Next, we express the 5th raw moment in terms of cumulants:

sol = RawToCumulant[5]

$$\mu_5' \to \kappa_1^5 + 10\,\kappa_1^3\,\kappa_2 + 15\,\kappa_1\,\kappa_2^2 + 10\,\kappa_1^2\,\kappa_3 + 10\,\kappa_2\,\kappa_3 + 5\,\kappa_1\,\kappa_4 + \kappa_5$$

which is an expression in κ_i, for $i = 1, \ldots, 5$. Here are the inverse relations:

inv = Table[CumulantToRaw[i], {i, 5}]; inv // TableForm

$$\kappa_1 \to \mu_1'$$
$$\kappa_2 \to -\mu_1'^2 + \mu_2'$$
$$\kappa_3 \to 2\,\mu_1'^3 - 3\,\mu_1'\,\mu_2' + \mu_3'$$
$$\kappa_4 \to -6\,\mu_1'^4 + 12\,\mu_1'^2\,\mu_2' - 3\,\mu_2'^2 - 4\,\mu_1'\,\mu_3' + \mu_4'$$
$$\kappa_5 \to 24\,\mu_1'^5 - 60\,\mu_1'^3\,\mu_2' + 30\,\mu_1'\,\mu_2'^2 + 20\,\mu_1'^2\,\mu_3' - 10\,\mu_2'\,\mu_3' - 5\,\mu_1'\,\mu_4' + \mu_5'$$

Substituting the inverse relations back into sol yields μ_5' again:

sol /. inv // Simplify

$$\mu_5' \to \mu_5'$$

Working 'about the mean' (*i.e.* taking $\kappa_1 = 0$) yields the CentralToCumulant conversions:

Table[CentralToCumulant[r], {r, 5}]

$$\{\mu_1 \to 0,\ \mu_2 \to \kappa_2,\ \mu_3 \to \kappa_3,\ \mu_4 \to 3\,\kappa_2^2 + \kappa_4,\ \mu_5 \to 10\,\kappa_2\,\kappa_3 + \kappa_5\}$$

The inverse relations are given by CumulantToCentral. Here is the 5th factorial moment $\mu'[5] = E\big[X(X-1)(X-2)(X-3)(X-4)\big]$ expressed in terms of raw moments:

FactorialToRaw[5]

$$\mu'[5] \to 24\,\mu_1' - 50\,\mu_2' + 35\,\mu_3' - 10\,\mu_4' + \mu_5'$$

This is easy to confirm by noting that:

x (x - 1) (x - 2) (x - 3) (x - 4) // Expand

$$24\,x - 50\,x^2 + 35\,x^3 - 10\,x^4 + x^5$$

The inverse relations are given by RawToFactorial:

RawToFactorial[5]

$$\mu_5' \to \mu'[1] + 15\,\mu'[2] + 25\,\mu'[3] + 10\,\mu'[4] + \mu'[5]$$

○ **The Converter Functions in Practice**

Sometimes, we know how to derive one class of moments (say raw moments), but not another (say cumulants). In these situations, the converter functions come to the rescue, for they enable us to derive the unknown moments in terms of the moments that can be calculated. This section illustrates how this can be done. The general approach is: First, express the desired moment (say κ_5) in terms of moments that we can calculate (say raw moments). Then, evaluate each raw moment μ'_i for the relevant distribution.

⊕ *Example 20:* Cumulants of $X \sim \text{Beta}(a, b)$

Let random variable $X \sim \text{Beta}(a, b)$ with pdf $f(x)$:

```
f = x^(a-1) (1 - x)^(b-1) / Beta[a, b]  ;  domain[f] = {x, 0, 1} && {a > 0, b > 0} ;
```

We wish to find the fourth cumulant. To do so, we can use the cumulant generating function approach, or the moment conversion approach.

(i) The cumulant generating function is:

```
cgf = Log[Expect[e^(t x), f]]
```

Log[Hypergeometric1F1[a, a + b, t]]

Then, the fourth cumulant is given by (2.24) as:

```
D[cgf, {t, 4}] /. t → 0 // FullSimplify
```

$$\frac{6\, a\, b\, (a^3 + a^2\, (1 - 2\, b) + b^2\, (1 + b) - 2\, a\, b\, (2 + b))}{(a + b)^4\, (1 + a + b)^2\, (2 + a + b)\, (3 + a + b)}$$

(ii) Moment conversion approach: Express the fourth cumulant in terms of raw moments:

```
sol = CumulantToRaw[4]
```

$\kappa_4 \to -6\, \mu'^4_1 + 12\, \mu'^2_1\, \mu'_2 - 3\, \mu'^2_2 - 4\, \mu'_1\, \mu'_3 + \mu'_4$

Here, each term μ'_r denotes $\mu'_r(X) = E[X^r]$, and hence can be evaluated with the Expect function. In the next input, we calculate each of the expectations that we require:

```
sol /. μ_r → Expect[x^r, f] // FullSimplify
```

$$\kappa_4 \to \frac{6\, a\, b\, (a^3 + a^2\, (1 - 2\, b) + b^2\, (1 + b) - 2\, a\, b\, (2 + b))}{(a + b)^4\, (1 + a + b)^2\, (2 + a + b)\, (3 + a + b)}$$

which is the same answer. ∎

2.5 Conditioning, Truncation and Censoring

2.5 A Conditional/Truncated Distributions

Let random variable X have pdf $f(x)$, with cdf $F(x) = P(X \le x)$. Further, let a and b be constants lying within the support of the domain. Then, the conditional density is

$$f(x \mid a < X \le b) = \frac{f(x)}{F(b) - F(a)} \qquad \text{Doubly truncated} \qquad (2.28)$$

$$f(x \mid X > a) = \frac{f(x)}{1 - F(a)} \quad (\text{let } b = \infty) \qquad \text{Truncated below} \qquad (2.29)$$

$$f(x \mid X \le b) = \frac{f(x)}{F(b)} \quad (\text{let } a = -\infty) \qquad \text{Truncated above} \qquad (2.30)$$

These conditional distributions are also sometimes known as *truncated distributions*. In each case, the conditional density on the left-hand side is expressed in terms of the unconditional (parent) pdf $f(x)$ on the right-hand side, which is adjusted by a scaling constant in the denominator so that the density still integrates to unity.

Proof of (2.30): The conditional probability that event Ω_1 occurs, given event Ω_2, is

$$P(\Omega_1 \mid \Omega_2) = \frac{P(\Omega_1 \cap \Omega_2)}{P(\Omega_2)} \text{ provided } P(\Omega_2) \ne 0.$$

$$\therefore P(X \le x \mid X \le b) = \frac{P(X \le x \cap X \le b)}{P(X \le b)} = \frac{P(X \le x)}{P(X \le b)} \text{ provided } x \le b.$$

$$\therefore F(x \mid X \le b) = \frac{F(x)}{F(b)}. \text{ Differentiating both sides with respect to } x \text{ yields (2.30).} \quad \square$$

⊕ **Example 21:** A 'Truncated Above' Standard Normal Distribution

```
ClearAll[f, F, g, b]
```

Let $X \sim N(0, 1)$ with pdf $f(x)$:

$$\mathtt{f} = \frac{e^{-\frac{x^2}{2}}}{\sqrt{2\pi}}; \qquad \mathtt{domain[f]} = \{\mathtt{x}, -\infty, \infty\};$$

and cdf $F(x)$:

```
F[x_] = Prob[x, f];
```

Let $g(x) = f(x \mid X \le b) = \dfrac{f(x)}{F(b)}$ denote a standard Normal pdf truncated above at b:

```
g = f/F[b];    domain[g] = {x, -∞, b} && {b ∈ Reals};
```

Figure 18 plots $g(x)$ at three different values of b.

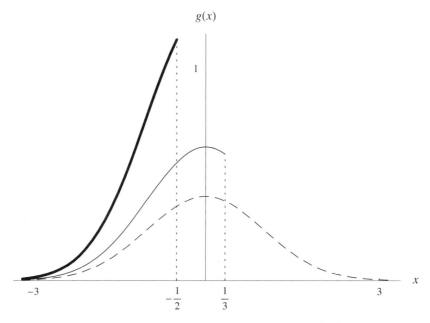

Fig. 18: A standard Normal pdf truncated above at $b = -\tfrac{1}{2}, \tfrac{1}{3}, \infty$

2.5 B Conditional Expectations

Let X have pdf $f(x)$. We wish to find the *conditional expectation* $E_f[u(X) \mid a < X \le b]$, where the notation $E_f[\,\cdot\,]$ indicates that the expectation is taken with respect to the random variable X whose pdf is $f(x)$. From (2.28), it follows that

$$E_f[u(X) \mid a < X \le b] = \frac{\int_a^b u(x)\, f(x)\, dx}{F(b) - F(a)}. \tag{2.31}$$

With **mathStatica**, an easier method is to first derive the conditional density via (2.28), say $g(x) = f(x \mid a < X \le b)$ with `domain[g] = {x, a, b}`. Then,

$$E_f[u(X) \mid a < X \le b] = E_g[u(X)]. \tag{2.32}$$

§2.5 B CONTINUOUS RANDOM VARIABLES

⊕ **Example 22:** Mean and Variance of a 'Truncated Above' Normal

Continuing *Example 21*, we have $X \sim N(0, 1)$ with pdf $f(x)$ (the parent distribution), and $g(x) = f(x \mid X \leq b)$ (a *truncated above* distribution). We wish to find $E_f[X \mid X \leq b]$. The solution is $E_g[X]$:

Expect[x, g]

$$-\frac{e^{-\frac{b^2}{2}} \sqrt{\frac{2}{\pi}}}{1 + \mathrm{Erf}\left[\frac{b}{\sqrt{2}}\right]}$$

Because $g(x)$ is 'truncated above' while $f(x)$ is not, it must always be the case that $E_g[X] < E_f[X]$. As b becomes 'large', the truncation becomes less severe, so $E_g[X] \to E_f[X]$. Thus, for our example, as $b \to \infty$, $E_g[X] \to 0$ from below, as per Fig. 19 (left panel). At the other extreme, as $b \to -\infty$, the 45° line forms an upper bound, since $E_g[X] \leq b$, if $X \leq b$.

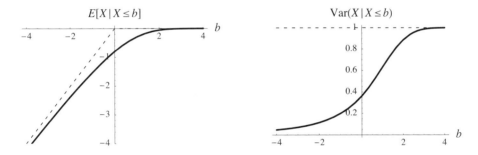

Fig. 19: Conditional mean (left) and variance (right) as a function of b

Similarly, the variance of a truncated distribution must always be smaller than the variance of its parent distribution, because the truncated distribution is a constrained version of the parent. As b becomes 'large', this constraint becomes insignificant, and so $\mathrm{Var}_g(X) \to \mathrm{Var}_f(X)$ from below. By contrast, as b tends toward the lower bound of the domain, truncation becomes more and more binding, causing the conditional variance to tend to 0, as per Fig. 19 (right panel). The conditional variance $\mathrm{Var}(X \mid X \leq b)$ is:

Var[x, g]

$$1 - \frac{2\,e^{-b^2}}{\pi \left(1 + \mathrm{Erf}\left[\frac{b}{\sqrt{2}}\right]\right)^2} - \frac{b\,e^{-\frac{b^2}{2}} \sqrt{\frac{2}{\pi}}}{1 + \mathrm{Erf}\left[\frac{b}{\sqrt{2}}\right]}$$

Finally, we Clear some symbols:

ClearAll[f, F, g]

… to prevent notational conflicts in future examples. ∎

2.5 C Censored Distributions

Consider the following examples:

(i) The demand for tickets to a concert is a random variable. Actual ticket sales, however, are bounded by the fixed capacity of the concert hall.
(ii) Similarly, electricity consumption (a random variable) is constrained above by the capacity of the grid.
(iii) The water level in a dam fluctuates randomly, but it can not exceed the physical capacity of the dam.
(iv) In some countries, foreign exchange rates are allowed to fluctuate freely within a band, but if they reach the edge of the band, the monetary authority intervenes to prevent the exchange rate from leaving the band.

Examples (i) and (ii) draw the distinction between *observed* data (*e.g.* ticket sales, electricity supply) and *unobserved* demand (some people may have been unable to purchase tickets). Examples (iii) and (iv) fall into the general class of stochastic processes that are bounded by reflecting (sticky) barriers; see Rose (1995). All of these examples (i–iv) can be modelled using censored distributions.

Let random variable X have pdf $f(x)$ and cdf $F(x)$, and let c denote a constant lying within the support of the domain. Then, Y has a *censored distribution*, censored below at point c, if

$$Y = \begin{cases} c & \text{if } X \le c \\ X & \text{if } X > c \end{cases} \qquad (2.33)$$

Figure 20 compares the pdf of X (the parent distribution) with the pdf of Y (the censored distribution). While X has a continuous pdf, the density of Y has both a discrete part and a continuous part. Here, all values of X smaller than c get compacted onto a single point c: thus, the point c occurs with positive probability $F(c)$.

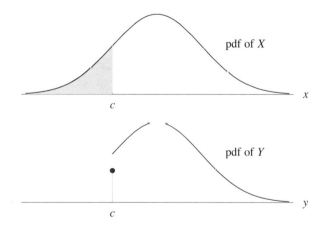

Fig. 20: Parent pdf (top) and censored pdf (bottom)

The definitions for a 'censored above' distribution, and a 'doubly censored' distribution (censored above and below) follow in similar fashion.

§2.5 C   CONTINUOUS RANDOM VARIABLES   69

⊕ *Example 23:* A 'Censored Below' Normal Distribution

ClearAll[f, c]

Let $X \sim N(0, 1)$ with pdf $f(x)$, and let $Y = \begin{cases} c & \text{if } X \le c \\ X & \text{if } X > c \end{cases}$. We enter all this as:

$$\mathbf{f} = \frac{e^{-\frac{x^2}{2}}}{\sqrt{2\pi}}; \quad \mathbf{domain[f]} = \{\mathbf{x}, -\infty, \infty\}; \quad \mathbf{y} = \mathbf{If[x \le c, c, x]};$$

Then, $E[Y]$ is:

Expect[y, f]

$$\frac{e^{-\frac{c^2}{2}}}{\sqrt{2\pi}} + \frac{1}{2} c \left(1 + \text{Erf}\left[\frac{c}{\sqrt{2}}\right]\right)$$

Note that this expression is equal to $f(c) + c F(c)$, where $F(c)$ is the cdf of X evaluated at the censoring point c. Similarly, Var(Y) is:

Var[y, f]

$$\frac{1}{4} \left(2 + c^2 - \frac{2 e^{-c^2}}{\pi} + \left(-2 - 2 c e^{-\frac{c^2}{2}} \sqrt{\frac{2}{\pi}}\right) \text{Erf}\left[\frac{c}{\sqrt{2}}\right] - c^2 \text{Erf}\left[\frac{c}{\sqrt{2}}\right]^2 \right)$$

Figure 21 plots $E[Y]$ and Var(Y) as a function of the censoring point c.

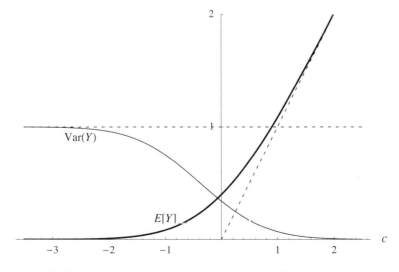

Fig. 21: The mean and variance of Y, plotted at different values of c

2.5 D   Option Pricing

Financial options are an interesting application of censored distributions. To illustrate, let time $t = 0$ denote today, let $\{S(t), t \geq 0\}$ denote the price of a stock at time t, and let S_T denote the stock price at a fixed future 'expiry' date $T > 0$. A European *call option* is a financial asset that gives its owner the right (but not the obligation) to buy stock at time T at a fixed price k (called the *strike price*). For example, if you own an Apple call option expiring on 19 July with strike $k = \$100$, it means you have the right to buy one share in Apple Computer at a price of $100 on July 19. If, on July 19, the stock price S_T is greater than $k = \$100$, the value of your option on the expiry date is $S_T - k$; however, if S_T is less than $100, it would not be worthwhile to purchase at $100, and so your option would have zero value. Thus, the value of a call option *at expiry* T is:

$$V_T = \begin{cases} S_T - k & \text{if } S_T > k \\ 0 & \text{if } S_T \leq k \end{cases} \quad (2.34)$$

We now know the value of an option at expiry — what then is the value of this option *today*, at $t = 0$, prior to expiry? At $t = 0$, the current stock price $S(0)$ is always known, while the future is of course unknown. That is, the future price S_T is a random variable whose pdf $f(s_T)$ is assumed known. Then, the value $V = V(0)$ of the option at $t = 0$ is simply the expected value of V_T, discounted for the time value of money between expiry ($t = T$) and today ($t = 0$):

$$V = V(0) = e^{-rT} E[V_T] \quad (2.35)$$

where r denotes the risk-free interest rate. This is the essence of option pricing, and we see that it rests crucially on censoring the distribution of future stock prices, $f(s_T)$.

⊕ **Example 24:** Black–Scholes Option Pricing (via Censored Distributions)

The Black–Scholes (1973) option pricing model is now quite famous, as acknowledged by the 1997 Nobel Memorial Prize in economics.[3] For our purposes, we just require the pdf of future stock prices $f(s_T)$. This, in turn, requires some stochastic calculus; readers unfamiliar with stochastic calculus can jump directly to (2.38) where $f(s_T)$ is stated, and proceed from there.

If investors are risk neutral,[4] and stock prices follow a geometric Brownian motion, then

$$\frac{dS}{S} = r\,dt + \sigma\,dz \quad (2.36)$$

with drift r and instantaneous standard deviation σ, where z is a Wiener process. By Ito's Lemma, this can be expressed as the ordinary Brownian motion

$$d\log(S) = \left(r - \frac{\sigma^2}{2}\right)dt + \sigma\,dz \quad (2.37)$$

so that $d\log(S_T) \sim N((r - \frac{\sigma^2}{2})T, \sigma^2 T)$. Expressing $d\log(S_T)$ as $\log(S_T) - \log(S(0))$, it then follows that

$$\log(S_T) \sim N(a, b^2) \quad \text{where} \quad \begin{cases} a = \log(S(0)) + \left(r - \frac{\sigma^2}{2}\right)T \\ b = \sigma\sqrt{T} \end{cases} \quad (2.38)$$

That is, $S_T \sim \text{Lognormal}(a, b^2)$, with pdf $f(s_T)$:

```
f = ─────────── Exp[- (Log[s_T] - a)² ];
    s_T b √(2 π)              2 b²

domain[f] = {s_T, 0, ∞} && {a ∈ Reals, b > 0};
```

The value of the option at expiry, V_T, may be entered via (2.34) as:

```
V_T = If[s_T > k, s_T - k, 0];
```

while the value $V = V(0)$ of a call option today is given by (2.35):

```
V = e^(-r T) Expect[V_T, f]
```

$$\frac{1}{2} e^{-rT} \left(-k \left(1 + \text{Erf}\left[\frac{a - \text{Log}[k]}{\sqrt{2}\, b}\right]\right) + e^{a + \frac{b^2}{2}} \left(1 + \text{Erf}\left[\frac{a + b^2 - \text{Log}[k]}{\sqrt{2}\, b}\right]\right) \right)$$

where a and b were defined in (2.38). This result is, in fact, identical to the Black–Scholes solution, though our derivation here via expectations is quite different (and much simpler) than the solution via partial differential equations used by Black and Scholes. Substituting in for a and b, and denoting today's stock price $S(0)$ by p, we have:

```
Value = V /. {a → Log[p] + (r - σ²/2) T, b → σ √T};
```

For example, if the current price of Apple stock is $p = S(0) = \$104$, the strike price is $k = \$100$, the interest rate is 5%, the volatility is 44% per annum ($\sigma = .44$), and there are 66 days left to expiry ($T = \frac{66}{365}$), then the value today (in \$) of the call option is:

```
Value /. {p → 104, k → 100, r → .05, σ → .44, T → 66/365}
```

```
10.2686
```

More generally, we can plot the value of our call option as a function of the current stock price p, as shown in Fig. 22.

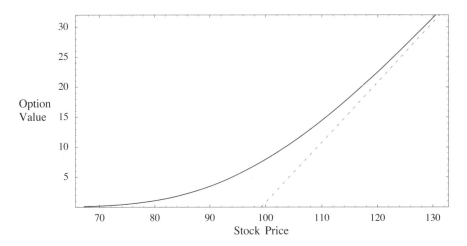

Fig. 22: Value of a call option as a function of today's stock price

As $p \to 0$, we become certain that $S_T < k$. Referring to (2.34), this means that as $p \to 0$, $V \to 0$, as Fig. 22 shows. By contrast, as $p \to \infty$, $P(S_T > k) \to 1$, so we become certain that $S_T > k$, and thus $V \to e^{-rT} E[S_T - k]$. The latter is equal to $p - e^{-rT} k$, as the reader can verify with `Expect[s_T - k, f]` and then substituting in for a and b. This explains the asymptotes in Fig. 22.

Many interesting comparative static calculations are now easily obtainable with *Mathematica*; for example, we can find the rate of change of option value with respect to σ as a symbolic entity with `D[Value,σ]//Simplify`. ∎

2.6 Pseudo-Random Number Generation

This section discusses different ways to generate pseudo-random drawings from a given distribution. If the distribution happens to be included in *Mathematica*'s Statistics package, the easiest approach is often to use the `Random[distribution]` function included in that package (§2.6 A). Of course, this is not a general solution, and it breaks down as soon as one encounters a distribution that is not in that package.

In the remaining parts of this section (§2.6 B–D), we discuss procedures that allow, in principle, any distribution to be sampled. We first consider the Inverse Method, which requires that both the cdf and inverse cdf can be computed, using either symbolic (§2.6 D) or numerical (§2.6 C) methods. Finally, §2.6 D discusses the Rejection Method, where neither the cdf nor the inverse cdf is required. Random number generation for discrete random variables is discussed in Chapter 3.

2.6 A *Mathematica*'s Statistics Package

The *Mathematica* statistics packages, `ContinuousDistributions`` and `NormalDistribution``, provide built-in pseudo-random number generation for well-known distributions such as the Normal, Gamma, and Cauchy. If we want to generate pseudo-random numbers from one of these well-known distributions, the simplest solution is to use these packages. They can be loaded as follows:

```
<< Statistics`
```

Suppose we want to generate pseudo-random drawings from a Gamma(a, b) distribution:

$$f = \frac{x^{a-1} e^{-x/b}}{\Gamma[a]\, b^a}; \qquad \text{domain}[f] = \{x, 0, \infty\}\ \&\&\ \{a > 0, b > 0\};$$

If $a = 2$ and $b = 3$, a single pseudo-random drawing is obtained as follows:

```
dist = GammaDistribution[2, 3];    Random[dist]

8.61505
```

while 10000 pseudo-random values can be generated with:

```
data = RandomArray[dist, 10000];
```

The **mathStatica** function, FrequencyPlot, can be used to compare this 'empirical' data with the true pdf $f(x)$:

```
FrequencyPlot[data, f /. {a → 2, b → 3}];
```

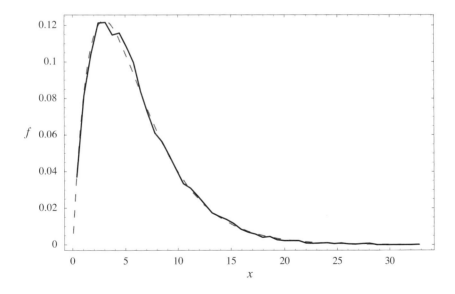

Fig. 23: The empirical pdf (———) and true pdf (– – –)

While it is certainly convenient to have pre-written code for special well-known distributions, this approach must, of course, break down as soon as we consider a distribution that is not in the package. Thus, more general methods are needed.

2.6 B   Inverse Method (Symbolic)

Let random variable X have pdf $f(x)$, cdf $p = F(x)$ and inverse cdf $x = F^{-1}(p)$, and let u be a pseudo-random drawing from Uniform(0, 1). Then a pseudo-random drawing from $f(x)$ is given by

$$x = F^{-1}(u) \tag{2.39}$$

In order for the *Inverse Method* to work efficiently, the inverse function $F^{-1}(\cdot)$ should be computationally tractable. Here is an example with the Levy distribution, with pdf $f(x)$:

$$f = \frac{e^{-\frac{1}{2x}}}{\sqrt{2\pi}\, x^{3/2}}; \quad \text{domain[f]} = \{x, 0, \infty\};$$

The cdf $F(x)$ is given by:

F = Prob[x, f]

$$1 - \text{Erf}\left[\frac{1}{\sqrt{2}\,\sqrt{x}}\right]$$

while the inverse cdf is:

inv = Solve[u == F, x] // Flatten

— Solve::ifun : Inverse functions are being used by Solve, so some solutions may not be found.

$$\left\{x \to \frac{1}{2\,\text{InverseErf}[0, 1-u]^2}\right\}$$

When u = Random[], this rule generates a pseudo-random Levy drawing x. More generally, if the inverse yields more than one possible solution, we would have to select the appropriate solution before proceeding. We now generate 10000 pseudo-random numbers from the Levy pdf, by replacing u with Random[]:

data = Table$\left[\dfrac{1}{2\,\text{InverseErf}[0, 1-\text{Random}[\,]]^2}, \{10000\}\right]$; // Timing

{2.36 Second, Null}

It is always a good idea to check the data set before continuing. The output here should only consist of positive real numbers. To check, here are the last 10 values:

Take[data, -10]

{6.48433, 0.229415, 3.70733, 4.53735, 0.356657,
0.646354, 1.09913, 0.443604, 1.17306, 0.532637}

§2.6 B   CONTINUOUS RANDOM VARIABLES   75

These numbers seem fine. We use the **mathStatica** function `FrequencyPlot` to inspect fit, and superimpose the parent density $f(x)$ on top:

```
FrequencyPlot[data, {0, 10, .1}, f];
```

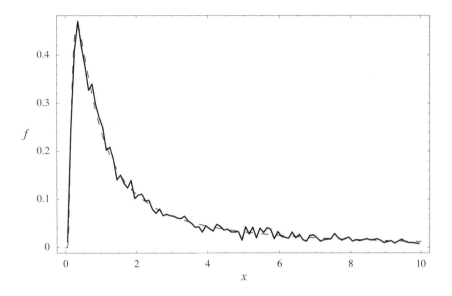

Fig. 24: The empirical pdf (——) and true pdf (– – –)

Some caveats: The Inverse Method can only work if we can determine both the cdf and its inverse. Inverse functions are tricky, and *Mathematica* may occasionally experience some difficulty in this regard. Also, since one ultimately has to work with a numerical density (*i.e.* numerical parameter values) when generating pseudo-random numbers, it is often best to specify parameter values at the very start — this makes it easier to calculate both the cdf and the inverse cdf.

2.6 C   Inverse Method (Numerical)

If it is difficult or impossible to find the inverse cdf symbolically, we can resort to doing so numerically. To illustrate, let random variable X have a half-Halo distribution with pdf $f(x)$:

```
f = 2/π √(1 - (x - 2)²);   domain[f] = {x, 1, 3};
```

with cdf $F(x)$:

```
F = Prob[x, f]
```

$$\frac{(-2 + x)\sqrt{-(-3 + x)(-1 + x)} + \text{ArcCos}[2 - x]}{\pi}$$

Mathematica cannot invert this cdf symbolically; that is, `Solve[u==F,x]` fails. Nevertheless, we can derive the inverse cdf using numerical methods. We do so by evaluating (F, x) at a finite number of different values of x, and then use interpolation to fill in the gaps in between these known points. How then do we decide at which values of x we should evaluate (F, x)? This is the same type of problem that *Mathematica*'s `Plot` function has to solve each time it makes a plot. So, following Abbott (1996), we use the `Plot` function to automatically select the values of x at which $(F(x), x)$ is to be constructed, and then record these values in a list called `lis`. The larger the number of `PlotPoints`, the more accurate will be the end result:

```
lis = {};
Plot[ (ss = F; AppendTo[lis, {ss, x}]; ss),  {x, 1, 3},
   PlotPoints → 2000,
   PlotRange → All, AxesLabel → {"x", "F"}];
```

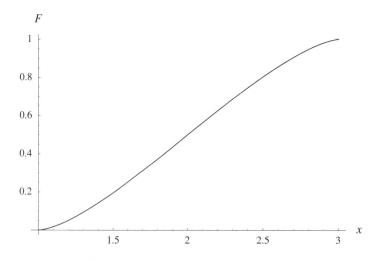

Fig. 25: The cdf $F(x)$ plotted as a function of x

Mathematica's `Interpolation` function is now used to fill in the gaps between the chosen points. We shall take the `Union` of `lis` so as to eliminate duplicate values that the `Plot` function can sometimes generate. Here, then, is our numerical inverse cdf function:

```
InverseCDF = Interpolation[Union[lis]]

InterpolatingFunction[{{1.89946 × 10⁻¹⁴, 1.}}, <>]
```

Here are 60000 pseudo-random drawings from the half-Halo distribution:

```
data = Table[ InverseCDF[ Random[] ], {60000}];
 // Timing

{1.1 Second, Null}
```

Figure 26 compares this pseudo-random data with the true pdf $f(x)$:

`FrequencyPlot[data, {1, 3, .02}, f];`

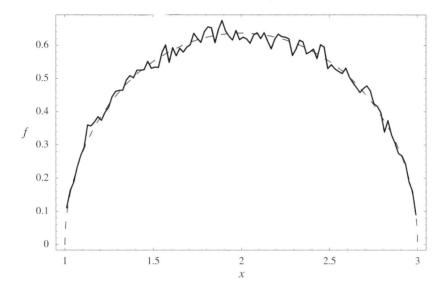

Fig. 26: The empirical pdf (———) and true half-Halo pdf (– – –)

2.6 D Rejection Method

Our objective is to generate pseudo-random numbers from some pdf $f(x)$. Sometimes, the Inverse Method may fail: typically, this happens because the cdf or the inverse cdf has an intractable functional form. In such cases, the Rejection Method can be very helpful — it provides a way to generate pseudo-random numbers from $f(x)$ (which we do *not* know how to do) by generating pseudo-random numbers from a density $g(x)$ (which we *do* know how to generate). Density $g(x)$ should have the following properties:

- $g(x)$ is defined over the same domain as $f(x)$, and

- there exists a constant $c > 0$ such that $\frac{f(x)}{g(x)} \le c$ for all x. That is, $c = \sup\left(\frac{f(x)}{g(x)}\right)$.

Let x_g denote a pseudo-random drawing from $g(x)$, and let u denote a pseudo-random drawing from the Unifom(0, 1) distribution. Then, the *Rejection Method* generates pseudo-random drawings from $f(x)$ in three steps:

The Rejection Method

(1) Generate x_g and u.

(2) If $u \le \frac{1}{c} \frac{f(x_g)}{g(x_g)}$, accept x_g as a random selection from $f(x)$.

(3) Else, return to step (1).

To illustrate, let $f(x)$ denote the pdf of a Birnbaum–Saunders distribution, with parameters α and β. This distribution has been used to represent the lifetime of components. We wish to generate pseudo-random drawings from $f(x)$ when say $\alpha = \frac{1}{2}$, $\beta = 4$:

```
f = e^(-(x-β)²/(2α² β x)) (x + β) / (2 α √(2 π β) x^(3/2))  /. {α → 1/2, β → 4};

domain[f] = {x, 0, ∞} && {α > 0, β > 0};
```

The Inverse Method will be of little help to us here, because *Mathematica* Version 4 cannot find the cdf of this distribution. Instead, we try the Rejection Method. We start by choosing a density $g(x)$. Suitable choices for $g(x)$ might include the Lognormal or the Levy (§2.6 B) or the Chi-squared(n), because each of these distributions has a similar shape to $f(x)$; this is easy to verify with a plot. We use Chi-squared(n) here, with $n = 4$:

```
g = x^(n/2-1) e^(-x/2) / (2^(n/2) Γ[n/2])  /. n → 4;   domain[g] = {x, 0, ∞};
```

Note that $g(x)$ is defined over the same domain as $f(x)$. Moreover, we can easily check whether $c = \sup\left(\frac{f(x)}{g(x)}\right)$ exists, by doing a quick plot of $\frac{f(x)}{g(x)}$.

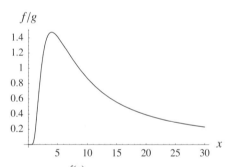

Fig. 27: $\frac{f(x)}{g(x)}$ plotted as a function of x

This suggests that c is roughly equal to 1.45. We can find the value of c more accurately using numerical methods:

```
c = FindMaximum[f/g, {x, 3, 6]][[1]]
```

1.4739

We can easily generate pseudo-random drawings x_g from $g(x)$ using *Mathematica*'s `Statistics` package:

```
<< Statistics`

dist = ChiSquareDistribution[4];   x_g = Random[dist]
```

18.8847

By step (2) of the Rejection Method, we accept x_g as a random selection from $f(x)$ if $u \le Q(x_g)$, where $Q(x_g) = \dfrac{1}{c} \dfrac{f(x_g)}{g(x_g)}$. We enter $Q(x)$ into *Mathematica* as follows:

```
Q[x_] = 1  f  // Simplify
       -  -
       c  g
```

$$\dfrac{29.5562 \, e^{-8/x} \, (4+x)}{x^{5/2}}$$

Steps (1) – (3) can now be modelled in just one line, by setting up a recursive function. In the following input, note how x_g (a pseudo-random Chi-squared drawing) is used to generate x_f (a pseudo-random Birnbaum–Saunders drawing):

```
x_f :=
  (x_g = Random[dist]; u = Random[]; If[u ≤ Q[x_g], x_g, x_f])
```

So, let us try it out ... here are 10000 pseudo-random Birnbaum–Saunders drawings:

```
data = Table[x_f, {10000}];
```

Check the fit:

```
FrequencyPlot[data, f];
```

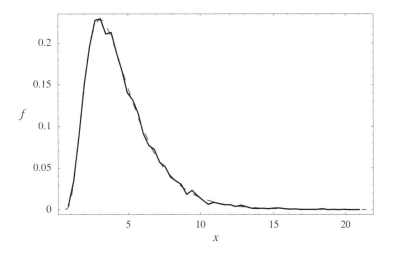

Fig. 28: The empirical pdf (——) and true pdf (– – –)

The Rejection Method is most useful when working with densities $f(x)$ that are not covered by *Mathematica*'s Statistics package, and for which the symbolic Inverse Method does not work. When using the Rejection Method, density $g(x)$ should be chosen so that it is easy to generate from, and is as similar in shape to $f(x)$ as possible. It is also worth checking that, at each stage of the process, output is numerical.

2.7 Exercises

1. Let continuous random variable X have a semi-Circular (half-Halo) distribution with pdf $f(x) = 2\sqrt{1-x^2}/\pi$ and domain of support $x \in (-1, 1)$. Plot the density $f(x)$. Find the cdf $P(X \leq x)$ and plot it. Find the mean and the variance of X.

2. Azzalini (1985) showed that if random variable X has a pdf $f(x)$ that is symmetric about zero, with cdf $F(x)$, then $2 f(x) F(\lambda x)$ is also a pdf, for parameter $\lambda \in \mathbb{R}$. In particular, when X is $N(0, 1)$, the density $g(x) = 2 f(x) F(\lambda x)$ is known as Azzalini's skew-Normal distribution. Find $g(x)$. Plot density $g(x)$ when $\lambda = 0, 1$ and 2. Find the mean and variance. Find upper and lower bounds on the variance.

3. Let $X \sim \text{Lognormal}(\mu, \sigma)$. Find the rth raw moment, the cdf, p^{th} quantile, and mode.

4. Let $f(x)$ denote a standard Normal pdf; further, let pdf $g(x) = (2\pi)^{-1}(1 + \cos(x))$, with domain of support $x \in (-\pi, \pi)$. Compare $f(x)$ with $g(x)$ by plotting both on a diagram. From the plot, which distribution has greater kurtosis? Verify your choice by calculating Pearson's measure of kurtosis.

5. Find the yth quantile for a standard Triangular distribution. Hence, find the median.

6. Let $X \sim \text{InverseGaussian}(\mu, \sigma)$ with pdf $f(x)$. Find the first 3 negative moments (i.e. $E[X^{-1}], E[X^{-2}], E[X^{-3}]$). Find the mgf, if it exists.

7. Let X have pdf $f(x) = \text{Sech}[x]/\pi$, $x \in \mathbb{R}$, which is known as the Hyperbolic Secant distribution. Derive the cf, and then the first 12 raw moments. Why are the odd-order moments zero?

8. Find the characteristic function of X^2, if $X \sim N(\mu, \sigma^2)$.

9. Find the cdf of the stable distribution $S(\frac{2}{3}, -1)$ as an exact symbolic entity.

10. The distribution of IQ in Class E2 at Rondebosch Boys High School is $X \sim N(\mu, \sigma^2)$. Mr Broster, the class teacher, decides to break the class into two streams: Stream 1 for those with IQ $> \omega$, and Stream 2 for those with IQ $\leq \omega$.
    (i) Find the average (expected) IQ in each stream, for any chosen value of ω.
    (ii) If $\mu = 100$ and $\sigma = 16$, plot (on one diagram) the average IQ in each stream as a function of ω.
    (iii) If $\mu = 100$ and $\sigma = 16$, how should Mr Broster choose ω if he wants:
    (a) the same number of students in each stream?
    (b) the average IQ of Stream 1 to be twice the average of Stream 2?
    For each case (a)–(b), find the average IQ in each stream.

11. Apple Computer is planning to host a live webcast of the next Macworld Conference. Let random variable X denote the number of people (measured in thousands) wanting to watch the live webcast, with pdf $f(x) = \frac{1}{144} e^{-x/12} x$, for $x > 0$. Find the expected number of people who want to watch the webcast. If Apple's web server can handle at most c simultaneous live streaming connections (measured in thousands), find the expected number of people who will be able to watch the webcast as a function of c. Plot the solution as a function of c.

12. Generate 20000 pseudo-random drawings from Azzalini's ($\lambda = 1$) skew-Normal distribution (see Exercise 2), using the exact inverse method (symbolic).

Chapter 3

Discrete Random Variables

3.1 Introduction

In this chapter, attention turns to random variables induced from experiments defined on countable, or enumerable, sample spaces. Such random variables are termed *discrete*; their values can be mapped in one-to-one correspondence with the set of integers. For example, the experiment of tossing a coin has two possible outcomes, a head and a tail, which can be mapped to 0 and 1, respectively. Accordingly, random variable X, taking values $x \in \{0, 1\}$, represents the experiment and is discrete.

The distinction between discrete and continuous random variables is made clearer by considering the *cumulative distribution function* (cdf). For a random variable X (discrete or continuous), its cdf $F(x)$, as a function of x, is defined as

$$F(x) = P(X \leq x), \quad \text{for all } x \in \mathbb{R}. \tag{3.1}$$

Now inspect the following cdf plots given in Fig. 1.

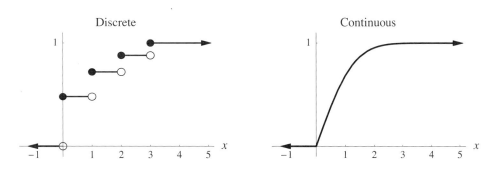

Fig. 1: Discrete and Continuous cumulative distribution functions

The left-hand panel depicts the cdf of a discrete random variable. It appears in the form of a step function. By contrast, the right-hand panel shows the cdf of a continuous random variable. Its cdf is everywhere continuous.

○ *List Form and Function Form*

The discrete random variable X depicted in Fig. 1 takes values 0, 1, 2, 3, with probability 0.48, 0.24, 0.16, 0.12, respectively. We can represent these details about X in two ways, namely *List Form* and *Function Form*. Table 1 gives List Form.

$P(X = x)$:	0.48	0.24	0.16	0.12
x:	0	1	2	3

Table 1: List Form

We enter List Form as:

```
f₁ = {0.48, 0.24, 0.16, 0.12};

domain[f₁] = {x, {0, 1, 2, 3}} && {Discrete};
```

Table 2 gives Function Form.

$$P(X = x) = \frac{12}{25(x+1)}; \quad x \in \{0, 1, 2, 3\}$$

Table 2: Function Form

We enter Function Form as:

```
f₂ = 12 / (25 (x + 1));

domain[f₂] = {x, 0, 3} && {Discrete};
```

Both List Form (f_1) and Function Form (f_2) express the same facts about X, and both are termed the *probability mass function* (pmf) of X. Notice especially the condition {Discrete} added to the domain statements. This is the device used to tell **mathStatica** that X is discrete. Importantly, appending the discreteness flag is *not* optional; if it is omitted, **mathStatica** will interpret the random variable as continuous.

The suite of **mathStatica** functions can operate on a pmf whether it is in List Form or Function Form—as we shall see repeatedly throughout this chapter. Here, for example, is a plot of the pmf of X from the List Form:

PlotDensity[f₁];

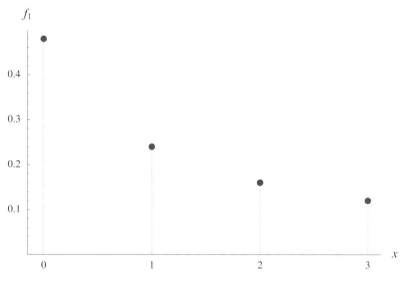

Fig. 2: The pmf of X

As a further illustration, here is the mean of X, using f_2:

Expect[x, f₂]

$\frac{23}{25}$

In general, the *expectation* of a function $u(X)$, where X is a discrete random variable, is defined as

$$E[u(X)] = \sum_x u(x) P(X = x) \qquad (3.2)$$

where summation is over all values x of X. For example, here is $E[\cos(X)]$ using f_1:

Expect[Cos[x], f₁]

0.42429

§3.2 examines aspects of probability through the experiment of 'throwing' a die. §3.3 details the more popular discrete distributions encountered in practice (see the range provided in **mathStatica**'s *Discrete* palette). Mixture distributions are examined in §3.4, for they provide a means to generate many further distributions. Finally, §3.5 discusses pseudo-random number generation of discrete random variables.

There exist many fine references on discrete random variables. In particular, Fraser (1958) and Hogg and Craig (1995) provide introductory material, while Feller (1968, 1971) and Johnson *et al.* (1993) provide advanced treatments.

3.2 Probability: 'Throwing' a Die

The study of probability is often motivated by experiments such as coin tossing, drawing balls from an urn, and throwing dice. Many fascinating problems have been posed using these simple, easily replicable physical devices. For example, Ellsberg (1961) used urn drawings to show that there are (at least) two different types of uncertainty: one can be described by (the usual concept of) probability, the other cannot (ambiguity). More recently, Walley (1991, 1996) illustrated his controversial notion of imprecise probability by using drawings from a bag of marbles. Probability also attracts widespread popular interest: it can be used to analyse games of chance, and it can help us analyse many intriguing paradoxes. For further discussion of many popular problems in probability, see Mosteller (1987). For discussion of probability theory, see, for example, Billingsley (1995) and Feller (1968, 1971).

In this, and the next two sections, we examine discrete random variables whose domain of support is the set (or subset) of the non-negative integers. For discrete random variables of this type, there exist generating functions that can be useful for analysing a variable's properties. For a discrete random variable X taking non-negative integer values, the *probability generating function* (pgf) is defined as

$$\Pi(t) = E[t^X] = \sum_{x=0}^{\infty} t^x P(X = x) \qquad (3.3)$$

which is a function of dummy variable t; it exists for any choice of $t \leq 1$. The pgf is similar to the moment generating function (mgf); indeed, subject to existence conditions (see §2.4 B), the mgf $M(t) = E[\exp(t X)]$ is equivalent to $\Pi(\exp(t))$. Likewise, $\Pi(\exp(i t))$ yields the characteristic function (cf). The pgf generates probabilities via the relation,

$$P(X = x) = \frac{1}{x!} \left. \frac{d^x \Pi(t)}{d t^x} \right|_{t=0} \qquad \text{for } x \in \{0, 1, 2, \ldots\}. \qquad (3.4)$$

⊕ **Example 1:** Throwing a Die

Consider the standard six-sided die with faces labelled 1 through 6. If X denotes the upmost face resulting from throwing the die onto a flat surface, such as a table-top, then X may be thought of as a discrete random variable with pmf given in Table 3.

$P(X = x)$:	$\frac{1}{6}$	$\frac{1}{6}$	$\frac{1}{6}$	$\frac{1}{6}$	$\frac{1}{6}$	$\frac{1}{6}$
x :	1	2	3	4	5	6

Table 3: The pmf of X

This pmf presupposes that the die is fair. The pmf of X may be entered in either List Form:

```
f = Table[ 1/6, {6}];

domain[f] = {x, Range[6]} && {Discrete};
```

... or Function Form:

$$g = \frac{1}{6};$$

$$\text{domain}[g] = \{x, 1, 6\} \,\&\&\, \{\text{Discrete}\};$$

The pgf of X may be derived from either representation of the pmf; for example, for the List Form representation:

pgf = Expect[t^x, f]

$$\frac{1}{6} t \,(1 + t + t^2 + t^3 + t^4 + t^5)$$

The probabilities can be recovered from the pgf using (3.4):

Table$\left[\frac{1}{x!} \text{D[pgf, \{t, x\}]}, \{x, 1, 6\}\right]$ /. t → 0

$$\left\{\frac{1}{6}, \frac{1}{6}, \frac{1}{6}, \frac{1}{6}, \frac{1}{6}, \frac{1}{6}\right\}$$

⊕ *Example 2:* The Sum of Two Die Rolls

Experiments involving more than one die have often been contemplated. For example, in 1693, Samuel Pepys wrote to Isaac Newton seeking an answer to a die roll experiment, apparently posed by a Mr Smith. Smith's question concerned the chances of throwing a minimum number of sixes with multiple dice: at least 1 six from a throw of a box containing 6 dice, at least 2 sixes from another box containing 12 dice, and 3 or more sixes from a third box filled with 18 dice. We leave this problem as an exercise for the reader to solve (see §3.3 B for some clues). The correspondence between Newton and Pepys, including Newton's solution, is given in Schell (1960).

The experiment we shall pose is the sum S obtained from tossing two fair dice, X_1 and X_2. The outcomes of X_1 and X_2 are independent, and their distribution is identical to that of X given in *Example 1*. We wish to derive the pmf of $S = X_1 + X_2$. In order to do so, consider its pgf:

$$\Pi_S(t) = E[t^S] = E[t^{X_1 + X_2}].$$

By independence $E[t^{X_1 + X_2}] = E[t^{X_1}] \, E[t^{X_2}]$ and by identicality $E[t^{X_1}] = E[t^{X_2}] = E[t^X]$, so $\Pi_S(t) = E[t^X]^2$. In *Mathematica*, the pgf of S is simply:

pgfS = pgf^2

$$\frac{1}{36} t^2 \,(1 + t + t^2 + t^3 + t^4 + t^5)^2$$

Now the domain of support of S is the integers from 2 to 12, so by (3.4) the pmf of S, say $h(s)$, in List Form, is:

```
h = Table[1/s! D[pgfS, {t, s}], {s, 2, 12}] /. t → 0
```

$$\left\{\frac{1}{36}, \frac{1}{18}, \frac{1}{12}, \frac{1}{9}, \frac{5}{36}, \frac{1}{6}, \frac{5}{36}, \frac{1}{9}, \frac{1}{12}, \frac{1}{18}, \frac{1}{36}\right\}$$

```
domain[h] = {s, Range[2, 12]} && {Discrete}
```

{s, {2, 3, 4, 5, 6, 7, 8, 9, 10, 11, 12}} && {Discrete}

⊕ ***Example 3:*** The Sum of Two Unfair Dice

Up until now, any die we have 'thrown' has been fair or, rather, assumed to be fair. It can be fun to consider the impact on an experiment when an unfair die is used! With an unfair die, the algebra of the experiment can rapidly become messy, but it is in such situations that *Mathematica* typically excels. There are two well-known methods of die corruption: loading it (attaching a weight to the inside of a face) and shaving it (slicing a thin layer from a face). In this example, we contemplate a shaved die. A shaved die is no longer a cube, and its total surface area is less than that of a fair die. Shaving upsets the relative equality in surface area of the faces. The shaved face, along with its opposing face, will have relatively more surface area than all the other faces. Consider, for instance, a die whose 1-face has been shaved. Then both the 1-face and the 6-face (opposing faces of a die sum to 7) experience no change in surface area, whereas the surface area of all the other faces is reduced.[1] Let us denote the increase in the probability of a 1 or 6 by δ. Then the probability of each of 2, 3, 4 and 5 must decrease by $\delta/2$ ($0 \le \delta < 1/3$). The List Form pmf of X, a 1-face shaved die, is thus:

```
f = {1/6 + δ, 1/6 - δ/2, 1/6 - δ/2, 1/6 - δ/2, 1/6 - δ/2, 1/6 + δ};
domain[f] = {x, Range[6]} && {Discrete};
```

We now repeat the experiment given in *Example 2*, only this time we use dice which are 1-face shaved. We may derive the List Form pmf of S exactly as before:

```
pgf = Expect[t^x, f];        pgfS = pgf^2;
h = Table[1/s! D[pgfS, {t, s}], {s, 2, 12}] /. t → 0 //
    Simplify
```

$$\left\{\frac{1}{36}(1 + 6\delta)^2, \frac{1}{18} + \frac{\delta}{6} - \delta^2, \frac{1}{12} - \frac{3\delta^2}{4}, \frac{1}{18}(2 - 3\delta - 9\delta^2),\right.$$
$$\frac{1}{36}(5 - 12\delta - 9\delta^2), \frac{1}{6} + 3\delta^2, \frac{1}{36}(5 - 12\delta - 9\delta^2),$$
$$\left.\frac{1}{18}(2 - 3\delta - 9\delta^2), \frac{1}{12} - \frac{3\delta^2}{4}, \frac{1}{18} + \frac{\delta}{6} - \delta^2, \frac{1}{36}(1 + 6\delta)^2\right\}$$

```
domain[h] = {s, Range[2, 12]} && {Discrete};
```

Figure 3 depicts the pmf of S using both fair and unfair dice. Both distributions are symmetric about their mean, 7, with a greater probability with shaved 1-face dice of sums of 2, 7 and 12. Moreover, as the distribution now appears fatter-tailed under shaved 1-face dice, we would expect the variability of its distribution to increase — a fact that can be verified by executing `Var[s, h /. δ → {0, 0.1}]`.

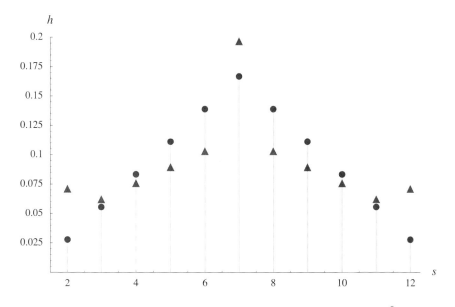

Fig. 3: The pmf of S for fair dice (●) and 1-face shaved dice (▲)

⊕ *Example 4:* The Game of Craps

The game of craps is a popular casino game. It involves a player throwing two dice, one or more times, until either a win or a loss occurs. The player wins on the first throw if the dice sum is 7 or 11. The player loses on the first throw if a sum of either 2, 3 or 12 occurs. If on the first throw the player neither wins nor loses, then the sum of the dice is referred to as the *point*. The game proceeds with the player throwing the dice until either the point occurs, in which case the player wins, or a sum of 7 occurs, in which case the player loses. When the dice are fair, it can be shown that the probability of the player winning is $244/495 \approx 0.49293$.

It is an interesting task to verify the probability of winning the game with *Mathematica*. However, for the purposes of this example, we use simulation methods to estimate the probability of winning. The following inputs combine to simulate the outcome of one game — returning 1 for a win, and 0 for a loss. First, here is a simple function that simulates the roll of a fair die:

```
TT := Random[Integer, {1, 6}]
```

Next is a function that simulates the first throw of the game, deciding whether to stop or simulate further throws:

```
Game := (s = TT + TT;
        Which[s == 7 || s == 11,       1,
   Ω           s == 2 || s == 3 || s == 12,   0,
              True,                    MoreThrows[s]])
```

Finally, if more than one throw is needed to complete the game:

```
MoreThrows[p_] := (s = TT + TT;
                  Which[s == p,  1,
                        s == 7,  0,
                        True,    MoreThrows[p]])
```

Notice that the `MoreThrows` function calls itself if a win or loss does not occur. In practice this will not result in an infinite recurrence because the probability that the game continues forever is zero. Let our estimator be the proportion of wins across a large number of games. Here is a simulated estimate of the probability of winning a game:

```
SampleMean[ Table[Game, {100000}] ] // N
```

0.49162

As a further illustration of simulation, suppose that a gambler starting with an initial fortune of \$5 repeatedly wagers \$1 against an infinitely rich opponent — the House — until his fortune is lost. Assuming that a win pays 1 to 1, the progress of his fortune from one game to the next can be represented by the function:

```
fortune[x_] := x - 1 + 2 Game
```

For example, here is one particular sequence of 10 games:

```
NestList[fortune, 5, 10]
```

{5, 4, 5, 4, 3, 4, 5, 6, 5, 4, 3}

After these games, his fortune has dropped to \$3, but as he is not yet ruined, he can still carry on gaming! Now suppose we wish to determine how many games the player can expect to play until ruin. To solve this, we take as our estimator the average length of a large number of matches. Here we simulate just 100 matches, and measure the length of each match:

```
matchLength = Table[
  NestWhileList[fortune, 5, Positive] // Length, {100}] - 1
```

{4059, 7, 37, 3193, 5, 5, 171, 45, 35, 15, 61, 573, 15, 125, 39, 67, 33,
13, 73, 11, 287, 27, 89, 49, 13, 3419, 2213, 4081, 11, 89, 697, 127,
179, 125, 33, 31, 9, 59, 973, 51, 5, 53, 613, 13, 13, 19, 19, 105, 53,
29, 163, 561, 107, 11, 25, 5, 435, 35, 7, 21, 27, 33, 19, 147, 61, 339,
101, 53, 239, 51, 23, 23, 403, 439, 6327, 7, 85, 5, 35, 107, 125, 49,
83, 33, 17, 439, 29, 15, 49, 9, 103, 13, 35, 43, 107, 145, 9, 45, 27, 81}

Then our estimate is:

SampleMean[matchLength] // N

334.16

In fact, the simulator estimator has performed reasonably well in this small trial, for the exact solution to the expected number of games until ruin can be shown to equal:

$$5 \Big/ \left(\frac{251}{495} - \frac{244}{495} \right) \text{ // N}$$

353.571

For details on the gambler's ruin problem see, for example, Feller (1968, Chapter 14). ∎

3.3 Common Discrete Distributions

This section presents a series of discrete distributions frequently applied in statistical practice: the Bernoulli, Binomial, Poisson, Geometric, Negative Binomial, and Hypergeometric distributions. Each distribution can be input into *Mathematica* with the **mathStatica** *Discrete* palette. The domain of support for all of these distributions is the set (or subset) of non-negative integers.

3.3 A   The Bernoulli Distribution

The Bernoulli distribution (named for Jacques Bernoulli, 1654–1705) is a fundamental building block in statistics. A Bernoulli distributed random variable has a two-point support, 0 and 1, with probability p that it takes the value 1, and probability $1-p$ that it is zero-valued. Experiments with binary outcomes induce a Bernoulli distributed random variable; for example, the ubiquitous coin toss can be coded 0 = tail and 1 = head, with probability one-half ($p = \frac{1}{2}$) assigned to each outcome if the coin is fair.

If X is a Bernoulli distributed random variable, its pmf is given by $P(X = x) = p^x(1-p)^{1-x}$, where $x \in \{0, 1\}$, and parameter p is such that $0 < p < 1$; p is often termed the success probability. From **mathStatica**'s *Discrete* palette:

f = p^x (1 - p)^{1-x};
domain[f] = {x, 0, 1} && {0 < p < 1} && {Discrete};

For example, the mean of X is:

Expect[x, f]

p

Although simple in structure, the Bernoulli distribution forms the backbone of many important statistical models encountered in practice.

⊕ **Example 5:** A Logit Model for the Bernoulli Response Probability

Suppose that sick patients are given differing amounts of a curative drug, and they respond to treatment after a fixed period of time as either 1 = cured or 0 = sick. Assume response $X \sim$ Bernoulli(p). Let y denote the amount of the drug given to a patient. Presumably the probability p that a patient is cured depends on y, all other factors held fixed. This probability is assumed to be governed, or modelled, by the logit relation:

$$p = \frac{1}{1 + e^{-(\alpha + \beta y)}};$$

Here, $\alpha \in \mathbb{R}$ and $\beta \in \mathbb{R}$ are unknown parameters whose values we wish to deduce or estimate. This will enable us to answer, for example, the following type of question: "If a patient receives a dose y^*, what is his chance of cure?". To illustrate, here is a set of artificial data on $n = 20$ patients:

| $x = 0$: | 7 | 17 | 14 | 3 | 15 | 19 | 11 | 6 | 20 | 12 |
| $x = 1$: | 46 | 33 | 19 | 32 | 43 | 34 | 51 | 16 | 35 | 30 |

Table 4: Dosage given (artificial data)

At the end of the treatment, 10 patients responded 0 = sick (the top row), while the remaining 10 patients were cured (the bottom row). The dosage y that each patient received appears in the body of the table. We may enter this data as follows:

```
dose = { 7, 17, 14,  3, 15, 19, 11,  6, 20, 12,
        46, 33, 19, 32, 43, 34, 51, 16, 35, 30};

response = { 0, 0, 0, 0, 0, 0, 0, 0, 0, 0,
             1, 1, 1, 1, 1, 1, 1, 1, 1, 1};
```

The observed log-likelihood function is (see Chapter 11 and Chapter 12 for further details):

```
obslogL = Log[Times @@ (f /. {y → dose, x → response})];
```

We use `FindMaximum` to find the maximum of the log-likelihood with respect to values for the unknown parameters:

```
sol = FindMaximum[obslogL, {α, 0}, {β, 0}][[2]]
```

$\{\alpha \to -7.47042, \beta \to 0.372755\}$

Given the data, the parameters of the model (α and β) have been estimated at the indicated values. The fitted model for the probability p of a cure as a function of dosage level y is given by:

p /. sol

$$\frac{1}{1 + e^{7.47042 - 0.372755\, y}}$$

The fitted p (the smooth curve) along with the data (the circled points) are plotted in Fig. 4.

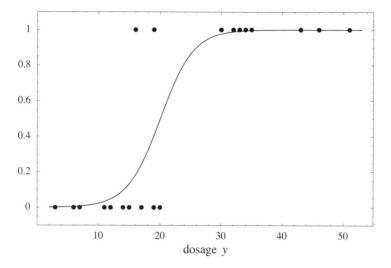

Fig. 4: Data and fitted p

Evidently, the fitted curve shows that if a patient receives a dosage of 20 units of the drug, then they have almost a 50% chance of cure (execute `Solve[(p/.sol)==0.5,y]`). Of course, room for error must exist when making a statement such as this, for we do not know the true values of the parameters α and β, nor whether the logistic formulation is the correct functional form.

Clear[p]

3.3 B   The Binomial Distribution

Let $X_1, X_2, \ldots, X_n$ be n mutually independent and identically distributed Bernoulli(p) random variables. The discrete random variable formed as the sum $X = \sum_{i=1}^{n} X_i$ is distributed as a Binomial random variable with index n and success probability p, written $X \sim \text{Binomial}(n, p)$; the domain of support of X is the integers $(0, 1, 2, \ldots, n)$. The pmf and its support may be entered directly from **mathStatica**'s *Discrete* palette:

f = Binomial[n, x] p^x (1 - p)^(n-x);

domain[f] = {x, 0, n} &&
{0 < p < 1, n > 0, n ∈ Integers} && {Discrete};

The Binomial derives its name from the expansion of $(p + q)^n$, where $q = 1 - p$. Here is the graph of the pmf, with p fixed at 0.4 and the index n taking values 10 (circles) and 20 (triangles):

```
PlotDensity[f /. {p → 0.4, n → {10, 20}}];
```

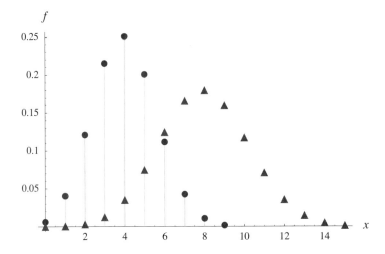

Fig. 5: Probability mass functions of X: $n = 10$, $p = 0.4$ (●); $n = 20$, $p = 0.4$ (▲)

The Binomial cdf, $P(X \leq x)$ for $x \in \mathbb{R}$, appears complicated:

```
Prob[x, f]
```

$$1 - \left((1-p)^{-1+n-\text{Floor}[x]} \, p^{1+\text{Floor}[x]} \, \Gamma[1+n] \, \text{Hypergeometric2F1}\left[1, 1-n+\text{Floor}[x], 2+\text{Floor}[x], \frac{p}{-1+p}\right] \right) \Big/ (\Gamma[n-\text{Floor}[x]] \, \Gamma[2+\text{Floor}[x]])$$

Figure 6 plots the cdf — it has the required step function appearance.

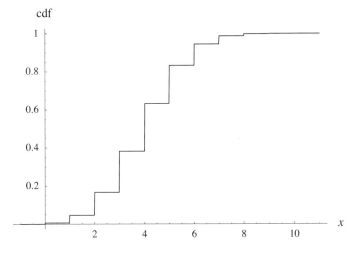

Fig. 6: The cdf of X: $n = 10$, $p = 0.4$

The mean, variance and other higher order moments of a Binomial random variable may be computed directly using Expect. For example, the mean $E[X]$ is:

μ = Expect[x, f]

n p

The variance of X is given by:

v = Var[x, f]

-n (-1 + p) p

Although the expression for the variance has a minus sign at the front, the variance is strictly positive because of the restriction on p.

Moments may also be obtained via a generating function method. Here, for example, is the central moment generating function $E[\exp(t(X - \mu))] = e^{-t\mu} E[\exp(tX)]$. In **mathStatica**:

mgfc = e$^{-t\mu}$ Expect[e^{tx}, f]

e^{-npt} (1 + (-1 + e^t) p)^n

Using mgfc, the ith central moment $\mu_i = E[(X - \mu)^i]$ is obtained by differentiation with respect to t (i times), and then setting t to zero. To illustrate, when computing Pearson's measure of kurtosis $\beta_2 = \mu_4 / \mu_2^2$:

$$\frac{D[\text{mgfc}, \{t, 4\}] /. t \to 0}{v^2} \text{ // FullSimplify}$$

$$\frac{-1 + 3 (-2 + n) (-1 + p) p}{n (-1 + p) p}$$

ClearAll[μ, v]

The Binomial distribution has a number of linkages to other statistical distributions. For example, if $X \sim$ Binomial(n, p) with mean $\mu = np$ and variance $\sigma^2 = np(1 - p)$, then the standardised discrete random variable

$$Y = \frac{X - np}{\sqrt{np(1-p)}}$$

has a limiting $N(0, 1)$ distribution, as n becomes large. In some settings, the Binomial distribution is itself a limiting distribution; *cf.* the Ehrenfest Urn. The Binomial distribution is also linked to another common discrete distribution — the Poisson distribution — which is discussed in §3.3 C.

⊕ **Example 6:** The Ehrenfest Urn

In physics, the Ehrenfest model describes the exchange of heat between two isolated bodies. In probabilistic terms, the model can be formulated according to urn drawings. Suppose there are two urns, labelled A and B, containing, in total, m balls. Starting at time $t = 0$ with some initial distribution of balls, the experiment proceeds at each $t \in \{1, 2, ...\}$ by randomly drawing a ball (from the entire collection of m balls) and then moving it from its present urn into the other. This means that if urn A contains $k \in \{0, 1, 2, ..., m\}$ balls (so B contains $m - k$ balls), and if the chosen ball is in urn A, then there are now $k - 1$ balls in A and $m - k + 1$ in B. On the other hand, if the chosen ball was in B, then there are now $k + 1$ balls in A and one fewer in B. Let X_t denote the number of balls in urn A at time t. Then, X_{t+1} depends only on X_t, its value being either one more or one less. Because each variable in the sequence $\{X_t\} = (X_1, X_2, X_3, ...)$ depends only on its immediate past, $\{X_t\}$ is said to form a Markov chain. The *conditional* pmf of X_{t+1}, given that $X_t = k$, appears Bernoulli-like, with support points $k + 1$ and $k - 1$.

When the chosen ball comes from urn B, we have

$$P(X_{t+1} = k + 1 \mid X_t = k) = \frac{m - k}{m} = 1 - \frac{k}{m} \qquad (3.5)$$

while if the chosen ball comes from urn A, we have

$$P(X_{t+1} = k - 1 \mid X_t = k) = \frac{k}{m}. \qquad (3.6)$$

The so-called limiting distribution of the sequence $\{X_t\}$ is often of interest; it is sometimes termed the long-run *unconditional* pmf of X_t.[2] It is given by the list of probabilities $p_0, p_1, p_2, ..., p_m$, and may be found by solving the simultaneous equation system,

$$p_k = \sum_{j=0}^{m} p_j P(X_{t+1} = k \mid X_t = j), \qquad k \in \{0, 1, 2, ..., m\} \qquad (3.7)$$

along with the adding-up condition,

$$p_0 + p_1 + p_2 + \cdots + p_m = 1. \qquad (3.8)$$

Substituting (3.5) and (3.6) into equations (3.7) yields, with some work, the equation system written as a function of m:

```
Ehrenfest[m_] := Join[
    Table[p_k == (1 - (k - 1)/m) p_{k-1} + (k + 1)/m p_{k+1}, {k, 1, m - 1}],
    {p_0 == p_1/m, p_m == p_{m-1}/m, Sum[p_i, {i, 0, m}] == 1}]
```

To illustrate, let $m = 5$ be the total number of balls distributed between the two urns. The long-run pmf is obtained as follows:

```
Solve[Ehrenfest[5], Table[p_i, {i, 0, 5}]]
```

$$\left\{\left\{p_0 \to \frac{1}{32},\ p_1 \to \frac{5}{32},\ p_2 \to \frac{5}{16},\ p_3 \to \frac{5}{16},\ p_4 \to \frac{5}{32},\ p_5 \to \frac{1}{32}\right\}\right\}$$

Now consider the Binomial$(m, \frac{1}{2})$ distribution, whose pmf is given by:

```
f = Binomial[m, x] (1/2)^m;

domain[f] = {x, 0, m} && {Discrete};
```

Computing all probabilities finds:

```
Table[f /. m → 5, {x, 0, 5}]
```

$$\left\{\frac{1}{32},\ \frac{5}{32},\ \frac{5}{16},\ \frac{5}{16},\ \frac{5}{32},\ \frac{1}{32}\right\}$$

which is equivalent to the limiting distribution of the Ehrenfest Urn when $m = 5$. In fact, for arbitrary m, the limiting distribution of the Ehrenfest Urn is Binomial$(m, \frac{1}{2})$. ∎

3.3 C   The Poisson Distribution

The Poisson distribution is an important discrete distribution, with vast numbers of applications in statistical practice. It is particularly relevant when the event of interest has a small chance of occurrence amongst a large population; for example, the daily number of automobile accidents in Los Angeles, where there are few accidents relative to the total number of trips undertaken. In fact, a link between the Binomial distribution and the Poisson can be made by allowing the Binomial index n to become large and the success probability p to become small, but simultaneously maintaining finiteness of the mean (see *Example 2* of Chapter 8). The Poisson often serves as an approximation to the Binomial distribution. For detailed material on the Poisson distribution see, amongst others, Haight (1967) and Johnson *et al.* (1993, Chapter 4).

A discrete random variable X with pmf:

```
f = e^-λ λ^x / x!;

domain[f] = {x, 0, ∞} && {λ > 0} && {Discrete};
```

is said to be Poisson distributed with parameter $\lambda > 0$; in short, $X \sim$ Poisson(λ). Figure 7 plots the pmf when $\lambda = 5$ and $\lambda = 10$.

```
PlotDensity[f /. λ → {5, 10}];
```

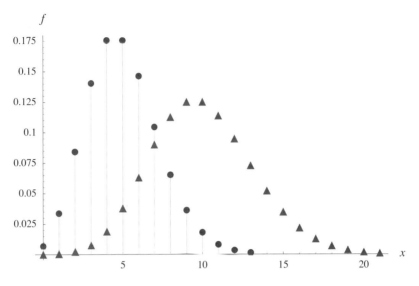

Fig. 7: Poisson pmf: $\lambda = 5$ (●), $\lambda = 10$ (▲)

From the graph, we see that the mass of the distribution shifts to the right as parameter λ increases. The Poisson mean is given by:

```
Expect[x, f]
```

λ

Curiously, the Poisson variance:

```
Var[x, f]
```

λ

... is identical to the mean. This feature alone serves to distinguish the Poisson from many other discrete distributions. Moreover, all cumulants of X are equal to λ, as can be shown by using the cumulant generating function:

```
cgf = Log[Expect[e^(t x), f]]
```

$\text{Log}[e^{(-1+e^t)\lambda}]$

To illustrate, here are the first 10 cumulants of X:

```
Table[D[cgf, {t, i}], {i, 10}] /. t → 0
```

$\{\lambda, \lambda, \lambda, \lambda, \lambda, \lambda, \lambda, \lambda, \lambda, \lambda\}$

⊕ **Example 7:** Probability Calculations

Let $X \sim \text{Poisson}(4)$ denote the number of ships arriving at a port each day. Determine:
(i) the probability that four or more ships arrive on a given day, and
(ii) repeat part (i) knowing that at least one ship arrives.

Solution: Begin by entering in X's details:

$$f = \frac{e^{-\lambda} \lambda^x}{x!} \; /. \; \lambda \to 4; \quad \text{domain}[f] = \{x, 0, \infty\} \; \&\& \; \{\text{Discrete}\};$$

(i) The required probability simplifies to $P(X \geq 4) = 1 - P(X \leq 3)$. Thus:

pp = 1 - Prob[3, f] // N

0.56653

(ii) We require the conditional probability $P(X \geq 4 \mid X \geq 1)$. For two events A and B, the conditional probability of A given B is defined as

$$P(A \mid B) = \frac{P(A \cap B)}{P(B)}, \quad \text{provided } P(B) > 0.$$

In our case, $A = \{X \geq 4\}$ and $B = \{X \geq 1\}$, so $A \cap B = \{X \geq 4\}$. We already have $P(X \geq 4)$, and $P(X \geq 1)$ may be found in the same manner. The conditional probability is thus:

$$\frac{pp}{1 - \text{Prob}[0, f]}$$

0.5771

⊕ **Example 8:** A Conditional Expectation

Suppose $X \sim \text{Poisson}(\lambda)$. Determine the conditional mean of X, given that X is odd-valued.

Solution: Enter in the details of X:

$$f = \frac{e^{-\lambda} \lambda^x}{x!}; \quad \text{domain}[f] = \{x, 0, \infty\} \; \&\& \; \{\lambda > 0\} \; \&\& \; \{\text{Discrete}\};$$

We require $E[X \mid X \in \{1, 3, 5, \ldots\}]$. This requires the pmf of $X \mid X \in \{1, 3, 5, \ldots\}$; namely, the distribution of X given that X is odd-valued, which is given by:

$$f1 = \frac{f}{\text{Sum}[\text{Evaluate}[f], \{x, 1, \infty, 2\}]};$$

$$\text{domain}[f1] = \{x, 1, \infty, 2\} \; \&\& \; \{\lambda > 0\} \; \&\& \; \{\text{Discrete}\};$$

Then, the required expectation is:

Expect[x, f1]

$\lambda \, \text{Coth}[\lambda]$

3.3 D   The Geometric and Negative Binomial Distributions

○ *The Geometric Distribution*

A Geometric experiment has similar properties to a Binomial experiment, *except* that the experiment is stopped when the first success is observed. Let p denote the probability of success in repeated independent Bernoulli trials. We are now interested in the probability that the *first* success occurs on the xth trial. Then X is said to be a Geometric random variable with pmf:

$$P(X = x) = p\,(1-p)^{x-1}, \quad x \in \{1, 2, 3, \ldots\}, \quad 0 < p < 1. \tag{3.9}$$

This can be entered with **mathStatica**'s *Discrete* palette:

f = p (1 - p)^(x-1);
domain[f] = {x, 1, ∞} && {0 < p < 1} && {Discrete};

Here, for example, is a plot of the pmf when $p = 0.6$:

PlotDensity[f /. p → .6, AxesOrigin → {3 / 4, 0}];

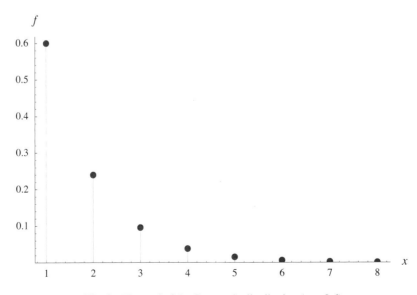

Fig. 8: The pmf of the Geometric distribution ($p = 0.6$)

§3.3 D DISCRETE RANDOM VARIABLES

○ *The Waiting-Time Negative Binomial Distribution*

A Waiting-Time Negative Binomial experiment has similar properties to the Geometric experiment, *except* that the experiment is now stopped when a *fixed* number of successes occur. As before, let p denote the probability of success in repeated independent Bernoulli trials. Of interest is the probability that the rth success occurs on the yth trial. Then Y is a Waiting-Time Negative Binomial random variable with pmf,

$$P(Y = y) = \binom{y-1}{r-1} p^r (1-p)^{y-r} \qquad (3.10)$$

for $y \in \{r, r+1, r+2, \ldots\}$ and $0 < p < 1$. We enter this as:

h = Binomial[y - 1, r - 1] p^r (1 - p)^{Y-r};
domain[h] = {y, r, ∞} && {0 < p < 1, r > 0} && {Discrete};

The mean $E[Y]$ and variance are, respectively:

Expect[y, h]

$\dfrac{r}{p}$

Var[y, h]

$\dfrac{r - p\, r}{p^2}$

○ *The Negative Binomial Distribution*

As its name would suggest, the Waiting-Time Negative Binomial distribution (3.10) is closely related to the Negative Binomial distribution. In fact, the latter may be obtained from the former by transforming $Y \to X$, such that $X = Y - r$:

f = Transform[x == y - r, h]

$(1 - p)^x\, p^r$ Binomial$[-1 + r + x, -1 + r]$

with domain:

domain[f] = TransformExtremum[x == y - r, h]

{x, 0, ∞} && {0 < p < 1, r > 0} && {Discrete}

as given in the *Discrete* palette. When r is an integer, the distribution is sometimes known as the Pascal distribution. Here is its pgf:

Expect[t^x, f]

$p^r\, (1 + (-1 + p)\, t)^{-r}$

3.3 E   The Hypergeometric Distribution

```
ClearAll[T, n, r, x]
```

Urn models in which balls are repeatedly drawn *without replacement* lead to the Hypergeometric distribution. This contrasts to sampling *with replacement* which leads to the Binomial distribution. To illustrate the former, suppose that an urn contains a total of T balls, r of which are red ($1 \le r < T$). The experiment proceeds by drawing one-by-one a sample of n balls from the urn without replacement ($1 \le n < T$).[3] Interest lies in determining the pmf of X, where X is the number of red balls drawn.

The domain of support of X is $x \in \{0, 1, \ldots, \min(n, r)\}$, where $\min(n, r)$ denotes the smaller of n and r. Next, consider the probability of a particular sequence of n draws, namely x red balls followed by $n - x$ other colours:

$$\left(\frac{r}{T} \times \frac{r-1}{T-1} \times \cdots \times \frac{r-x+1}{T-x+1} \right) \left(\frac{T-r}{T-x} \times \frac{T-r-1}{T-x-1} \times \cdots \times \frac{T-r-(n-x-1)}{T-x-(n-x-1)} \right)$$

$$= \frac{r!}{(r-x)!} \frac{(T-r)!}{(T-r-n+x)!} \frac{(T-n)!}{T!}$$

$$= \binom{T-n}{r-x} \Big/ \binom{T}{r}.$$

In total, there are $\binom{n}{x}$ arrangements of x red balls amongst the n drawn, each having the above probability. Hence, the pmf of X is

$$f(x) = \binom{n}{x}\binom{T-n}{r-x} \Big/ \binom{T}{r}$$

where $x \in \{0, 1, \ldots, \min(n, r)\}$. We may enter the pmf of X as:

```
       Binomial[n, x] Binomial[T - n, r - x]
f  =  ─────────────────────────────────────── ;
                 Binomial[T, r]

domain[f] = {x, 0, n} && {Discrete};
```

We have set `domain[f]={x,0,n}`, rather than `{x,0,Min[n,r]}`, because **mathStatica** does not support the latter. This alteration does not affect the pmf.[4]

The Hypergeometric distribution gets its name from the appearance of the Gaussian hypergeometric function in its pgf:

```
pgf = Expect[t^x, f]
```

   (Γ[1 - n + T] Hypergeometric2F1Regularized[-n,
       -r, 1 - n - r + T, t]) / (Binomial[T, r] Γ[1 + r])

Here are the mean and variance of X:

Expect[x, f]

$$\frac{n\,r}{T}$$

Var[x, f]

$$\frac{n\,r\,(n-T)\,(r-T)}{(-1+T)\,T^2}$$

⊕ *Example 9:* The Number of Ace Cards

Obtain the pmf of the distribution of the number of ace cards in a poker hand. Then plot it.

Solution: In this example, the 'urn' is the deck of $T = 52$ playing cards, and the 'red balls' are the ace cards, so $r = 4$. There are $n = 5$ cards in a hand. Therefore, the pmf of the number of ace cards in a poker hand is given by:

Table[f /. {T → 52, n → 5, r → 4}, {x, 0, 4}]

$$\left\{ \frac{35673}{54145},\ \frac{3243}{10829},\ \frac{2162}{54145},\ \frac{94}{54145},\ \frac{1}{54145} \right\}$$

which we may plot as:

**PlotDensity[f /. {T → 52, n → 5, r → 4}, {x, 0, 4},
  AxesOrigin → {-0.25, 0}];**

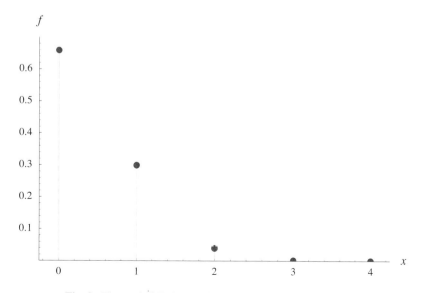

Fig. 9: The pmf of X, the number of ace cards in a poker hand

3.4 Mixing Distributions

At times, it may be necessary to use distributions with unusual characteristics, such as long-tailed behaviour or multimodality. Unfortunately, it can be difficult to write down from scratch the pdf/pmf of a distribution with the desired characteristic. Fortunately, progress can usually be made with the method of *mixing distributions*. Two prominent approaches to mixing are presented here: component-mixing (§3.4 A) and parameter-mixing (§3.4 B). The first type, component-mixing, forms distributions from linear combinations of other distributions. It is a method well-suited for generating random variables with multimodal distributions. The second type, parameter-mixing, relaxes the assumption of fixed parameters, allowing them to vary according to some specified distribution.

3.4 A Component-Mix Distributions

Component mix distributions are formed from linear combinations of distributions. To fix notation, let the pmf of a discrete random variable X_i be $f_i(x) = P(X_i = x)$ for $i = 1, \ldots, n$, and let ω_i be a constant such that $0 < \omega_i < 1$ and $\sum_{i=1}^{n} \omega_i = 1$. The linear combination of the component random variables defines the n-component-mix random variable,

$$X \sim \omega_1 X_1 + \omega_2 X_2 + \cdots + \omega_n X_n \tag{3.11}$$

and its pmf is given by the weighted average

$$f(x) = \sum_{i=1}^{n} \omega_i f_i(x). \tag{3.12}$$

Importantly, the domain of support of X is taken to be all points x contained in the union of support points of the component distributions.[5] Titterington *et al.* (1985) deals extensively with distributions formed from component-mixes.

⊕ ***Example 10:*** A Poisson Two-Component-Mix

Let $X_1 \sim$ Poisson(2) and $X_2 \sim$ Poisson(10) be independent, and set $\omega_1 = \omega_2 = \frac{1}{2}$. Plot the pmf of the two-component-mix $X \sim \omega_1 X_1 + \omega_2 X_2$.

Solution: The general form of the pmf of X can be entered directly from (3.12):

```
        e⁻ᶿ θˣ              e⁻λ λˣ
f₁ = ─────── ;       f₂ = ─────── ;       f = ω₁ f₁ + ω₂ f₂ ;
          x !                 x !
```

As both X_1 and X_2 are supported on the set of non-negative integers, then this is also the domain of support of X. As the parameter restrictions are unimportant in this instance, the domain of support of X may be entered into *Mathematica* simply as:

```
domain[f] = {x, 0, ∞} && {Discrete};
```

The plot we require is:

PlotDensity$\left[\texttt{f} \;/.\; \left\{\theta \to 2,\; \lambda \to 10,\; \omega_1 \to \frac{1}{2},\; \omega_2 \to \frac{1}{2}\right\}\right]$;

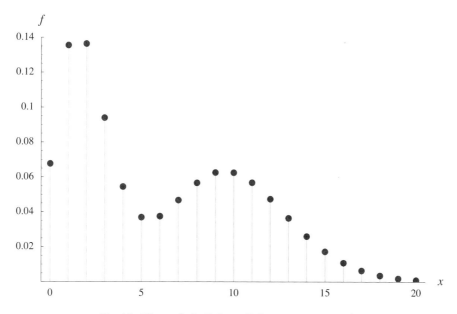

Fig. 10: The pmf of a Poisson–Poisson component-mix

For our chosen mixing weights, the pmf of X is bimodal, a feature not shared by either of the components. We return to the Poisson two-component-mix distribution in §12.6, where maximum likelihood estimation of its parameters is considered. ∎

○ *Zero-Inflated Distributions*

A survey of individual consumption patterns can often return an excessively large number of zero observations on consumption of items such as cigarettes. *Zero-Inflated distributions* can be used to model such variables. They are just a special case of (3.11), and are formed from the two-component-mix,

$$X \sim \omega_1 X_1 + \omega_2 X_2 = (1 - \omega) X_1 + \omega X_2 \qquad (3.13)$$

where, because $\omega_1 + \omega_2 = 1$, we can express the mix with a single weight ω. In this component-mix, zero-inflated distributions correspond to nominating X_1 as a degenerate distribution with all its mass at the origin; that is, $P(X_1 = 0) = 1$. The distribution of X is therefore a modification of the distribution of X_2. If the domain of support of X_2 does not include zero, then this device serves to add zero to the domain of support of X. On the other hand, if X_2 can take value zero, then $P(X = 0) > P(X_2 = 0)$ because ω is such that $0 < \omega < 1$. In both scenarios, the probability of obtaining a zero is boosted.

⊕ **Example 11:** The Zero-Inflated Poisson Distribution

Consider the two-component-mix (3.13) with $P(X_1 = 0) = 1$, and $X_2 \sim \text{Poisson}(\lambda)$. In this case, X has the so-called Zero-Inflated Poisson distribution, or ZIP for short. The pmf of X is

$$P(X = x) = \begin{cases} 1 - \omega + \omega e^{-\lambda} & \text{if } x = 0 \\ \omega \dfrac{e^{-\lambda} \lambda^x}{x!} & \text{if } x \in \{1, 2, \ldots\} \end{cases}$$

where $0 < \omega < 1$ and $\lambda > 0$. To obtain the pgf, we only require the pgf of X_2, denoted $\Pi_2(t)$, since

$$\begin{aligned} \Pi(t) &= \sum_{x=0}^{\infty} t^x P(X = x) \\ &= (1 - \omega) + \omega \sum_{x=0}^{\infty} t^x P(X_2 = x) \\ &= (1 - \omega) + \omega \Pi_2(t). \end{aligned} \quad (3.14)$$

For our example, the pgf of $X_2 \sim \text{Poisson}(\lambda)$ is:

    f₂ = $\dfrac{e^{-\lambda} \lambda^x}{x!}$;

    domain[f₂] = {x, 0, ∞} && {λ > 0} && {Discrete};

    pgf₂ = Expect[tˣ, f₂]

    $e^{(-1+t)\lambda}$

Then, by (3.14), the pgf of X is:

    pgf = (1 - ω) + ω pgf₂

    $1 - \omega + e^{(-1+t)\lambda} \omega$

Taking, for example, $\omega = 0.5$ and $\lambda = 5$, $P(X = 0)$ is quite substantial.

    pgf /. {ω → 0.5, λ → 5, t → 0}

    0.503369

… when compared to the same chance for its Poisson component alone:

    pgf₂ /. {λ → 5, t → 0} // N

    0.00673795

3.4 B Parameter-Mix Distributions

When the distribution of a random variable X depends upon a parameter θ, the (unknown) true value of θ is usually assumed fixed in the population. In some instances, however, an argument can be made for relaxing parameter fixity, which yields our second type of mixing distribution: parameter-mix distributions.

Two key components are required to form a parameter-mix distribution, namely the conditional distribution of the random variable given the parameter, and the distribution of the parameter itself. Let $f(x\,|\,\Theta = \theta)$ denote the density of $X\,|\,(\Theta = \theta)$, and let $g(\theta)$ denote the density of Θ. With this notation, the so-called '$g(\theta)$ parameter-mix of $f(x\,|\,\Theta = \theta)$' is written as

$$f(x\,|\,\Theta = \theta) \bigwedge_{\Theta} g(\theta) \tag{3.15}$$

and is equal to

$$E_{\Theta}\!\left[f(x\,|\,\Theta = \theta)\right] \tag{3.16}$$

where $E_{\Theta}[\,]$ is the usual expectation operator, with its subscript indicating that the expectation is taken with respect to the distribution of Θ. The solution to (3.16) is the unconditional distribution of X, which is the statistical model of interest. For instance,

$$\text{Binomial}(N,\,p) \bigwedge_{N} \text{Poisson}(\lambda)$$

denotes a Binomial(N, p) distribution in which parameter N (instead of being fixed) has a Poisson(λ) distribution. In this fashion, many distributions can be created using a parameter-mix approach; indeed the parameter-mix approach is often used as a device in its own right for developing new distributions. Table 5 lists five parameter-mix distributions (only the first three are discrete distributions).

$$\text{Negative Binomial }(r,\,p) = \text{Poisson}(L) \bigwedge_{L} \text{Gamma}\!\left(r,\,\tfrac{1-p}{p}\right)$$

$$\text{Holla }(\mu,\,\lambda) = \text{Poisson}(L) \bigwedge_{L} \text{InverseGaussian}(\mu,\,\lambda)$$

$$\text{Pólya–Aeppli }(b,\,\lambda) = \text{Poisson}(\Theta) \bigwedge_{\Theta} \text{Gamma}(A,\,b) \bigwedge_{A} \text{Poisson}(\lambda)$$

$$\text{Student's }t(n) = \text{Normal}(0,\,S^2) \bigwedge_{S^2} \text{InverseGamma}\!\left(\tfrac{n}{2},\,\tfrac{2}{n}\right)$$

$$\text{Noncentral Chi-squared }(n,\,\lambda) = \text{Chi-squared}(n + 2K) \bigwedge_{K} \text{Poisson}\!\left(\tfrac{\lambda}{2}\right)$$

Table 5: Parameter-mix distributions

For extensive details on parameter-mixing, see Johnson *et al.* (1993, Chapter 8). The following examples show how to construct parameter-mix distributions with **mathStatica**.

⊕ **Example 12:** A Binomial–Poisson Mixture

Find the distribution of X, when it is formed as $\text{Binomial}(N, p) \bigwedge_{N} \text{Poisson}(\lambda)$.

Solution: Of the two Binomial parameters, index N is permitted to vary according to a Poisson(λ) distribution, while the success probability p remains fixed. Begin by entering the key components. The first distribution, say $f(x)$, is the conditional distribution $X \mid (N = n) \sim \text{Binomial}(n, p)$:

```
f = Binomial[n, x] p^x (1 - p)^(n-x);

domain[f] = {x, 0, n} &&
            {0 < p < 1, n > 0, n ∈ Integers} && {Discrete};
```

The second is the parameter distribution $N \sim \text{Poisson}(\lambda)$:

$$g = \frac{e^{-\lambda} \lambda^n}{n!};$$

```
domain[g] = {n, 0, ∞} && {λ > 0} && {Discrete};
```

From (3.16), we require the expectation $E_N[\text{Binomial}(N, p)]$. The pmf of the parameter-mix distribution is then found by entering:

```
Expect[f, g]
```

$$\frac{e^{-p\lambda} (p\lambda)^x}{x!}$$

The mixing distribution is discrete and has a Poisson form: $X \sim \text{Poisson}(p\lambda)$. ∎

⊕ **Example 13:** A Binomial–Beta Mixture: The Beta–Binomial Distribution

Consider a Beta parameter-mix of the success probability of a Binomial distribution:

$$\text{Binomial}(n, P) \bigwedge_{P} \text{Beta}(a, b).$$

The conditional distribution $X \mid (P = p) \sim \text{Binomial}(n, p)$ is:

```
f = Binomial[n, x] p^x (1 - p)^(n-x);

domain[f] = {x, 0, n} &&
            {0 < p < 1, n > 0, n ∈ Integers} && {Discrete};
```

The distribution of the parameter $P \sim \text{Beta}(a, b)$ is:

```
g = p^(a-1) (1 - p)^(b-1) / Beta[a, b];    domain[g] = {p, 0, 1} && {a > 0, b > 0};
```

§3.4 B   DISCRETE RANDOM VARIABLES   107

We obtain the parameter-mix distribution by evaluating $E_P[f(x \mid P = p)]$ as per (3.16):

Expect[f, g]

$$\frac{\text{Binomial}[n, x] \, \Gamma[b + n - x] \, \Gamma[a + x]}{\text{Beta}[a, b] \, \Gamma[a + b + n]}$$

This is a Beta–Binomial distribution, with domain of support on the set of integers $\{0, 1, 2, \ldots, n\}$. The distribution is listed in the *Discrete* palette. ∎

⊕ *Example 14:* A Geometric–Exponential Mixture: The Yule Distribution

Consider an Exponential parameter-mix of a Geometric distribution:

$$\text{Geometric}(e^{-W}) \bigwedge_W \text{Exponential}\left(\frac{1}{\lambda}\right).$$

For a fixed value w of W, the conditional distribution is Geometric with the set of positive integers as the domain of support. The Geometric's success probability parameter p is coded as $p = e^{-w}$, which will lie between 0 and 1 provided $w > 0$. Here is the conditional distribution $X \mid (W = w) \sim \text{Geometric}(e^{-w})$:

f = p (1 - p)^(x-1) /. p → e^-w;
domain[f] = {x, 1, ∞} && {w > 0} && {Discrete};

Parameter W is such that $W \sim \text{Exponential}\left(\frac{1}{\lambda}\right)$:

g = λ e^(-λw);
domain[g] = {w, 0, ∞} && {λ > 0};

The parameter-mix distribution is found by evaluating:

Expect[f, g]

$$\frac{\lambda \, \Gamma[x] \, \Gamma[1 + \lambda]}{\Gamma[1 + x + \lambda]}$$

This is a Yule distribution, with domain of support on the set of integers $\{1, 2, 3, \ldots\}$, with parameter $\lambda > 0$. The Yule distribution is also given in **mathStatica**'s *Discrete* palette. The Yule distribution has been applied to problems in linguistics. Another distribution with similar areas of application is the Riemann Zeta distribution. It too may be entered from **mathStatica**'s *Discrete* palette. The Riemann Zeta distribution has also been termed the Zipf distribution, and it may be viewed as the discrete analogue of the continuous Pareto(a, 1) distribution; see Johnson *et al.* (1993, Chapter 11) for further details. ∎

⊕ *Example 15:* Modelling the Change in the Price of a Security (Stocks, Options, etc.)

Let the continuous random variable Y denote the change in the price of a security (measured in natural logarithms) using daily data. In economics and finance, it is common practice to assume that $Y \sim N(0, \sigma^2)$. Alas, empirically, the Normal model tends to under-predict both large and small price changes. That is, many empirical densities of price changes appear to be both more peaked and have fatter tails than a Normal pdf with the same variance; see, for instance, Merton (1990, p.59). In light of this, we need to replace the Normal model with another model that exhibits the desired behaviour. Let there be t transactions in any given day, and let $Y_i \sim N(0, \omega^2)$, $i \in \{1, \ldots, t\}$, represent the change in price on the ith transaction.[6] Thus, the daily change in price is obtained as $Y = Y_1 + Y_2 + \cdots + Y_t$, a sum of t random variables. For Y_i independent of Y_j ($i \neq j$), we now have $Y \sim N(0, t\omega^2)$, with pdf $f(y)$:

```
        1               y²
f = ─────────── Exp[- ─────── ];
    √t ω √2 π         2 t ω²

domain[f] = {y, -∞, ∞} && {ω > 0, t > 0, t ∈ Integers};
```

Parameter-mixing provides a resolution to the deficiency of the Normal model. Instead of treating t as a fixed parameter, we are now going to treat it as a discrete random variable $T = t$. Then, Y is a random-length sum of T random variables, and is in fact a member of the Stopped-Sum class of distributions; see Johnson *et al.* (1993, Chapter 9). In parameter-mix terms, f is the conditional model $Y \mid (T = t)$. For the purposes of this example, let the parameter distribution $T \sim \text{Geometric}(p)$, with density $g(t)$:

```
g = p (1 - p)^(t-1);

domain[g] = {t, 1, ∞} && {0 < p < 1} && {Discrete};
```

The desired mixture is

$$N(0, T\omega^2) \bigwedge_{T} \text{Geometric}(p) = E_T\left[f(y \mid T = t)\right]$$

which we can attempt to solve as:

```
Expect[f, g]
```

$$\sum_{t=1}^{\infty} \frac{e^{-\frac{y^2}{2t\omega^2}} (1-p)^{-1+t} p}{\sqrt{2\pi} \sqrt{t} \omega}$$

This does not evaluate further, in this case, as there is no closed form solution to the sum. However, we can proceed by using numerical methods.[7] Figure 11 illustrates. In the left panel, we see that the parameter-mix pdf (the solid line) is more peaked in the neighbourhood of the origin than a Normal pdf (the dashed line). In the right panel (which zooms-in on the distribution's right tail), it is apparent that the tails of the pdf are fatter than a Normal pdf. The parameter-mix distribution exhibits the attributes observed in empirical practice.

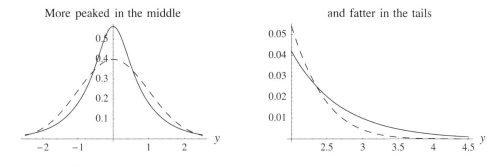

Fig. 11: Parameter-mix pdf (——) versus Normal pdf (– – –)

If a closed form solution is desired, we could simply select a different model for $g(t)$; for example, a Gamma or a Rayleigh distribution works nicely. While the latter are continuous densities, they yield the same qualitative results. For other models of changes in security prices see, for example, Fama (1965) and Clark (1973). ■

3.5 Pseudo-Random Number Generation

3.5 A Introducing `DiscreteRNG`

Let a discrete random variable X have domain of support $\Omega = \{x: x_0, x_1, ...\}$, with cdf $F(x) = P(X \le x)$ and pmf $f(x) = P(X = x)$ such that $\sum_{x \in \Omega} f(x) = 1$. This section tackles the problem of generating pseudo-random copies of X. One well-known approach is the inverse method: if u is a pseudo-random drawing from the Uniform(0, 1), the (continuous) uniform distribution defined on the unit interval, then $x = F^{-1}(u)$ is a pseudo-random copy of X. Of course, this method is only desirable if the inverse function of F is computationally tractable, and this, unfortunately, rarely occurs. In this section, we present a discrete pseudo-random number generator entitled `DiscreteRNG` that is virtuous in two respects. First, it is universal—it applies in principle to any discrete univariate distribution without alteration. This is achieved by constructing $F^{-1}(u)$ as a lookup table, instead of trying to do so symbolically. Second, this approach is surprisingly efficient. Given that pluralism and efficiency are usually mutually incompatible, the attainment of both goals is particularly pleasing. Detailed discussion of the function appears in Rose and Smith (1997).

The **mathStatica** function `DiscreteRNG[n, f]` generates n pseudo-random copies of a discrete random variable X, with pmf f. It allows f to take either Function Form or List Form. We illustrate its use with both input types by example.

⊕ **Example 16:** The Poisson Distribution

Suppose that $X \sim \text{Poisson}(6)$. Then, in Function Form, its pmf $f(x)$ is:

```
f = e^-λ λ^x / x! /. λ → 6;

domain[f] = {x, 0, ∞} && {Discrete};
```

Here, then, are 10 pseudo-random copies of X:

```
DiscreteRNG[10, f]
```

{5, 6, 9, 3, 5, 9, 2, 8, 5, 7}

and here are a few more:

```
data = DiscreteRNG[50000, f]; // Timing
```

{0.38 Second, Null}

Notice that it took `DiscreteRNG` a fraction of a second to produce 50000 Poisson(6) pseudo-random numbers!

In order to check how effective `DiscreteRNG` is in replicating the true distribution, we contrast the relative empirical distribution of the generated data with the true distribution of X using the **mathStatica** function `FrequencyPlotDiscrete`. The two distributions are overlaid as follows:

```
FrequencyPlotDiscrete[data, f];
```

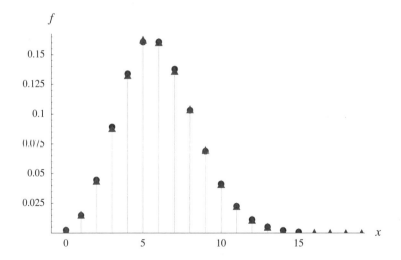

Fig. 12: Comparison of the empirical pmf (▲) to the Poisson(6) pmf (●)

The triangles give the generated empirical pmf, while the circles represent the true Poisson(6) pmf. The fit is superb.   ■

§3.5 A DISCRETE RANDOM VARIABLES

⊕ *Example 17:* A Discrete Distribution in List Form

The previous example dealt with Function Form input. DiscreteRNG can also be used for List Form input. Suppose that random variable X is distributed as follows:

$P(X = x)$:	0.1	0.4	0.3	0.2
x:	-1	$3/2$	π	4.4

Table 6: The pmf of X

X's details in List Form are:

```
f = {0.1, 0.4, 0.3, 0.2};
domain[f] = {x, {-1, 3/2, π, 4.4}} && {Discrete};
```

Here are eight pseudo-random numbers from the distribution:

```
DiscreteRNG[8, f]
```

$\{1.5, 3.14159, 1.5, 4.4, 3.14159, 4.4, 4.4, 3.14159\}$

And here are a few more:

```
data = DiscreteRNG[50000, f]; // Timing
```

$\{0.39 \text{ Second}, \text{Null}\}$

The empirical pmf overlaid with the true pmf is given by:

```
FrequencyPlotDiscrete[data, f];
```

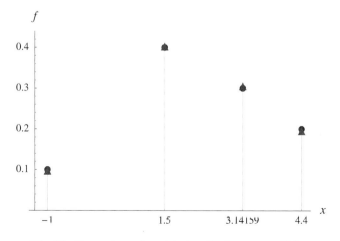

Fig. 13: Comparison of empirical pmf (▲) to true pmf (●)

Once again, the fit is superb.

⊕ *Example 18:* Holla's Distribution

Because `DiscreteRNG` is a general solution, it can generate random numbers, in principle, from any discrete distribution, not just the limited number of distributions that have been pre-programmed into *Mathematica*'s Statistics package. Consider, for example, Holla's distribution (see Table 5 for its parameter-mix derivation):

$$f = \frac{1}{x!}\left(e^{\lambda/\mu}\sqrt{\frac{2}{\pi}}\sqrt{\lambda}\left(\frac{2}{\lambda}+\frac{1}{\mu^2}\right)^{\frac{1}{4}(1-2x)} \text{BesselK}\left[\frac{1}{2}-x, \sqrt{\lambda\left(2+\frac{\lambda}{\mu^2}\right)}\right]\right);$$

$$\text{domain}[f] = \{x, 0, \infty\} \,\&\&\, \{\mu > 0, \lambda > 0\} \,\&\&\, \{\text{Discrete}\};$$

It would be a substantial undertaking to attempt to generate pseudo-random numbers from Holla's distribution using the inverse method. However, for given values of μ and λ, `DiscreteRNG` has no trouble in performing the task. Here is the code to produce 50000 pseudo-random copies:

```
data = DiscreteRNG[50000, f /. {μ → 1, λ → 4}]; // Timing
```

{0.39 Second, Null}

We again compare the empirical distribution to the true distribution:

```
FrequencyPlotDiscrete[data, f /. {μ → 1, λ → 4}];
```

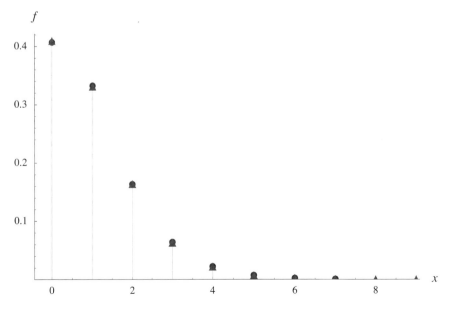

Fig. 14: Comparison of empirical pmf (▲) to Holla's distribution (●)

○ *Computationl Efficiency*

Mathematica's `Statistics`DiscreteDistributions`` package includes the Poisson distribution used in *Example 16*, so we can compare the computational efficiency of *Mathematica*'s generator to **mathStatica**'s generator. After loading the Statistics add-on:

    <<Statistics`

generate 50000 copies from Poisson(6) using the customised routine contained in this package:

    dist = PoissonDistribution[6];
    RandomArray[dist, {50000}]; // Timing

    {77.67 Second, Null}

By contrast, **mathStatica** takes just 0.38 seconds (see *Example 16*) to generate the same number of copies. Thus, for this example, DiscreteRNG is around 200 times faster than *Mathematica*'s Statistics package, even though DiscreteRNG is a general solution that has not been specially optimised for the Poisson. In further comparative experiments against the small range of discrete distributions included in the *Mathematica* Statistics package, Rose and Smith (1997) report complete efficiency dominance for DiscreteRNG.

3.5 B Implementation Notes

DiscreteRNG works by internally constructing a numerical lookup table of the specified discrete random variable's cdf.[8] When generating many pseudo-random numbers from a particular discrete distribution, it therefore makes sense to ask for all the desired pseudo-random numbers in one go, rather than repeatedly constructing the lookup table. The contrast in performance is demonstrated by the following timings for a Riemann Zeta distribution:

$$f = \frac{x^{-(\rho+1)}}{\text{Zeta}[1+\rho]} \, /. \, \rho \to 3;$$

    domain[f] = {x, 1, ∞} && {Discrete};

The first input below calls DiscreteRNG 1000 times, whereas the second generates all 1000 in just one call and is clearly far more efficient:

    Table[DiscreteRNG[1, f], {1000}]; // Timing

    {12.31 Second, Null}

    DiscreteRNG[1000, f]; // Timing

    {0. Second, Null}

A numerical lookup table is naturally limited to a finite number of elements. Thus, if the discrete random variable has an infinite number of points of support (as in the Riemann Zeta case), then the tail(s) of the distribution must be censored. The default in DiscreteRNG is to automatically censor the left and right tails of the distribution in such a way that less than $\varepsilon = 10^{-6}$ of the density mass is disturbed at either tail. The related **mathStatica** function RNGBounds identifies the censoring points and calculates the probability mass that is theoretically affected by censoring. For example, for $X \sim$ Riemann Zeta(3):

RNGBounds[f]

- The density was not censored below.
- Censored above at x = 68. This can affect 9.58084×10^{-7} of the density mass.

The printed output tells us that when DiscreteRNG is used at its default settings, it can generate copies of X from the set $\Omega^* = \{1, \ldots, 68\}$. By censoring at 68, outcomes $\Omega_* = \{69, 70, 71, \ldots\}$ are reported as 68. Thus, the censored mass is not lost; it is merely shifted to the censoring point. The density mass shifted in this way corresponds to $P(X \in \Omega_*)$ which is equal to:

1 - Prob[68, f] // N

9.58084×10^{-7}

as reported by RNGBounds above.

If censoring at $x = 68$ is not desirable, tighter (or weaker) tolerance levels can be set. RNGBounds[f, $\underline{\varepsilon}$, $\overline{\varepsilon}$] can be used to inspect the effect of arbitrary tolerance settings, while DiscreteRNG[n, f, $\underline{\varepsilon}$, $\overline{\varepsilon}$] imposes those settings on the generator; $\underline{\varepsilon}$ is the tolerance setting for the left tail, and $\overline{\varepsilon}$ is the setting for the right tail. For example:

RNGBounds[f, 10^{-8}, 10^{-8}]

- The density was not censored below.
- Censored above at x = 313. This can affect 9.99556×10^{-9} of the density mass.

Thus, DiscreteRNG[n, f, 10^{-8}, 10^{-8}] will generate n copies of $X \sim$ Riemann Zeta(3), with outcomes restricted to the integers in $\Omega^* = \{1, 2, \ldots, 313\}$; censoring occurs on the right at 313 which results in just under 10^{-8} of the density mass being shifted to that point. The reason the censoring point has 'blown out' to 313 is because the Riemann Zeta distribution is long-tailed.

DiscreteRNG and RNGBounds are defined for tolerance settings $\varepsilon \geq 10^{-15}$. Setting ε outside this interval is not meaningful and may cause problems. (It is also assumed that $\varepsilon < 0.25$, although this constraint should never be binding.) In List Form examples, the distribution is never censored, so RNGBounds does not apply and, by design, will not evaluate. For Function Form examples, we recommend that whenever DiscreteRNG is applied, the printed output from RNGBounds should also be inspected.

Finally, by constructing a lookup table, `DiscreteRNG` trades off a small fixed cost in return for a lower marginal cost. This trade-off will be particularly beneficial if a large number of pseudo-random numbers are required. If only a few are needed, it may not be worthwhile. The fixed cost is itself proportional to the size of the lookup table. For instance, a Discrete Uniform such as $f = 10^{-6}$ defined on $\Omega = \{1, ..., 10^6\}$ will require a huge lookup table. Here, a technique such as `Random[Integer,{1,10^6}]` is clearly more appropriate.

3.6 Exercises

1. Let random variable X take values 1, 2, 3, 4, 5, with probability $\frac{1}{55}, \frac{4}{55}, \frac{9}{55}, \frac{16}{55}, \frac{25}{55}$, respectively.

    (i) Enter the pmf of X in List Form, plot the pmf, and then evaluate $E[X]$.

    (ii) Enter the pmf of X in Function Form, and evaluate $E[X]$.

    (iii) Repeat (i) and (ii) when X takes values 1, 3, 5, with probability $\frac{1}{35}, \frac{9}{35}, \frac{25}{35}$, respectively.

2. Enter the Binomial(n, p) pmf from **mathStatica**'s *Discrete* palette. Express the pmf in List Form when $n = 10$ and $p = 0.4$.

3. Derive the mean, variance, cdf, mgf and pgf for the following distributions whose pmf may be entered from **mathStatica**'s *Discrete* palette: (i) Geometric, (ii) Hypergeometric, (iii) Logarithmic, and (iv) Yule.

4. Using the shaved 1-face dice described in *Example 3*, plot the probability of winning Craps against δ.

5. A gambler aims to increase his initial capital of $5 to $10 by playing Craps, betting $1 per game. Using simulation, estimate the probability that the gambler can, before ruin (*i.e.* his balance is depleted to $0), achieve his goal.

6. In a large population of n individuals, each person must submit a blood sample for test. Let p denote the probability that an individual returns a positive test, and assume that p is small. The test designer suggests pooling samples of blood from m individuals, testing the pooled sample with a single test. If a negative test is returned, then this one test indicates that all m individuals are negative. However, if a positive test is returned, then the test is carried out on each individual in the pool. For this sampling design, determine μ (the expected number of tests), and the optimal value of m when $p = 0.01$. Assume all individuals in the population are mutually independent, and that p is the same across all individuals.

7. What are the chances of throwing: (i) at least 1 six from a throw of a box containing 6 dice, (ii) at least 2 sixes from another box containing 12 dice, and (iii) 3 or more sixes from a third box filled with 18 dice?

8. An urn contains 20 balls, 4 of which are coloured red. A sample of 5 balls is drawn one-by-one from the urn. What is the probability that one of the balls drawn is red:

    (i) if each ball that is drawn is returned to the urn?

    (ii) if each ball that is drawn is set aside?

9. Experience indicates that a firm will, on average, fire 3 workers per year. Assuming that the number of employees fired per year is Poisson distributed, what is the probability that in the coming year the firm will: (i) not fire any workers, and (ii) fire at least 4 workers?

10. Let a random variable $X \sim \text{Poisson}(\lambda)$. Determine the smallest value of λ such that $P(X \le 1) \le 0.05$.

11. Determine the pmf of the following parameter-mixes, and plot it at the indicated values of the parameters:

   (i) $\text{Binomial}(N, p) \bigwedge_{N} \text{Binomial}(m, q)$. Plot for $p = \frac{3}{4}, q = \frac{1}{2}, m = 10$.

   (ii) $\text{Negative Binomial}(R, p) \bigwedge_{R} \text{Geometric}(q)$. Plot for $p = \frac{1}{4}, q = \frac{2}{3}$.

   (iii) $\text{Poisson}(\Theta) \bigwedge_{\Theta} \text{Lindley}(\delta)$. Plot for $\delta = 1$.

12. (i) Use `DiscreteRNG` to generate 20000 pseudo-random drawings from the Geometric(0.1) distribution. Then use `FrequencyPlotDiscrete` to plot the empirical distribution, with the true distribution superimposed on top.

   (ii) Repeat (i), this time using *Mathematica*'s Statistics package pseudo-random number generator:

   `RandomArray[GeometricDistribution[0.1], 20000] + 1`

   (the "+1" is required because the Geometric distribution hardwired in the Statistics package includes 0 in its domain of support).

   (iii) Report on any discrepancies you observe between the empirical and true distributions.

13. (i) Generate 20000 pseudo-random numbers from a Zero-Inflated Poisson distribution (parameters ω and λ; see *Example 11*) when $\omega = 0.6$ and $\lambda = 4$. Compare the empirical distribution to the theoretical distribution.

   (ii) Generate 20000 pseudo-random numbers from a Poisson two-component-mix distribution (parameters ω, λ and θ; see *Example 10*) when $\omega = 0.6$, $\lambda = 9$ and $\theta = 3$. Compare the empirical distribution to the theoretical distribution.

Chapter 4

Distributions of Functions of Random Variables

4.1 Introduction

This chapter is concerned with the following problem, which we state here in its simplest form:

> Let X be a random variable with density $f(x)$.
> What is the distribution of $Y = u(X)$, where $u(X)$ denotes some function of X?

This problem is of interest for several reasons. First, it is crucial to an understanding of statistical *distribution theory*: for instance, this chapter derives (from first principles) distributions such as the Lognormal, Pareto, Extreme Value, Rayleigh, Chi-squared, Student's t, Fisher's F, noncentral Chi-squared, noncentral F, Triangular and Laplace, amongst many others. Second, it is important in *sampling theory*: the chapter discusses ways to find the exact sampling distribution of statistics such as the sample sum, the sample mean, and the sample sum of squares. Third, it is of practical importance: for instance, a gold mine may have a profit function $u(x)$ that depends on the gold price X (a random variable). The firm is interested to know the distribution of its profits, given the distribution of X.

In statistics, there are two standard methods for solving these problems:

- The *Transformation Method*: this only applies to one-to-one transformations.

- The *MGF Method*: this is less restrictive, but can be more difficult to solve. It is based on the Uniqueness Theorem relating moment generating functions to densities.

§4.2 discusses the Transformation Method, while §4.3 covers the MGF Method. These two methodologies are then applied to some important examples in §4.4 (products and ratios of random variables) and §4.5 (sums and differences of random variables).

4.2 The Transformation Method

This section discusses the Transformation Method: §4.2 A discusses transformations of a single random variable, §4.2 B extends the analysis to the multivariate case, while §4.2 C considers transformations that are not strictly one-to-one, as well as manual methods.

4.2 A Univariate Cases

A *one-to-one transformation* implies that each value x is related to one (and only one) value $y = u(x)$, and that each value y is related to one (and only one) value $x = u^{-1}(y)$. Any univariate monotonic increasing or decreasing function yields a one-to-one transformation. Figure 1, for instance, shows two transformations.

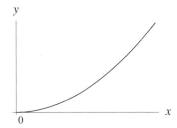
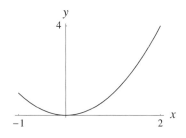

Fig. 1:   (i) $y = x^2$, for $x \in \mathbb{R}_+$ (ii) $y = x^2$, for $x \in (-1, 2)$

Case (i): Even though $y = x^2$ has two solutions, namely:

Solve[y == x², x]

$\{\{x \to -\sqrt{y}\}, \{x \to \sqrt{y}\}\}$

only the latter solution is valid for the given domain ($x \in \mathbb{R}_+$). Therefore, *over the given domain*, the function is monotonically increasing, and thus case (i) is a one-to-one transformation.

Case (ii): Here, for some values of y, there exists more than one corresponding value of x; there are now two valid solutions, neither of which can be excluded. Thus, case (ii) is *not* a one-to-one transformation. Fortunately, a theorem exists to deal with such cases. see §4.2 C.

Theorem 1: Let X be a *continuous* random variable with pdf $f(x)$, and let $Y = u(X)$ define a one-to-one transformation between the values of X and Y. Then the pdf of Y, say $g(y)$, is

$$g(y) = f(u^{-1}(y)) |J| \qquad (4.1)$$

where $x = u^{-1}(y)$ is the inverse function of $y = u(x)$, and $J = \dfrac{d u^{-1}(y)}{dy}$ denotes the Jacobian of the transformation; u^{-1} is assumed to be differentiable.

Proof: We will only sketch the proof.[1] To aid intuition, suppose $Y = u(X)$ defines a one-to-one *increasing* transformation between the values of X and Y. Then $P(Y \le y) = P(X \le x)$, or equivalently in terms of their respective cdf's, $G(y) = F(x)$. Then, by the chain rule of differentiation:

$$g(y) = \frac{dG(y)}{dy} = \frac{dF(x)}{dx}\frac{dx}{dy} = f(x)\frac{dx}{dy} \qquad \text{where } x = u^{-1}(y).$$

Remark: If X is a *discrete* random variable, then (4.1) becomes:

$$g(y) = f\left(u^{-1}(y)\right)$$

The **mathStatica** function, `Transform[eqn, f]` finds the density of $Y = u(X)$, where X has density $f(x)$, for both continuous and discrete random variables, while `TransformExtremum[eqn, f]` calculates the domain of Y, if it can do so. As per Theorem 1, `Transform` and `TransformExtremum` should only be used on transformations that are one-to-one. The `Transform` function is best illustrated by example ...

⊕ **Example 1:** Derivation of the Cauchy Distribution

Let X have Uniform density $f(x) = \frac{1}{\pi}$, defined on $\left(-\frac{\pi}{2}, \frac{\pi}{2}\right)$:

```
f = 1/π;    domain[f] = {x, -π/2, π/2};
```

Then, the density of $Y = \tan(X)$ is derived as follows:

```
Transform[y == Tan[x], f]
```

$$\frac{1}{\pi + \pi y^2}$$

with domain of support:

```
TransformExtremum[y == Tan[x], f]
```

$\{y, -\infty, \infty\}$

This is the pdf of a Cauchy distributed random variable. Note the double equal sign in the transformation equation: `y == Tan[x]`. If, by mistake, we enter `y = Tan[x]` with a single equal sign (or if `y` was previously given some value), we would need to `Clear[y]` before trying again.   ∎

⊕ **Example 2:** Standardising a $N(\mu, \sigma^2)$ Random Variable

Let $X \sim N(\mu, \sigma^2)$ with density $f(x)$:

$$\texttt{f} = \frac{e^{-\frac{(x-\mu)^2}{2\sigma^2}}}{\sigma\sqrt{2\pi}}; \quad \texttt{domain[f] = \{x, -\infty, \infty\} \&\& \{\mu \in Reals, \sigma > 0\};}$$

Then, the density of $Z = \frac{X-\mu}{\sigma}$, denoted $g(z)$ is:

$$\texttt{g = Transform}\left[\texttt{z} == \frac{x-\mu}{\sigma}, \texttt{f}\right]$$
$$\texttt{domain[g] = TransformExtremum}\left[\texttt{z} == \frac{x-\mu}{\sigma}, \texttt{f}\right]$$

$$\frac{e^{-\frac{z^2}{2}}}{\sqrt{2\pi}}$$

$\{\texttt{z}, -\infty, \infty\}$

That is, Z is a $N(0, 1)$ random variable. ∎

⊕ **Example 3:** Derivation of the Lognormal Distribution

Let $X \sim N(\mu, \sigma^2)$ with density $f(x)$, as entered above in *Example 2*. Then, the density of $Y = e^X$, denoted $g(y)$, is:

$$\texttt{g = Transform[y == }e^x\texttt{, f]}$$
$$\texttt{domain[g] = TransformExtremum[y == }e^x\texttt{, f]}$$

$$\frac{e^{-\frac{(\mu-\text{Log}[y])^2}{2\sigma^2}}}{\sqrt{2\pi}\, y\, \sigma}$$

$\{\texttt{y}, 0, \infty\}\ \&\&\ \{\mu \in \texttt{Reals}, \sigma > 0\}$

This is a Lognormal distribution, so named because $\log(Y)$ has a Normal distribution. Figure 2 plots the Lognormal pdf, when $\mu = 0$ and $\sigma = 1$.

PlotDensity[g /. {μ → 0, σ → 1}];

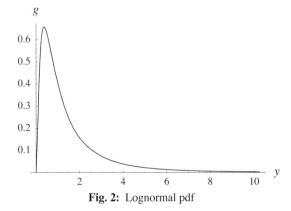

Fig. 2: Lognormal pdf

⊕ **Example 4:** Derivation of Uniform, Pareto, Extreme Value and Rayleigh Distributions

Let X have a standard Exponential distribution with density $f(x)$:

```
f = e^-x;      domain[f] = {x, 0, ∞};
```

We shall consider the following simple transformations:

(i) $Y = e^{-X}$ (ii) $Y = e^{X}$ (iii) $Y = -\log(X)$ (iv) $Y = \sqrt{X}$

(i) When $Y = e^{-X}$, we obtain the standard Uniform distribution:

```
g = Transform[y == e^-x, f]
domain[g] = TransformExtremum[y == e^-x, f]
```

1

{y, 0, 1}

(ii) When $Y = e^{X}$, we obtain a Pareto distribution:

```
g = Transform[y == e^x, f]
domain[g] = TransformExtremum[y == e^x, f]
```

$$\frac{1}{y^2}$$

{y, 1, ∞}

More generally, if $X \sim \text{Exponential}(\frac{1}{a})$, then $Y = b\, e^{X}$ ($b > 0$) yields the Pareto density with pdf $a\, b^a\, y^{-(a+1)}$, defined for $y > b$.[2] This is often used in economics to model the distribution of income, and is named after the economist Vilfredo Pareto (1848–1923).

(iii) When $Y = -\log(X)$, we obtain the standard Extreme Value distribution:

```
g = Transform[y == -Log[x], f]
domain[g] = TransformExtremum[y == -Log[x], f]
```

$e^{-e^{-y} - y}$

{y, -∞, ∞}

(iv) When $Y = \sqrt{X}$, we obtain a Rayleigh distribution:

```
g = Transform[y == √x , f]
domain[g] = TransformExtremum[y == √x , f]
```

$2 e^{-y^2} y$

$\{y, 0, \infty\}$

as given in the *Continuous* palette (simply replace σ with $\sqrt{1/2}$ to get the same result). More generally, if $X \sim$ Exponential(λ), then $Y = \sqrt{X} \sim$ Rayleigh(σ) with $\sigma = \sqrt{\lambda/2}$. This distribution is often used in engineering to model the life of electronic components. ∎

⊕ **Example 5:** Transformations of the Uniform Distribution

Let $X \sim$ Uniform(α, β) with density $f(x)$, where $0 < \alpha < \beta < \infty$:

```
f = 1/(β - α) ;   domain[f] = {x, α, β} && {0 < α < β};
```

We seek the distributions of: (i) $Y = 1 + X^2$ and (ii) $Y = (1 + X)^{-1}$.

Solution: Let $g(y)$ denote the pdf of Y. Then the solution to (i) is:

```
g = Transform[y == 1 + x² , f]
domain[g] = TransformExtremum[y == 1 + x² , f]
```

$\dfrac{1}{\sqrt{-1 + y}\,(-2\alpha + 2\beta)}$

$\{y, 1 + \alpha^2, 1 + \beta^2\}$ && $\{0 < \alpha < \beta\}$

while the solution to the second part is:

```
g = Transform[y == (1 + x)⁻¹ , f]
domain[g] = TransformExtremum[y == (1 + x)⁻¹ , f]
```

$\dfrac{1}{-y^2 \alpha + y^2 \beta}$

$\left\{y, \dfrac{1}{1+\beta}, \dfrac{1}{1+\alpha}\right\}$ && $\{0 < \alpha < \beta\}$

Generally, transformations involving parameters pose no problem, provided we remember to attach the appropriate assumptions to the original `domain[f]` statement at the very start. ∎

4.2 B Multivariate Cases

Thus far, we have considered the distribution of a transformation of a single random variable. This section extends the analysis to more than one random variable. The concepts discussed in the univariate case carry over to the multivariate case with the appropriate modifications.

Theorem 2: Let X_1 and X_2 be *continuous* random variables with joint pdf $f(x_1, x_2)$. Let $Y_1 = u_1(X_1, X_2)$ and $Y_2 = u_2(X_1, X_2)$ define a one-to-one transformation between the values of (X_1, X_2) and (Y_1, Y_2). Then the joint pdf of Y_1 and Y_2 is

$$g(y_1, y_2) = f\big(u_1^{-1}(y_1, y_2),\ u_2^{-1}(y_1, y_2)\big)\ |J| \qquad (4.2)$$

where $u_i^{-1}(y_1, y_2)$ is the inverse function of $Y_i = u_i(X_1, X_2)$, and

$$J = \det \begin{pmatrix} \dfrac{\partial x_1}{\partial y_1} & \dfrac{\partial x_1}{\partial y_2} \\ \dfrac{\partial x_2}{\partial y_1} & \dfrac{\partial x_2}{\partial y_2} \end{pmatrix}$$

is the Jacobian of the transformation, with $\dfrac{\partial x_i}{\partial y_j}$ denoting the partial derivative of $x_i = u_i^{-1}(y_1, y_2)$ with respect to y_j, and $\det(\cdot)$ denotes the determinant of the matrix. Transformations in higher dimensional systems follow in similar fashion.

Proof: The proof is analogous to Theorem 1; see Tjur (1980, §3.1) for more detail.

Remark: If the X_i are *discrete* random variables, (4.2) becomes:

$$g(y_1, y_2) = f\big(u_1^{-1}(y_1, y_2),\ u_2^{-1}(y_1, y_2)\big)$$

The **mathStatica** function, `Transform`, may also be used in multivariate settings. Of course, by Theorem 2, it should only be used to solve transformations that are one-to-one.

The transition from univariate to multivariate transformations raises two new issues:

(i) How many random variables?

The Transformation Method requires that there are as many 'new' variables Y_i as there are 'old' variables X_i. Suppose, for instance, that X_1, X_2 and X_3 have joint pdf $f(x_1, x_2, x_3)$, and that we seek the pdf of $Y_1 = u_1(X_1, X_2, X_3)$. This problem involves three steps. *First*, we must create two additional random variables, $Y_2 = u_2(X_1, X_2, X_3)$ and $Y_3 = u_3(X_1, X_2, X_3)$, and we must do so in such a way that there is one-to-one transformation from the values of (X_1, X_2, X_3) to (Y_1, Y_2, Y_3). This could, for example, be done by setting $Y_2 = X_2$, and $Y_3 = X_3$. *Second*, we can then find the joint pdf of (Y_1, Y_2, Y_3). *Third*, we can then derive the desired marginal pdf of Y_1 from the joint pdf of (Y_1, Y_2, Y_3) by integrating out Y_2 and Y_3. *Example 7* illustrates this procedure.

(ii) Non-rectangular domains

Let (X_1, X_2) have joint pdf $f(x_1, x_2)$. Let $Y_1 = u_1(X_1, X_2)$ and $Y_2 = u_2(X_1, X_2)$ define a one-to-one transformation from the values of (X_1, X_2) to the values of (Y_1, Y_2), and let $g(y_1, y_2)$ denote the joint pdf of (Y_1, Y_2). Finally, let $\mathcal{A}$ denote the space where $f(x_1, x_2) > 0$, and let $\mathcal{B}$ denote the space where $g(y_1, y_2) > 0$; $\mathcal{A}$ and $\mathcal{B}$ are therefore the domains of support. Then, the transformation is said to map space $\mathcal{A}$ (in the x_1-x_2 plane) onto space $\mathcal{B}$ (in the y_1-y_2 plane). If the domain of a joint pdf does *not* depend on any of its constituent random variables, then we say the domain defines an *independent product space*. For instance, the domain $\mathcal{A} = \{(x_1, x_2) : \frac{1}{2} < x_1 < 3, \ 1 < x_2 < 4\}$ is an independent product space, because the domain of X_1 does not depend on the domain of X_2, and vice versa. If plotted in x_1-x_2 space, this domain would appear rectangular, as the left panel in Fig. 3 illustrates.

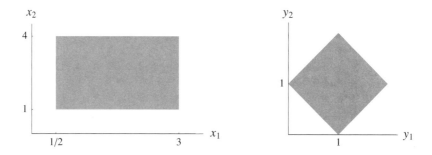

Fig. 3: Rectangular (left) and non-rectangular (right) domains

In this vein, we refer to domains as being either *rectangular* or *non-rectangular*. Even though space $\mathcal{A}$ is rectangular, it is important to realise that a multivariate transformation will often create dependence in space $\mathcal{B}$. To see this, consider the following example:

⊕ **Example 6:** A Non-Rectangular Domain

Let X_1 and X_2 be defined on the unit interval with joint pdf $f(x_1, x_2) = 1$:

```
f = 1;  domain[f] = {{x₁, 0, 1}, {x₂, 0, 1}};
```

Let $Y_1 = X_1 + X_2$ and $Y_2 = X_1 - X_2$. Then, we have:

```
eqn = {y₁ == x₁ + x₂, y₂ == x₁ - x₂};
```

Note the bracketing on the transformation equation — it takes the same form as *Mathematica*'s Solve function. Then the joint pdf of Y_1 and Y_2, denoted $g(y_1, y_2)$, is:

```
g = Transform[eqn, f]
```

$\frac{1}{2}$

The **mathStatica** function DomainPlot[*eqn*, *f*] illustrates set $\mathcal{B}$, denoting the space in the y_1-y_2 plane where $g(y_1, y_2) = \frac{1}{2}$.

DomainPlot[eqn, f];

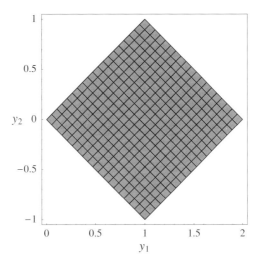

Fig. 4: Space in the y_1-y_2 plane where $g(y_1, y_2) = \frac{1}{2}$

The domain here is $\mathcal{B} = \{(y_1, y_2) : 0 < y_1 + y_2 < 2, -2 < y_2 - y_1 < 0\}$. This is clearly a non-rectangular domain, indicating that Y_1 and Y_2 are dependent.

Notes:

(i) In the multivariate case, TransformExtremum does not derive the domain itself; instead it calculates the extremities of the domain:

TransformExtremum[eqn, f]

{{y₁, 0, 2}, {y₂, -1, 1}}

This is sometimes helpful to verify ones working. However, as this example shows, extremities and domains are *not* always the same, and care must be taken not to confuse them.

(ii) For more information on DomainPlot, see the **mathStatica Help** file.

(iii) It is worth noting that even though Y_1 and Y_2 are dependent, they are uncorrelated:

Corr[{x₁ + x₂, x₁ - x₂}, f]

0

It follows that zero correlation does *not* imply independence. ∎

⊕ **Example 7:** Product of Uniform Random Variables

Let $X_1 \sim$ Uniform(0, 1) be independent of $X_2 \sim$ Uniform(0, 1), and let $Y = X_1 X_2$. Find $P(Y \le \frac{1}{4})$.

Solution: Due to independence, the joint pdf of X_1 and X_2, say $f(x_1, x_2)$, is just the pdf of X_1 multiplied by the pdf of X_2:

```
f = 1;     domain[f] = {{x₁, 0, 1}, {x₂, 0, 1}};
```

Take $Y = X_1 X_2$, and let $Z = X_2$, so that the number of 'new' variables is equal to the number of 'old' ones. Then, the transformation equation is:

```
eqn = {y == x₁ x₂, z == x₂};
```

Let $g(y, z)$ denote the joint pdf of (Y, Z):

```
g = Transform[eqn, f]
```

$$\frac{1}{z}$$

Since X_1 and X_2 are $U(0, 1)$, and $Y = X_1 X_2$ and $Z = X_2$, it follows that $0 < y < z < 1$. To see this visually, evaluate `DomainPlot[eqn, f]`. We enter $0 < y < z < 1$ as follows:

```
domain[g] = {{y, 0, z}, {z, y, 1}};
```

Then the marginal pdf of $Y = X_1 X_2$ is:

```
h = Marginal[y, g]
```

$$-\text{Log}[y]$$

with domain of support:

```
domain[h] = {y, 0, 1};
```

Finally, we require $P(Y \le \frac{1}{4})$. This is given by:

```
Prob[ 1/4 , h]
```

$$\frac{1}{4} (1 + \text{Log}[4])$$

which is approximately 0.5966. It can be helpful, sometimes, to check that one's symbolic workings make sense by using an alternative methodology. For instance, we can use simulation to estimate $P(X_1 X_2 \le \frac{1}{4})$. Here, then, are 10000 drawings of $Y = X_1 X_2$:

```
data = Table[ Random[] Random[], {10000}];
```

We now count how many copies of Y are smaller than (or equal to) $\frac{1}{4}$, and divide by 10000 to get our estimate of $P(Y \leq \frac{1}{4})$:

$$\frac{\texttt{Count[data, y\_ /; y} \leq \tfrac{1}{4}\texttt{]}}{\texttt{10000.}}$$

0.5952

which is close to the exact result derived above. ∎

4.2 C Transformations That Are *Not* One-to-One; Manual Methods

In §4.2 A, we considered the transformation $Y = X^2$ defined on $x \in (-1, 2)$. This is *not* a one-to-one transformation, because for some values of Y there are two corresponding values of X. This section discusses how to undertake such transformations.

Theorem 3: Let X be a *continuous* random variable with pdf $f(x)$, and let $Y = u(X)$ define a transformation between the values of X and Y that is *not* one-to-one. Thus, if $\mathcal{A}$ denotes the space where $f(x) > 0$, and $\mathcal{B}$ denotes the space where $g(y) > 0$, then there exist points in $\mathcal{B}$ that correspond to more than one point in $\mathcal{A}$. However, if set $\mathcal{A}$ can be partitioned into k sets, $\mathcal{A}_1, \ldots, \mathcal{A}_k$, such that u defines a one-to-one transformation of each $\mathcal{A}_i$ onto $\mathcal{B}_i$ (the image of $\mathcal{A}_i$ under u), then the pdf of Y is

$$g(y) = \sum_{i=1}^{k} \delta_i(y) f\!\left(u_i^{-1}(y)\right) |J_i| \qquad \text{for } i = 1, \ldots, k \tag{4.3}$$

where $\delta_i(y) = 1$ if $y_i \in \mathcal{B}_i$ and 0 otherwise, $x = u_i^{-1}(y)$ is the inverse function of $Y = u(X)$ in partition i, and $J_i = \dfrac{d u_i^{-1}(y)}{dy}$ denotes the Jacobian of the transformation in partition i.[3]

All this really means is that, for each region i, we simply work as we did before with Theorem 1; we then add up all the parts $i = 1, \ldots, k$.

⊕ **Example 8:** A Transformation That Is *Not* One-to-One

Let X have pdf $f(x) = \dfrac{e^x}{e^2 - e^{-1}}$ defined on $x \in (-1, 2)$, and let $Y = X^2$. We seek the pdf of Y. We have:

$$\texttt{f = }\frac{\texttt{e}^{\texttt{x}}}{\texttt{e}^2 - \texttt{e}^{-1}}\texttt{;} \qquad \texttt{domain[f] = \{x, -1, 2\};} \qquad \texttt{eqn = \{y == x}^2\texttt{\};}$$

Solution: The transformation from X to Y is not one-to-one over the given domain. We can, however, partition the domain into two sets of points that are both one-to-one. We do this as follows:

```
f₁ = f;  domain[f₁] = {x, -1, 0};
f₂ = f;  domain[f₂] = {x,  0, 2};
```

Let $g_1(y)$ denote the density of Y corresponding to when $x \leq 0$, and similarly, let $g_2(y)$ denote the density of Y corresponding to $x > 0$:

```
{g₁ = Transform[eqn, f₁], TransformExtremum[eqn, f₁]}
{g₂ = Transform[eqn, f₂], TransformExtremum[eqn, f₂]}
```

$$\left\{ \frac{e^{1-\sqrt{y}}}{(-2 + 2\,e^3)\,\sqrt{y}},\ \{y, 0, 1\} \right\}$$

$$\left\{ \frac{e^{1+\sqrt{y}}}{(-2 + 2\,e^3)\,\sqrt{y}},\ \{y, 0, 4\} \right\}$$

By (4.3), it follows that

$$g(y) = \begin{cases} g_1 + g_2 & 0 < y \leq 1 \\ g_2 & 1 < y < 4 \end{cases}$$

which we enter, using **mathStatica**, as:

```
g = If[y ≤ 1, g₁ + g₂, g₂];  domain[g] = {y, 0, 4};
```

Figure 5 plots the pdf.

```
PlotDensity[g, PlotRange → {0, .5}];
```

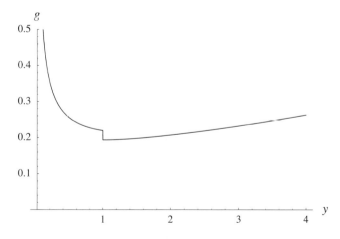

Fig. 5: The pdf of $Y = X^2$, with discontinuity at $y = 1$

Despite the discontinuity of the pdf at $y = 1$, **mathStatica** functions such as `Prob` and `Expect` will still work perfectly well. For instance, here is the cdf $P(Y \le y)$:

cdf = Prob[y, g]

$$\text{If}\left[y \le 1, \; \frac{2\,e\,\text{Sinh}\left[\sqrt{y}\right]}{-1 + e^3}, \; \frac{-1 + e^{1+\sqrt{y}}}{-1 + e^3}\right]$$

This can be easily illustrated with `Plot[cdf,{y,0,4}]`. ∎

⊕ **Example 9:** The Square of a Normal Random Variable: The Chi-squared Distribution

Let $X \sim N(0, 1)$ with density $f(x)$. We seek the distribution of $Y = X^2$. Thus, we have:

f = $\dfrac{e^{-\frac{x^2}{2}}}{\sqrt{2\pi}}$;   domain[f] = {x, -∞, ∞};   eqn = {y == x^2};

Solution: The transformation equation here is *not* one-to-one over the given domain. By Theorem 3, we can, however, partition the domain into two disjoint sets of points that are both one-to-one:

f$_1$ = f;   domain[f$_1$] = {x, -∞, 0};
f$_2$ = f;   domain[f$_2$] = {x, 0, ∞};

Let $g_1(y)$ denote the density of Y corresponding to when $x \le 0$, and similarly, let $g_2(y)$ denote the density of Y corresponding to when $x > 0$:

{g$_1$ = Transform[eqn, f$_1$], TransformExtremum[eqn, f$_1$]}
{g$_2$ = Transform[eqn, f$_2$], TransformExtremum[eqn, f$_2$]}

$$\left\{ \frac{e^{-y/2}}{2\sqrt{2\pi}\sqrt{y}}, \; \{y, 0, \infty\} \right\}$$

$$\left\{ \frac{e^{-y/2}}{2\sqrt{2\pi}\sqrt{y}}, \; \{y, 0, \infty\} \right\}$$

By Theorem 3, it follows that

$$g(y) = \begin{cases} g_1 + g_2 & 0 < y < \infty \\ 0 & \text{otherwise} \end{cases}$$

where $g_1 + g_2$ is:

g$_1$ + g$_2$

$$\frac{e^{-y/2}}{\sqrt{2\pi}\sqrt{y}}$$

This is the pdf of a Chi-squared random variable with 1 degree of freedom. ∎

○ **Manual Methods**

In all the examples above, we have always posed the transformation problem as:

Q. Let X be a random variable with pdf $f(x)$. What is the pdf of $Y = e^X$?
A. `Transform[y == e^x, f]`

But what if the same problem is posed as follows?

Q. Let X be a random variable with pdf $f(x)$. What is the pdf of Y, given $X = \log(Y)$?
A. `Transform[x == Log[y], f]` will fail, as this syntax is not supported.

We are now left with two possibilities:

(i) We could simply invert the transformation equation manually in *Mathematica* with `Solve[x == Log[y], f]`, and then derive the solution automatically with `Transform[y == e^x, f]`. Unfortunately, *Mathematica* may not always be able to neatly invert the transformation equation into the desired form, and we are then stuck.

(ii) Alternatively, we could adopt a manual approach by implementing either Theorem 1 (§4.2 A) or Theorem 2 (§4.2 B) ourselves in *Mathematica*. In a univariate setting, the basic approach would be to define:

`g = (f /. x → Log[y]) * Jacob[x /. x → Log[y], y]`

where the **mathStatica** function `Jacob` calculates the Jacobian of the transformation in absolute value. A multivariate example of a manual step-by-step transformation is given in Chapter 6 (see *Example 20, §6.4 A*).

4.3 The MGF Method

The moment generating function (mgf) method is based on the Uniqueness Theorem (§2.4 D) which states that there is a one-to-one correspondence between the mgf and the pdf of a random variable (if the mgf exists). Thus, if two mgf's are the same, then they must share the same density. As before, let X have density $f(x)$, and consider the transformation to $Y = u(X)$. We seek the pdf of Y, say $g(y)$. Two steps are involved:

Step 1: Find the mgf of Y.

Step 2: Hence, find the pdf of Y. This is normally done by matching the functional form of the mgf of Y with well-known moment generating functions. One usually does this armed with a textbook that lists the mgf's for well-known distributions, unless one has a fine memory for such things. If we can find a match, then the pdf is identified by the Uniqueness Theorem. Unfortunately, this matching process is often neither easy nor obvious. Moreover, if the pdf of Y is not well-known, then matching may not be possible. The mgf method is particularly well-suited to deriving the distribution of sample sums and sample means. This is discussed in §4.5 B, which provides further examples.

⊕ **Example 10:** The Square of a Normal Random Variable (again)

Let random variable $X \sim N(0, 1)$ with pdf $f(x)$:

```
f = e^(-x²/2) / √(2π);   domain[f] = {x, -∞, ∞};
```

We seek the distribution of $Y = X^2$.

Solution: The mgf of $Y = X^2$ is given by $E[e^{tX^2}]$:

```
mgf_Y = Expect[e^(t x²), f]
```

    — This further assumes that: $\{t < \frac{1}{2}\}$

$$\frac{1}{\sqrt{1 - 2t}}$$

By referring to a listing of mgf's, we see that this output is identical to the mgf of a Chi-squared random variable with 1 degree of freedom, confirming what was found in *Example 9*. Hence, if $X \sim N(0, 1)$, then X^2 is Chi-squared with 1 degree of freedom.

Using Characteristic Functions

The Uniqueness Theorem applies to both the moment generating function and the characteristic function (cf). As such, instead of deriving the mgf of Y, we could just as well have derived the characteristic function. Indeed, using the cf has two advantages. First, for many densities, the mgf does not exist, whereas the cf does. Second, once we have the cf, we can (in theory) derive the pdf that is associated with it by means of the Inversion Theorem (§2.4 D), rather than trying to match it with a known cf in a textbook appendix. This is particularly important if the derived cf is not of a standard (or common) form.

In this vein, we now obtain the pdf of Y directly by the Inversion Theorem. To start, we need the cf. Since we already know the mgf (derived above), we can easily derive the cf by simply replacing the argument t with it, as follows:

```
cf = mgf_Y /. t → i t
```

$$\frac{1}{\sqrt{1 - 2it}}$$

and then apply the Inversion Theorem (as per §2.4 D) to yield the pdf:

```
pdf = InverseFourierTransform[cf, t, y,
              FourierParameters → {1, 1}]
```

$$\frac{(1 + \text{Sign}[y])(\text{Cosh}[\frac{y}{2}] - \text{Sinh}[\frac{y}{2}])}{2\sqrt{2\pi}\,(y^2)^{1/4}}$$

which simplifies further if we note that Y is always positive:

FullSimplify[pdf, y > 0]

$$\frac{e^{-y/2}}{\sqrt{2\pi}\ \sqrt{y}}$$

which is the pdf we obtained in *Example 9*. Although inverting the cf is much more attractive than matching mgf's with textbook appendices, the inversion process is computationally difficult (even with *Mathematica*) and success is not that common in practice. ∎

⊕ *Example 11:* Product of Two Normals

Let X_1 and X_2 be independent $N(0, 1)$ random variables. We wish to find the density of the product $Y = X_1 X_2$ using the mgf/cf method.

Solution: The joint pdf $f(x_1, x_2)$ is:

$$f = \frac{e^{-\frac{x_1^2}{2}}}{\sqrt{2\pi}}\ \frac{e^{-\frac{x_2^2}{2}}}{\sqrt{2\pi}};$$

domain[f] = {{x₁, -∞, ∞}, {x₂, -∞, ∞}};

The cf of Y is given by $E[e^{itY}] = E[e^{itX_1 X_2}]$:

cf = Expect[e^(i t x₁ x₂), f]

— This further assumes that: $\{t^2 > -1\}$

$$\frac{1}{\sqrt{1 + t^2}}$$

Inverting the cf yields the pdf of Y:

**pdf = InverseFourierTransform[cf, t, y,
        FourierParameters → {1, 1}]**

$$\frac{\text{BesselK}[0, y\ \text{Sign}[y]]}{\pi}$$

where `BesselK` denotes the modified Bessel function of the second kind. Figure 6 contrasts the pdf of Y with that of the Normal pdf.

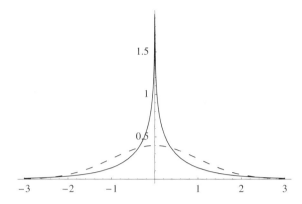

Fig. 6: The pdf of the product of two Normals (——) compared to a Normal pdf (– – –)

4.4 Products and Ratios of Random Variables

This section discusses random variables that are formed as products or ratios of other random variables.

⊕ *Example 12:* Product of Two Normals (again)

Let X_1 and X_2 be two independent standard Normal random variables. In *Example 11*, we found the pdf of the product $X_1 X_2$ using the MGF Method. We now do so using the Transformation Method.

Solution: Let $f(x_1, x_2)$ denote the joint pdf of X_1 and X_2. Due to independence, $f(x_1, x_2)$ is just the pdf of X_1 multiplied by the pdf of X_2:

$$f = \frac{e^{-\frac{x_1^2}{2}}}{\sqrt{2\pi}} \frac{e^{-\frac{x_2^2}{2}}}{\sqrt{2\pi}}; \quad \texttt{domain[f] = \{\{x}_1\texttt{, -}\infty\texttt{, }\infty\texttt{\}, \{x}_2\texttt{, -}\infty\texttt{, }\infty\texttt{\}\};}$$

Let $Y_1 = X_1 X_2$ and $Y_2 = X_2$. Then, the joint pdf of (Y_1, Y_2), say $g(y_1, y_2)$, is:

```
         g = Transform[{y₁ == x₁ x₂, y₂ == x₂}, f];
domain[g] = {{y₁, -∞, ∞}, {y₂, -∞, ∞}};
```

In the interest of brevity, we have suppressed the output of g here by putting a semi-colon at the end of each line of the input. Nevertheless, one should always inspect the solution for g by removing the semi-colon, before proceeding further. Given $g(y_1, y_2)$, the marginal pdf of Y_1 is:

Marginal[y₁, g]

$$\frac{\texttt{BesselK[0, Abs[y}_1\texttt{]]}}{\pi}$$

as per *Example 11*. ∎

⊕ **Example 13:** Ratio of Two Normals: The Cauchy Distribution

Let X_1 and X_2 be two independent standard Normal random variables. We wish to find the pdf of the ratio X_1/X_2.

Solution: The joint pdf $f(x_1, x_2)$ was entered in *Example 12*. Let $g(y_1, y_2)$ denote the joint pdf of $Y_1 = X_1/X_2$ and $Y_2 = X_2$. Then:

```
g = Transform[{y₁ == x₁/x₂, y₂ == x₂}, f];
domain[g] = {{y₁, -∞, ∞}, {y₂, -∞, ∞}};
```

Again, one should inspect the solution to g by removing the semi-colons. The pdf of Y_1 is:

```
Marginal[y₁, g]
```

$$\frac{1}{\pi + \pi y_1^2}$$

where Y_1 has domain of support $(-\infty, \infty)$. That is, the ratio of two independent $N(0, 1)$ random variables has a Cauchy distribution. ∎

⊕ **Example 14:** Derivation of Student's *t* Distribution

Let $X \sim N(0, 1)$ be independent of $Y \sim$ Chi-squared(n). We seek the density of the (scaled) ratio $T = \dfrac{X}{\sqrt{Y/n}}$.

Solution: Due to independence, the joint pdf of (X, Y), say $f(x, y)$, is the pdf of X multiplied by the pdf of Y:

```
f = (e^(-x²/2)/√(2π)) * (y^(n/2-1) e^(-y/2) / (2^(n/2) Γ[n/2]));
domain[f] = {{x, -∞, ∞}, {y, 0, ∞}} && {n > 0};
```

Let $T = \dfrac{X}{\sqrt{Y/n}}$ and $Z = Y$. Then, the joint pdf of (T, Z), say $g(t, z)$, is obtained with:

```
g = Transform[{t == x/√(y/n), z == y}, f];
domain[g] = {{t, -∞, ∞}, {z, 0, ∞}} && {n > 0};
```

Then, the pdf of T is:

```
Marginal[t, g]
```

$$\frac{n^{n/2} \, (n + t^2)^{\frac{1}{2}(-1-n)} \, \Gamma[\frac{1+n}{2}]}{\sqrt{\pi} \, \Gamma[\frac{n}{2}]}$$

where T has domain of support $(-\infty, \infty)$. This is the pdf of a random variable distributed according to Student's *t* distribution with n degrees of freedom. ∎

§4.4 FUNCTIONS OF RANDOM VARIABLES

⊕ **Example 15:** Derivation of Fisher's F Distribution

Let $X_1 \sim \chi_a^2$ be independent of $X_2 \sim \chi_b^2$, where χ_a^2 and χ_b^2 are Chi-squared distributions with degrees of freedom a and b, respectively. We seek the distribution of the (scaled) ratio $R = \dfrac{X_1/a}{X_2/b}$.

Solution: Due to independence, the joint pdf of (X_1, X_2), say $f(x_1, x_2)$, is just the pdf of X_1 multiplied by the pdf of X_2:

```
f  =  ( x₁^(a/2 - 1) e^(-x₁/2) / (2^(a/2) Γ[a/2]) ) * ( x₂^(b/2 - 1) e^(-x₂/2) / (2^(b/2) Γ[b/2]) );

domain[f] = {{x₁, 0, ∞}, {x₂, 0, ∞}} && {a > 0, b > 0};
```

Let $Z = X_2$. Then, the joint pdf of (R, Z), say $g(r, z)$, is obtained with:

```
g  =  Transform[{r == (x₁/a)/(x₂/b), z == x₂}, f];

domain[g] = {{r, 0, ∞}, {z, 0, ∞}} && {a > 0, b > 0};
```

Then, the pdf of random variable R is:

```
Marginal[r, g]
```

$$\frac{\left(\frac{ar}{b}\right)^{a/2} \left(1 + \frac{ar}{b}\right)^{\frac{1}{2}(-a-b)} \Gamma\!\left[\frac{a+b}{2}\right]}{r\, \Gamma\!\left[\frac{a}{2}\right] \Gamma\!\left[\frac{b}{2}\right]}$$

with domain of support $(0, \infty)$. This is the pdf of a random variable with Fisher's F distribution, with parameters a and b denoting the numerator and denominator degrees of freedom, respectively. ∎

⊕ **Example 16:** Derivation of Noncentral F Distribution

Let $X_1 \sim \chi_a^2(\lambda)$ be independent of $X_2 \sim \chi_b^2$, where $\chi_a^2(\lambda)$ denotes a noncentral Chi-squared distribution with noncentrality parameter λ. We seek the distribution of the (scaled) ratio $R = \dfrac{X_1/a}{X_2/b}$.

Solution: Let $f(x_1, x_2)$ denote the joint pdf of X_1 and X_2. Due to independence, $f(x_1, x_2)$ is just the pdf of X_1 multiplied by the pdf of X_2. As usual, the *Continuous* palette can be used to help enter the densities:

```
f = ( 2^(-a/2) e^(-(x₁+λ)/2) x₁^((a-2)/2)  *

     HypergeometricOF1Regularized[ a/2, (x₁ λ)/4 ] ) ( x₂^(b/2 - 1) e^(-x₂/2) / (2^(b/2) Γ[b/2]) );

domain[f] = {{x₁, 0, ∞}, {x₂, 0, ∞}} && {a > 0, b > 0, λ > 0};
```

With $Z = X_2$, the joint pdf of (R, Z), say $g(r, z)$, is obtained with:

```
g = Transform[{r == (x₁/a)/(x₂/b), z == x₂}, f];
domain[g] = {{r, 0, ∞}, {z, 0, ∞}} && {a > 0, b > 0, λ > 0};
```

Then, the pdf of random variable R is:

```
Marginal[r, g]
```

$$\frac{1}{r\,\Gamma[\frac{b}{2}]} \left(e^{-\lambda/2} \left(\frac{a\,r}{b}\right)^{a/2} \left(1 + \frac{a\,r}{b}\right)^{\frac{1}{2}(-a-b)} \Gamma\!\left[\frac{a+b}{2}\right] \right.$$
$$\left. \text{Hypergeometric1F1Regularized}\!\left[\frac{a+b}{2},\ \frac{a}{2},\ \frac{a\,r\,\lambda}{2\,b + 2\,a\,r}\right] \right)$$

with domain of support $(0, \infty)$. This is the pdf of a random variable with a noncentral F distribution with noncentrality parameter λ, and degrees of freedom a and b. ∎

4.5 Sums and Differences of Random Variables

This section discusses random variables that are formed as sums or differences of other random variables. §4.5 A applies the Transformation Method, while §4.5 B applies the MGF Method which is particularly well-suited to dealing with sample sums and sample means.

4.5 A Applying the Transformation Method

⊕ *Example 17:* Sum of Two Exponential Random Variables

Let X_1 and X_2 be independent random variables, each distributed Exponentially with parameter λ. We wish to find the density of $X_1 + X_2$.

Solution: Let $f(x_1, x_2)$ denote the joint pdf of (X_1, X_2):

```
f = (e^(-x₁/λ))/λ * (e^(-x₂/λ))/λ;  domain[f] = {{x₁, 0, ∞}, {x₂, 0, ∞}};
```

Let $Y = X_1 + X_2$ and $Z = X_2$. Since X_1 and X_2 are positive, it follows that $0 < z < y < \infty$. Then the joint pdf of (Y, Z), say $g(y, z)$, is obtained with:

```
g = Transform[{y == x₁ + x₂, z == x₂}, f];
domain[g] = {{y, z, ∞}, {z, 0, y}};
```

Then, the pdf of $Y = X_1 + X_2$ is:

Marginal[y, g]

$$\frac{e^{-\frac{y}{\lambda}} y}{\lambda^2}$$

with domain of support $(0, \infty)$, which is the pdf of a random variable with a Gamma distribution with shape parameter $a = 2$, and scale parameter $b = \lambda$. This is easy to verify using **mathStatica**'s *Continuous* palette. ∎

⊕ *Example 18:* Sum of Poisson Random Variables

Let $X_1 \sim \text{Poisson}(\lambda_1)$ be independent of $X_2 \sim \text{Poisson}(\lambda_2)$. We seek the distribution of the sum $X_1 + X_2$.

Solution: Let $f(x_1, x_2)$ denote the joint pmf of (X_1, X_2):

$$f = \frac{e^{-\lambda_1} \lambda_1^{x_1}}{x_1!} \frac{e^{-\lambda_2} \lambda_2^{x_2}}{x_2!};$$
$$\text{domain}[f] = \{\{x_1, 0, \infty\}, \{x_2, 0, \infty\}\} \text{ \&\& } \{\text{Discrete}\};$$

Let $Y = X_1 + X_2$ and $Z = X_2$. Then the joint pmf of (Y, Z), say $g(y, z)$, is:

g = Transform[{y == x₁ + x₂, z == x₂}, f]

$$\frac{e^{-\lambda_1 - \lambda_2} \lambda_1^{y-z} \lambda_2^z}{(y - z)! \, z!}$$

where $0 \leq z \leq y < \infty$. We seek the pmf of Y, and so sum out the values of Z:

$$\text{sol} = \sum_{z=0}^{y} \text{Evaluate}[g]$$

$$\frac{e^{-\lambda_1 - \lambda_2} \lambda_1^y \left(\frac{\lambda_1 + \lambda_2}{\lambda_1}\right)^y}{\Gamma[1 + y]}$$

which simplifies further:

FullSimplify[sol, y ∈ Integers]

$$\frac{e^{-\lambda_1 - \lambda_2} (\lambda_1 + \lambda_2)^y}{\Gamma[1 + y]}$$

This is the pmf of a Poisson$(\lambda_1 + \lambda_2)$ random variable. Thus, the sum of independent Poisson variables is itself Poisson distributed. This result is particularly important in the following scenario: consider the sample sum comprised of n independent Poisson(λ) variables. Then, $\sum_{i=1}^{n} X_i \sim \text{Poisson}(n\lambda)$. ∎

⊕ **Example 19:** Sum of Two Uniform Random Variables: A Triangular Distribution

Let $X_1 \sim$ Uniform(0, 1) be independent of $X_2 \sim$ Uniform(0, 1). We seek the density of $Y = X_1 + X_2$.

Solution: Let $f(x_1, x_2)$ denote the joint pdf of (X_1, X_2):

```
f = 1;     domain[f] = {{x₁, 0, 1}, {x₂, 0, 1}};
```

Let $Y = X_1 + X_2$ and $Z = X_2$. Then the joint pdf of (Y, Z), say $g(y, z)$, is:

```
eqn = {y == x₁ + x₂, z == x₂};     g = Transform[eqn, f]

1
```

Deriving the domain of this joint pdf is a bit more tricky, but can be assisted by using `DomainPlot`, which again plots the space in the *y-z* plane where $g(y, z) > 0$:

```
DomainPlot[eqn, f];
```

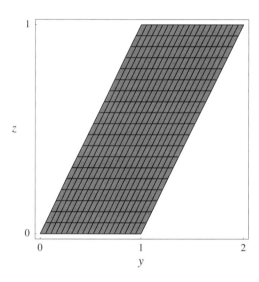

Fig. 7: The space in the *y-z* plane where $g(y, z) > 0$

We see that the domain (the shaded region) can be defined as follows:

When $y < 1$: $0 < z < y < 1$
When $y > 1$: $1 < y < 1 + z < 2$, or equivalently, $0 < y - 1 < z < 1$

The density of Y, say $h(y)$, is then obtained by integrating out Z in each part of the domain. This is easiest to do manually here:

$$h = \text{If}\left[y < 1, \text{Evaluate}\left[\int_0^y g \, dz\right], \text{Evaluate}\left[\int_{y-1}^1 g \, dz\right]\right]$$

```
If[y < 1, y, 2 - y]
```

with domain of support:

domain[h] = {y, 0, 2};

Figure 8 plots the pdf of Y.

PlotDensity[h];

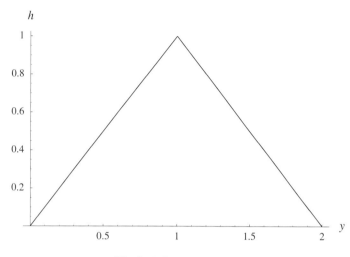

Fig. 8: Triangular pdf

This is known as a Triangular distribution. More generally, if $X_1, \ldots, X_n$ are independent Uniform(0,1) random variables, the distribution of $S_n = \sum_{i=1}^{n} X_i$ is known as the Irwin–Hall distribution (see *Example 18* of Chapter 2). By contrast, the distribution of S_n/n is known as Bates's distribution (*cf. Example 6* of Chapter 8). ∎

⊕ **Example 20:** Difference of Exponential Random Variables: The Laplace Distribution

Let X_1 and X_2 be independent random variables, each distributed Exponentially with parameter $\lambda = 1$. We seek the density of $Y = X_1 - X_2$.

Solution: Let $f(x_1, x_2)$ denote the joint pdf of X_1 and X_2. Due to independence:

f = e⁻ˣ¹ * e⁻ˣ² ;   domain[f] = {{x₁, 0, ∞}, {x₂, 0, ∞}};

Let $Z = X_2$. Then the joint pdf of (Y, Z), say $g(y, z)$, is:

eqn = {y == x₁ - x₂, z == x₂};   g = Transform[eqn, f]

e^{-y-2z}

Deriving the domain of support of Y and Z is a bit more tricky. To make things clearer, we again use `DomainPlot` to plot the space in the y-z plane where $g(y, z) > 0$. Because x_1

and x_2 are unbounded above, we need to manually specify the plot bounds; we use $\{x_1, 0, 100\}$, $\{x_2, 0, 100\}$ here:

```
DomainPlot[eqn, f, {x₁, 0, 100}, {x₂, 0, 100},
    PlotRange → {{-2, 2}, {-1, 3}}];
```

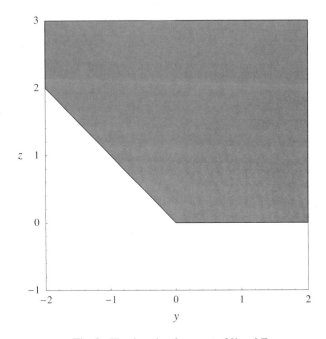

Fig. 9: The domain of support of Y and Z

This suggests that the domain (the shaded region in Fig. 9) can be defined as follows:

When $y < 0$: $\quad 0 < -y \le z < \infty$
When $y > 0$: $\quad \{0 < y < \infty, \; 0 < z < \infty\}$

The density of Y, say $h(y)$, is then obtained by integrating out Z in each part of the domain. This is done manually here:

$$h = \text{If}\left[y < 0, \; \text{Evaluate}\left[\int_{-y}^{\infty} g \, dz\right], \; \text{Evaluate}\left[\int_{0}^{\infty} g \, dz\right]\right]$$

$$\text{If}\left[y < 0, \; \frac{e^y}{2}, \; \frac{e^{-y}}{2}\right]$$

with domain of support:

```
domain[h] = {y, -∞, ∞};
```

This is often expressed in texts as $h(y) = \frac{1}{2} e^{-|y|}$, for $y \in \mathbb{R}$. This is the pdf of a random variable with a standard Laplace distribution (also known as the Double Exponential distribution). ∎

4.5 B   Applying the MGF Method

The MGF Method is especially well-suited to finding the distribution of the sum of independent and identical random variables. Let $(X_1, \ldots, X_n)$ denote a random sample of size n drawn from a random variable X whose mgf is $M_X(t)$. Further, let:

$$s_1 = \sum_{i=1}^{n} X_i \quad \text{(sample sum)}$$

$$s_2 = \sum_{i=1}^{n} X_i^2 \quad \text{(sample sum of squares)}$$

(4.4)

Then, the following results are a special case of the MGF Theorem of Chapter 2:

mgf of s_1: $M_{s_1}(t) = \prod_{i=1}^{n} M_X(t) = \{M_X(t)\}^n = \left(E[e^{tX}]\right)^n$

mgf of $\overline{X} = \frac{s_1}{n}$: $M_{\overline{X}}(t) = M_{s_1}(\frac{t}{n}) = \{M_X(\frac{t}{n})\}^n = \left(E[e^{\frac{t}{n}X}]\right)^n$ (4.5)

mgf of s_2: $M_{s_2}(t) = \prod_{i=1}^{n} M_{X^2}(t) = \{M_{X^2}(t)\}^n = \left(E[e^{tX^2}]\right)^n$

We shall make use of these relations in the following examples.

⊕ **Example 21:** Sum of n Bernoulli Random Variables: The Binomial Distribution

Suppose that the discrete random variable X is Bernoulli distributed with parameter p. That is, $X \sim$ Bernoulli(p), where $P(X=1) = p$, $P(X=0) = 1-p$, and $0 < p < 1$.

```
        g = p^x (1 - p)^(1-x);
domain[g] = {x, 0, 1} && {0 < p < 1} && {Discrete};
```

For a random sample of size n on X, the mgf of the sample sum s_1 is derived from (4.5) as:

$$\text{mgf}_{s_1} = \text{Expect}[e^{tx}, g]^n$$

$$(1 + (-1 + e^t) p)^n$$

This is equivalent to the mgf of a Binomial(n, p) variable, as the reader can easily verify (use the *Discrete* palette to enter the Binomial pmf). Therefore, if $X \sim$ Bernoulli(p), then $s_1 \sim$ Binomial(n, p).   ∎

⊕ **Example 22:** Sum of n Exponential Random Variables: The Gamma Distribution

Let $X \sim$ Exponential(λ). For a random sample of size n, $(X_1, \ldots, X_n)$, we wish to find the distribution of the sample sum $s_1 = \sum_{i=1}^{n} X_i$.

Solution: Let $f(x)$ denote the pdf of X:

```
f = (1/λ) e^(-x/λ) ;        domain[f] = {x, 0, ∞} && {λ > 0};
```

By (4.5), the mgf of the sample sum s_1 is:

$$\text{mgf}_{s_1} = \text{Expect}[e^{tx}, f]^n$$

$$\left(\frac{1}{1-t\lambda}\right)^n$$

This is identical to the mgf of a Gamma(a, b) random variable with parameter $a = n$, and $b = \lambda$, as we now verify:

$$g = \frac{x^{a-1} e^{-x/b}}{\Gamma[a]\, b^a}; \quad \text{domain}[g] = \{x, 0, \infty\}\;\&\&\;\{a > 0, b > 0\};$$

$$\text{Expect}[e^{tx}, g]$$

$$(1 - b\,t)^{-a}$$

Thus, if $X \sim$ Exponential(λ), then $s_1 \sim$ Gamma(n, λ). ∎

⊕ **Example 23:** Sum of n Chi-squared Random Variables

Let $X \sim \chi_v^2$, a Chi-squared random variable with v degrees of freedom, and let $(X_1, \ldots, X_n)$ denote a random sample of size n drawn from X. We wish to find the distribution of the sample sum $s_1 = \sum_{i=1}^{n} X_i$.

Solution: Let $f(x)$ denote the pdf of X:

$$f = \frac{x^{v/2-1} e^{-x/2}}{2^{v/2}\, \Gamma[\tfrac{v}{2}]}; \quad \text{domain}[f] = \{x, 0, \infty\}\;\&\&\;\{v > 0\};$$

The mgf of X is:

$$\text{mgf} = \text{Expect}[e^{tx}, f]$$

— This further assumes that: $\{t < \tfrac{1}{2}\}$

$$(1 - 2\,t)^{-v/2}$$

By (4.5), the mgf of the sample sum s_1 is:

$$\text{mgf}_{s_1} = \text{mgf}^n \;//\; \text{PowerExpand}$$

$$(1 - 2\,t)^{-\frac{nv}{2}}$$

which is the mgf of a Chi-squared random variable with nv degrees of freedom. Thus, if $X \sim \chi_v^2$, then $s_1 \sim \chi_{nv}^2$. ∎

⊕ *Example 24:* Distribution of the Sample Mean for a Normal Random Variable

If $X \sim N(\mu, \sigma^2)$, find the distribution of the sample mean, for a random sample of size n.

Solution: Let $f(x)$ denote the pdf of X:

```
f = e^(-(x-μ)²/(2σ²)) / (σ √(2π)) ;   domain[f] = {x, -∞, ∞} && {μ ∈ Reals, σ > 0};
```

Then the mgf of the sample mean, $\overline{X}$, is given by (4.5) as $\left(E\left[e^{\frac{t}{n}X}\right]\right)^n$:

```
Expect[e^(t/n x), f]^n   // PowerExpand // Simplify
```

$$e^{t\mu + \frac{t^2 \sigma^2}{2n}}$$

which is the mgf of a $N\!\left(\mu, \frac{\sigma^2}{n}\right)$ variable. Therefore, $\overline{X} \sim N\!\left(\mu, \frac{\sigma^2}{n}\right)$. ∎

⊕ *Example 25:* Distribution of the Sample Mean for a Cauchy Random Variable

Let X be a Cauchy random variable. We wish to find the distribution of the sample mean, $\overline{X}$, for a random sample of size n.

Solution: Let $f(x)$ denote the pdf of X:

```
f = 1 / (π (1 + x²)) ;   domain[f] = {x, -∞, ∞};
```

The mgf of a Cauchy random variable does not exist, so we shall use the characteristic function (cf) instead, as the latter always exists. Recall that the cf of X is $E[e^{itX}]$:

```
cf = Expect[e^(i t x), f]
```

— This further assumes that: {Im[t] == 0}

$$e^{-t\,\text{Sign}[t]}$$

By (4.5), the cf of $\overline{X}$ is given by:

```
cf_X̄ = (cf /. t → t/n)^n // Simplify[#, {n > 0, n ∈ Integers}] &
```

$$e^{-t\,\text{Sign}[t]}$$

Note that the cf of $\overline{X}$ is identical to the cf of X. Therefore, if X is Cauchy, then $\overline{X}$ has the same distribution. ∎

⊕ **Example 26:** Distribution of the Sample Sum of Squares for $X_i \sim N(\mu, 1)$
→ Derivation of a Noncentral Chi-squared Distribution

Let $(X_1, \ldots, X_n)$ be independent random variables, with $X_i \sim N(\mu, 1)$. We wish to find the density of the sample sum of squares $s_2 = \sum_{i=1}^n X_i^2$ using the mgf method.

Solution: Let $X \sim N(\mu, 1)$ have pdf $f(x)$:

```
f = e^(-1/2 (x-μ)²) / √(2π) ;        domain[f] = {x, -∞, ∞};
```

By (4.5), the mgf of s_2 is $(E[e^{tX^2}])^n$:

```
mgf = Expect[e^(t x²), f]^n  // PowerExpand
```

— This further assumes that: $\{t < \frac{1}{2}\}$

$$e^{\frac{n t \mu^2}{1-2t}} (1 - 2t)^{-n/2}$$

This expression is equivalent to the mgf of a noncentral Chi-squared variable $\chi_n^2(\lambda)$ with n degrees of freedom and noncentrality parameter $\lambda = n\mu^2$. To demonstrate this, we use **mathStatica**'s *Continuous* palette to input the $\chi_n^2(\lambda)$ pdf, and match its mgf to the one derived above:

```
f = HypergeometricOF1Regularized[n/2, xλ/4] / (2^(n/2) e^((x+λ)/2) x^(-(n-2)/2)) ;

domain[f] = {x, 0, ∞} && {n > 0, λ > 0};
```

Its mgf is given by:

```
Expect[e^(t x), f]
```

— This further assumes that: $\{t < \frac{1}{2}\}$

$$e^{\frac{t \lambda}{1-2t}} (1 - 2t)^{-n/2}$$

We see that the mgf's are equivalent provided $\lambda = n\mu^2$, as claimed. Thus, if $X \sim N(\mu, 1)$, then $s_2 = \sum_{i=1}^n X_i^2 \sim \chi_n^2(n\mu^2)$. If $\mu = 0$, the noncentrality parameter disappears, and we revert to the familiar Chi-squared(n) pdf:

```
f /. λ → 0

2^(-n/2) e^(-x/2) x^(1/2 (-2+n)) / Γ[n/2]
```

Figure 10 illustrates the noncentral Chi-squared pdf $\chi_{n=4}^2(\lambda)$, at different values of λ.

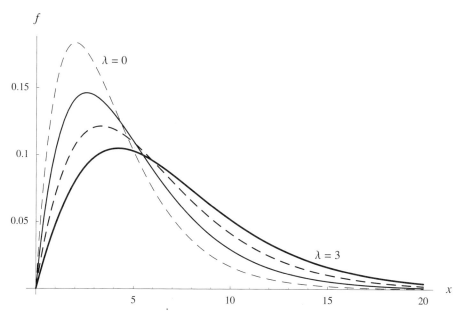

Fig. 10: Noncentral Chi-squared pdf when $n = 4$ and $\lambda = 0, 1, 2, 3$

⊕ **Example 27:** Distribution of the Sample Sum of Squares About the Mean

Let $(X_1, \ldots, X_n)$ be independent random variables, with $X_i \sim N(0, 1)$. We wish to find the density of the sum of squares about the sample mean; i.e. $SS = \sum_{i=1}^{n} (X_i - \overline{X})^2$ where $\overline{X} = \frac{1}{n} \sum_{i=1}^{n} X_i$. Unlike previous examples, the random variable SS is not listed in (4.5). Nevertheless, we can find the solution by first applying a transformation known as *Helmert's transformation* and then applying a result obtained above with the mgf method. Helmert's transformation is given by:

$$\begin{aligned}
Y_1 &= (X_1 - X_2)/\sqrt{2} \\
Y_2 &= (X_1 + X_2 - 2X_3)/\sqrt{6} \\
Y_3 &= (X_1 + X_2 + X_3 - 3X_4)/\sqrt{12} \\
&\vdots \\
Y_{n-1} &= (X_1 + X_2 + \cdots + X_{n-1} - (n-1)X_n)/\sqrt{n(n-1)} \\
Y_n &= (X_1 + X_2 + \cdots + X_n)/\sqrt{n}
\end{aligned} \quad (4.6)$$

For our purposes, the Helmert transformation has two important features:

(i) If each X_i is independent $N(0, 1)$, then each Y_i is also independent $N(0, 1)$.

(ii) $SS = \sum_{i=1}^{n} (X_i - \overline{X})^2 = \sum_{i=1}^{n-1} Y_i^2$.

The rest is easy: we know from *Example 26* that if $Y_i \sim N(0, 1)$, then $\sum_{i=1}^{n-1} Y_i^2$ is Chi-squared with $n - 1$ degrees of freedom. Therefore, for a random sample of size n on a standard Normal random variable, $\sum_{i=1}^{n} (X_i - \overline{X})^2 \sim \chi_{n-1}^2$.

To illustrate properties (i) and (ii), we can implement the Helmert transformation (4.6) in *Mathematica*:

```
Helmert[n_Integer] := Append[
    Table[y_j == (∑_{i=1}^{j} x_i - j x_{j+1})/√(j (j + 1)), {j, n - 1}], y_n == (∑_{i=1}^{n} x_i)/√n ]
```

When, say, $n = 4$, we have:

```
x⃗ = Table[x_i, {i, 4}];
y⃗ = Table[y_i, {i, 4}];
eqn = Helmert[4]
```

$$\left\{ y_1 == \frac{x_1 - x_2}{\sqrt{2}},\ y_2 == \frac{x_1 + x_2 - 2 x_3}{\sqrt{6}},\ y_3 == \frac{x_1 + x_2 + x_3 - 3 x_4}{2\sqrt{3}},\ y_4 == \frac{1}{2}(x_1 + x_2 + x_3 + x_4) \right\}$$

Let $f(\vec{x})$ denote the joint pdf of the X_i:

```
f = ∏_{i=1}^{n} e^{-x_i²/2}/√(2π)  /. n → 4;   domain[f] = Thread[{x⃗, -∞, ∞}];
```

and let $g(\vec{y})$ denote the joint pdf of the Y_i:

```
g = Transform[eqn, f]
domain[g] = Thread[{y⃗, -∞, ∞}];
```

$$\frac{e^{\frac{1}{2}(-y_1^2 - y_2^2 - y_3^2 - y_4^2)}}{4\pi^2}$$

Property (i) states that if the X_i are $N(0, 1)$, then the Y_i are also independent $N(0, 1)$. This is easily verified — the marginal distributions of each of Y_1, Y_2, Y_3 and Y_4:

```
Map[ Marginal[#, g] &, y⃗ ]
```

$$\left\{ \frac{e^{-y_1^2/2}}{\sqrt{2\pi}},\ \frac{e^{-y_2^2/2}}{\sqrt{2\pi}},\ \frac{e^{-y_3^2/2}}{\sqrt{2\pi}},\ \frac{e^{-y_4^2/2}}{\sqrt{2\pi}} \right\}$$

… are all $N(0, 1)$, while independence follows since the joint pdf $g(\vec{y})$ is equal to the product of the marginals.

Property (ii) states that $\sum_{i=1}^{n} (X_i - \overline{X})^2 = \sum_{i=1}^{n-1} Y_i^2$. To show this, we first find the inverse of the transformation equations:

```
inv = Solve[eqn, x⃗][[1]]
```

$$\left\{ x_4 \to \frac{1}{2}\left(-\sqrt{3}\, y_3 + y_4\right), \right.$$
$$x_3 \to \frac{1}{6}\left(-2\sqrt{6}\, y_2 + \sqrt{3}\, y_3 + 3 y_4\right),$$
$$x_1 \to \frac{1}{6}\left(3\sqrt{2}\, y_1 + \sqrt{6}\, y_2 + \sqrt{3}\, y_3 + 3 y_4\right),$$
$$\left. x_2 \to \frac{1}{6}\left(-3\sqrt{2}\, y_1 + \sqrt{6}\, y_2 + \sqrt{3}\, y_3 + 3 y_4\right) \right\}$$

and then examine the sum $\sum_{i=1}^{n} (X_i - \overline{X})^2$, given the transformation of X to Y:

```
∑_{i=1}^{n} (x_i - (1/n) ∑_{i=1}^{n} x_i)^2  /. n → 4  /. inv  // Simplify
```

$$y_1^2 + y_2^2 + y_3^2$$

One final point is especially worth noting: since SS is a function of (Y_1, Y_2, Y_3), and since each of these variables is independent of Y_4, it follows that SS is independent of Y_4 or any function of it, including $Y_4/\sqrt{n}$, which is equal to $\overline{X}$, by (4.6). Hence, in Normal samples, SS is independent of $\overline{X}$. This applies not only when $n = 4$, but also quite generally for arbitrary n. The independence of SS and $\overline{X}$ in Normal samples is an important property that is useful when constructing statistics for hypothesis testing. ∎

4.6 Exercises

1. Let $X \sim \text{Uniform}(0, 1)$. Show that the distribution of $Y = \log\left(\frac{X}{1-X}\right)$ is standard Logistic.

2. Let $X \sim N(\mu, \sigma^2)$. Find the distribution of $Y = \exp(\exp(X))$.

3. Find the pdf of $Y = 1/X$:
   (i) if $X \sim \text{Gamma}(a, b)$; ($Y$ has an InverseGamma(a, b) distribution).
   (ii) if $X \sim \text{PowerFunction}(a, c)$; ($Y$ has a Pareto distribution).
   (iii) if $X \sim \text{InverseGaussian}(\mu, \lambda)$; ($Y$ has a Random Walk distribution).
   Plot the Random Walk pdf when $\mu = 1$ and $\lambda = 1, 4$ and 16.

4. Let X have a Maxwell–Boltzmann distribution. Find the distribution of $Y = X^2$ using both the Transformation Method and the MGF Method.

5. Let X_1 and X_2 have joint pdf $f(x_1, x_2) = 4 x_1 x_2$, $0 < x_1 < 1$, $0 < x_2 < 1$. Find the joint pdf of $Y_1 = X_1^2$ and $Y_2 = X_1 X_2$. Plot the domain of support of Y_1 and Y_2.

6. Let X_1 and X_2 be independent standard Cauchy random variables. Find the distribution of $Y = X_1 X_2$ and plot it.

7. Let X_1 and X_2 be independent Gamma variates with the same scale parameter b. Find the distribution of $Y = \dfrac{X_1}{X_1 + X_2}$.

8. Let $X \sim \text{Geometric}(p)$ and $Y \sim \text{Geometric}(q)$ be independent random variables. Find the distribution of $Z = Y - X$. Plot the pmf of Z when (i) $p = q = \frac{1}{2}$, (ii) $p = \frac{1}{2}$, $q = \frac{1}{8}$.

9. Find the sum of n independent Gamma(a, b) random variables.

Chapter 5

Systems of Distributions

5.1 Introduction

This chapter discusses three systems of distributions: (i) the Pearson family, §5.2, which defines a density in terms of its slope; (ii) the Johnson system, §5.3, which describes a density in terms of transformations of the standard Normal; and (iii) a Gram–Charlier expansion, §5.4, which represents a density as a series expansion of the standard Normal density.

The Pearson system, in particular, is of interest in its own right because it nests many common distributions such as the Gamma, Normal, Student's t, and Beta as special cases. The family of stable distributions is discussed in Chapter 2. Non-parametric kernel density estimation is briefly discussed in §5.5, while the method of moments estimation technique (used throughout the chapter) is covered in §5.6.

5.2 The Pearson Family

5.2 A Introduction

The Pearson system is the family of solutions $p(x)$ to the differential equation

$$\frac{dp(x)}{dx} = -\frac{a+x}{c_0 + c_1 x + c_2 x^2} p(x) \tag{5.1}$$

that yield well-defined density functions. The shape of the resulting distribution will clearly depend on the Pearson parameters (a, c_0, c_1, c_2). As we shall see later, these parameters can be expressed in terms of the first four moments of the distribution (§5.2 D). Thus, if we know the first four moments, we can construct a density function that is consistent with those moments. This provides a rather neat way of constructing density functions that approximate a given set of data. Karl Pearson grouped the family into a number of *types* (§5.2 C). These *types* can be classified in terms of β_1 and β_2 where

$$\beta_1 = \frac{\mu_3^2}{\mu_2^3} \quad \text{and} \quad \beta_2 = \frac{\mu_4}{\mu_2^2}. \tag{5.2}$$

The value of $\sqrt{\beta_1}$ is often used as a measure of *skewness*, while β_2 is often used as a measure of *kurtosis*. Figure 1 illustrates this classification system in (β_1, β_2) space.

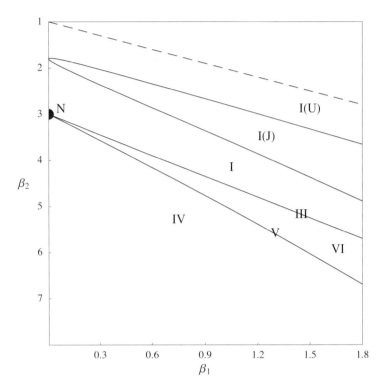

Fig. 1: The β_1, β_2 chart for the Pearson system

The classification consists of several types, as listed in Table 1.

Main types :	*Type I* including $I(U)$ and $I(J)$, *Type IV* and *Type VI*
Transition types :	*Type III* (a line), *Type V* (a line)
Symmetrical types :	If the distribution is symmetrical, then $\mu_3 = 0$, so $\beta_1 = 0$. This yields three special cases : • The N at (0, 3) denotes the Normal distribution. • *Type II* (not labelled) occurs when $\beta_1 = 0$ and $\beta_2 < 3$, and is thus just a special case of *Type I*. • *Type VII* occurs when $\beta_1 = 0$ and $\beta_2 > 3$ (a special case of *Type IV*).

Table 1: Pearson types

The dashed line denotes the upper limit for all distributions. The vertical axis is 'upside-down'. This has become an established (though rather peculiar) convention which we follow. *Type I*, *I(U)* and *I(J)* all share the same functional form—they are all *Type I*. However, they differ in appearance: *Type I(U)* yields a U-shaped density, while *Type I(J)* yields a J-shaped density.[1] The electronic notebook version of this chapter provides an animated tour of the Pearson system here:

5.2 B Fitting Pearson Densities

This section illustrates how to construct a Pearson distribution that is consistent with a set of data whose first four moments are known. With **mathStatica**, this is a two step process:

(i) Use PearsonPlot[$\{\mu_2, \mu_3, \mu_4\}$] to ascertain which Pearson *Type* is consistent with the data.
(ii) If it is say *Type III*, then PearsonIII[μ, $\{\mu_2, \mu_3, \mu_4\}$, x] yields the desired density function $f(x)$ (and its domain).

The full set of functions is:

    PearsonI   PearsonII   PearsonIII   PearsonIV
    PearsonV   PearsonVI   PearsonVII

In the following examples, we categorise data as follows:

- Is it *population* data or *sample* data?
- Is it *raw* data or *grouped* data?

⊕ **Example 1:** Fitting a Pearson Density to *Raw Population* Data

The marks.dat data set lists the final marks of all 891 first year students in the Department of Econometrics at the University of Sydney in 1996. It is raw data because it has not been grouped or altered in any way, and may be thought of as population data (as opposed to sample data) because the entire population's results are listed in the data set. To proceed, we first load the data set into *Mathematica*:

    data = ReadList["marks.dat"];

and then find its mean:

    mean = SampleMean[data] // N

    58.9024

We can use the **mathStatica** function FrequencyPlot to get an intuitive visual perspective on this data set:

    FrequencyPlot[data];

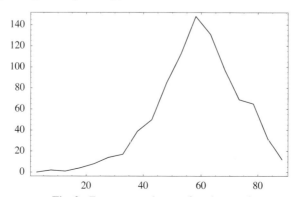

Fig. 2: Frequency polygon of student marks

The x-axis in Fig. 2 represents the range of possible marks from 0 to 100, while the y-axis plots frequency. Of course, there is nothing absolute about the shape of this plot, because the shape varies with the chosen bandwidth c. To see this, evaluate `FrequencyPlot[data, {0, 100, c}]` at different values of c, changing the bandwidth from, say, 4 to 12. Although the shape changes, this empirical pdf nevertheless does give a rough idea of what our Pearson density will look like. Alternatively, see the non-parametric kernel density estimator in §5.5.

Next, we need to find the population central moments μ_2, μ_3, μ_4. Since we have population data, we can use the `CentralMoment` function in *Mathematica*'s `Statistics`DescriptiveStatistics`` package, which we load as follows:

<< Statistics`

μ_{234} = Table[CentralMoment[data, r], {r, 2, 4}] // N

Step (i): `PearsonPlot[`μ_{234}`]` calculates β_1 and β_2 from μ_{234}, and then indicates which Pearson *Type* is appropriate for this data set by plotting a large black dot at the point (β_1, β_2):

PearsonPlot[μ_{234}];

$\{\beta_1 \to 0.173966, \beta_2 \to 3.55303\}$

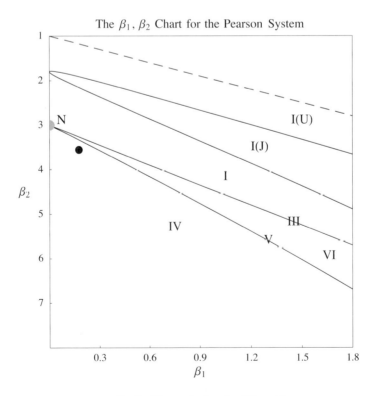

Fig. 3: The marks data is of *Type IV*

§5.2 B   SYSTEMS OF DISTRIBUTIONS

Step (ii): The large black dot is within the *Type IV* zone (the most feared of them all!), so the fitted Pearson density $f(x)$ and its domain are given by:

{f, domain[f]} = PearsonIV[mean, μ_{234}, x]

$$\left\{ \frac{1.14587 \times 10^{25} \, e^{13.4877 \, \text{ArcTan}[1.55011 - 0.0169455 \, x]}}{(448.276 - 6.92074 \, x + 0.0378282 \, x^2)^{13.2177}}, \, \{x, -\infty, \infty\} \right\}$$

The FrequencyPlot function can now be used to compare the empirical pdf (——) with the fitted Pearson pdf (– – –):

p1 = FrequencyPlot[data, f];

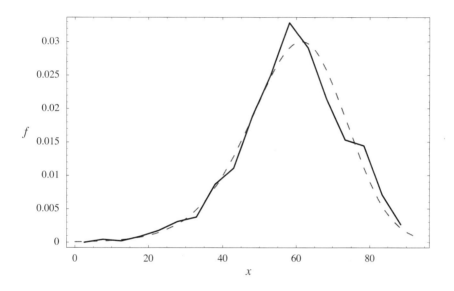

Fig. 4: The empirical pdf (——) and fitted Pearson pdf (– – –) for the marks data

⊕ **Example 2:** Fitting a Pearson Density to *Raw Sample* Data

The file grain.dat contains data that measures the yield from 1500 different rows of wheat. The data comes from Andrews and Herzberg (1985) and StatLib. We shall treat it as raw sample data. To proceed, we first load the data set into *Mathematica*:

data = ReadList["grain.dat"];

and find its sample mean:

mean = SampleMean[data] // N

587.722

Because this is sample data, the population central moments μ_2, μ_3, μ_4 are unknown. We shall not use the CentralMoment function from *Mathematica*'s Statistics package to estimate the population central moments, because the CentralMoment function is a biased estimator. Instead, we shall use **mathStatica**'s UnbiasedCentralMoment function, as discussed in Chapter 7, because it is an unbiased estimator of population central moments (and has many other desirable properties). As it so happens, the bias from using the CentralMoment function will be small in this example because the sample size is large, but that may not always be the case. Here, then, is our estimate of the vector (μ_2, μ_3, μ_4):

$\hat{\mu}_{234}$ = **Table[UnbiasedCentralMoment[data, r], {r, 2, 4}]**

{9997.97, 576417., 3.39334 × 10⁸}

PearsonPlot$[\hat{\mu}_{234}]$ shows that this is close to *Type III*, so we fit a Pearson density, $f(x)$, to *Type III*:

{f, domain[f]} = PearsonIII[mean, $\hat{\mu}_{234}$, x]

$\{2.39465 \times 10^{-35}\ e^{-0.0324339\,x}\ (-7601.05 + 30.832\,x)^{10.0661}$,
$\{x, 246.531, \infty\}\}$

Once again, the FrequencyPlot function compares the empirical pdf (———) with the fitted Pearson pdf (– – –):

FrequencyPlot[data, f];

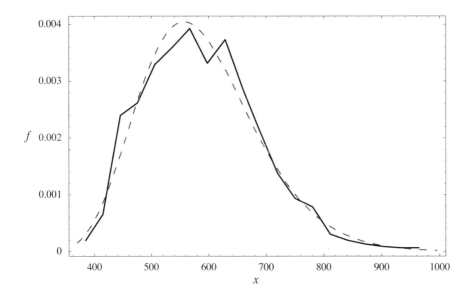

Fig. 5: The empirical pdf (———) and fitted Pearson pdf (– – –) for wheat yield data

⊕ **Example 3:** Fitting a Pearson Density to *Grouped* Data

Table 2 stems from Elderton and Johnson (1969, p. 5):

age X	freq
< 19	34
20–24	145
25–29	156
30–34	145
35–39	123
40–44	103
45–49	86
50–54	71
55–59	55
60–64	37
65–69	21
70–74	13
75–79	7
80–84	3
85–89	1

Table 2: The number of sick people at different ages (in years)

Here, ages 20–24 includes those aged from $19\frac{1}{2}$ up to $24\frac{1}{2}$, and so on. Let X denote the mid-point of each class interval of ages (note that these are equally spaced), while freq denotes the frequency of each interval. Finally, let τ denote the relative frequency. The mid-point of the first class is taken to be 17 to ensure equal bandwidths. Then:

```
X = {17, 22, 27, 32, 37, 42, 47, 52, 57, 62, 67, 72, 77, 82, 87};
freq = {34, 145, 156, 145, 123, 103, 86, 71, 55, 37, 21, 13, 7, 3, 1};
τ = freq / (Plus @@ freq);
```

The **mathStatica** function FrequencyGroupPlot provides a 'histogram' of this grouped data:

FrequencyGroupPlot[{X, freq}];

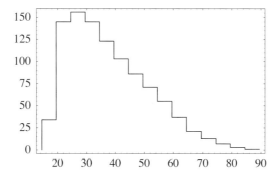

Fig. 6: 'Histogram' of the number of sick people at different ages

which gives some idea of what the fitted Pearson density should look like.

When data is given in table form, the mean is conventionally taken as $\sum_{i=1}^{k} X_i \tau_i$, where X_i is the mid-point of each interval, and τ_i is the relative frequency of each interval, over the k class intervals. Thus:

mean = X.τ // N

37.875

A quick and slightly dirty [2] (though widely used) estimator of the rth central moment for grouped data is given by:

DirtyMu[r_] := (X - mean)^r.τ

Then our estimates of (μ_2, μ_3, μ_4) are:

$\hat{\mu}_{234}$ = {DirtyMu[2], DirtyMu[3], DirtyMu[4]}

{191.559, 1888.36, 107703.}

which is *Type I*, as PearsonPlot[$\hat{\mu}_{234}$] will verify. Then, the fitted Pearson density is:

{f, domain[f]} = PearsonI[mean, $\hat{\mu}_{234}$, x]

{9.70076 × 10^{-8} (94.3007 − 1. x)$^{2.77976}$ (−16.8719 + 1. x)$^{0.406924}$, {x, 16.8719, 94.3007}}

Of course, the density $f(x)$ should be consistent with the central moments that generated it. Thus, if we calculated the first few central moments of $f(x)$, we should obtain $\{\mu_2 \to 191.559, \mu_3 \to 1888.36, \mu_4 \to 107703\}$, as above. A quick check verifies these results:

Expect[(x - mean)$^{\{2, 3, 4\}}$, f]

{191.559, 1888.36, 107703.}

The FrequencyGroupPlot function can now be used to compare the 'histogram' with the smooth fitted Pearson pdf:

FrequencyGroupPlot[{X, freq}, f];

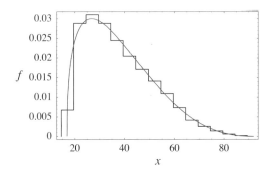

Fig. 7: The 'histogram' and the fitted Pearson pdf (smooth)

5.2 C Pearson Types

Recall that the Pearson family is defined as the set of solutions to

$$\frac{dp(x)}{dx} = -\frac{a+x}{c_0 + c_1 x + c_2 x^2} p(x).$$

In *Mathematica*, the solution to this differential equation can be expressed as:

$$\text{Pearson} := \text{DSolve}\left[\text{p}'[\text{x}] == -\frac{(\text{a}+\text{x})\,\text{p}[\text{x}]}{\text{c0}+\text{c1}\,\text{x}+\text{c2}\,\text{x}^2}, \text{p}[\text{x}], \text{x}\right]$$

Since $\frac{dp}{dx} = 0$ when $x = -a$, the latter defines the mode, while the shape of the density will depend on the roots of the quadratic $c_0 + c_1 x + c_2 x^2$. The various Pearson *Types* correspond to the different forms this quadratic may take. We briefly consider the main seven types, in no particular order. Before doing so, we set up `MrClean` to ensure that we start our analysis of each *Type* with a clean slate:

$$\text{MrClean} := \text{ClearAll}[\text{a}, \text{c0}, \text{c1}, \text{c2}, \text{p}, \text{x}];$$

Type IV occurs when $c_0 + c_1 x + c_2 x^2$ does not have real roots. In *Mathematica*, this is equivalent to finding the solution to the differential equation without making any special assumption at all about the roots. This works because *Mathematica* typically finds the most general solution, and does not assume the roots are real:

$$\text{MrClean}; \quad \text{Pearson // Simplify}$$

$$\left\{\left\{\text{p}[\text{x}] \to e^{\frac{(\text{c1}-2\,\text{a}\,\text{c2})\,\text{ArcTan}\left[\frac{\text{c1}+2\,\text{c2}\,\text{x}}{\sqrt{-\text{c1}^2+4\,\text{c0}\,\text{c2}}}\right]}{\text{c2}\,\sqrt{-\text{c1}^2+4\,\text{c0}\,\text{c2}}}} (\text{c0} + \text{x}\,(\text{c1}+\text{c2}\,\text{x}))^{-\frac{1}{2\,\text{c2}}} \text{C}[1]\right\}\right\}$$

The domain is $\{x, -\infty, \infty\}$. Under *Type IV*, numerical integration is usually required to find the constant of integration `C[1]`.

Type VII is the special symmetrical case of *Type IV*, and it occurs when $c_1 = a = 0$. This nests Student's t distribution:

$$\% \,/.\, \{\text{c1} \to 0,\, \text{a} \to 0\}$$

$$\left\{\left\{\text{p}[\text{x}] \to (\text{c0} + \text{c2}\,\text{x}^2)^{-\frac{1}{2\,\text{c2}}} \text{C}[1]\right\}\right\}$$

Type III (Gamma distribution) occurs when $c_2 = 0$:

$$\text{MrClean}; \quad \text{c2} = 0; \quad \text{Pearson // Simplify}$$

$$\left\{\left\{\text{p}[\text{x}] \to e^{-\frac{x}{\text{c1}}} (\text{c0} + \text{c1}\,\text{x})^{\frac{\text{c0}-\text{a}\,\text{c1}}{\text{c1}^2}} \text{C}[1]\right\}\right\}$$

In order for this solution to be a well-defined pdf, we require $p(x) > 0$. Thus, if $c_1 > 0$, the domain is $x > -c_0/c_1$; if $c_1 < 0$, the domain is $x < -c_0/c_1$.

Type V occurs when the quadratic $c_0 + c_1 x + c_2 x^2$ has one real root. This occurs when $c_1^2 - 4 c_0 c_2 = 0$. Hence:

MrClean; c0 = $\frac{c1^2}{4\, c2}$; Pearson // Simplify

$$\left\{\left\{p[x] \to e^{\frac{-c1+2\, a\, c2}{c2\, (c1+2\, c2\, x)}} (c1 + 2\, c2\, x)^{-1/c2}\, C[1]\right\}\right\}$$

The Normal distribution is obtained when $c_1 = c_2 = 0$:

MrClean; c1 = 0; c2 = 0; Pearson

$$\left\{\left\{p[x] \to e^{-\frac{a\,x}{c0} - \frac{x^2}{2\,c0}}\, C[1]\right\}\right\}$$

Completing the square allows us to write this as:

$$p[x] = k\, e^{-\frac{(x+a)^2}{2\, c0}}\, ; \quad \texttt{domain}[p[x]] = \{x, -\infty, \infty\};$$

where, in order to be a well-defined density, constant k must be such that the density integrates to unity; that is, that $P(X < \infty) = 1$:

Solve[Prob[∞, p[x]] == 1, k]

— This further assumes that: $\{c0 > 0\}$

$$\left\{\left\{k \to \frac{1}{\sqrt{c0}\, \sqrt{2\, \pi}}\right\}\right\}$$

The result is thus Normal with mean $-a$, and variance $c_0 > 0$.

That leaves *Type I*, *Type II* and *Type VI*. These cases occur if $c_0 + c_1 x + c_2 x^2 = 0$ has two *real* roots, r_1 and r_2. In particular, *Type I* occurs if $r_1 < 0 < r_2$ (roots are of *opposite* sign), with domain $r_1 < x < r_2$. This nests the Beta distribution. *Type II* is identical to *Type I*, except that we now further assume that $r_1 = -r_2$. This yields a symmetrical curve with $\beta_1 = 0$. *Type VI* occurs if r_1 and r_2 are the *same* sign; the domain is $x > r_2$ if $0 < r_1 < r_2$, or $x < r_2$ if $r_2 < r_1 < 0$. In the case of *Type VI*, with two real roots of the same sign, one can express $c_0 + c_1 x + c_2 x^2$ as $c_2(x - r_1)(x - r_2)$. The family of solutions is then:

MrClean;

DSolve$\left[p'[x] == -\frac{a+x}{c2\, (x-r1)\, (x-r2)}\, p[x],\, p[x],\, x\right]$ // Simplify

$$\left\{\left\{p[x] \to (-r1 + x)^{\frac{a+r1}{-c2\, r1+c2\, r2}} (-r2 + x)^{\frac{a+r2}{c2\, r1-c2\, r2}}\, C[1]\right\}\right\}$$

where the constant of integration can now be solved for the relevant domain.

5.2 D Pearson Coefficients in Terms of Moments

```
ClearAll[a, c0, c1, c2, eqn, μ]
```

It is possible to express the Pearson coefficients a, c_0, c_1 and c_2 in terms of the first four raw moments $\acute{\mu}_r$ ($r = 1, 4$). To do so, we first multiply both sides of (5.1) by x^r and integrate over the domain of X:

$$\int_{-\infty}^{\infty} x^r (c_0 + c_1 x + c_2 x^2) \frac{dp(x)}{dx} dx = -\int_{-\infty}^{\infty} x^r (a + x) p(x) dx. \tag{5.3}$$

If we integrate the left-hand side by parts,

$$\int_{-\infty}^{\infty} f g' dx = f g \Big]_{-\infty}^{\infty} - \int_{-\infty}^{\infty} f' g \, dx \qquad \text{with } g' = \frac{dp(x)}{dx}$$

and break the right-hand side into two, then (5.3) becomes

$$x^r (c_0 + c_1 x + c_2 x^2) p(x) \Big]_{-\infty}^{\infty}$$

$$- \int_{-\infty}^{\infty} \{ r c_0 x^{r-1} + (r+1) c_1 x^r + (r+2) c_2 x^{r+1} \} p(x) \, dx \tag{5.4}$$

$$= -\int_{-\infty}^{\infty} a x^r p(x) \, dx - \int_{-\infty}^{\infty} x^{r+1} p(x) \, dx$$

If we assume that $x^r p(x) \to 0$ at the extremum of the domain, then the first expression on the left-hand side vanishes, and after substituting raw moments $\acute{\mu}$ for integrals, we are left with

$$-r c_0 \acute{\mu}_{r-1} - (r+1) c_1 \acute{\mu}_r - (r+2) c_2 \acute{\mu}_{r+1} = -a \acute{\mu}_r - \acute{\mu}_{r+1}. \tag{5.5}$$

This recurrence relation defines any moment in terms of lower moments. Further, since the density must integrate to unity, we have the boundary condition $\acute{\mu}_0 = 1$. In *Mathematica* notation, we write this relation as:

```
eqn[r_] :=
    (-r c0 μ́_{r-1} - (r + 1) c1 μ́_r - (r + 2) c2 μ́_{r+1} == -a μ́_r - μ́_{r+1})
      /. μ́_0 → 1
```

We wish to find a, c_0, c_1 and c_2 in terms of $\acute{\mu}_r$. Putting $r = 0, 1, 2$ and 3 yields the required 4 equations (for the 4 unknowns) which we now solve simultaneously to yield the solution:

```
Z = Solve[Table[eqn[r], {r, 0, 3}], {a, c0, c1, c2}]
    // Simplify
```

$$a \to \frac{20\,\mu_1'^2\,\mu_2'\,\mu_3' - 12\,\mu_1'^3\,\mu_4' - \mu_3'\left(3\,\mu_2'^2 + \mu_4'\right) + \mu_1'\left(-9\,\mu_2'^3 - 8\,\mu_3'^2 + 13\,\mu_2'\,\mu_4'\right)}{2\left(9\,\mu_2'^3 + 4\,\mu_1'^3\,\mu_3' - 16\,\mu_1'\,\mu_2'\,\mu_3' + 6\,\mu_3'^2 - 5\,\mu_2'\,\mu_4' + \mu_1'^2\left(-3\,\mu_2'^2 + 5\,\mu_4'\right)\right)}$$

$$c_0 \to \frac{\mu_1'\,\mu_3'\left(\mu_2'^2 + \mu_4'\right) + \mu_2'\left(3\,\mu_3'^2 - 4\,\mu_2'\,\mu_4'\right) + \mu_1'^2\left(-4\,\mu_3'^2 + 3\,\mu_2'\,\mu_4'\right)}{2\left(9\,\mu_2'^3 + 4\,\mu_1'^3\,\mu_3' - 16\,\mu_1'\,\mu_2'\,\mu_3' + 6\,\mu_3'^2 - 5\,\mu_2'\,\mu_4' + \mu_1'^2\left(-3\,\mu_2'^2 + 5\,\mu_4'\right)\right)}$$

$$c_1 \to \frac{8\,\mu_1'^2\,\mu_2'\,\mu_3' - 6\,\mu_1'^3\,\mu_4' - \mu_3'\left(3\,\mu_2'^2 + \mu_4'\right) + \mu_1'\left(-3\,\mu_2'^3 - 2\,\mu_3'^2 + 7\,\mu_2'\,\mu_4'\right)}{2\left(9\,\mu_2'^3 + 4\,\mu_1'^3\,\mu_3' - 16\,\mu_1'\,\mu_2'\,\mu_3' + 6\,\mu_3'^2 - 5\,\mu_2'\,\mu_4' + \mu_1'^2\left(-3\,\mu_2'^2 + 5\,\mu_4'\right)\right)}$$

$$c_2 \to \frac{6\,\mu_2'^3 + 4\,\mu_1'^3\,\mu_3' - 10\,\mu_1'\,\mu_2'\,\mu_3' + 3\,\mu_3'^2 - 2\,\mu_2'\,\mu_4' + \mu_1'^2\left(-3\,\mu_2'^2 + 2\,\mu_4'\right)}{2\left(9\,\mu_2'^3 + 4\,\mu_1'^3\,\mu_3' - 16\,\mu_1'\,\mu_2'\,\mu_3' + 6\,\mu_3'^2 - 5\,\mu_2'\,\mu_4' + \mu_1'^2\left(-3\,\mu_2'^2 + 5\,\mu_4'\right)\right)}$$

If we work *about the mean*, then $\mu_1' = 0$, and $\mu_r' = \mu_r$ for $r \geq 2$. The formulae then become:

```
Z /. {μ1 → 0, μ → μ}
```

$$a \to -\frac{\mu_3\left(3\,\mu_2^2 + \mu_4\right)}{2\left(9\,\mu_2^3 + 6\,\mu_3^2 - 5\,\mu_2\,\mu_4\right)}$$

$$c_0 \to \frac{\mu_2\left(3\,\mu_3^2 - 4\,\mu_2\,\mu_4\right)}{2\left(9\,\mu_2^3 + 6\,\mu_3^2 - 5\,\mu_2\,\mu_4\right)}$$

$$c_1 \to -\frac{\mu_3\left(3\,\mu_2^2 + \mu_4\right)}{2\left(9\,\mu_2^3 + 6\,\mu_3^2 - 5\,\mu_2\,\mu_4\right)}$$

$$c_2 \to \frac{6\,\mu_2^3 + 3\,\mu_3^2 - 2\,\mu_2\,\mu_4}{2\left(9\,\mu_2^3 + 6\,\mu_3^2 - 5\,\mu_2\,\mu_4\right)}$$

Note that a and c_1 are now equal; this only applies when one works *about the mean*. With these definitions, the Pearson *Types* of §5.2 C can now be expressed in terms of the first 4 moments, instead of parameters a, c_0, c_1 and c_2. This is, in fact, how the various automated Pearson fitting functions are constructed (§5.2 B).

5.2 E   Higher Order Pearson-Style Families

Instead of basing the Pearson system upon the quadratic $c_0 + c_1 x + c_2 x^2$, one can instead consider using higher order polynomials as the foundation stone. If the moments of the population are known, then this endeavour must unambiguously yield a better fit. If, however, the observed data is a random sample drawn from the population, there is a trade-off: a higher order polynomial implies that higher order moments are required, and the estimates of the latter may be unreliable (have high variance), unless the sample size is 'large'.

In this section, we consider a Pearson-style system based upon a cubic polynomial. This will be the family of solutions $p(x)$ to the differential equation

$$\frac{dp(x)}{dx} = -\frac{a+x}{c_0 + c_1 x + c_2 x^2 + c_3 x^3} p(x). \tag{5.6}$$

Adopting the method introduced in §5.2 D once again yields a recurrence relation, but now with one extra term on the left-hand side. Equation (5.5) now becomes

$$-r c_0 \acute{\mu}_{r-1} - (r+1) c_1 \acute{\mu}_r - (r+2) c_2 \acute{\mu}_{r+1} - (r+3) c_3 \acute{\mu}_{r+2} = -a \acute{\mu}_r - \acute{\mu}_{r+1}. \tag{5.7}$$

Given the boundary condition $\acute{\mu}_0 = 1$, we enter this recurrence relation into *Mathematica* as:

```
eqn2[r_] :=
   (-r c0 µ´_{r-1} - (r + 1) c1 µ´_r - (r + 2) c2 µ´_{r+1} - (r + 3) c3 µ´_{r+2} ==
   -a µ´_r - µ´_{r+1}) /. µ´_0 → 1
```

Our objective is to find a, c_0, c_1, c_2 and c_3 in terms of $\acute{\mu}_r$. Putting $r = 0, 1, 2, 3, 4$ yields the required 5 equations (for the 5 unknowns) which we now solve simultaneously:

```
Z1 = Solve[Table[eqn2[r], {r, 0, 4}], {a, c0, c1, c2, c3}];
```

The solution is rather long, so we will not print it here. However, if we work *about the mean*, taking $\acute{\mu}_1 = 0$, and $\acute{\mu}_r = \mu_r$ for $r \geq 2$, the solution reduces to:

```
Z2 = Z1[[1]] /. {µ´_1 → 0, µ´ → µ} // Simplify;
Z2 // TableForm
```

$a \rightarrow \dfrac{-117 \mu_3^2 \mu_3 \mu_4 - 16 \mu_3^3 \mu_4 + 81 \mu_2^4 \mu_5 + 5 \mu_4^2 \mu_5 + \mu_2 \mu_3 (25 \mu_4^2 + 24 \mu_3 \mu_5) + 3 \mu_2^2 (16 \mu_3^3 - 18 \mu_4 \mu_5 + 7 \mu_3 \mu_6) + \mu_3 (-12 \mu_5^2 + 7 \mu_4 \mu_6)}{2 (96 \mu_3^4 - 27 \mu_2^4 \mu_4 - 50 \mu_4^3 + 93 \mu_3 \mu_4 \mu_5 + 15 \mu_2^2 (7 \mu_4^2 + 9 \mu_3 \mu_5) + 9 \mu_2^3 (2 \mu_3^3 - 7 \mu_6) - 42 \mu_3^2 \mu_6 + \mu_2 (-272 \mu_3^2 \mu_4 - 36 \mu_5^2 + 35 \mu_4 \mu_6))}$

$c_0 \rightarrow \dfrac{-3 \mu_2^3 (4 \mu_4^2 - 3 \mu_3 \mu_5) + 8 \mu_3^3 (-2 \mu_4^2 + 3 \mu_3 \mu_5) + \mu_2 (40 \mu_4^3 - 77 \mu_3 \mu_4 \mu_5 + 21 \mu_3^2 \mu_6) + \mu_2^2 (3 \mu_3^2 \mu_4 + 36 \mu_5^2 - 28 \mu_4 \mu_6)}{2 (-96 \mu_3^4 + 27 \mu_2^4 \mu_4 + 50 \mu_4^3 - 93 \mu_3 \mu_4 \mu_5 - 15 \mu_2^2 (7 \mu_4^2 + 9 \mu_3 \mu_5) + 42 \mu_3^2 \mu_6 + \mu_2^3 (-18 \mu_3^2 + 63 \mu_6) + \mu_2 (272 \mu_3^2 \mu_4 + 36 \mu_5^2 - 35 \mu_4 \mu_6))}$

$c_1 \rightarrow \dfrac{-33 \mu_2^2 \mu_3 \mu_4 - 16 \mu_3^3 \mu_4 + 27 \mu_2^4 \mu_5 + 5 \mu_4^2 \mu_5 + \mu_2 \mu_3 (37 \mu_4^2 + 6 \mu_3 \mu_5) + 3 \mu_2^2 (4 \mu_3^3 - 16 \mu_4 \mu_5 + 7 \mu_3 \mu_6) + \mu_3 (-12 \mu_5^2 + 7 \mu_4 \mu_6)}{2 (96 \mu_3^4 - 27 \mu_2^4 \mu_4 - 50 \mu_4^3 + 93 \mu_3 \mu_4 \mu_5 + 15 \mu_2^2 (7 \mu_4^2 + 9 \mu_3 \mu_5) + 9 \mu_2^3 (2 \mu_3^3 - 7 \mu_6) - 42 \mu_3^2 \mu_6 + \mu_2 (-272 \mu_3^2 \mu_4 - 36 \mu_5^2 + 35 \mu_4 \mu_6))}$

$c_2 \rightarrow \dfrac{-48 \mu_3^4 + 18 \mu_2^4 \mu_4 + 20 \mu_4^3 - 39 \mu_3 \mu_4 \mu_5 - 3 \mu_2^2 (22 \mu_4^2 + 23 \mu_3 \mu_5) - 6 \mu_3^2 (2 \mu_3^2 - 7 \mu_6) + 21 \mu_3^2 \mu_6 + \mu_2 (143 \mu_3^2 \mu_4 + 12 \mu_5^2 - 14 \mu_4 \mu_6)}{2 (-96 \mu_3^4 + 27 \mu_2^4 \mu_4 + 50 \mu_4^3 - 93 \mu_3 \mu_4 \mu_5 - 15 \mu_2^2 (7 \mu_4^2 + 9 \mu_3 \mu_5) + 42 \mu_3^2 \mu_6 + \mu_2^3 (-18 \mu_3^2 + 63 \mu_6) + \mu_2 (272 \mu_3^2 \mu_4 + 36 \mu_5^2 - 35 \mu_4 \mu_6))}$

$c_3 \rightarrow \dfrac{14 \mu_2^2 \mu_3 \mu_4 - 9 \mu_3^2 \mu_5 + \mu_3 (2 \mu_4^2 - 3 \mu_3 \mu_5) + \mu_2 (-6 \mu_3^3 + \mu_4 \mu_5)}{-96 \mu_3^4 + 27 \mu_2^4 \mu_4 + 50 \mu_4^3 - 93 \mu_3 \mu_4 \mu_5 - 15 \mu_2^2 (7 \mu_4^2 + 9 \mu_3 \mu_5) + 42 \mu_3^2 \mu_6 + \mu_2^3 (-18 \mu_3^2 + 63 \mu_6) + \mu_2 (272 \mu_3^2 \mu_4 + 36 \mu_5^2 - 35 \mu_4 \mu_6)}$

which is comparatively compact (for a more legible rendition, see the electronic notebook). Whereas the second-order (quadratic) Pearson family can be expressed in terms of the first 4 moments, the third-order (cubic) Pearson-style family requires the first 6 moments. Note that a and c_1 are no longer equal.

⊕ *Example 4:* Fitting a Third-Order (Cubic) Pearson-Style Density

In this example, we fit a third-order (cubic) Pearson-style density to the data set: marks.dat. *Example 1* fitted the standard second-order (quadratic) Pearson distribution to this data set. It will be interesting to see how a third-order Pearson-style distribution compares. First, we load the required data set into *Mathematica*, if this has not already been done:

```
data = ReadList["marks.dat"];
```

The population central moments $\mu_2, \mu_3, \mu_4, \mu_5$ and μ_6 are given by:

```
<< Statistics`
μ⃗ = Table[μ_r → CentralMoment[data, r] // N, {r, 2, 6}]
```

$\{\mu_2 \to 193.875, \mu_3 \to -1125.94, \mu_4 \to 133550.,$
$\mu_5 \to -2.68578 \times 10^6, \mu_6 \to 1.77172 \times 10^8\}$

In the quadratic system, this data was of *Type IV* (the most general form). Consequently, in the cubic system, we will once again try the most general solution (*i.e.* without making any assumptions about the roots of the cubic polynomial). The solution then is:

$$\text{DSolve}\left[p'[x] == -\frac{a+x}{c0 + c1\,x + c2\,x^2 + c3\,x^3}\,p[x], p[x], x\right]$$

$$\left\{\left\{p[x] \to e^{-\text{RootSum}\left[c0+c1\,\#1+c2\,\#1^2+c3\,\#1^3\,\&,\,\frac{a\,\text{Log}[x-\#1]+\text{Log}[x-\#1]\,\#1}{c1+2\,c2\,\#1+3\,c3\,\#1^2}\,\&\right]}\,C[1]\right\}\right\}$$

Mathematica provides the solution in terms of a RootSum object. If we now replace the Pearson coefficients $\{a, c_0, c_1, c_2, c_3\}$ with central moments $\{\mu_2, \mu_3, \mu_4, \mu_5, \mu_6\}$ via Z2 derived above, and then replace the latter with the empirical $\vec{\mu}$, we obtain:

```
sol = e^{-RootSum[c0+c1 #1+c2 #1²+c3 #1³ &, (a Log[x-#1]+Log[x-#1] #1)/(c1+2 c2 #1+3 c3 #1²) &]} /.
      Z2 /. μ⃗ // Simplify
```

$((-31.6478 - 52.712\,\mathrm{i}) + x)^{-9.86369+6.66825\,\mathrm{i}}$
$((-31.6478 + 52.712\,\mathrm{i}) + x)^{-9.86369-6.66825\,\mathrm{i}}\,(556.021 + x)^{19.7274}$

while the constant of integration over, say, $\{x, -100, 100\}$ is:

```
cn = NIntegrate[sol, {x, -100, 100}]
```

4.22732×10^{32}

A quick plot illustrates:

Plot[sol / cn, {x, -50, 50}];

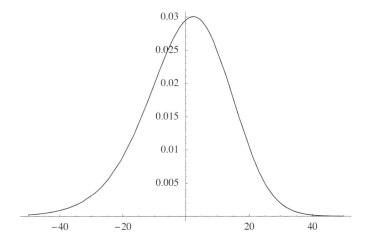

Fig. 8: Cubic Pearson fit for the marks data set

This looks identical to the plot of *f* derived in *Example 1*, except the origin is now at zero, rather than at the mean. If *f* from *Example 1* is derived with zero mean, one can then Plot[f-sol/cn,{x,-50,50}] to see the difference between the two solutions. Doing so yields Fig. 9.

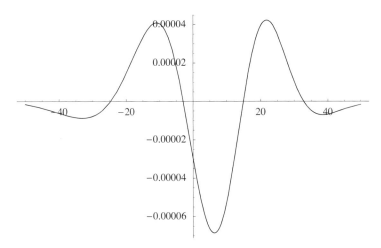

Fig. 9: The difference between the quadratic and cubic Pearson fit

The difference between the plots is remarkably small (note the scale on the vertical axis). This outcome is rather reassuring for those who prefer to use the much simpler quadratic Pearson system. ∎

5.3 Johnson Transformations

5.3 A Introduction

Recall that the Pearson family provides a unique distribution for every possible (β_1, β_2) combination. The Johnson family provides the same feature, and does so by using a set of three transformations of the standard Normal. In particular, if $Z \sim N(0, 1)$ with density $\phi(z)$, and Y is a transform of Z, then the Johnson family is given by:

(1) S_L (Lognormal) $Y = \exp(\frac{Z-\gamma}{\delta})$ $\Leftrightarrow$ $Z = \gamma + \delta \log(Y)$ $(0 < y < \infty)$

(2) S_U (Unbounded) $Y = \sinh(\frac{Z-\gamma}{\delta})$ $\Leftrightarrow$ $Z = \gamma + \delta \sinh^{-1}(Y)$ $(-\infty < y < \infty)$

(3) S_B (Bounded) $Y = \frac{1}{1+\exp\left(-\frac{Z-\gamma}{\delta}\right)}$ $\Leftrightarrow$ $Z = \gamma + \delta \log(\frac{Y}{1-Y})$ $(0 < y < 1)$

Applying a second transform $X = \xi + \lambda Y$ (or equivalently $Y = \frac{X-\xi}{\lambda}$) expands the system from two parameters (γ, δ) to the full set of four $(\gamma, \delta, \xi, \lambda)$, where δ and λ are taken to be positive. Since $X = \xi + \lambda Y$, the shape of the distribution of X will be the same as that of Y. Hence, the parameters may be interpreted as follows: γ and δ determine the shape of the distribution of X; λ is a scale factor; and ξ is a location factor. Figure 10 illustrates the classification system in (β_1, β_2) space.

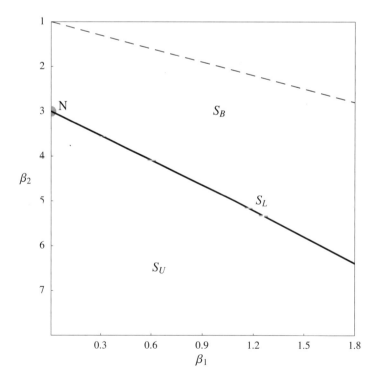

Fig. 10: The β_1, β_2 chart for the Johnson system

Several points are of note:

(i) The classification consists of two *main* types, namely S_U and S_B. These are separated by a *transition* type, the S_L line, which corresponds to the family of Lognormal distributions. The N at $(\beta_1, \beta_2) = (0, 3)$ once again denotes the Normal distribution, which may be thought of as a limiting form of the three systems as $\delta \to \infty$.

(ii) The S_U system is termed *unbounded* because the domain here is $\{y : y \in \mathbb{R}\}$. The S_B system is termed *bounded* because the domain for this system is $\{y : 0 < y < 1\}$.

(iii) The dashed line represents the bound on all distributions, and is given by $\beta_2 - \beta_1 = 1$.

Whereas the Pearson system can be easily 'automated' for fitting purposes, the Johnson system requires some hands-on fine tuning. We consider each system in turn: S_L (§5.3 B); S_U (§5.3 C); and S_B (§5.3 D).

5.3 B S_L System (Lognormal)

Let $Z \sim N(0, 1)$ with density $\phi(z)$:

```
ϕ = e^(-z²/2) / √(2π);    domain[ϕ] = {z, -∞, ∞} && {γ ∈ Reals, δ > 0};
```

The S_L system is defined by the transformation $Y = \exp\left(\frac{Z-\gamma}{\delta}\right)$. Then, the density of Y, say $g(y)$, is:

```
g          = Transform[y == e^((z-γ)/δ), ϕ]
domain[g]  = TransformExtremum[y == e^((z-γ)/δ), ϕ]
```

$$\frac{e^{-\frac{1}{2}(\gamma + \delta \log[y])^2} \delta}{\sqrt{2\pi}\, y}$$

```
{y, 0, ∞} && {γ ∈ Reals, δ > 0}
```

The Lognormal density is positively skewed, though as δ increases, the curve tends to symmetry. In Fig. 11, the density on the far left corresponds to a 'small' δ, while each successive density to the right corresponds to a doubling of δ.

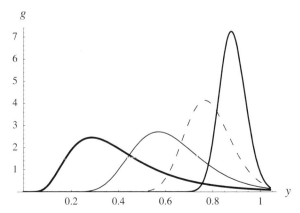

Fig. 11: The Lognormal pdf $g(y)$ when $\gamma = 2$, and $\delta = 2, 4, 8$ and 16

Since $Y = \exp\left(\frac{Z-\gamma}{\delta}\right)$, and Z has density $\phi(z)$, the rth raw moment $E[Y^r]$ can be expressed as:

$$\Omega = \text{Expect}\left[e^{\frac{(z-\gamma)\,r}{\delta}},\ \phi\right]$$

$$e^{\frac{r\,(r-2\,\gamma\,\delta)}{2\,\delta^2}}$$

Thus, the first 4 raw moments (rm) are:

$$\text{rm} = \text{Table}\left[\acute{\mu}_r \to \Omega,\ \{r, 4\}\right]$$

$$\left\{\acute{\mu}_1 \to e^{\frac{1-2\,\gamma\,\delta}{2\,\delta^2}},\ \acute{\mu}_2 \to e^{\frac{2-2\,\gamma\,\delta}{\delta^2}},\ \acute{\mu}_3 \to e^{\frac{3\,(3-2\,\gamma\,\delta)}{2\,\delta^2}},\ \acute{\mu}_4 \to e^{\frac{2\,(4-2\,\gamma\,\delta)}{\delta^2}}\right\}$$

This can be expressed in terms of central moments (cm), as follows:

```
cm = Table[
        CentralToRaw[r] /. rm // Simplify,
        {r, 2, 4}];

cm // TableForm
```

$$\mu_2 \to e^{\frac{1-2\,\gamma\,\delta}{\delta^2}}\left(-1 + e^{\frac{1}{\delta^2}}\right)$$

$$\mu_3 \to e^{\frac{3-6\,\gamma\,\delta}{2\,\delta^2}}\left(-1 + e^{\frac{1}{\delta^2}}\right)^2\left(2 + e^{\frac{1}{\delta^2}}\right)$$

$$\mu_4 \to e^{\frac{2-4\,\gamma\,\delta}{\delta^2}}\left(-1 + e^{\frac{1}{\delta^2}}\right)^2\left(-3 + 3\,e^{\frac{2}{\delta^2}} + 2\,e^{\frac{3}{\delta^2}} + e^{\frac{4}{\delta^2}}\right)$$

Then β_1 and β_2 can be expressed as:

$$\beta_1 = \frac{\mu_3^2}{\mu_2^3}\ /.\ \text{cm}\ //\ \text{Simplify}$$

$$\left(-1 + e^{\frac{1}{\delta^2}}\right)\left(2 + e^{\frac{1}{\delta^2}}\right)^2$$

and

$$\beta_2 = \frac{\mu_4}{\mu_2^2}\ /.\ \text{cm}\ //\ \text{Simplify}$$

$$-3 + 3\,e^{\frac{2}{\delta^2}} + 2\,e^{\frac{3}{\delta^2}} + e^{\frac{4}{\delta^2}}$$

These equations define the Lognormal curve parametrically in (β_1, β_2) space, as δ increases from 0 to ∞, as Fig. 12 illustrates. In *Mathematica*, one can use `ParametricPlot` to derive this curve.

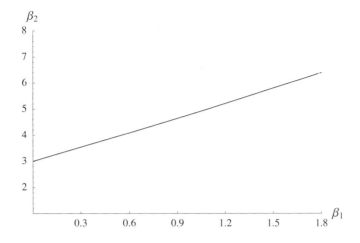

Fig. 12: The Lognormal curve in (β_1, β_2) space

This is identical to the S_L curve shown in Fig. 10 (The Johnson Plot), except that the vertical axis is not inverted here. Despite appearances, the curve in Fig. 12 is not linear; this is easy to verify with a ruler. In the limit, as $\delta \to \infty$, β_1 and β_2 tend to 0 and 3, respectively:

Limit[{β_1, β_2}, $\delta \to \infty$]

{0, 3}

so that the Normal distribution is obtained as a limit case of the Lognormal.

Given an empirical value for β_1 (or β_2), we can now 'solve' for δ. This is particularly easy since γ is not required. For instance, if $\hat{\beta}_1 = 0.829$:

Solve[β_1 == 0.829, δ]

— Solve::ifun : Inverse functions are being
   used by Solve, so some solutions may not be found.

{ {$\delta \to$ -3.46241},
  {$\delta \to$ -0.457213 - 0.354349 i},
  {$\delta \to$ -0.457213 + 0.354349 i},
  {$\delta \to$ 0.457213 - 0.354349 i},
  {$\delta \to$ 0.457213 + 0.354349 i},
  {$\delta \to$ 3.46241} }

Since we require δ to be both real and positive, only the last of these solutions is feasible. One can now find γ by comparing μ_2 (derived above) with its empirical estimate $\hat{\mu}_2$.

5.3 C S_U System (Unbounded)

Once again, let $Z \sim N(0, 1)$ with density $\phi(z)$:

$$\phi = \frac{e^{-\frac{z^2}{2}}}{\sqrt{2\pi}}; \quad \text{domain}[\phi] = \{z, -\infty, \infty\} \,\&\&\, \{\gamma \in \text{Reals}, \delta > 0\};$$

The S_U system is defined by the transformation $Y = \sinh\!\left(\frac{Z-\gamma}{\delta}\right)$. Hence, the density of Y, say $g(y)$, is:

$$g = \text{Transform}\!\left[y == \text{Sinh}\!\left[\frac{z-\gamma}{\delta}\right], \phi\right]$$

$$\text{domain}[g] = \text{TransformExtremum}\!\left[y == \text{Sinh}\!\left[\frac{z-\gamma}{\delta}\right], \phi\right]$$

$$\frac{e^{-\frac{1}{2}(\gamma + \delta \,\text{ArcSinh}[y])^2}\,\delta}{\sqrt{2\pi}\,\sqrt{1+y^2}}$$

$$\{y, -\infty, \infty\} \,\&\&\, \{\gamma \in \text{Reals}, \delta > 0\}$$

Figure 13 indicates shapes that are typical in the S_U family.

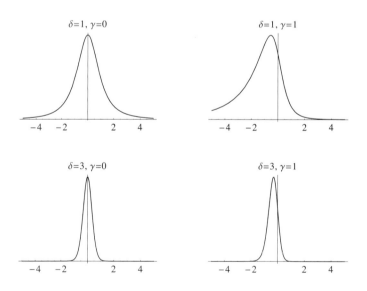

Fig. 13: Typical pdf shapes in the S_U family

Since $Y = \sinh\!\left(\frac{Z-\gamma}{\delta}\right)$, and Z has density $\phi(z)$, the rth moment $E[Y^r]$ can be expressed:

$$\Omega := \text{Expect}\!\left[\text{Sinh}\!\left[\frac{z-\gamma}{\delta}\right]^r, \phi\right]\,/\!/\,\text{ExpToTrig}\,/\!/\,\text{FullSimplify}$$

This time, *Mathematica* cannot find the solution as a function of r, which is why we use a delayed evaluation (:=) instead of an immediate evaluation (=).

The first 4 raw moments (rm) are now given by:

```
rm = Table[μ'ᵣ → Ω, {r, 4}];        rm // TableForm
```

$\mu'_1 \to -e^{\frac{1}{2\delta^2}} \sinh[\frac{\gamma}{\delta}]$

$\mu'_2 \to \frac{1}{2}\left(-1 + e^{\frac{2}{\delta^2}} \cosh[\frac{2\gamma}{\delta}]\right)$

$\mu'_3 \to -\frac{1}{4} e^{\frac{1}{2\delta^2}} \left(-3 \sinh[\frac{\gamma}{\delta}] + e^{\frac{4}{\delta^2}} \sinh[\frac{3\gamma}{\delta}]\right)$

$\mu'_4 \to \frac{1}{8}\left(3 - 4 e^{\frac{2}{\delta^2}} \cosh[\frac{2\gamma}{\delta}] + e^{\frac{8}{\delta^2}} \cosh[\frac{4\gamma}{\delta}]\right)$

This can be expressed in terms of central moments (cm), as follows:[3]

```
cm = Table[CentralToRaw[r] /. rm // FullSimplify, {r, 2, 4}]
```

$\{\mu_2 \to \frac{1}{2}\left(-1 + e^{\frac{1}{\delta^2}}\right)\left(1 + e^{\frac{1}{\delta^2}} \cosh[\frac{2\gamma}{\delta}]\right),$

$\mu_3 \to -\frac{1}{4} e^{\frac{1}{2\delta^2}} \left(-1 + e^{\frac{1}{\delta^2}}\right)^2 \left(3 \sinh[\frac{\gamma}{\delta}] + e^{\frac{1}{\delta^2}}\left(2 + e^{\frac{1}{\delta^2}}\right) \sinh[\frac{3\gamma}{\delta}]\right),$

$\mu_4 \to \frac{1}{8}\left(3 + e^{\frac{2}{\delta^2}}\left(e^{\frac{6}{\delta^2}} \cosh[\frac{4\gamma}{\delta}] + 4 \cosh[\frac{2\gamma}{\delta}]\left(-1 + 6 e^{\frac{1}{\delta^2}} \sinh[\frac{\gamma}{\delta}]^2\right) - 8 \sinh[\frac{\gamma}{\delta}]\left(3 \sinh[\frac{\gamma}{\delta}]^3 + e^{\frac{3}{\delta^2}} \sinh[\frac{3\gamma}{\delta}]\right)\right)\right)\}$

Then β_1 and β_2 can be expressed as:

```
β₁ = μ₃²/μ₂³ /. cm // Simplify
```

$\dfrac{e^{\frac{1}{\delta^2}}\left(-1 + e^{\frac{1}{\delta^2}}\right)\left(3 \sinh[\frac{\gamma}{\delta}] + e^{\frac{1}{\delta^2}}\left(2 + e^{\frac{1}{\delta^2}}\right) \sinh[\frac{3\gamma}{\delta}]\right)^2}{2\left(1 + e^{\frac{1}{\delta^2}} \cosh[\frac{2\gamma}{\delta}]\right)^3}$

```
β₂ = μ₄/μ₂² /. cm // Simplify
```

$\dfrac{3 + e^{\frac{2}{\delta^2}}\left(e^{\frac{6}{\delta^2}} \cosh[\frac{4\gamma}{\delta}] + 4 \cosh[\frac{2\gamma}{\delta}]\left(-1 + 6 e^{\frac{1}{\delta^2}} \sinh[\frac{\gamma}{\delta}]^2\right) - 8 \sinh[\frac{\gamma}{\delta}]\left(3 \sinh[\frac{\gamma}{\delta}]^3 + e^{\frac{3}{\delta^2}} \sinh[\frac{3\gamma}{\delta}]\right)\right)}{2\left(-1 + e^{\frac{1}{\delta^2}}\right)^2\left(1 + e^{\frac{1}{\delta^2}} \cosh[\frac{2\gamma}{\delta}]\right)^2}$

○ *Fitting the S_U System*

To fit the S_U system, we adopt the following steps:

(i) Given values for (β_1, β_2), solve for (δ, γ), noting that $\delta > 0$, and that the sign of γ is opposite to that of μ_3.

(ii) This gives us $g(y | \gamma, \delta)$. Given the transform $X = \xi + \lambda Y$, solve for ξ, and $\lambda > 0$.

⊕ **Example 5:** Fit a Johnson Density to the `marks.dat` Population Data Set

First, load the data set, if this has not already been done:

    `data = ReadList["marks.dat"];`

The mean of this data set is:

    `mean = SampleMean[data] // N`

    `58.9024`

Empirical values for μ_2, μ_3 and μ_4 are once again given by:

    `<< Statistics`` `

    μ_{234} = `Table[ CentralMoment[data, r], {r, 2, 4}] // N`

    `{193.875, -1125.94, 133550.}`

If we were working with sample data, we would replace the `CentralMoment` function with `UnbiasedCentralMoment` (just cut and paste). Just as `PearsonPlot` was used in *Example 1* to indicate the appropriate Pearson *Type*, we now use `JohnsonPlot` to indicate which of the Johnson systems is suitable for this data set:

    `JohnsonPlot[`μ_{234}`];`

    $\{\beta_1 \to 0.173966,\ \beta_2 \to 3.55303\}$

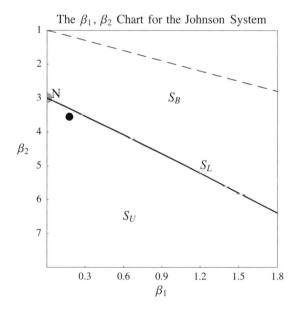

Fig. 14: The marks data lies in the S_U system

The black dot, depicting (β_1, β_2) for this data set, lies in the S_U system. We derived β_1 and β_2 in terms of δ and γ above. Thus, given values $\{\beta_1 \to 0.173966,\ \beta_2 \to 3.55303\}$, it is

§5.3 C SYSTEMS OF DISTRIBUTIONS 171

now possible to 'solve' for (δ, γ). The `FindRoot` function simultaneously solves the two equations for δ and γ:

```
sol = FindRoot[
         { β₁ == 0.17396604431160143`,
           β₂ == 3.5530347934625883`}, {δ, 2}, {γ, 2}]
```

$\{\delta \to 3.74767, \gamma \to 2.0016\}$

Note that `FindRoot` is a numerical technique that returns the first solution it finds, so different starting points may yield different solutions. In evaluating the solution, it helps to note that δ should be positive, while γ should be opposite in sign to μ_3. Johnson (1949, p. 164) and Johnson *et al.* (1994, p. 36) provide a diagram known as an *abac* that provides a rough estimate of γ and δ, given values for β_1 and β_2. These rough estimates make an excellent starting point for the `FindRoot` function. In a similar vein, see Bowman and Shenton (1980).

The full 4-parameter $(\gamma, \delta, \xi, \lambda)$ Johnson S_U system is obtained by applying the further transformation $X = \xi + \lambda Y$ or equivalently $Y = \frac{X-\xi}{\lambda}$. Since we are adding two new parameters, we shall add some assumptions about them:

```
domain[g] = domain[g] && {ξ ∈ Reals, λ > 0};
```

Then the density of $X = \xi + \lambda Y$, say $f(x)$, is:

```
f         = Transform[x == ξ + λ y, g]
domain[f] = TransformExtremum[x == ξ + λ y, g]
```

$$\frac{e^{-\frac{1}{2}(\gamma + \delta \operatorname{ArcSinh}[\frac{x-\xi}{\lambda}])^2} \delta}{\sqrt{2\pi} \lambda \sqrt{1 + \frac{(x-\xi)^2}{\lambda^2}}}$$

$\{x, -\infty, \infty\}$ && $\{\gamma \in \text{Reals}, \delta > 0, \xi \in \text{Reals}, \lambda > 0\}$

where γ and δ have already been found. Since $X = \xi + \lambda Y$, $\operatorname{Var}(X) = \lambda^2 \operatorname{Var}(Y)$. Here, $\operatorname{Var}(Y)$ was found above as $\mu_2(\gamma, \delta)$ (part of cm), while $\operatorname{Var}(X)$ is taken to be the empirical variance 193.875 of the data set. Thus, at the fitted values, the equation $\operatorname{Var}(X) = \lambda^2 \operatorname{Var}(Y)$ becomes:

```
193.875 == λ² μ₂ /. cm /. sol
```

$193.875 == 0.101355 \lambda^2$

Solving for λ yields:

```
λ̂ = Solve[%, λ]
```

$\{\{\lambda \to -43.7359\}, \{\lambda \to 43.7359\}\}$

Since we require $\lambda > 0$, the second solution is the desired one. That leaves ξ ...

Since $X = \xi + \lambda Y$, $E[X] = \xi + \lambda E[Y]$. Here, $E[Y]$ was found above as $\mu'_1(\gamma, \delta)$ (part of rm), while $E[X]$ is taken to be the empirical mean of the data set. Thus, at the fitted values, $E[X] = \xi + \lambda E[Y]$ becomes:

mean == $\xi + \lambda \mu'_1$ /. rm /. sol /. $\hat{\lambda}$[[2]]

58.9024 == -25.3729 + ξ

Solving for ξ yields:

$\hat{\xi}$ = Solve[%, ξ]

{{$\xi \to 84.2752$}}

The desired fitted density $f(x)$ is thus:

f = f /. sol /. $\hat{\lambda}$[[2]] /. $\hat{\xi}$[[1]]

$$\frac{0.0341848 \, e^{-\frac{1}{2}(2.0016+3.74767\,\text{ArcSinh}[0.0228645\,(-84.2752+x)])^2}}{\sqrt{1+0.000522787\,(-84.2752+x)^2}}$$

which has an unbounded domain, like all S_U distributions.

As in *Example 1*, the **mathStatica** function FrequencyPlot allows one to compare the fitted density with the empirical pdf of the data:

p2 = FrequencyPlot[data, f];

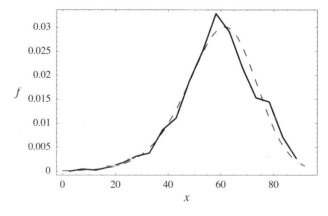

Fig. 15: The empirical pdf (———) and the fitted Johnson S_U pdf (– – –)

This Johnson S_U fitted density appears almost identical to the PearsonIV fit derived in *Example 1*. The final diagram in *Example 1* was labelled p1. If p1 is still in memory, the command Show[p1/.Hue[__] → Hue[.4], p2] shows both plots together, but now with the fitted Pearson curve in green rather than red, enabling a visual comparison (note that Hue[__] contains two _ characters). The curves are so similar that only a tiny tinge of green would be visible on screen. ∎

5.3 D S_B System (Bounded)

Once again, let $Z \sim N(0, 1)$ with density $\phi(z)$:

$$\phi = \frac{e^{-\frac{z^2}{2}}}{\sqrt{2\pi}}; \quad \text{domain}[\phi] = \{z, -\infty, \infty\} \,\&\&\, \{\gamma \in \text{Reals}, \delta > 0\};$$

The S_B (bounded) system is defined by the transformation $Y = (1 + \exp(-\frac{Z-\gamma}{\delta}))^{-1}$. Then, the density of Y, say $g(y)$, is:

$$g = \text{Transform}\left[y == \left(1 + e^{-\frac{z-\gamma}{\delta}}\right)^{-1}, \phi\right]$$
$$\text{domain}[g] = \text{TransformExtremum}\left[y == \left(1 + e^{-\frac{z-\gamma}{\delta}}\right)^{-1}, \phi\right]$$

$$\frac{e^{-\frac{1}{2}(\gamma-\delta \log[-1+\frac{1}{y}])^2} \delta}{\sqrt{2\pi} (y - y^2)}$$

$$\{y, 0, 1\} \,\&\&\, \{\gamma \in \text{Reals}, \delta > 0\}$$

The full 4-parameter $(\gamma, \delta, \xi, \lambda)$ Johnson S_B system is obtained by applying the further transformation $X = \xi + \lambda Y$ or equivalently $Y = \frac{X-\xi}{\lambda}$. Since we are adding two new parameters, we shall add some assumptions about them:

$$\text{domain}[g] = \text{domain}[g] \,\&\&\, \{\xi \in \text{Reals}, \lambda > 0\};$$

Then the density of X, say $f(x)$, is:

$$f = \text{Transform}[\{x == \xi + \lambda y\}, g]$$
$$\text{domain}[f] = \text{TransformExtremum}[\{x == \xi + \lambda y\}, g]$$

$$\frac{e^{-\frac{1}{2}(\gamma-\delta \log[-1+\frac{\lambda}{x-\xi}])^2} \delta \lambda}{\sqrt{2\pi} (x - \xi) (-x + \lambda + \xi)}$$

$$\{x, \xi, \lambda + \xi\} \,\&\&\, \{\gamma \in \text{Reals}, \delta > 0, \xi \in \text{Reals}, \lambda > 0\}$$

Figure 16 shows some plots from the S_B (γ, δ) family.

The moments of the S_B system are extremely complicated. Johnson (1949) obtained a solution for μ'_1, though this does not have a closed form; nor can it be implemented usefully in *Mathematica*. As such, the method of moments is not generally used for fitting S_B systems. Instead, a method of percentile points is used, which equates percentile points of the observed and fitted curves. This approach is not an exact methodology, and we refer the interested reader to Johnson (1949) or Elderton and Johnson (1969, p. 131). Alternatively, one can always use the automated Pearson fitting functions as a substitute, which is inevitably a much simpler strategy.

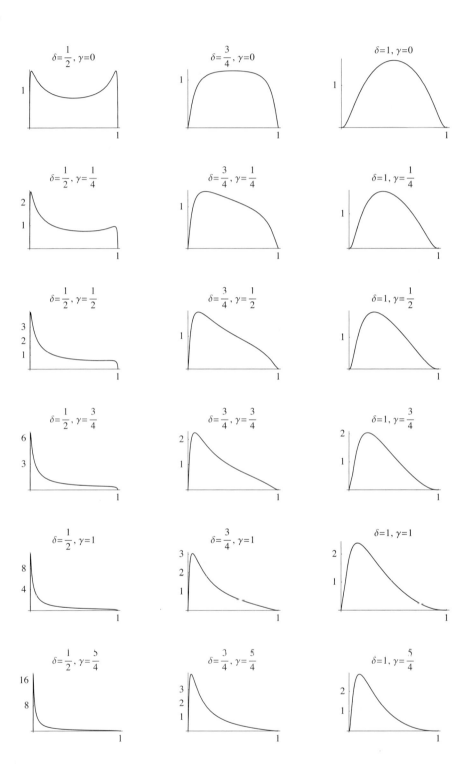

Fig. 16: Some pdf shapes in the S_B family

5.4 Gram–Charlier Expansions

5.4 A Definitions and Fitting

Let $\phi(z)$ denote a standard Normal density:

$$\phi = \frac{e^{-\frac{z^2}{2}}}{\sqrt{2\pi}}; \quad \text{domain}[\phi] = \{z, -\infty, \infty\};$$

and let $\psi(z)$ denote an arbitrary pdf that has been standardised so that its mean is 0 and variance is 1. If $\psi(z)$ can be expanded as a series of derivatives of $\phi(z)$, then

$$\psi(z) = \sum_{j=0}^{\infty} c_j (-1)^j \frac{d^j \phi(z)}{d z^j}. \tag{5.8}$$

This assumes the expansion is convergent — Stuart and Ord (1994, Section 6.22) provide conditions in this regard. Further, let $H_j(z) = \frac{(-1)^j}{\phi(z)} \frac{d^j \phi(z)}{d z^j}$; $H_j(z)$ is known as a Hermite polynomial and has a number of interesting properties (see §5.4 B). Then (5.8) may be written as

$$\psi(z) = \phi(z) \sum_{j=0}^{\infty} c_j H_j(z). \tag{5.9}$$

Then, for sufficiently large t, $\psi(z) \simeq \phi(z) \sum_{j=0}^{t} c_j H_j(z)$. In *Mathematica*, we explicitly model this as a function of t:

$$\psi[\texttt{t\_}] := \phi \sum_{j=0}^{t} c[j] \, H[j]$$

This has two components: (i) $H_j(z)$ and (ii) c_j.

(i) The Hermite polynomial $H_j(z)$ is defined by:[4]

$$H[\texttt{j\_}] := \frac{(-1)^j}{\phi} \partial_{\{z,j\}} \phi \quad // \text{ Expand}$$

Then the first few Hermite polynomials are:

$$\texttt{Table}[H_j \to H[j], \{j, 0, 10\}]$$
$$\quad // \text{ TableForm } // \text{ TraditionalForm}$$

$H_0 \to 1$

$H_1 \to z$

$H_2 \to z^2 - 1$

$H_3 \to z^3 - 3z$

$H_4 \to z^4 - 6z^2 + 3$

$H_5 \to z^5 - 10z^3 + 15z$

$H_6 \to z^6 - 15z^4 + 45z^2 - 15$

$H_7 \to z^7 - 21z^5 + 105z^3 - 105z$

$H_8 \to z^8 - 28z^6 + 210z^4 - 420z^2 + 105$

$H_9 \to z^9 - 36z^7 + 378z^5 - 1260z^3 + 945z$

$H_{10} \to z^{10} - 45z^8 + 630z^6 - 3150z^4 + 4725z^2 - 945$

(ii) The c_j terms are formally derived in §5.4 B where it is shown that c_j is a function of the first j moments of $\psi(z)$. Since we are basing the expansion on $\phi(z)$ (a standardised Normal), c_j is given here in terms of standardised moments (*i.e.* assuming $\mu'_1 = \mu_1 = 0$, $\mu_2 = 1$). The solution takes a similar functional form to $H_j(x)$, which we can exploit in *Mathematica* through pattern matching:

```
c[j_] := H[j]/j! /. z^i_. :> μᵢ /. {μ₁ → 0, μ₂ → 1}
```

The first few c_j terms are given by:

```
Table[cⱼ → c[j], {j, 0, 10}] // TableForm
```

$c_0 \to 1$

$c_1 \to 0$

$c_2 \to 0$

$c_3 \to \frac{\mu_3}{6}$

$c_4 \to \frac{1}{24}(-3 + \mu_4)$

$c_5 \to \frac{1}{120}(-10\mu_3 + \mu_5)$

$c_6 \to \frac{1}{720}(30 - 15\mu_4 + \mu_6)$

$c_7 \to \frac{105\mu_3 - 21\mu_5 + \mu_7}{5040}$

$c_8 \to \frac{-315 + 210\mu_4 - 28\mu_6 + \mu_8}{40320}$

$c_9 \to \frac{-1260\mu_3 + 378\mu_5 - 36\mu_7 + \mu_9}{362880}$

$c_{10} \to \frac{3780 - 3150\mu_4 + 630\mu_6 - 45\mu_8 + \mu_{10}}{3628800}$

§5.4 A   SYSTEMS OF DISTRIBUTIONS   177

We can now evaluate the *Mathematica* function ψ[t] for arbitrarily large t, as a function of the first t (standardised) moments of $\psi(z)$. Here is an example with $t = 7$:

ψ[7]

$$\frac{1}{\sqrt{2\pi}}\left(e^{-\frac{z^2}{2}}\left(1 + \frac{1}{6}(-3z + z^3)\mu_3 + \frac{1}{24}(3 - 6z^2 + z^4)(-3 + \mu_4) + \frac{1}{120}(15z - 10z^3 + z^5)(-10\mu_3 + \mu_5) + \frac{1}{720}(-15 + 45z^2 - 15z^4 + z^6)(30 - 15\mu_4 + \mu_6) + \frac{(-105z + 105z^3 - 21z^5 + z^7)(105\mu_3 - 21\mu_5 + \mu_7)}{5040}\right)\right)$$

⊕ *Example 6:* Fit a Gram–Charlier Density to the marks.dat Population Data

First, load the data if this has not already been done:

data = ReadList["marks.dat"];

Once again, its mean is:

mean = SampleMean[data] // N

58.9024

Evaluating the first 6 central moments (cm) yields:

<< Statistics`

cm = Table[CentralMoment[data, r] // N, {r, 1, 6}]

{0., 193.875, -1125.94,
  133550., -2.68578 × 10^6, 1.77172 × 10^8}

(Once again, if we were working with sample data, we would replace the CentralMoment function with UnbiasedCentralMoment in the line above.) To obtain standardised moments, note that $\mu_i^{\text{standardised}} = \mu_i / \mu_2^{i/2}$. Then, empirical values for the first 6 standardised moments (sm) are:

sm = Table$\left[\mu_i \to \frac{\text{cm}[\![i]\!]}{\text{cm}[\![2]\!]^{i/2}},\ \{i,\ 1,\ 6\}\right]$

$\{\mu_1 \to 0.,\ \mu_2 \to 1.,\ \mu_3 \to -0.417092,$
$\mu_4 \to 3.55303,\ \mu_5 \to -5.13177,\ \mu_6 \to 24.3125\}$

Evaluating ψ[6] at these values yields:

```
ψ₆ = ψ[6] /. sm  // Simplify
domain[ψ₆] = {z, -∞, ∞};
```

$0.000563511 \, e^{-\frac{z^2}{2}} (-5.24309 + z) (-3.14529 + z)$
$(8.28339 - 1.45564 \, z + z^2) (5.43111 + 4.17537 \, z + z^2)$

The above gives the density in standardised units. To find the density in original units, say $f(x)$, transform from $Z = \frac{X-\mu}{\sigma}$ to $X = \mu + \sigma Z$:

```
eqn = {x == mean + √cm〚2〛 z};

      f = Transform[eqn, ψ₆]
domain[f] = TransformExtremum[eqn, ψ₆]
```

$5.55363 \times 10^{-12} \, e^{-0.00257898 \, (-58.9024 + x)^2}$
$(-131.907 + x) (-102.697 + x)$
$(6269.27 - 138.073 \, x + x^2) (1098.01 - 59.6673 \, x + x^2)$

$\{x, -\infty, \infty\}$

Once again, FrequencyPlot allows one to compare the empirical pdf with the fitted density:

```
p3 = FrequencyPlot[data, f];
```

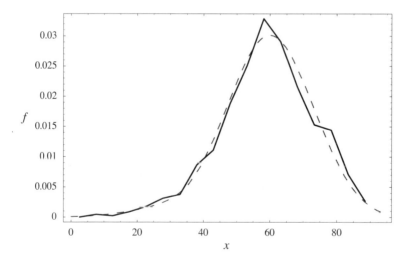

Fig. 17: The empirical pdf (——) and the fitted Gram–Charlier pdf (– – –)

This fitted Gram–Charlier density is actually very similar to the previous Johnson and PearsonIV results. The final Pearson fit was labelled p1. If it is still in memory, the command Show[p1/.Hue[__] → Hue[.4], p3] shows both plots together, but now with the fitted Pearson curve in green rather than red, enabling a visual comparison (note that Hue[__] contains two _ characters). On screen, the difference is apparent, but very slight. ∎

Some Advantages and Disadvantages of Gram–Charlier Expansions

By construction, Pearson densities must be unimodal; this follows from equation (5.1), since $dp/dx = 0$ at $x = -a$. Given bimodal data, Pearson densities may yield a very poor fit. In the Johnson family, both the S_L and S_U systems are unimodal. Although the S_B system can produce bimodal densities under certain conditions, the latter is not pleasant to work with. By contrast, Gram–Charlier expansions can produce mildly multimodal densities. On the downside, however, Gram–Charlier expansions have an undesirable tendency to sometimes produce small negative frequencies, particularly in the tails. In an ideal world, these negatives frequencies could be avoided by taking higher order expansions. This in turn requires higher order moments, which in turn have high variance and may be unreliable unless the sample size is sufficiently large. Finally, from a practical viewpoint, Gram–Charlier expansions are often 'unstable' in the sense that adding an extra $(t+1^{th})$ term may actually yield a worse fit, so some care is required in choosing an appropriate value for t.

5.4 B Hermite Polynomials; Gram–Charlier Coefficients

Let j denote the degree of the polynomial $P_j(z)$. Then, the family of polynomials $P_j(z)$, $j = 0, 1, 2, \ldots$, is said to be *orthogonal* to the weight function $w(z)$ if

$$\int_{-\infty}^{\infty} P_i(z) P_j(z) w(z) \, dz = 0 \quad \text{for } i \neq j. \tag{5.10}$$

Hermite polynomials are orthogonal to the weight function $w(z) = e^{-z^2/2}$. They are defined by

$$H_j(z) = \frac{(-1)^j}{w(z)} \frac{d^j w(z)}{dz^j} = \frac{(-1)^j}{\phi(z)} \frac{d^j \phi(z)}{dz^j} \tag{5.11}$$

and have the property that

$$\int_{-\infty}^{\infty} H_i(z) H_j(z) \phi(z) \, dz = \begin{cases} 0 & \text{if } i \neq j \\ j! & \text{if } i = j \end{cases} \tag{5.12}$$

To illustrate the point, compare: (*Note*: H[j] and ϕ were inputted in §5.4 A)

$$\int_{-\infty}^{\infty} \texttt{H[2] H[3]} \, \phi \, dz$$

0

with

$$\int_{-\infty}^{\infty} \texttt{H[3] H[3]} \, \phi \, dz$$

6

Multiplying both sides of (5.9) by $H_i(z)$ yields

$$H_i(z)\,\psi(z) = \sum_{j=0}^{\infty} c_j\, H_i(z)\, H_j(z)\, \phi(z). \tag{5.13}$$

Integrating both sides yields, by the orthogonal property (5.12),

$$\int_{-\infty}^{\infty} H_i(z)\,\psi(z)\,dz = c_i\, i! \tag{5.14}$$

Thus,

$$c_i = \frac{1}{i!}\, E[H_i(z)] \tag{5.15}$$

where the expectation is carried out with respect to $\psi(z)$. We already know the form of the Hermite polynomials. For instance, $H_6(z)$ is:

H[6]

$-15 + 45\, z^2 - 15\, z^4 + z^6$

It immediately follows that $E[H_6(z)] = \left(-15 + 45\,\acute\mu_2 - 15\,\acute\mu_4 + \acute\mu_6\right)$ where $\acute\mu_i$ denotes the ith raw moment of $\psi(z)$. In *Mathematica*, this conversion from z^i to $\acute\mu_i$ can be neatly achieved through pattern matching:

H[6] /. $z^{\text{i}\_.}$:> $\acute\mu_{\text{i}}$

$-15 + 45\,\acute\mu_2 - 15\,\acute\mu_4 + \acute\mu_6$

Finally, since we have assumed that $\psi(z)$ is a standardised density, replace $\acute\mu$ with μ, and let $\mu_1 = 0$ and $\mu_2 = 1$. Then c_6 reduces to $(30 - 15\mu_4 + \mu_6)/6!$. These substitutions accord with the definition of the c[j] function in §5.4 A, and so c[6] yields:

c[6]

$\dfrac{1}{720}\,(30 - 15\,\mu_4 + \mu_6)$

Finally, the nomenclature 'Gram–Charlier Expansion of *Type A*' suggests other types of expansions also exist. Indeed, just as *Type A* uses the standard Normal $\phi(z)$ as a generating function, Charlier's '*Type B*' uses the Poisson weight function $e^{-\lambda}\lambda^x/x!$ as its generating function, defined for $x = 0, 1, 2, \ldots$. This has the potential to perform better than the standard Normal when approximating skew densities. However, it assumes a discrete ordinate system and perhaps for this reason is rarely used.

5.5 Non-Parametric Kernel Density Estimation

Kernel density estimation does not typically belong in a chapter on *Systems of Distributions*. However, just as a Pearson curve gives an impression of the distribution of the underlying population, so too does kernel density estimation, which helps explain why it is included here.

One of the virtues of working with families of distributions, rather than a specific distribution, is that it reduces the chance of making the wrong parametric assumption about the distribution's correct form. Instead of assuming a particular functional form, one assumes a particular family, which is more general. If our assumption is correct, then our estimates should be efficient. However, assumptions do not always hold, and by locking our analysis into an incorrect assumptional framework, we can end up doing rather poorly. As such, it is usually wise to conduct a preliminary investigation of the data based upon minimal assumptions. Smoothing methods serve to do this, as density smoothness is all that is imposed. The so-called *kernel density estimator* is

$$\hat{f}(y) = \frac{1}{nc} \sum_{i=1}^{n} K\left(\frac{y - Y_i}{c}\right) \qquad (5.16)$$

where $(Y_1, ..., Y_n)$ is a random sample of size n collected on a random variable Y. The function K is known as the kernel and is specified by the analyst; it is often chosen to be a density function with zero mean and finite variance. Parameter $c > 0$ is known as the *bandwidth* and it too is specified by the analyst; small values of c produce a rough estimate, while large values produce a very smooth estimate. For further details on kernel density estimation, see Silverman (1986) and Simonoff (1996); Stine (1996) gives an implementation under *Mathematica* Version 2.2.

⊕ *Example 7:* Non-Parametric Kernel Density Estimation

In practice, the kernel density estimate is presented in the form of a plot, and this is exactly the output produced by the **mathStatica** function NPKDEPlot (non-parametric kernel density estimator). To illustrate its use, we apply it to Parzen's (1979) yearly 'Snowfall in Buffalo' data (63 data points collected from 1910 to 1972, and measured in inches):

```
data = ReadList["snowfall.dat"];
```

Two steps are required:

(i) Specify the kernel K
(ii) Choose the bandwidth c

We can then use NPKDEPlot to plot the kernel density estimate.

Step (i): In this example, we select K to be of form

$$K(u) = \frac{(2r+1)!!}{r!\, 2^{r+1}} (1-u^2)^r, \qquad -1 \le u \le 1 \qquad (5.17)$$

where $r = 1, 2, 3, \ldots$ denotes the weight of the kernel, and !! is the double factorial function. The $r = 1$ case yields the Epanechnikov kernel (ep):

$$\texttt{ep} = \frac{3}{4}\,(1 - u^2)\texttt{;} \quad \texttt{domain[ep]} = \{u, -1, 1\}\texttt{;}$$

Other common choices for K include the bi-weight kernel ($r = 2$), the tri-weight kernel ($r = 3$), and the Gaussian kernel $(2\pi)^{-1/2} \exp(-u^2/2)$ which is defined everywhere on the real line.

Step (ii): Next, we select the bandwidth c. This is most important, and experimenting with different values of c is advisable. A number of methods exist to automate bandwidth choice; **mathStatica** implements both the Silverman (1986) approach (default) and the more sophisticated (but much slower) Sheather and Jones (1991) method. They can be used as stand-alone bandwidth selectors, or, better still, as a starting point for experimentation. For the snowfall data set, the Sheather–Jones optimal bandwidth (using the Epanechnikov kernel) is:

c = Bandwidth[data, ep, Method → SheatherJones]

37.2621

Since K and c have now been specified, we can plot the smoothed *non-parametric kernel density estimate* using the NPKDEPlot[*data*, *K*, *c*] function:

NPKDEPlot[data, ep, c];

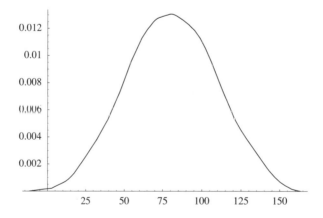

Fig. 18: Plot of the non-parametric kernel density estimate, snowfall data ($c = 37.26$)

This estimate has produced a distinct mode for snowfall of around 80 inches. Suppose we keep the same kernel, but choose a smaller bandwidth with $c = 10$:

```
NPKDEPlot[data, ep, 10];
```

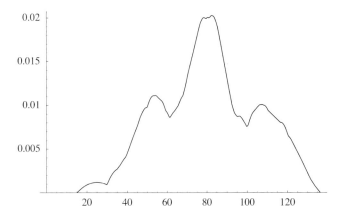

Fig. 19: Plot of the non-parametric kernel density estimate, snowfall data ($c = 10$)

Our new estimate exposes two lesser modes on either side of the 80-inch mode, at around 53 inches and 108 inches. A comparison of the two estimates suggests that the Sheather–Jones bandwidth is too large for this data set and has over-smoothed. This observation is in line with Parzen (1979, p.114) who reports that a trimodal shape for this data is "the more likely answer". This serves to highlight the importance of the experimentation process. Clicking the 'View Animation' button in the electronic notebook brings up an animation in which the bandwidth c varies from 4 to 25 in step sizes of $1/4$. This provides a rather neat way to visualise how the shape of the estimate changes with c.

5.6 The Method of Moments

The *method of moments* is employed throughout this chapter to estimate unknown parameters. This technique essentially equates sample moments with population moments. The latter are generally functions of unknown parameters, and are then solved for those parameters.

To be specific, suppose the random variable Y has density $f(y; \theta)$, where θ is a $(k \times 1)$ vector containing all unknown parameters. Now construct the first r raw moments of Y. That is, construct $\mu'_i = E[Y^i]$ for $i = 1, \ldots, r$ and $r \geq k$ (in all our examples, it suffices to set $r = k$). Generally, each moment will depend (often non-linearly) upon the parameters, so $\mu'_i = \mu'_i(\theta)$. Now let $(Y_1, \ldots, Y_n)$ denote a random sample of size n collected on Y. We then construct the sample raw moments $\hat{m}_i = \frac{1}{n} \sum_{j=1}^{n} Y_j^i$ for each i. The method of moments estimator, denoted by $\hat{\theta}$, solves the set of k equations $\mu'_i(\hat{\theta}) = \hat{m}_i$ for $\hat{\theta}$. The estimator is defined by equating the population moment with the sample moment, even though population moments and sample moments are generally not equal; that is, $\mu'_i(\theta) \neq \hat{m}_i$. This immediately questions the validity of the method of moments estimator. While not pursuing the answer in any detail here, we shall merely assert that the estimator may be justified using asymptotic arguments; for further discussion, see Mittelhammer (1996). Asymptotic theory is considered in detail in Chapter 8.

⊕ *Example 8:* The Bernoulli Distribution

Let $Y \sim \text{Bernoulli}(\theta)$, where $\theta = P(Y = 1)$, with pmf $g(y)$:

```
g        = θ^y (1 - θ)^(1-y);
domain[g] = {y, 0, 1} && {0 < θ < 1} && {Discrete};
```

The population mean of Y is easily derived as:

$\acute{\mu}_1$ = **Expect[y, g]**

θ

For a random sample of size n, the method of moments estimator is defined as the solution to $\acute{\mu}_1(\hat{\theta}) = \acute{m}_1$, which needs no further effort in this case: $\hat{\theta} = \acute{m}_1$. ∎

⊕ *Example 9:* The Gamma Distribution

Let $Y \sim \text{Gamma}(a, b)$ denote the Gamma distribution with parameter $\theta = \begin{pmatrix} a \\ b \end{pmatrix}$ and pdf $f(y)$:

```
f = y^(a-1) e^(-y/b) / (Γ[a] b^a);    domain[f] = {y, 0, ∞} && {a > 0, b > 0};
```

To estimate θ using the method of moments, we require the first two population raw moments:

$\acute{\mu}_1$ = **Expect[y, f]**
$\acute{\mu}_2$ = **Expect[y^2, f]**

a b

a (1 + a) b²

Then, the method of moments estimator of parameters a and b is obtained via:

Solve[{$\acute{\mu}_1$ == $\acute{m}_1$, $\acute{\mu}_2$ == $\acute{m}_2$}, {a, b}]

$\left\{\left\{a \rightarrow -\frac{\acute{m}_1^2}{\acute{m}_1^2 - \acute{m}_2}, \; b \rightarrow \frac{-\acute{m}_1^2 + \acute{m}_2}{\acute{m}_1}\right\}\right\}$

Mathematica gives the solution as a replacement rule for a and b. Note that the symbols $\acute{\mu}_1$ and $\acute{\mu}_2$ are 'reserved' for use by **mathStatica**'s moment converter functions. To avoid any confusion, it is best to Unset them:

$\acute{\mu}_1$ = .; $\acute{\mu}_2$ =.;

... prior to leaving this section. ∎

5.7 Exercises

1. Identify where each of the following distributions will be found on a Pearson diagram:
   (i) Exponential(λ)
   (ii) standard Logistic
   (iii) Azzalini's skew-Normal distribution with $\lambda > 0$ (see Chapter 2, Exercise 2).

2. The data "stock.dat" provides monthly US stock market returns from 1834 to 1925, yielding a sample of 1104 observations. The data is the same as that used in Pagan and Ullah (1999, Section 2.10).[5]
   (i) Fit a Pearson density to this data.
   (ii) Estimate the density of stock market returns using a non-parametric kernel density estimator, with a Gaussian kernel.
   (iii) Compare the Pearson fit to the kernel density estimate.

   To load the data, use: ReadList["stock.dat"].

3. Derive the equation describing the *Type III* and *Type V* lines in the Pearson diagram. [Hint: use the recurrence relation (5.5) to solve the moments $\left(\mu'_1, \mu'_2, \mu'_3, \mu'_4\right)$ as a function of the Pearson coefficients (a, c_0, c_1, c_2). Hence, find β_1 and β_2 in terms of (a, c_0, c_1, c_2). Then impose the parameter assumptions that define *Type III* and *Type V*, and find the relation between β_1 and β_2.] *

4. Exercise 3 derived the formulae describing the *Type III* and *Type V* lines, respectively, as:

   Type III: $\quad \beta_2 = \frac{3}{2} \beta_1 + 3$

   Type V: $\quad \beta_2 = \dfrac{3(-16 - 13\beta_1 - 2(4+\beta_1)^{3/2})}{\beta_1 - 32}$

   Use these results to show that a Gamma distribution defines the *Type III* line in a Pearson diagram, and that an Inverse Gamma distribution defines the *Type V* line.

5. Let random variable $X \sim \text{Beta}(a, 1)$ with density $f(x) = a x^{a-1}$, for $0 < x < 1$; this is also known as a Power Function distribution. Show that this distribution defines the *Type I(J)* line(s) on a Pearson diagram, as parameter a varies.

6. Let random variable X have a standard Extreme Value distribution. Find μ and $\{\mu_2, \mu_3, \mu_4\}$. Fit a Pearson density to these moments. Compare the true pdf (Extreme Value) with the Pearson fit.

7. Recall that the Johnson family is based on transformations of $Z \sim N(0, 1)$. In similar vein, a Johnson-style family can be constructed using transformations of $Z \sim$ Logistic (Tadikamalla and Johnson (1982)). Thus, if $Z \sim$ Logistic, find the pdf of $Y = \sinh\left(\frac{Z-\gamma}{\delta}\right)$, $\gamma \in \mathbb{R}$, $\delta > 0$. Plot the pdf when $\gamma = 0$ and $\delta = 1, 2$ and 3. Find the first 4 raw moments of random variable Y.

8. Construct a non-parametric kernel density estimator plot of the "sd.dat" data set (which measures the diagonal length of 100 forged Swiss bank notes and 100 real Swiss bank notes) using a Logistic kernel and the Silverman optimal bandwidth.

Chapter 6

Multivariate Distributions

6.1 Introduction

Thus far, we have considered the distribution of a single random variable. This chapter extends the analysis to a collection of random variables $\vec{X} = (X_1, X_2, \ldots, X_m)$. When $m = 2$, we have a bivariate setting; when $m = 3$, a trivariate ... and so on. Although the transition from univariate to multivariate analysis is 'natural', it does introduce some new concepts, in particular: joint densities §6.1 A, non-rectangular domains §6.1 B, joint distribution functions §6.1 C, marginal distributions §6.1 D, and conditional distributions §6.1 E. Multivariate expectations, product moments, generating functions and multivariate moment conversion functions are discussed in §6.2. Next, §6.3 examines the properties of independence and dependence. §6.4 is devoted to the multivariate Normal, §6.5 discusses the multivariate t and the multivariate Cauchy, while §6.6 looks at the Multinomial distribution and the bivariate Poisson distribution.

6.1 A Joint Density Functions

○ *Continuous Random Variables*

Let $\vec{X} = (X_1, \ldots, X_m)$ denote a collection of m random variables defined on a domain of support $\Lambda \subset \mathbb{R}^m$, where we assume Λ is an open set in $\mathbb{R}^m$. Then a function $f : \Lambda \to \mathbb{R}_+$ is a joint *probability density function* (pdf) if it has the following properties:

$$f(x_1, \ldots, x_m) > 0, \quad \text{for } (x_1, \ldots, x_m) \in \Lambda$$

$$\int \cdots \int_\Lambda f(x_1, \ldots, x_m) \, dx_1 \cdots dx_m = 1 \tag{6.1}$$

⊕ *Example 1:* Joint pdf

Consider the function $f(x, y)$ with domain of support $\Lambda = \{(x, y) : 0 < x < \infty, 0 < y < \infty\}$:

```
       -1-x
      e  y   x
f = ─────────── ;    domain[f] = {{x, 0, ∞}, {y, 0, ∞}};
         y⁴
```

Clearly, f is positive over its domain, and it integrates to unity over the domain:

```
Integrate[f, {x,0,∞}, {y,0,∞}]
```

1

Thus, $f(x, y)$ may represent the joint pdf of a pair of random variables. Figure 1 plots $f(x, y)$ over part of its support.

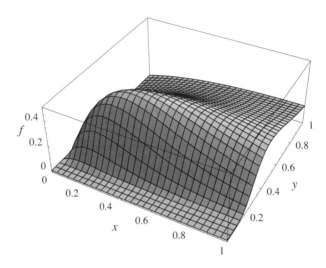

Fig. 1: The joint pdf $f(x, y)$

A contour plot allows one to pick out specific contours along which $z = f(x, y)$ is constant. That is, each contour joins points on the surface that have the same height z. Figure 2 plots all combinations of x and y such that $f(x, y) = \frac{1}{30}$. The edge of the dark-shaded region is the contour line.

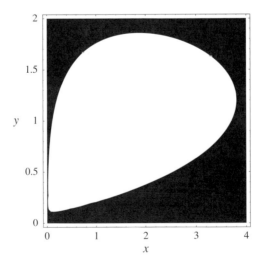

Fig. 2: The contour $f(x, y) = \frac{1}{30}$

Discrete Random Variables

Let $\vec{X} = (X_1, \ldots, X_m)$ denote a collection of m random variables defined on a domain of support $\Lambda \subset \mathbb{R}^m$. Then a function $f: \Lambda \to \mathbb{R}_+$ is a joint *probability mass function* (pmf) if it has the following properties:

$$f(x_1, \ldots, x_m) = P(X_1 = x_1, \ldots, X_m = x_m) > 0, \text{ for } (x_1, \ldots, x_m) \in \Lambda$$

$$\sum_\Lambda \cdots \sum f(x_1, \ldots, x_m) = 1$$

(6.2)

⊕ **Example 2:** Joint pmf

Let random variables X and Y have joint pmf $h(x, y) = \frac{x+1-y}{54}$ with domain of support $\Lambda = \{(x, y) : x \in \{3, 5, 7\}, y \in \{0, 1, 2, 3\}\}$, as per Table 1.

	$Y = 0$	$Y = 1$	$Y = 2$	$Y = 3$
$X = 3$	$\frac{4}{54}$	$\frac{3}{54}$	$\frac{2}{54}$	$\frac{1}{54}$
$X = 5$	$\frac{6}{54}$	$\frac{5}{54}$	$\frac{4}{54}$	$\frac{3}{54}$
$X = 7$	$\frac{8}{54}$	$\frac{7}{54}$	$\frac{6}{54}$	$\frac{5}{54}$

Table 1: Joint pmf of $h(x, y) = \frac{x+1-y}{54}$

In *Mathematica*, this pmf may be entered as:

```
pmf = Table[ x + 1 - y
             ─────────, {x, 3, 7, 2}, {y, 0, 3}]
                54
```

$$\begin{pmatrix} \frac{2}{27} & \frac{1}{18} & \frac{1}{27} & \frac{1}{54} \\ \frac{1}{9} & \frac{5}{54} & \frac{2}{27} & \frac{1}{18} \\ \frac{4}{27} & \frac{7}{54} & \frac{1}{9} & \frac{5}{54} \end{pmatrix}$$

This is a well-defined pmf since all the probabilities are positive, and they sum to 1:

```
Plus @@ Plus @@ pmf
```

1

The latter can also be evaluated with:

```
Plus @@ (pmf // Flatten)
```

1

Figure 3 interprets the joint pmf in the form of a three-dimensional bar chart.

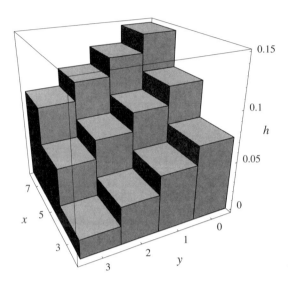

Fig. 3: Joint pmf of $h(x, y) = \frac{x+1-y}{54}$

6.1 B   Non-Rectangular Domains

If the domain of a joint pdf does not depend on any of its constituent random variables, then we say the domain defines an independent product space. For instance, the domain $\{(x, y): \frac{1}{2} < x < 3,\ 1 < y < 4\}$ is an independent product space, because the domain of X does not depend on the domain of Y, and vice versa. We enter such domains into **mathStatica** as:

$$\text{domain}[\text{f}] = \left\{\left\{\text{x},\ \frac{1}{2},\ 3\right\},\ \{\text{y},\ 1,\ 4\}\right\}$$

If plotted, this domain would appear rectangular, as Fig. 4 illustrates. In this vein, we refer to such domains as being *rectangular*.

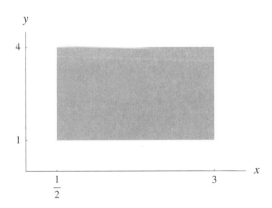

Fig. 4: A rectangular domain

Sometimes, the domain itself may depend on random variables. We refer to such domains as being *non-rectangular*. Examples include:

(i) $\{(x, y): 0 < x < y < \infty\}$. This would appear triangular in the two-dimensional plane. We can enter this domain into **mathStatica** as:

domain[f] = {{x, 0, y}, {y, x, ∞}}

(ii) $\{(x, y): x^2 + y^2 < 1\}$. This would appear circular in the two-dimensional plane. At present, **mathStatica** does not support such domains. However, this feature is planned for a future version of **mathStatica**, once *Mathematica* itself can support multiple integration over inequality defined regions.

6.1 C Probability and **Prob**

○ *Continuous Random Variables*

Given some joint pdf $f(x_1, \ldots, x_m)$, the joint *cumulative distribution function* (cdf) is given by:

$$P(X_1 \leq x_1, \ldots, X_m \leq x_m) = \int_{-\infty}^{x_m} \cdots \int_{-\infty}^{x_1} f(w_1, \ldots, w_m) \, dw_1 \cdots dw_m. \quad (6.3)$$

The **mathStatica** function Prob[$\{x_1, \ldots, x_m\}, f$] calculates $P(X_1 \leq x_1, \ldots, X_m \leq x_m)$. The position of each element $\{x_1, x_2, \ldots\}$ in Prob[$\{x_1, \ldots, x_m\}, f$] is important, and must correspond to the ordering specified in the domain statement.

⊕ *Example 3:* Joint cdf

Consider again the joint pdf given in *Example 1*:

$$f = \frac{e^{-\frac{1+x}{y}} x}{y^4}; \quad \text{domain[f] = \{\{x, 0, \infty\}, \{y, 0, \infty\}\};}$$

Here is the cdf $F(x, y) = P(X \leq x, Y \leq y)$:

F = Prob[{x, y}, f]

$$e^{-1/y} \left(1 - \frac{e^{-\frac{x}{y}} (x + x^2 + y + 2 x y)}{(1 + x)^2 \, y}\right)$$

Since $F(x, y)$ may be viewed as the anti-derivative of $f(x, y)$, differentiating F yields the original joint pdf $f(x, y)$:

D[F, x, y] // Simplify

$$\frac{e^{-\frac{1+x}{y}} x}{y^4}$$

Figure 5 plots the joint cdf.

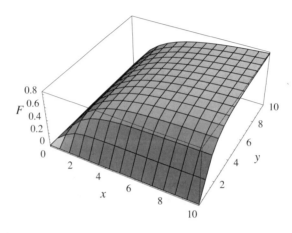

Fig. 5: The joint cdf $F(x, y)$

The surface approaches 1 asymptotically, which it reaches in the limit:

Prob[{∞, ∞}, f]

1

⊕ *Example 4:* Probability Content of a Region — Introducing `MrSpeedy`

Let $\vec{X} = (X_1, X_2, X_3)$ have joint pdf $g(x_1, x_2, x_3)$:

g = k e^x₁ x₁ (x₂ + 1) / x₃²;
domain[g] = {{x₁, 0, 1}, {x₂, 2, 4}, {x₃, 3, 5}};

where the constant $k > 0$ is defined such that g integrates to unity over its domain. The cdf of g is:

Clear[G];

G[x1_, x2_, x3_] = Prob[{x1, x2, x3}, g]

$$\frac{k\,(1 + e^{x1}\,(-1 + x1))\,(-2 + x2)\,(4 + x2)\,(-3 + x3)}{6\,x3}$$

Note that we have set up G as a *Mathematica* function of x1 through x3, and can thus apply it as a function in the standard way. Here, we find k by evaluating G at the upper boundary of the domain:

G[1, 4, 5]

$$\frac{16\,k}{15}$$

This requires $k = \frac{15}{16}$ in order for g to be a well-defined pdf. If we require the probability content of a region within the domain, we could just type in the whole integral. For instance, the probability of being within the region

$$S = \{(x_1, x_2, x_3): \ 0 < x_1 < \tfrac{1}{2}, \ 3 < x_2 < \tfrac{7}{2}, \ 4 < x_3 < \tfrac{9}{2}\}$$

is given by:

$$\int_4^{\frac{9}{2}} \int_3^{\frac{7}{2}} \int_0^{\frac{1}{2}} g \, dx_1 \, dx_2 \, dx_3$$

$$\frac{17}{288} \left(1 - \frac{\sqrt{e}}{2}\right) k$$

While this is straightforward, it is by no means the fastest solution. In particular, the probability content of a region within the domain can be found purely by using the function G[] (which we have already found) *and* the boundaries of that region, without any need for further integration. *Note:* the solution is *not* $G[\tfrac{1}{2}, \tfrac{7}{2}, \tfrac{9}{2}] - G[0, 3, 4]$. Rather, one must evaluate the cdf at every possible extremum defined by set S. The **mathStatica** function MrSpeedy[cdf, S] does this.

   ? MrSpeedy

      MrSpeedy[cdf, S] calculates the probability content
      of a region defined by set S, by making use of the
      known distribution function cdf[x1, x2, ..., xm].

For our example:

   $S = \{\{0, \tfrac{1}{2}\}, \{3, \tfrac{7}{2}\}, \{4, \tfrac{9}{2}\}\};$

   MrSpeedy[G, S]

   $\dfrac{17}{288} \left(1 - \dfrac{\sqrt{e}}{2}\right) k$

MrSpeedy typically provides at least a 20-fold speed increase over direct integration. To see the calculations MrSpeedy performs, replace G with say Φ:

   MrSpeedy[Φ, S]

   $-\Phi[0, 3, 4] + \Phi[0, 3, \tfrac{9}{2}] + \Phi[0, \tfrac{7}{2}, 4] - \Phi[0, \tfrac{7}{2}, \tfrac{9}{2}] +$
   $\Phi[\tfrac{1}{2}, 3, 4] - \Phi[\tfrac{1}{2}, 3, \tfrac{9}{2}] - \Phi[\tfrac{1}{2}, \tfrac{7}{2}, 4] + \Phi[\tfrac{1}{2}, \tfrac{7}{2}, \tfrac{9}{2}]$

MrSpeedy evaluates the cdf at each of these points. Note that this approach applies to any m-variate distribution. ∎

○ **Discrete Random Variables**

Given some joint pmf $f(x_1, \ldots, x_m)$, the joint cdf is

$$P(X_1 \leq x_1, \ldots, X_m \leq x_m) = \sum_{w_1 \leq x_1} \cdots \sum_{w_m \leq x_m} f(w_1, \ldots, w_m). \tag{6.4}$$

Note that the `Prob` function does not operate on multivariate *discrete* domains.

⊕ **Example 5:** Joint cdf

In *Example 2*, we considered the bivariate pmf $h(x, y) = \frac{x+1-y}{54}$ with domain of support $\Lambda = \{(x, y): x \in \{3, 5, 7\}, y \in \{0, 1, 2, 3\}\}$. The cdf, $H(x, y) = P(X \leq x, Y \leq y)$, can be defined in *Mathematica* as follows:

```
H[x_, y_] = Sum[ (w1 + 1 - w2)/54 , {w1, 3, x, 2}, {w2, 0, y}]
```

$$\frac{1}{108}\left(8 + 7y - y^2 + 10\,\text{Floor}\left[\tfrac{1}{2}(-3+x)\right] + 9y\,\text{Floor}\left[\tfrac{1}{2}(-3+x)\right] - y^2\,\text{Floor}\left[\tfrac{1}{2}(-3+x)\right] + 2\,\text{Floor}\left[\tfrac{1}{2}(-3+x)\right]^2 + 2y\,\text{Floor}\left[\tfrac{1}{2}(-3+x)\right]^2\right)$$

Then, for instance, $P(X \leq 5, Y \leq 3)$ is:

```
H[5, 3]
```

$$\frac{14}{27}$$

Figure 6 plots the joint cdf as a three-dimensional bar chart.

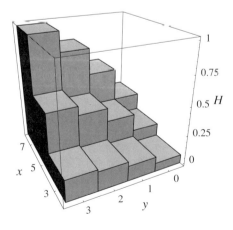

Fig. 6: The joint cdf $H(x, y)$

6.1 D Marginal Distributions

○ *Continuous Random Variables*

Let the *continuous* random variables X_1 and X_2 have joint pdf $f(x_1, x_2)$. Then the *marginal pdf* of X_1 is $f_1(x_1)$, where

$$f_1(x_1) = \int_{x_2} f(x_1, x_2)\, dx_2. \tag{6.5}$$

More generally, if $(X_1, \ldots, X_m)$ have joint pdf $f(x_1, \ldots, x_m)$, then the marginal pdf of a group $r < m$ of these random variables is obtained by 'integrating out' the $(m - r)$ variables that are not of interest. The **mathStatica** function, Marginal$[\vec{x}_r, f]$, derives the marginal joint pdf of the variable(s) specified in $\vec{x}_r$. If there is more than one variable in $\vec{x}_r$, then it must take the form of a list. The ordering of the variables in this list does not matter.

⊕ *Example 6:* Marginal

Let the continuous random variables $\vec{X} = (X_1, X_2, X_3, X_4)$ have joint pdf $f(x_1, x_2, x_3, x_4)$:

```
f = k e^x1 x1 (x2 + 1) (x3 - 3)^2 / x4^2;
domain[f] = {{x1, 0, 1}, {x2, 1, 2}, {x3, 2, 3}, {x4, 3, 4}};
```

where k is a constant. The marginal bivariate distribution of X_2 *and* X_4 is given by:

Marginal[{x2, x4}, f]

$$\frac{k\,(1 + x_2)}{3\,x_4^2}$$

The resulting marginal density depends only on values of X_2 and X_4, since X_1 and X_3 have been integrated out. Similarly, the marginal distribution of X_4 does not depend on values of X_1, X_2 or X_3:

Marginal[x4, f]

$$\frac{5\,k}{6\,x_4^2}$$

We can use Marginal to determine k, by letting $\vec{x}_r$ be an empty set. Then all the random variables are 'integrated out':

Marginal[{}, f]

$$\frac{5\,k}{72}$$

Thus, in order for f to be a well-defined density function, k must equal $\frac{72}{5}$. ∎

○ **Discrete Random Variables**

In a discrete world, the $\int$ symbol in (6.5) is replaced by the summation symbol Σ. Thus, if the discrete random variables X_1 and X_2 have joint pmf $f(x_1, x_2)$, then the *marginal pmf* of X_1 is $f_1(x_1)$, where

$$f_1(x_1) = \sum_{x_2} f(x_1, x_2). \tag{6.6}$$

The `Marginal` function only operates on continuous domains; it is not currently implemented for discrete domains.

⊕ **Example 7:** Discrete Marginal

Recall, from *Example 2*, the joint pmf $h(x, y) = \frac{x+1-y}{54}$ with domain of support $\{(x, y) : x \in \{3, 5, 7\}, y \in \{0, 1, 2, 3\}\}$:

$$\texttt{pmf = Table}\Big[\frac{\texttt{x + 1 - y}}{54}, \texttt{\{x, 3, 7, 2\}, \{y, 0, 3\}}\Big];$$

By (6.6), the marginal pmf of Y is:

$$\texttt{pmf}_\texttt{Y} = \texttt{Sum}\Big[\frac{\texttt{x + 1 - y}}{54}, \texttt{\{x, 3, 7, 2\}}\Big] \texttt{ // Simplify}$$

$$\frac{6-y}{18}$$

where Y may take values of 0, 1, 2 or 3. That is:

$$\texttt{pmf}_\texttt{Y} \texttt{ /. y} \to \texttt{\{0, 1, 2, 3\}}$$

$$\left\{\frac{1}{3}, \frac{5}{18}, \frac{2}{9}, \frac{1}{6}\right\}$$

Alternatively, we can derive the same result directly, by finding the sum of each column of Table 1:

$$\texttt{Plus @@ pmf}$$

$$\left\{\frac{1}{3}, \frac{5}{18}, \frac{2}{9}, \frac{1}{6}\right\}$$

The sum of each row can be found with:

$$\texttt{Plus @@ Transpose[pmf]}$$

$$\left\{\frac{5}{27}, \frac{1}{3}, \frac{13}{27}\right\}$$

Further examples of discrete multivariate distributions are given in §6.6. ∎

6.1 E Conditional Distributions

○ *Continuous Random Variables*

Let the continuous random variables X_1 and X_2 have joint pdf $f(x_1, x_2)$. Then the *conditional pdf* of X_1 given $X_2 = x_2$ is denoted by $f(x_1 \mid X_2 = x_2)$ or, for short, $f(x_1 \mid x_2)$. It is defined by

$$f(x_1 \mid x_2) = \frac{f(x_1, x_2)}{f_2(x_2)}, \qquad \text{provided } f_2(x_2) > 0 \qquad (6.7)$$

where $f_2(x_2)$ denotes the marginal pdf of X_2 evaluated at $X_2 = x_2$. More generally, if $(X_1, \ldots, X_m)$ have joint pdf $f(x_1, \ldots, x_m)$, the joint conditional pdf of a group of r of these random variables (given that the remaining $m - r$ variables are fixed) is the joint pdf of the m variables divided by the joint marginal pdf of the $m - r$ fixed variables.

Since the conditional pdf $f(x_1 \mid x_2)$ is a well-defined pdf, we can use it to calculate probabilities and expectations. For instance, if $u(X_1)$ is a function of X_1, then the *conditional expectation* $E[u(X_1) \mid X_2 = x_2]$ is given by

$$E[u(X_1) \mid x_2] = \int_{x_1} u(x_1)\, f(x_1 \mid x_2)\, dx_1. \qquad (6.8)$$

With **mathStatica**, conditional expectations are easily calculated by first deriving the conditional density, say $f_{\text{con}}(x_1) = f(x_1 \mid x_2)$ and domain[f_{con}]. The desired conditional expectation is then given by Expect[u, f_{con}]. Two particular examples of conditional expectations are the conditional mean $E[X_1 \mid x_2]$, which is known as the *regression function* of X_1 on X_2, and the conditional variance $\text{Var}(X_1 \mid x_2)$, which is known as the *scedastic function*.

⊕ **Example 8:** Conditional

The **mathStatica** function, Conditional[$\vec{x}_r$, f], derives the conditional pdf of $\vec{x}_r$ variable(s), given that the remaining variables are fixed. As above, if there is more than one variable in $\vec{x}_r$, then it must take the form of a list; it does not matter how the variables in this list are sorted. To eliminate any confusion, a message clarifies what is (and what is not) being conditioned on. For density $f(x_1, x_2, x_3, x_4)$, defined in *Example 6*, the joint conditional pdf of X_2 and X_4, given $X_1 = x_1$ and $X_3 = x_3$ is:

    Conditional[{x₂, x₄}, f]

– Here is the conditional pdf f (x₂ , x₄ | x₁ , x₃):

$$\frac{24\,(1 + x_2)}{5\,x_4^2}$$

Note that this output is the same as the first Marginal example above (given $k = \frac{72}{5}$). This is because (X_1, X_2, X_3, X_4) are mutually stochastically independent (see §6.3 A). ∎

⊕ ***Example 9:*** Conditional Expectation (Continuous)

Let X_1 and X_2 have joint pdf $f(x_1, x_2) = x_1 + x_2$, supported on the unit rectangle $\{(x_1, x_2) : 0 < x_1 < 1, 0 < x_2 < 1\}$:

```
f = x₁ + x₂;    domain[f] = {{x₁, 0, 1}, {x₂, 0, 1}};
```

as illustrated below in Fig. 7. Derive the conditional mean and conditional variance of X_1, given $X_2 = x_2$.

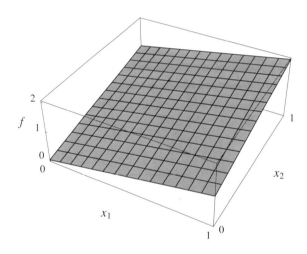

Fig. 7: The joint pdf $f(x_1, x_2) = x_1 + x_2$

Solution: The conditional pdf $f(x_1 \mid x_2)$, denoted f_{con}, is:[1]

```
f_con = Conditional[x₁, f]
```

— Here is the conditional pdf $f(x_1 \mid x_2)$:

$$\frac{x_1 + x_2}{\frac{1}{2} + x_2}$$

In order to apply **mathStatica** functions to the conditional pdf f_{con}, we need to declare the domain over which it is defined. This is because **mathStatica** will only recognise f_{con} as a pdf if its domain has been specified. Since random variable X_2 is now fixed at x_2, the domain of f_{con} is:

```
domain[f_con] = {x₁, 0, 1};
```

The required conditional mean is:

```
Expect[x₁, f_con]
```

$$\frac{2 + 3 x_2}{3 + 6 x_2}$$

The conditional variance is:

Var[x₁, f_con]

$$\frac{1 + 6 x_2 + 6 x_2^2}{18 (1 + 2 x_2)^2}$$

As this result depends on X_2, the conditional variance is heteroscedastic. ∎

○ **Discrete Random Variables**

The transition to a discrete world is once again straightforward: if the discrete random variables, X_1 and X_2, have joint pmf $f(x_1, x_2)$, then the *conditional pmf* of X_2 given $X_1 = x_1$ is denoted by $f(x_2 \mid X_1 = x_1)$ or, for short, $f(x_2 \mid x_1)$. It is defined by

$$f(x_2 \mid x_1) = \frac{f(x_1, x_2)}{f_1(x_1)}, \qquad \text{provided } f_1(x_1) > 0 \qquad (6.9)$$

where $f_1(x_1)$ denotes the marginal pmf of X_1, evaluated at $X_1 = x_1$, as defined in (6.6). Note that **mathStatica**'s Conditional function only operates on continuous domains; it is not implemented for discrete domains. As above, the conditional pmf $f(x_2 \mid x_1)$ can be used to calculate probabilities and expectations. Thus, if $u(X_2)$ is a function of X_2, the *conditional expectation* $E[u(X_2) \mid X_1 = x_1]$ is given by

$$E[u(X_2) \mid x_1] = \sum_{x_2} u(x_2) f(x_2 \mid x_1). \qquad (6.10)$$

⊕ **Example 10:** Conditional Mean (Discrete)

Find the conditional mean of X, given $Y = y$, for the pmf $h(x, y) = \frac{x+1-y}{54}$ with domain of support $\{(x, y): x \in \{3, 5, 7\}, y \in \{0, 1, 2, 3\}\}$.

Solution: We require $E[X \mid Y = y] = \sum_x x\, h(x \mid y) = \sum_x x\, \frac{h(x, y)}{h_y(y)}$. In *Example 7*, we found that the marginal pmf of Y was $h_y(y) = \frac{6-y}{18}$. Hence, the solution is:

```
sol = Sum[x (x + 1 - y)/54 / (6 - y)/18, {x, 3, 7, 2}] // Simplify
```

$$\frac{98 - 15 y}{18 - 3 y}$$

This depends, of course, on $Y = y$. Since we can assign four possible values to y, the four possible conditional expectations $E[X \mid Y = y]$ are:

```
sol /. y → {0, 1, 2, 3}
```

$$\left\{ \frac{49}{9}, \frac{83}{15}, \frac{17}{3}, \frac{53}{9} \right\}$$

6.2 Expectations, Moments, Generating Functions

6.2 A Expectations

Let the collection of m random variables $(X_1, \ldots, X_m)$ have joint density function $f(x_1, \ldots, x_m)$. Then the *expectation* of some function u of the random variables, $u(X_1, \ldots, X_m)$, is

$$E[u(X_1, \ldots, X_m)] = \begin{cases} \int_{x_m} \cdots \int_{x_1} u(x_1, \ldots, x_m) f(x_1, \ldots, x_m) \, dx_1 \cdots dx_m \\ \sum_{x_1} \cdots \sum_{x_m} u(x_1, \ldots, x_m) f(x_1, \ldots, x_m) \end{cases} \quad (6.11)$$

corresponding to the continuous and discrete cases, respectively. **mathStatica**'s Expect function generalises neatly to a multivariate continuous setting. For instance, in §6.1 D, we considered the following pdf $g(x_1, x_2, x_3, x_4)$:

```
g = 72/5 e^x1 x1 (x2 + 1) (x3 - 3)^2 / x4^2;
domain[g] = {{x1, 0, 1}, {x2, 1, 2}, {x3, 2, 3}, {x4, 3, 4}};
```

We now find both $E[X_1(X_4^2 - X_2)]$ and $E[X_4]$:

```
Expect[x1 (x4^2 - x2), g]
```

$$\frac{157}{15} (-2 + e)$$

```
Expect[x4, g]
```

$$12 \, \text{Log}\left[\frac{4}{3}\right]$$

6.2 B Product Moments, Covariance and Correlation

Multivariate moments are a special type of multivariate expectation. To illustrate, let X_1 and X_2 have joint bivariate pdf $f(x_1, x_2)$. Then, the bivariate *raw moment* $\mu'_{r,s}$ is

$$\mu'_{r,s} = E[X_1^r X_2^s]. \quad (6.12)$$

With $s = 0$, $\mu'_{r,0}$ denotes the rth raw moment of X_1. Similarly, with $r = 0$, $\mu'_{0,s}$ denotes the sth raw moment of X_2. More generally, $\mu'_{r,s}$ is known as a *product* raw moment or joint raw moment. These definitions extend in the obvious way to higher numbers of variables.

The bivariate *central moment* $\mu_{r,s}$ is defined as

$$\mu_{r,s} = E\bigl[(X_1 - E[X_1])^r (X_2 - E[X_2])^s\bigr]. \tag{6.13}$$

The *covariance* of X_i and X_j, denoted $\mathrm{Cov}(X_i, X_j)$, is defined by

$$\mathrm{Cov}(X_i, X_j) = E\bigl[(X_i - E[X_i])(X_j - E[X_j])\bigr]. \tag{6.14}$$

When $i = j$, $\mathrm{Cov}(X_i, X_j)$ is equivalent to $\mathrm{Var}(X_i)$. More generally, the *variance-covariance* matrix of $\vec{X} = (X_1, X_2, \ldots, X_m)$ is the $(m \times m)$ symmetric matrix:

$$\mathrm{Varcov}(\vec{X}) = E\bigl[(\vec{X} - E[\vec{X}])(\vec{X} - E[\vec{X}])^{\mathrm{T}}\bigr]$$

$$= E\left[\begin{pmatrix} X_1 - EX_1 \\ X_2 - EX_2 \\ \vdots \\ X_m - EX_m \end{pmatrix} \bigl((X_1 - EX_1),\ (X_2 - EX_2),\ \ldots,\ (X_m - EX_m)\bigr)\right]$$

$$= E\left[\begin{pmatrix} (X_1 - EX_1)^2 & (X_1 - EX_1)(X_2 - EX_2) & \cdots & (X_1 - EX_1)(X_m - EX_m) \\ (X_2 - EX_2)(X_1 - EX_1) & (X_2 - EX_2)^2 & \cdots & (X_2 - EX_2)(X_m - EX_m) \\ \vdots & \vdots & \ddots & \vdots \\ (X_m - EX_m)(X_1 - EX_1) & (X_m - EX_m)(X_2 - EX_2) & \cdots & (X_m - EX_m)^2 \end{pmatrix}\right]$$

$$= \begin{pmatrix} \mathrm{Var}(X_1) & \mathrm{Cov}(X_1, X_2) & \cdots & \mathrm{Cov}(X_1, X_m) \\ \mathrm{Cov}(X_2, X_1) & \mathrm{Var}(X_2) & \cdots & \mathrm{Cov}(X_2, X_m) \\ \vdots & \vdots & \ddots & \vdots \\ \mathrm{Cov}(X_m, X_1) & \mathrm{Cov}(X_m, X_2) & \cdots & \mathrm{Var}(X_m) \end{pmatrix}$$

It follows from (6.14) that $\mathrm{Cov}(X_i, X_j) = \mathrm{Cov}(X_j, X_i)$, and thus that the variance-covariance matrix is symmetric. In the notation of (6.13), one could alternatively express $\mathrm{Varcov}(\vec{X})$ as follows:

$$\mathrm{Varcov}(\vec{X}) = \begin{pmatrix} \mu_{2,0,0,\ldots,0} & \mu_{1,1,0,\ldots,0} & \cdots & \mu_{1,0,\ldots,0,1} \\ \mu_{1,1,0,\ldots,0} & \mu_{0,2,0,\ldots,0} & \cdots & \mu_{0,1,\ldots,0,1} \\ \vdots & \vdots & \ddots & \vdots \\ \mu_{1,0,\ldots,0,1} & \mu_{0,1,0,\ldots,1} & \cdots & \mu_{0,0,\ldots,0,2} \end{pmatrix} \tag{6.15}$$

which again highlights its symmetry.

Finally, the *correlation* between X_i and X_j is defined as

$$\rho(X_i, X_j) = \rho_{ij} = \frac{\mathrm{Cov}(X_i, X_j)}{\sqrt{\mathrm{Var}(X_i)\mathrm{Var}(X_j)}} \tag{6.16}$$

where it can be shown that $-1 \leq \rho_{ij} \leq 1$. If X_i and X_j are mutually stochastically independent (§6.3 A), then $\rho_{ij} = 0$; the converse does not always hold (see *Example 16*).

⊕ **Example 11:** Product Moments, `Cov, Varcov, Corr`

Let the continuous random variables X, Y and Z have joint pdf $f(x, y, z)$:

$$f = \frac{1}{\sqrt{2\pi\lambda}} e^{-\frac{x^2}{2} - \frac{z}{\lambda}} \left(1 + \alpha (2y - 1) \operatorname{Erf}\left[\frac{x}{\sqrt{2}}\right]\right);$$

$$\text{domain}[f] = \{\{x, -\infty, \infty\}, \{y, 0, 1\}, \{z, 0, \infty\}\}$$
$$\&\& \{-1 < \alpha < 1, \lambda > 0\};$$

The mean vector is $\vec{\mu} = E[(X, Y, Z)]$:

`Expect[{x, y, z}, f]`

$$\left\{0, \frac{1}{2}, \lambda\right\}$$

Here is the product raw moment $\mu'_{3,2,1} = E[X^3 \, Y^2 \, Z]$:

`Expect[x³ y² z, f]`

$$\frac{5 \alpha \lambda}{12 \sqrt{\pi}}$$

Here is the product central moment $\mu_{2,0,2} = E\left[(X - E[X])^2 (Z - E[Z])^2\right]$:

`Expect[ (x - Expect[x, f])² (z - Expect[z, f])², f]`

$$\lambda^2$$

$\operatorname{Cov}(X, Y)$ is given by:

`Cov[{x, y}, f]`

$$\frac{\alpha}{6 \sqrt{\pi}}$$

More generally, the variance-covariance matrix is:

`Varcov[f]`

$$\begin{pmatrix} 1 & \frac{\alpha}{6\sqrt{\pi}} & 0 \\ \frac{\alpha}{6\sqrt{\pi}} & \frac{1}{12} & 0 \\ 0 & 0 & \lambda^2 \end{pmatrix}$$

The correlation between X and Y is:

`Corr[{x, y}, f]`

$$\frac{\alpha}{\sqrt{3\pi}}$$

6.2 C Generating Functions

The multivariate *moment generating function* (mgf) is a natural extension to the univariate case defined in Chapter 2. Let $\vec{X} = (X_1, \ldots, X_m)$ denote an m-variate random variable, and let $\vec{t} = (t_1, \ldots, t_m) \in \mathbb{R}^m$ denote a vector of dummy variables. Then the mgf $M_{\vec{X}}(\vec{t})$ is a function of $\vec{t}$; when no confusion is possible, we denote $M_{\vec{X}}(\vec{t})$ by $M(\vec{t})$. It is defined by

$$M(\vec{t}) = E\left[e^{\vec{t} \cdot \vec{X}}\right] = E[e^{t_1 X_1 + \cdots + t_m X_m}] \qquad (6.17)$$

provided the expectation exists for all $t_i \in (-c, c)$, for some constant $c > 0$, $i = 1, \ldots, m$. If it exists, the mgf can be used to generate the product raw moments. In, say, a bivariate setting, the product raw moment $\mu'_{r,s} = E[X_1^r X_2^s]$ may be obtained from $M(\vec{t})$ as follows:

$$\mu'_{r,s} = E[X_1^r X_2^s] = \left.\frac{\partial^{r+s} M(\vec{t})}{\partial t_1^r \partial t_2^s}\right|_{\vec{t}=\vec{0}} . \qquad (6.18)$$

The *central moment generating function* may be obtained from the mgf (6.17) as follows:

$$E\left[e^{\vec{t} \cdot (\vec{X} - \vec{\mu})}\right] = e^{-\vec{t} \cdot \vec{\mu}} M(\vec{t}), \qquad \text{where } \vec{\mu} = E[\vec{X}]. \qquad (6.19)$$

The *cumulant generating function* is the natural logarithm of the mgf. The multivariate *characteristic function* is similar to (6.17) and given by

$$C(\vec{t}) = E\left[\exp(i\vec{t} \cdot \vec{X})\right] = E\left[\exp(i(t_1 X_1 + t_2 X_2 + \cdots + t_m X_m))\right] \qquad (6.20)$$

where i denotes the unit imaginary number.

Given discrete random variables defined on subsets of the non-negative integers $\{0, 1, 2, \ldots\}$, the multivariate *probability generating function* (pgf) is

$$\Pi(\vec{t}) = E[t_1^{X_1} t_2^{X_2} \cdots t_m^{X_m}] . \qquad (6.21)$$

The pgf provides a way to determine the probabilities. For instance, in the bivariate case,

$$P(X_1 = r, X_2 = s) = \frac{1}{r!\,s!} \left.\frac{\partial^{r+s} \Pi(\vec{t})}{\partial t_1^r \partial t_2^s}\right|_{\vec{t}=\vec{0}} . \qquad (6.22)$$

The pgf can also be used as a *factorial moment generating function*. For instance, in a bivariate setting, the product factorial moment,

$$\mu'[r, s] = E[X_1^{[r]} X_2^{[s]}]$$
$$= E[X_1(X_1 - 1) \cdots (X_1 - r + 1) \times X_2(X_2 - 1) \cdots (X_2 - s + 1)] \qquad (6.23)$$

may be obtained from $\Pi(\vec{t})$ as follows:

$$\mu'[r, s] = E[X_1^{[r]} X_2^{[s]}] = \left.\frac{\partial^{r+s} \Pi(\vec{t})}{\partial t_1^r \partial t_2^s}\right|_{\vec{t}=\vec{1}} . \qquad (6.24)$$

Note that $\vec{t}$ is set here to $\vec{1}$ and not $\vec{0}$. To then convert from factorial moments to product raw moments, see the `FactorialToRaw` function of §6.2 D.

⊕ *Example 12:* Working with Generating Functions

Gumbel (1960) considered a bivariate Exponential distribution with cdf given by:

```
F = 1 - e^-x - e^-y + e^-(x+y+θ x y) ;
```

for $0 \le \theta \le 1$. Because X and Y are continuous random variables, the joint pdf $f(x, y)$ may be obtained by differentiation:

```
        f = D[F, x, y] // Simplify
domain[f] = {{x, 0, ∞}, {y, 0, ∞}} && {0 < θ < 1};
```

$$e^{-x-y-xy\theta}\,(1 + (-1 + x + y)\,\theta + x\,y\,\theta^2)$$

This is termed a bivariate Exponential distribution because its marginal distributions are standard Exponential. For instance:

```
Marginal[x, f]
```

$$e^{-x}$$

Here is the mgf (this takes about 100 seconds on our reference machine):

```
t = {t₁, t₂};   v = {x, y};   mgf = Expect[e^(t.v), f]
```

— This further assumes that: $\left\{t_1 < 1,\ \text{Arg}\left[\frac{-1 + t_2}{\theta}\right] \ne 0\right\}$

$$-\frac{t_1}{-1 + t_1} + \frac{1}{1 - t_2} + \frac{1}{\theta^2}\left(e^{\frac{(-1+t_1)(-1+t_2)}{\theta}}\left(\text{MeijerG}[\{\{\}, \{1\}\}, \{\{0, 0\}, \{\}\}, \frac{(-1 + t_1)(-1 + t_2)}{\theta}\right](-1 + t_1)(1 + (-1 + \theta)\,t_2) + \text{ExpIntegralE}\left[1, \frac{(-1 + t_1)(-1 + t_2)}{\theta}\right](1 - t_1 + (-1 + \theta + t_1)\,t_2)\right)\right)$$

where the condition $\text{Arg}[\frac{-1+t_2}{\theta}] \ne 0$ is just *Mathematica*'s way of saying $t_2 < 1$. We can now obtain any product raw moment $\mu'_{r,s} = E[X_1^r X_2^s]$ from the mgf, as per (6.18). For instance, $\mu'_{3,4} = E[X_1^3 X_2^4]$ is given by:

```
D[ mgf, {t₁, 3}, {t₂, 4}] /. t_ → 0   // FullSimplify
```

$$\frac{12\,\theta\,(1 + \theta\,(5 + 2\,\theta)) - 12\,e^{\frac{1}{\theta}}\,(1 + 6\,\theta\,(1 + \theta))\,\text{Gamma}[0, \frac{1}{\theta}]}{\theta^6}$$

If we plan to do many of these calculations, it is convenient to write a little *Mathematica* function, Moment$[r, s] = E[X^r Y^s]$, to automate this calculation:

```
Moment[r_, s_] :=
   D[mgf, {t₁, r}, {t₂, s}] /. t_ → 0   // FullSimplify
```

Then $\mu'_{3,4}$ is now given by:

Moment[3, 4]

$$\frac{12\,\theta\,(1 + \theta\,(5 + 2\,\theta)) - 12\,e^{\frac{1}{\theta}}\,(1 + 6\,\theta\,(1 + \theta))\,\text{Gamma}[0, \frac{1}{\theta}]}{\theta^6}$$

Just as we derived the 'mgf about the origin' above, we can also derive the 'mgf about the mean' (*i.e.* the central mgf). To do so, we first need the mean vector $\vec{\mu} = (E[X], E[Y])$, given by:

$\vec{\mu}$ = {Moment[1, 0], Moment[0, 1]}

{1, 1}

Then, by (6.19), the centralised mgf is:

mgfc = $e^{-\vec{t}\cdot\vec{\mu}}$ mgf;

Just as differentiating the mgf yields raw moments, differentiating the centralised mgf yields central moments. In particular, the variances and the covariance of X and Y can be obtained using the following function:

MyCov[i_, j_] := D[mgfc, t$_i$, t$_j$] /. t_ → 0 // FullSimplify

which we apply as follows:

Array[MyCov, {2, 2}]

$$\begin{pmatrix} 1 & -1 + \frac{e^{\frac{1}{\theta}}\,\text{Gamma}[0, \frac{1}{\theta}]}{\theta} \\ -1 + \frac{e^{\frac{1}{\theta}}\,\text{Gamma}[0, \frac{1}{\theta}]}{\theta} & 1 \end{pmatrix}$$

To see how this works, evaluate:

Array[σ, {2, 2}]

$$\begin{pmatrix} \sigma[1, 1] & \sigma[1, 2] \\ \sigma[2, 1] & \sigma[2, 2] \end{pmatrix}$$

We could, of course, alternatively derive the variance-covariance matrix directly with Varcov[f], which takes roughly 6 seconds to evaluate on our reference machine. ∎

6.2 D Moment Conversion Formulae

The moment converter functions introduced in Chapter 2 extend naturally to a multivariate setting. Using these functions, one can express any multivariate moment (μ', μ or κ) in terms of any other moment (μ', μ or κ). The supported conversions are:

function	description
RawToCentral[{$r, s, \ldots$}]	not implemented
RawToCumulant[{$r, s, \ldots$}]	$\mu'_{r,s,\ldots}$ in terms of $\kappa_{i,j,\ldots}$
CentralToRaw[{$r, s, \ldots$}]	$\mu_{r,s,\ldots}$ in terms of $\mu'_{i,j,\ldots}$
CentralToCumulant[{$r, s, \ldots$}]	$\mu_{r,s,\ldots}$ in terms of $\kappa_{i,j,\ldots}$
CumulantToRaw[{$r, s, \ldots$}]	$\kappa_{r,s,\ldots}$ in terms of $\mu'_{i,j,\ldots}$
CumulantToCentral[{$r, s, \ldots$}]	$\kappa_{r,s,\ldots}$ in terms of $\mu_{i,j,\ldots}$
and	
RawToFactorial[{$r, s, \ldots$}]	$\mu'_{r,s,\ldots}$ in terms of $\mu'[i, j, \ldots]$
FactorialToRaw[{$r, s, \ldots$}]	$\mu'[r, s]$ in terms of $\mu'_{i,j}$

Table 2: Multivariate moment conversion functions

⊕ *Example 13:* Express Cov(X, Y) in terms of Raw Moments

Solution: By (6.13), the covariance between X and Y is the central moment $\mu_{1,1}(X, Y)$. Thus, to express the covariance in terms of raw moments, we use the function `CentralToRaw[{1,1}]`:

CentralToRaw[{1, 1}]

$\mu_{1,1} \to -\mu'_{0,1} \mu'_{1,0} + \mu'_{1,1}$

This is just the well-known result that $\mu_{1,1} = E[XY] - E[Y]E[X]$. ∎

Cook (1951) gives *raw* → *cumulant* conversions and *central* → *cumulant* conversions, as well as the inverse relations *cumulant* → *raw* and *cumulant* → *central*, all in a bivariate world with $r + s \le 6$; see also Stuart and Ord (1994, Section 3.29). With **mathStatica**, we can derive these relations on the fly. Here is the bivariate raw moment $\mu'_{3,2}$ expressed in terms of bivariate cumulants:

RawToCumulant[{3, 2}]

$\mu'_{3,2} \to \kappa_{0,1}^2 \kappa_{1,0}^3 + \kappa_{0,2} \kappa_{1,0}^3 + 6 \kappa_{0,1} \kappa_{1,0}^2 \kappa_{1,1} + 6 \kappa_{1,0} \kappa_{1,1}^2 +$
$\quad 3 \kappa_{1,0}^2 \kappa_{1,2} + 3 \kappa_{0,1}^2 \kappa_{1,0} \kappa_{2,0} + 3 \kappa_{0,2} \kappa_{1,0} \kappa_{2,0} +$
$\quad 6 \kappa_{0,1} \kappa_{1,1} \kappa_{2,0} + 3 \kappa_{1,2} \kappa_{2,0} + 6 \kappa_{0,1} \kappa_{1,0} \kappa_{2,1} + 6 \kappa_{1,1} \kappa_{2,1} +$
$\quad 3 \kappa_{1,0} \kappa_{2,2} + \kappa_{0,1}^2 \kappa_{3,0} + \kappa_{0,2} \kappa_{3,0} + 2 \kappa_{0,1} \kappa_{3,1} + \kappa_{3,2}$

Working 'about the mean' (*i.e.* set $\kappa_{1,0} = \kappa_{0,1} = 0$) yields the `CentralToCumulant` conversions. Here is:

CentralToCumulant[{3, 2}]

$$\mu_{3,2} \to 3\,\kappa_{1,2}\,\kappa_{2,0} + 6\,\kappa_{1,1}\,\kappa_{2,1} + \kappa_{0,2}\,\kappa_{3,0} + \kappa_{3,2}$$

The inverse relations are given by `CumulantToRaw` and `CumulantToCentral`. Here, for instance, is the trivariate cumulant $\kappa_{2,1,1}$ expressed in terms of trivariate raw moments:

CumulantToRaw[{2, 1, 1}]

$$\kappa_{2,1,1} \to -6\,\mu'_{0,0,1}\,\mu'_{0,1,0}\,{\mu'}^2_{1,0,0} + 2\,\mu'_{0,1,1}\,{\mu'}^2_{1,0,0} + \\ 4\,\mu'_{0,1,0}\,\mu'_{1,0,0}\,\mu'_{1,0,1} + 4\,\mu'_{0,0,1}\,\mu'_{1,0,0}\,\mu'_{1,1,0} - \\ 2\,\mu'_{1,0,1}\,\mu'_{1,1,0} - 2\,\mu'_{1,0,0}\,\mu'_{1,1,1} + 2\,\mu'_{0,0,1}\,\mu'_{0,1,0}\,\mu'_{2,0,0} - \\ \mu'_{0,1,1}\,\mu'_{2,0,0} - \mu'_{0,1,0}\,\mu'_{2,0,1} - \mu'_{0,0,1}\,\mu'_{2,1,0} + \mu'_{2,1,1}$$

The converter functions extend to any arbitrarily large variate system, of any weight. Here is the input for a 4-variate cumulant $\kappa_{3,1,3,1}$ of weight 8 expressed in terms of central moments:

CumulantToCentral[{3, 1, 3, 1}]

The same expression in raw moments is about 5 times longer and contains 444 different terms. It takes less than a second to evaluate:

Length[CumulantToRaw[{3, 1, 3, 1}]⟦2⟧] // Timing

{0.383333 Second, 444}

Factorial moments were discussed in §6.2 C, and are applied in §6.6 B. David and Barton (1957, p. 144) list multivariate *factorial* → *raw* conversions up to weight 4, along with the inverse relation *raw* → *factorial*. With **mathStatica**, we can again derive these relations on the fly. Here is the bivariate factorial moment $\mu[3, 2]$ expressed in terms of bivariate raw moments:

FactorialToRaw[{3, 2}]

$$\mu[3, 2] \to -2\,\mu'_{1,1} + 2\,\mu'_{1,2} + 3\,\mu'_{2,1} - 3\,\mu'_{2,2} - \mu'_{3,1} + \mu'_{3,2}$$

and here is a trivariate `RawToFactorial` conversion of weight 7:

RawToFactorial[{4, 1, 2}]

$$\mu'_{4,1,2} \to \mu[1, 1, 1] + \mu[1, 1, 2] + 7\,\mu[2, 1, 1] + 7\,\mu[2, 1, 2] + \\ 6\,\mu[3, 1, 1] + 6\,\mu[3, 1, 2] + \mu[4, 1, 1] + \mu[4, 1, 2]$$

○ *The Converter Functions in Practice*

Sometimes, one might know how to derive one class of moments (say raw moments) but not another (say cumulants), or vice versa. In such situations, the converter functions come to the rescue, for they enable one to derive any moment (μ', μ or κ), provided one class of moments can be calculated. This section illustrates how this can be done. The general approach is as follows: first, we express the desired moment (say $\kappa_{2,1}$) in terms of moments that we can calculate (say raw moments):

```
CumulantToRaw[{2, 1}]
```

$$\kappa_{2,1} \to 2\,\mu'_{0,1}\,{\mu'_{1,0}}^2 - 2\,\mu'_{1,0}\,\mu'_{1,1} - \mu'_{0,1}\,\mu'_{2,0} + \mu'_{2,1}$$

and then we evaluate each raw moment $\mu'_{\_}$ for the relevant distribution. This can be done in two ways:

Method (i):   derive $\mu'_{\_}$ from a known mgf
Method (ii): derive $\mu'_{\_}$ directly using the `Expect` function.

Examples 14 and *15* illustrate the two approaches, respectively.

⊕ **Example 14:** Method (i)

Find $\mu_{2,1,2}$ for Cheriyan and Ramabhadran's multivariate Gamma distribution.

Solution: Kotz *et al.* (2000, p.456) give the joint mgf of Cheriyan and Ramabhadran's *m*-variate Gamma distribution as follows:

```
GammaMGF[m_] :=
```
$$\left(1 - \sum_{j=1}^{m} t_j\right)^{-\theta_0} \prod_{j=1}^{m} (1 - t_j)^{-\theta_j}$$

So, for a trivariate system, the mgf is:

```
mgf = GammaMGF[3]
```

$$(1 - t_1)^{-\theta_1}\,(1 - t_2)^{-\theta_2}\,(1 - t_3)^{-\theta_3}\,(1 - t_1 - t_2 - t_3)^{-\theta_0}$$

The desired central moment $\mu_{2,1,2}$ can be expressed in terms of raw moments:

```
sol = CentralToRaw[{2, 1, 2}]
```

$$\mu_{2,1,2} \to 4\,{\mu'_{0,0,1}}^2\,\mu'_{0,1,0}\,{\mu'_{1,0,0}}^2 -$$
$$\mu'_{0,0,2}\,\mu'_{0,1,0}\,{\mu'_{1,0,0}}^2 - 2\,\mu'_{0,0,1}\,\mu'_{0,1,1}\,{\mu'_{1,0,0}}^2 + \mu'_{0,1,2}\,{\mu'_{1,0,0}}^2 -$$
$$4\,\mu'_{0,0,1}\,\mu'_{0,1,0}\,\mu'_{1,0,0}\,\mu'_{1,0,1} + 2\,\mu'_{0,1,0}\,\mu'_{1,0,0}\,\mu'_{1,0,2} -$$
$$2\,{\mu'_{0,0,1}}^2\,\mu'_{1,0,0}\,\mu'_{1,1,0} + 4\,\mu'_{0,0,1}\,\mu'_{1,0,0}\,\mu'_{1,1,1} -$$
$$2\,\mu'_{1,0,0}\,\mu'_{1,1,2} - {\mu'_{0,0,1}}^2\,\mu'_{0,1,0}\,\mu'_{2,0,0} + 2\,\mu'_{0,0,1}\,\mu'_{0,1,0}\,\mu'_{2,0,1} -$$
$$\mu'_{0,1,0}\,\mu'_{2,0,2} + {\mu'_{0,0,1}}^2\,\mu'_{2,1,0} - 2\,\mu'_{0,0,1}\,\mu'_{2,1,1} + \mu'_{2,1,2}$$

Here, each term $\acute{\mu}_{r,s,v}$ denotes $\acute{\mu}_{r,s,v}(X, Y, Z) = E[X^r Y^s Z^v]$, which we can, in turn, find by differentiating the mgf. Since we wish to do this many times, let us write a little *Mathematica* function, Moment$[r, s, v] = E[X^r Y^s Z^v]$, to automate this calculation:

```
Moment[r_, s_, v_] :=
  D[mgf, {t₁, r}, {t₂, s}, {t₃, v}] /. t_ → 0
```

Then, the solution is:

```
sol /. μ́ₖ_ :→ Moment[k] // Simplify
```

$\mu_{2,1,2} \to 2\,\Theta_0\,(12 + 10\,\Theta_0 + \Theta_1 + \Theta_3)$

An alternative solution to this particular problem, without using the converter functions, is to first find the mean vector $\vec{\mu} = \{E[X], E[Y], E[Z]\}$:

```
μ⃗ = {Moment[1, 0, 0], Moment[0, 1, 0], Moment[0, 0, 1]}
```

$\{\Theta_0 + \Theta_1,\ \Theta_0 + \Theta_2,\ \Theta_0 + \Theta_3\}$

Second, find the central mgf, by (6.19):

```
t⃗ = {t₁, t₂, t₃};   mgfc = e^(-t⃗·μ⃗) mgf
```

$e^{-t_1(\Theta_0+\Theta_1)-t_2(\Theta_0+\Theta_2)-t_3(\Theta_0+\Theta_3)}\,(1-t_1)^{-\Theta_1}$
$(1-t_2)^{-\Theta_2}\,(1-t_3)^{-\Theta_3}\,(1-t_1-t_2-t_3)^{-\Theta_0}$

Then, differentiating the central mgf yields the desired central moment $\mu_{2,1,2}$ again:

```
D[mgfc, {t₁, 2}, {t₂, 1}, {t₃, 2}] /. t_ → 0 // Simplify
```

$2\,\Theta_0\,(12 + 10\,\Theta_0 + \Theta_1 + \Theta_3)$

⊕ ***Example 15:*** Method (ii)

Let random variables X and Y have joint density $f(x, y)$:

$$f = \frac{1}{\sqrt{2\pi}}\, e^{-\frac{x^2}{2} - 2y} \left(e^y + \alpha\,(e^y - 2)\,\text{Erf}\!\left[\frac{x}{\sqrt{2}}\right] \right);$$

domain$[f] = \{\{x, -\infty, \infty\}, \{y, 0, \infty\}\}$ && $\{-1 < \alpha < 1\}$;

For the given density, find the product cumulant $\kappa_{2,2}$.

Solution: If we knew the mgf, we could immediately derive the cumulant generating function. Unfortunately, *Mathematica* Version 4 can not derive the mgf; nor is it likely to be listed in any textbook, because this is not a common distribution. To resolve this problem, we will make use of the moment conversion formulae. The desired solution, $\kappa_{2,2}$, expressed in terms of raw moments, is:

sol = CumulantToRaw[{2, 2}]

$$\kappa_{2,2} \to -6\,\mu'^2_{0,1}\,\mu'^2_{1,0} + 2\,\mu'_{0,2}\,\mu'^2_{1,0} + 8\,\mu'_{0,1}\,\mu'_{1,0}\,\mu'_{1,1} - 2\,\mu'^2_{1,1} -$$
$$2\,\mu'_{1,0}\,\mu'_{1,2} + 2\,\mu'^2_{0,1}\,\mu'_{2,0} - \mu'_{0,2}\,\mu'_{2,0} - 2\,\mu'_{0,1}\,\mu'_{2,1} + \mu'_{2,2}$$

Here, each term $\mu'_{r,s}$ denotes $\mu'_{r,s}(X, Y) = E[X^r Y^s]$, and so can be evaluated with the Expect function. In the next input, we calculate each of the expectations that we require:

sol /. $\mu'_{r\_,s\_}$:→ Expect[x^r y^s, f] // Simplify

$$\kappa_{2,2} \to -\frac{\alpha^2}{2\pi}$$

The calculation takes about 6 seconds on our reference machine. ∎

6.3 Independence and Dependence

6.3 A Stochastic Independence

Let random variables $\vec{X} = (X_1, \ldots, X_m)$ have joint pdf $f(x_1, \ldots, x_m)$, with marginal density functions $f_1(x_1), \ldots, f_m(x_m)$. Then $(X_1, \ldots, X_m)$ are said to be *mutually stochastically independent* if and only if

$$f(x_1, \ldots, x_m) = f_1(x_1) \times \cdots \times f_m(x_m). \tag{6.25}$$

That is, the joint pdf is equal to the product of the marginal pdf's. A number of well-known theorems apply to mutually stochastically independent random variables, which we state here without proof. In particular:

	If $(X_1, \ldots, X_m)$ are mutually stochastically independent, then:
(i)	$P(a \leq X_1 \leq b, \ldots, c \leq X_m \leq d) = P(a \leq X_1 \leq b) \times \cdots \times P(c \leq X_m \leq d)$
(ii)	$E[u_1(X_1) \cdots u_m(X_m)] = E[u_1(X_1)] \times \cdots \times E[u_m(X_m)]$ for arbitrary functions $u_i(\cdot)$
(iii)	$M(t_1, \ldots, t_m) = M(t_1) \times \cdots \times M(t_m)$ mgf of the joint distribution = product of the mgf's of the marginal distributions
(iv)	$\text{Cov}(X_i, X_j) = 0$ for all $i \neq j$ However, zero covariance does *not* imply independence.

Table 3: Properties of mutually stochastic independent random variables

⊕ ***Example 16:*** Stochastic Dependence and Correlation

Let the random variables X, Y and Z have joint pdf $h(x, y, z)$:

```
         Exp[- 1/2 (x² + y² + z²)] (1 + x y z Exp[- 1/2 (x² + y² + z²)])
h = ─────────────────────────────────────────────────────────────────── ;
                                  (2 π)^(3/2)
domain[h] = {{x, -∞, ∞}, {y, -∞, ∞}, {z, -∞, ∞}};
```

Since the product of the marginal pdf's:

```
Marginal[x, h]  Marginal[y, h]  Marginal[z, h]
```

$$\frac{e^{-\frac{x^2}{2} - \frac{y^2}{2} - \frac{z^2}{2}}}{2\sqrt{2}\,\pi^{3/2}}$$

... is *not* equal to the joint pdf $h(x, y, z)$, it follows by (6.25) that X, Y and Z are mutually stochastically *dependent*. Even though X, Y and Z are mutually dependent, their correlations ρ_{ij} ($i \neq j$) are all zero:

```
Varcov[h]
```

$$\begin{pmatrix} 1 & 0 & 0 \\ 0 & 1 & 0 \\ 0 & 0 & 1 \end{pmatrix}$$

Clearly, zero correlation does not imply independence. ∎

6.3 B Copulae

Copulae provide a method for constructing multivariate distributions from known marginal distributions. We shall only consider the bivariate case here. For more detail, see Joe (1997) and Nelsen (1999).

Let the continuous random variable X have pdf $f(x)$ and cdf $F(x)$; similarly, let the continuous random variable Y have pdf $g(y)$ and cdf $G(y)$. We wish to create a bivariate distribution $H(x, y)$ from these marginals. The joint distribution function $H(x, y)$ is given by

$$H(x, y) = C(F, G) \tag{6.26}$$

where C denotes the copula function. Then, the joint pdf $h(x, y)$ is given by

$$h(x, y) = \frac{\partial^2 H(x, y)}{\partial x \, \partial y}. \tag{6.27}$$

Table 4 lists some examples of copulae.

copula	formula	restrictions
Independent	$C = FG$	
Morgenstern	$C = FG(1 + \alpha(1-F)(1-G))$	$-1 < \alpha < 1$
Ali–Mikhail–Haq	$C = \dfrac{FG}{1 - \alpha(1-F)(1-G)}$	$-1 \leq \alpha \leq 1$
Frank	$C = -\dfrac{1}{\alpha} \log\left[1 + \dfrac{(e^{-\alpha F} - 1)(e^{-\alpha G} - 1)}{e^{-\alpha} - 1}\right]$	$\alpha \neq 0$

Table 4: Copulae

With the exception of the independent case, each copula in Table 4 includes parameter α. This term induces a new parameter into the joint bivariate distribution $h(x, y)$, which gives added flexibility. In each case, setting parameter $\alpha = 0$ (or taking the limit $\alpha \to 0$, in the Frank case) yields the independent copula $C = FG$ as a special case. When $\alpha = 1$, the Ali–Mikhail–Haq copula simplifies to $C = \dfrac{FG}{F + G - FG}$, as used in Exercise 8.

In the following two examples, we shall work with the Morgenstern (1956) copula.[2] We enter it as follows:

```
ClearAll[F, G]
Copula := F G (1 + α (1 - F) (1 - G))
```

⊕ *Example 17:* Bivariate Uniform (à la Morgenstern)

Let $X \sim$ Uniform(0, 1) with pdf $f(x)$ and cdf $F(x)$, and let $Y \sim$ Uniform(0, 1) with pdf $g(y)$ and cdf $G(y)$:

```
f = 1;   domain[f] = {x, 0, 1};   F = Prob[x, f];
g = 1;   domain[g] = {y, 0, 1};   G = Prob[y, g];
```

Let $h(x, y)$ denote the bivariate Uniform obtained via a Morgenstern copula. Then:

```
h = D[Copula, x, y] // Simplify

1 + (-1 + 2 x) (-1 + 2 y) α
```

with domain of support:

```
domain[h] = {{x, 0, 1}, {y, 0, 1}} && {-1 < α < 1};
```

Figure 8 plots the joint pdf $h(x, y)$ when $\alpha = \frac{1}{2}$. Clicking the 'View Animation' button in the electronic notebook brings up an animation of $h(x, y)$, allowing parameter α to vary from -1 to 1 in step sizes of $\frac{1}{10}$. This provides a rather neat way to visualise positive and negative correlation.

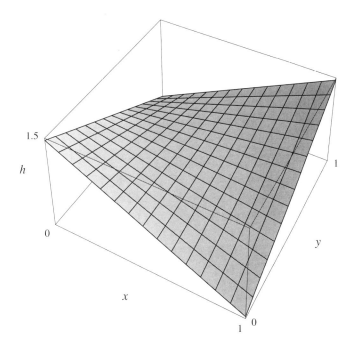

Fig. 8: Bivariate Uniform joint pdf $h(x, y)$ when $\alpha = \frac{1}{2}$

We already know the joint cdf $H(x, y) = P(X \le x, Y \le y)$, which is just the copula function:

Copula

$$x\,y\,(1 + (1 - x)\,(1 - y)\,\alpha)$$

The variance-covariance matrix is given by:

Varcov[h]

$$\begin{pmatrix} \frac{1}{12} & \frac{\alpha}{36} \\ \frac{\alpha}{36} & \frac{1}{12} \end{pmatrix}$$

⊕ *Example 18:* Normal–Uniform Bivariate Distribution (à la Morgenstern)

Let $X \sim N(0, 1)$ with pdf $f(x)$ and cdf $F(x)$, and let $Y \sim \text{Uniform}(0, 1)$ with pdf $g(y)$ and cdf $G(y)$:

```
f = e^(-x²/2) / √(2π);    domain[f] = {x, -∞, ∞};    F = Prob[x, f];

g = 1;                    domain[g] = {y, 0, 1};     G = Prob[y, g];
```

Let $h(x, y)$ denote the bivariate distribution obtained via a Morgenstern copula. Then:

```
h = D[Copula, x, y]   // Simplify
```

$$\frac{e^{-\frac{x^2}{2}} \left(1 + (-1 + 2y)\, \alpha\, \mathrm{Erf}\!\left[\frac{x}{\sqrt{2}}\right]\right)}{\sqrt{2\pi}}$$

with domain of support:

```
domain[h] = {{x, -∞, ∞}, {y, 0, 1}}  &&  {-1 ≤ α ≤ 1};
```

Figure 9 plots the joint pdf $h(x, y)$ when $\alpha = 0$.

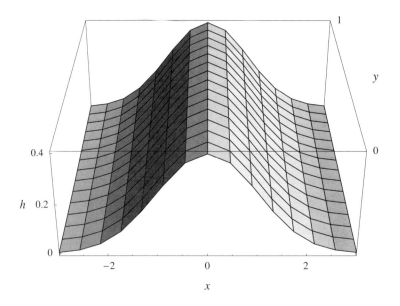

Fig. 9: Normal–Uniform joint pdf $h(x, y)$ when $\alpha = 0$

The joint cdf $H(x, y) = P(X \le x, Y \le y)$ is the copula function:

```
Copula // Simplify
```

$$\tfrac{1}{2} y \left(1 + \tfrac{1}{2}(-1+y)\, \alpha \left(-1 + \mathrm{Erf}\!\left[\tfrac{x}{\sqrt{2}}\right]\right)\right)\left(1 + \mathrm{Erf}\!\left[\tfrac{x}{\sqrt{2}}\right]\right)$$

We can confirm that the marginal distributions are in fact Normal and Uniform, respectively:

```
Marginal[x, h]
Marginal[y, h]
```

$$\frac{e^{-\frac{x^2}{2}}}{\sqrt{2\pi}}$$

1

§6.3 B MULTIVARIATE DISTRIBUTIONS

The variance-covariance matrix is:

Varcov[h]

$$\begin{pmatrix} 1 & \frac{\alpha}{6\sqrt{\pi}} \\ \frac{\alpha}{6\sqrt{\pi}} & \frac{1}{12} \end{pmatrix}$$

Let $h_c(y)$ denote the conditional density function of Y, given $X = x$:

h_c = Conditional[y, h]

— Here is the conditional pdf $h(y \mid x)$:

$$1 + (-1 + 2y) \, \alpha \, \text{Erf}\left[\frac{x}{\sqrt{2}}\right]$$

with domain:

domain[h_c] = {y, 0, 1} && {-1 ≤ α ≤ 1};

Then, the conditional mean $E[Y \mid X = x]$ is:

Expect[y, h_c]

$$\frac{1}{6}\left(3 + \alpha \, \text{Erf}\left[\frac{x}{\sqrt{2}}\right]\right)$$

and the conditional variance $\text{Var}(Y \mid X = x)$ is:

Var[y, h_c]

$$\frac{1}{36}\left(3 - \alpha^2 \, \text{Erf}\left[\frac{x}{\sqrt{2}}\right]^2\right)$$

Figure 10 plots the conditional mean and the conditional variance, when X and Y are correlated ($\alpha = 1$) and uncorrelated ($\alpha = 0$).

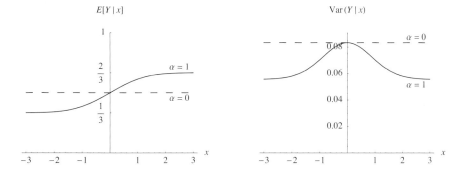

Fig. 10: Conditional mean and variance

6.4 The Multivariate Normal Distribution

The *Mathematica* package, `Statistics`MultinormalDistribution``, has several functions that are helpful throughout this section. We load this package as follows:

```
<< Statistics`
```

The multivariate Normal distribution is pervasive throughout statistics, so we devote an entire section to it and to some of its properties. Given $\vec{X} = (X_1, \ldots, X_m)$, we denote the m-variate *multivariate Normal distribution* by $N(\vec{\mu}, \Sigma)$, with mean vector $\vec{\mu} = (\mu_1, \ldots, \mu_m) \in \mathbb{R}^m$, variance-covariance matrix Σ, and joint pdf

$$f(\vec{x}) = (2\pi)^{-m/2} |\Sigma|^{-1/2} \exp\left(-\tfrac{1}{2}(\vec{x} - \vec{\mu})^T \Sigma^{-1} (\vec{x} - \vec{\mu})\right) \qquad (6.28)$$

where $\vec{x} = (x_1, \ldots, x_m) \in \mathbb{R}^m$, and Σ is a symmetric, positive definite $(m \times m)$ matrix. When $m = 1$, (6.28) simplifies to the univariate Normal pdf.

6.4 A The Bivariate Normal

Let random variables X_1 and X_2 have a bivariate Normal distribution, with zero mean vector, and variance-covariance matrix $\Sigma = \begin{pmatrix} 1 & \rho \\ \rho & 1 \end{pmatrix}$. Here, ρ denotes the correlation coefficient between X_1 and X_2. That is:

```
𝐗⃗ = {x₁, x₂};   μ⃗ = {0, 0};   Σ = ( 1  ρ )
                                    ( ρ  1 );
dist2 = MultinormalDistribution[μ⃗, Σ];
```

Then, we enter our bivariate Normal pdf $f(x_1, x_2)$ as:

```
f = PDF[dist2, 𝐗⃗] // Simplify
domain[f] = Thread[{𝐗⃗, -∞, ∞}] && {-1 < ρ < 1}
```

$$\frac{e^{\frac{x_1^2 - 2\rho x_1 x_2 + x_2^2}{-2 + 2\rho^2}}}{2\pi\sqrt{1-\rho^2}}$$

$$\{\{x_1, -\infty, \infty\}, \{x_2, -\infty, \infty\}\} \&\& \{-1 < \rho < 1\}$$

where the `PDF` and `MultinormalDistribution` functions are defined in *Mathematica*'s `Statistics` package.

When $\rho = 0$, the cdf can be expressed in terms of the built-in error function as:[3]

```
F₀ = Prob[{x₁, x₂}, f /. ρ → 0]
```

$$\tfrac{1}{4}\left(1 + \mathrm{Erf}\left[\tfrac{x_1}{\sqrt{2}}\right]\right)\left(1 + \mathrm{Erf}\left[\tfrac{x_2}{\sqrt{2}}\right]\right)$$

○ *Diagrams*

Figure 11 plots the zero correlation pdf and cdf.

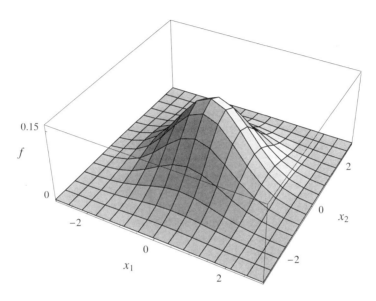

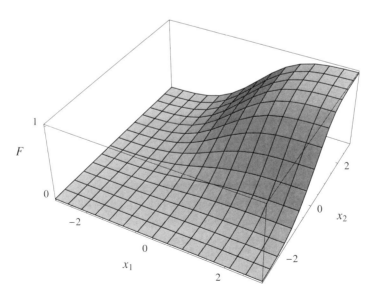

Fig. 11: The bivariate Normal joint pdf f (top) and joint cdf F (bottom), when $\rho = 0$

The shape of the contours of $f(x_1, x_2)$ depends on ρ, as Fig. 12 illustrates with a set of contour plots.

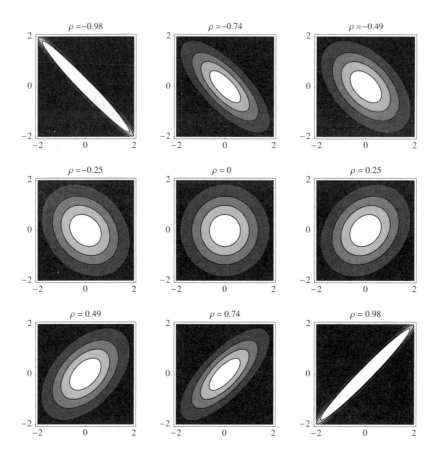

Fig. 12: Contour plots of the bivariate Normal pdf, for different values of ρ

Each plot corresponds to a specific value of ρ. In the top left corner, $\rho = -0.98$ (almost perfect negative correlation), whereas in the bottom right corner, $\rho = 0.98$ (almost perfect positive correlation). The middle plot corresponds to the case of zero correlation. In any given plot, the edge of each shaded region represents the contour line, and each contour is a two-dimensional ellipse along which f is constant. The ellipses are aligned along the $x_1 = x_2$ line when $\rho > 0$, or the $x_1 = -x_2$ line when $\rho < 0$.

We can even plot the specific ellipse that encloses $q\%$ of the distribution by using the `EllipsoidQuantile[dist, q]` function in *Mathematica*'s `Statistics` package. This is illustrated in Fig. 13, which plots the ellipses that enclose 15% (bold), 90% (dashed) and 99% (plain) of the distribution, respectively, when ρ is 0.6. Figure 14 superimposes 1000 pseudo-random drawings from this distribution on top of Fig. 13. On average, we would expect around 1% of the simulated data to lie outside the 99% quantile. For this particular set of simulated data, there are 11 such points (the large dots in Fig. 14).

§6.4 A   MULTIVARIATE DISTRIBUTIONS   219

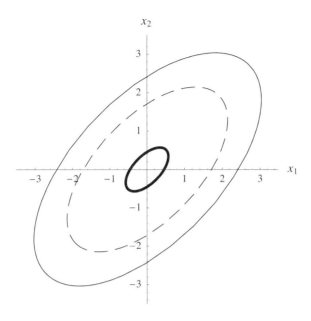

Fig. 13: Quantiles: 15% (bold), 90% (dashed) and 99% (plain)

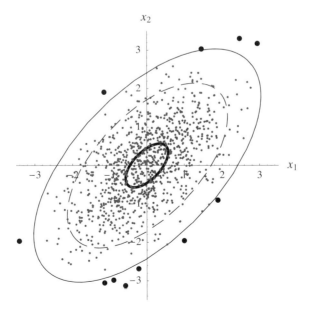

Fig. 14: Quantiles plotted with 1000 pseudo-random drawings

○ *Applying the* **mathStatica** *Toolset*

We can try out the **mathStatica** toolset on density f. The marginal distribution of X_1 is well known to be $N(0, 1)$, as we confirm with:

Marginal[x$_1$, f]

$$\frac{e^{-\frac{x_1^2}{2}}}{\sqrt{2\pi}}$$

The variance-covariance matrix is, of course, equal to Σ:

Varcov[f]

$$\begin{pmatrix} 1 & \rho \\ \rho & 1 \end{pmatrix}$$

The conditional distribution of X_1 given $X_2 = x_2$ is $N(\rho x_2, 1 - \rho^2)$, as we confirm with:

Conditional[x$_1$, f]

— Here is the conditional pdf f (x$_1$ | x$_2$):

$$\frac{e^{\frac{(x_1 - \rho x_2)^2}{2(-1+\rho^2)}}}{\sqrt{2\pi}\sqrt{1-\rho^2}}$$

Here is the product moment $E[X_1^2 X_2^2]$:

Expect[x$_1^2$ x$_2^2$, f]

$1 + 2\rho^2$

The moment generating function is given by:

t̃ = {t$_1$, t$_2$}; mgf = Expect[e$^{\vec{t}\cdot\vec{x}}$, f]

$$e^{\frac{1}{2}(t_1^2 + 2\rho t_1 t_2 + t_2^2)}$$

Here, again, is the product moment $E[X_1^2 X_2^2]$, but now derived from the mgf:

D[mgf, {t$_1$, 2}, {t$_2$, 2}] /. t_ → 0

$1 + 2\rho^2$

If the mgf is known, this approach to deriving moments is much faster than the direct Expect approach. However, in higher variate (or more general) examples, *Mathematica* may not always be able to find the mgf, nor the cf. In the special case of the multivariate Normal distribution, this is not necessarily a problem since *Mathematica*'s Statistics package 'knows' the solution. Of course, this concept of 'knowledge' is somewhat

§6.4 A   MULTIVARIATE DISTRIBUTIONS   221

artificial—*Mathematica*'s Statistics package does not derive the solution, but rather regurgitates the answer just like a textbook appendix does. In this vein, the Statistics package and a textbook appendix both work the same way: someone typed the answer in. For instance, for our example, the cf is immediately outputted (*not* derived) by the Statistics package as:

CharacteristicFunction[dist2, {t₁, t₂}]

$e^{\frac{1}{2}(-t_2(\rho t_1+t_2)-t_1(t_1+\rho t_2))}$

While this works well here, the regurgitation approach unfortunately breaks down as soon as one veers from the chosen path, as we shall see in *Example 21*.

⊕ **Example 19:** The Normal Linear Regression Model

Let us suppose that the random variables Y and X are jointly distributed, and that the conditional mean of Y given $X = x$ can be expressed as

$$E[Y \mid X = x] = \alpha_1 + \alpha_2 x \qquad (6.29)$$

where α_1 and α_2 are unknown but fixed parameters. The conditional mean, being linear in the parameters, is called a *linear regression function*. We may write

$$Y = \alpha_1 + \alpha_2 x + U \qquad (6.30)$$

where the random variable $U = Y - E[Y \mid X = x]$ is referred to as the *disturbance*, and has, by construction, a conditional mean equal to zero; that is, $E[U \mid X = x] = 0$. If Y is conditionally Normally distributed, then by linearity so too is U conditionally Normal, in which case we have the **Normal** linear regression model. This model can arise from a setting in which (Y, X) are jointly Normally distributed. To see this, let (Y, X) have joint bivariate pdf $N(\vec{\mu}, \Sigma)$ where:

$\vec{\mu}$ = {μ_Y, μ_X}; Σ = $\begin{pmatrix} \sigma_Y^2 & \sigma_Y \sigma_X \rho \\ \sigma_Y \sigma_X \rho & \sigma_X^2 \end{pmatrix}$;
cond = {σ_Y > 0, σ_X > 0, -1 < ρ < 1};
dist = MultinormalDistribution[$\vec{\mu}$, Σ];

Let $f(y, x)$ denote the joint pdf:

f = Simplify[PDF[dist, {y, x}], cond]
domain[f] = {{y, -∞, ∞}, {x, -∞, ∞}} && cond

$$\frac{e^{-\frac{(y-\mu_Y)^2 \sigma_X^2 - 2\rho(x-\mu_X)(y-\mu_Y)\sigma_X\sigma_Y + (x-\mu_X)^2 \sigma_Y^2}{2(-1+\rho^2)\sigma_X^2 \sigma_Y^2}}}{2\pi\sqrt{1-\rho^2}\,\sigma_X\,\sigma_Y}$$

{{y, -∞, ∞}, {x, -∞, ∞}} && {σ_Y > 0, σ_X > 0, -1 < ρ < 1}

The regression function $E[Y \mid X = x]$ can be derived in two steps (as per *Example 9*):

(i) We first determine the conditional pdf of Y given $X = x$:

```
f_con = Conditional[y, f]
```

— Here is the conditional pdf $f(y \mid x)$:

$$\frac{e^{\frac{((y-\mu_Y)\sigma_X + \rho(-x+\mu_X)\sigma_Y)^2}{2(-1+\rho^2)\sigma_X^2 \sigma_Y^2}}}{\sqrt{2\pi}\sqrt{1-\rho^2}\,\sigma_Y}$$

where the domain of the conditional distribution is:

```
domain[f_con] = {y, -∞, ∞} && cond;
```

(ii) We can now find $E[Y \mid X = x]$:

```
regf = Expect[y, f_con]
```

$$\mu_Y + \frac{\rho(x - \mu_X)\sigma_Y}{\sigma_X}$$

This expression is of form $\alpha_1 + \alpha_2 x$. To see this, we can use the `CoefficientList` function to obtain the parameters α_1 and α_2:

```
CoefficientList[regf, x]
```

$$\left\{ \mu_Y - \frac{\rho\,\mu_X\,\sigma_Y}{\sigma_X},\ \frac{\rho\,\sigma_Y}{\sigma_X} \right\}$$

In summary, if (Y, X) are jointly bivariate Normal, then the regression function $E[Y \mid X = x]$ is linear in the parameters, of form $\alpha_1 + \alpha_2 x$, where $\alpha_1 = \mu_Y - \alpha_2 \mu_X$ and $\alpha_2 = \frac{\rho\sigma_Y}{\sigma_X}$, which is what we set out to show. Finally, inspection of `f_con` reveals that the conditional distribution of $Y \mid (X = x)$ is Normal. Joint Normality therefore determines a Normal linear regression model. ∎

⊕ **Example 20:** Robin Hood

Robin Hood has entered the coveted Nottingham Forest Archery competition, where contestants shoot arrows at a vertical target. For Mr Hood, it is known that the distribution of horizontal and vertical deviations from the centre of the target is bivariate Normal, with zero means, equal variances σ^2 and correlation ρ. What is the probability that he gets a bull's-eye, if the latter has unit radius?

Solution: We begin by setting up the appropriate bivariate Normal distribution:

```
x⃗ = {x_1, x_2};   μ⃗ = {0, 0};   Σ = σ^2 ( 1  ρ
                                            ρ  1 );
dist = MultinormalDistribution[μ⃗, Σ];
cond = {σ > 0, -1 < ρ < 1, r > 0, 0 < θ < 2π};
```

Let $f(x_1, x_2)$ denote the joint pdf:

```
f = Simplify[PDF[dist, x⃗], cond]
domain[f] = {{x₁, -∞, ∞}, {x₂, -∞, ∞}} && cond;
```

$$\frac{e^{-\frac{x_1^2 - 2\rho x_1 x_2 + x_2^2}{2(-1+\rho^2)\sigma^2}}}{2\pi\sqrt{1-\rho^2}\,\sigma^2}$$

The solution requires a transformation to polar co-ordinates. Thus:

```
Ω = {x₁ → r Cos[θ], x₂ → r Sin[θ]};
```

Here, $R = \sqrt{X_1^2 + X_2^2}$ represents the distance of (X_1, X_2) from the origin, while $\Theta = \arctan(X_2/X_1)$ represents the angle of (X_1, X_2) with respect to the X_1 axis. Thus, $R = r \in \mathbb{R}_+$ and $\Theta = \theta \in \{\theta : 0 < \theta < 2\pi\}$. We seek the joint pdf of R and Θ. We thus apply the transformation method (Chapter 4). We do so manually (see §4.2 C), because there are two solutions, differing only in respect to sign. The desired joint density is $g(r, \theta)$:

```
g = Simplify[(f /. Ω) Jacob[x⃗ /. Ω, {r, θ}], cond]
domain[g] = {{r, 0, ∞}, {θ, 0, 2π}} && cond;
```

$$\frac{e^{-\frac{r^2(-1+\rho \sin[2\theta])}{2(-1+\rho^2)\sigma^2}}\,r}{2\pi\sqrt{1-\rho^2}\,\sigma^2}$$

The probability of hitting the bull's-eye is given by $P(R \leq 1)$. In the simple case of zero correlation ($\rho = 0$), this is:

```
pr = Prob[{1, 2π}, g /. ρ → 0]
```

$$1 - e^{-\frac{1}{2\sigma^2}}$$

As expected, this probability is decreasing in the standard deviation σ, as Fig. 15 illustrates.

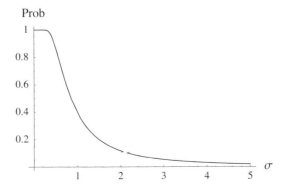

Fig. 15: Probability that Robin Hood hits a bull's-eye, as a function of σ

More generally, in the case of non-zero correlation ($\rho \neq 0$), *Mathematica* cannot determine this probability exactly. This is not surprising as the solution does not have a convenient closed form. Nevertheless, given values of the parameters σ and ρ, one can use numerical integration. For instance, if $\sigma = 2$, and $\rho = 0.7$, the probability of a bull's-eye is:

```
NIntegrate[g /. {σ → 2, ρ → 0.7}, {r, 0, 1}, {θ, 0, 2π}]
```

0.155593

which contrasts with a probability of 0.117503 when $\rho = 0$. More generally, it appears that a contestant whose shooting is 'elliptical' ($\rho \neq 0$) will hit the bull's-eye more often than an 'uncorrelated' ($\rho = 0$) contestant! ∎

⊕ **Example 21:** Truncated Bivariate Normal

Let $(X, Y) \sim N(\vec{0}, \Sigma)$ with joint pdf $f(x, y)$ and cdf $F(x, y)$, with $\Sigma = \begin{pmatrix} 1 & \rho \\ \rho & 1 \end{pmatrix}$, where we shall assume that $0 < \rho < 1$. Corresponding to $f(x, y)$, let $g(x, y)$ denote the pdf of a truncated distribution with Y restricted to the positive real line ($Y > 0$). We wish to find the pdf of the truncated distribution $g(x, y)$, the marginal distributions $g_X(x)$ and $g_Y(y)$, and the new variance-covariance matrix.

Solution: Since the truncated distribution is not a 'textbook' Normal distribution, *Mathematica*'s Multinormal package is not designed to answer such questions. By contrast, **mathStatica** adopts a general approach and so can solve such problems. Given:

```
V = {x, y};   μ⃗ = {0, 0};   Σ = ( 1  ρ
                                  ρ  1 );   cond = {0 < ρ < 1};
```

Then, the parent pdf $f(x, y)$ is:

```
f = Simplify[PDF[MultinormalDistribution[μ⃗, Σ], V], cond];
domain[f] = {{x, -∞, ∞}, {y, -∞, ∞}} && cond;
```

By familiar truncation arguments (§2.5 A):

$$g(x, y) = \frac{f(x, y)}{1 - F(\infty, 0)} = 2 f(x, y), \quad \text{for } x \in \mathbb{R}, y \in \mathbb{R}_+$$

which we enter as:

```
g = 2 f;
domain[g] = {{x, -∞, ∞}, {y, 0, ∞}} && cond;
```

The marginal pdf of Y, when Y is truncated below at zero, is $g_Y(y)$:

```
g_Y = Marginal[y, g]
```

$$e^{-\frac{y^2}{2}} \sqrt{\frac{2}{\pi}}$$

This is the pdf of a half-Normal random variable, as illustrated in Fig. 16.

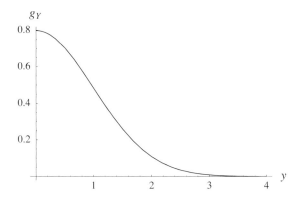

Fig. 16: The marginal pdf of Y, when Y is truncated below at zero

By contrast, the marginal pdf of X, when Y is truncated below at 0, is given by $g_X(x)$:

g_X = Marginal[x, g]

$$\frac{e^{-\frac{x^2}{2}} \left(1 + \text{Erf}\left[\frac{x\rho}{\sqrt{2-2\rho^2}}\right]\right)}{\sqrt{2\pi}}$$

which is Azzalini's skew-Normal(λ) pdf with $\lambda = \rho/\sqrt{1-\rho^2}$ (see Chapter 2, Exercise 2). Even though X is not itself truncated, $g_X(x)$ *is* affected by the truncation of Y, because X is correlated with Y. Now consider the two extremes: if $\rho = 0$, X and Y are uncorrelated, so $g_X(\cdot) = f_X(\cdot)$, and we obtain a standard Normal pdf; at the other extreme, if $\rho = 1$, X and Y are perfectly correlated, so $g_X(\cdot) = g_Y(\cdot)$, and we obtain a half-Normal pdf. For $0 < \rho < 1$, we obtain a result between these two extremes. This can be seen from Fig. 17, which plots both extremes, and three cases in between.

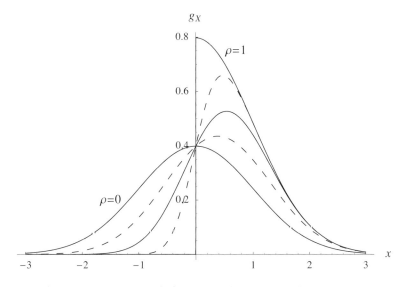

Fig. 17: The marginal pdf of X, when Y is truncated below at zero.

The mean vector, when Y is truncated below at zero, is:

Expect[{x, y}, g]

$$\left\{ \sqrt{\frac{2}{\pi}}\, \rho,\ \sqrt{\frac{2}{\pi}} \right\}$$

The variance-covariance matrix for (X, Y), when Y is truncated below at zero, is:

Varcov[g]

$$\begin{pmatrix} 1 - \frac{2\rho^2}{\pi} & \frac{(-2+\pi)\,\rho}{\pi} \\ \frac{(-2+\pi)\,\rho}{\pi} & \frac{-2+\pi}{\pi} \end{pmatrix}$$

This illustrates that, in a mutually dependent setting, the truncation of one random variable effects all the random variables (not just the truncated variable). ∎

6.4 B The Trivariate Normal

The trivariate Normal distribution for (X, Y, Z) is fully specified by the (3×1) vector of means and the (3×3) variance-covariance matrix. When the mean vector is $\vec{0}$ and the variances are all equal to unity, we have:

$$V = \{x, y, z\};\quad \vec{\mu} = \{0, 0, 0\};\quad \Sigma = \begin{pmatrix} 1 & \rho_{xy} & \rho_{xz} \\ \rho_{xy} & 1 & \rho_{yz} \\ \rho_{xz} & \rho_{yz} & 1 \end{pmatrix};$$

dist3 = MultinormalDistribution[$\vec{\mu}$, Σ];

cond = {-1 < ρ_{xy} < 1, -1 < ρ_{xz} < 1, -1 < ρ_{yz} < 1, Det[Σ] > 0};

where ρ_{ij} denotes the correlation between variable i and variable j, and the condition Det[Σ] > 0 reflects the fact that the variance-covariance matrix is positive definite. Let $g(x, y, z)$ denote the joint pdf:

g = PDF[dist3, V] // Simplify

$$\frac{e^{\frac{x^2 + y^2 + z^2 - z^2 \rho_{xy}^2 - y^2 \rho_{xz}^2 - 2 y z \rho_{yz} - x^2 \rho_{yz}^2 - 2 \rho_{xz}\,(z - y\,\rho_{yz}) + 2\rho_{xy}\,(-x y + z\,\rho_{xz} + x z\,\rho_{yz})}{2\,(-1 + \rho_{xy}^2 + \rho_{xz}^2 - 2\rho_{xy}\rho_{xz}\rho_{yz} + \rho_{yz}^2)}}}{2\sqrt{2}\,\pi^{3/2}\,\sqrt{1 - \rho_{xy}^2 - \rho_{xz}^2 + 2\rho_{xy}\rho_{xz}\rho_{yz} - \rho_{yz}^2}}$$

with domain:

domain[g] = Thread[{V, -∞, ∞}] && cond

{{x, -∞, ∞}, {y, -∞, ∞}, {z, -∞, ∞}} && {-1 < ρ_{xy} < 1, -1 < ρ_{xz} < 1, -1 < ρ_{yz} < 1, 1 - ρ_{xy}^2 - ρ_{xz}^2 + 2 $\rho_{xy}\rho_{xz}\rho_{yz}$ - ρ_{yz}^2 > 0}

Here, for example, is $E[X\,Y\,e^Z]$; the calculation takes about 70 seconds on our reference computer:

Expect[x y e^z, g]

$\sqrt{e}\;(\rho_{xy} + \rho_{xz}\,\rho_{yz})$

Figure 12, above, illustrated that a contour plot of a bivariate Normal pdf yields an ellipse, or a circle given zero correlation. Figure 18 illustrates a specific contour of the trivariate pdf $g(x, y, z)$, when $\rho_{xy} \to 0.2$, $\rho_{yz} \to 0.3$, $\rho_{xz} \to 0.4$, and $g(x, y, z) = 0.05$. Once again, the symmetry of the plot will be altered by the choice of correlation coefficients. Whereas the bivariate Normal yields elliptical contours (or a circle given zero correlation), the trivariate case yields the intuitive 3D equivalent, namely the surface of an ellipsoid (or that of a sphere given zero correlations). Here, parameter ρ_{xy} alters the 'orientation' of the ellipsoid in the x-y plane, just as ρ_{yz} does in the y-z plane, and ρ_{xz} does in the x-z plane.

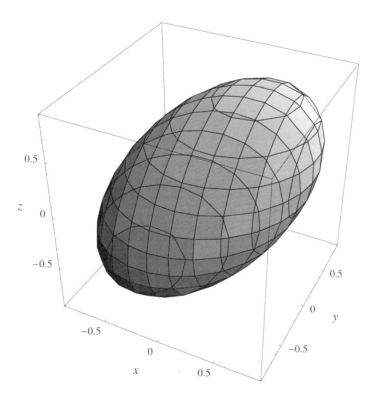

Fig. 18: The contour $g(x, y, z) = 0.05$ for the trivariate Normal pdf

Just as in the 2D case, we can plot the specific ellipsoid that encloses $q\%$ of the distribution by using the function `EllipsoidQuantile[dist, q]`. This is illustrated in Fig. 19 below, which plots the ellipsoids that enclose 60% (solid) and 90% (wireframe) of the distribution, respectively, given $\rho_{xy} \to 0.01$, $\rho_{yz} \to 0.01$, $\rho_{xz} \to 0.4$. Ideally, one would plot the 90% ellipsoid using translucent graphics. Unfortunately, *Mathematica* Version 4 does not support translucent graphics, so we use a `WireFrame` instead.

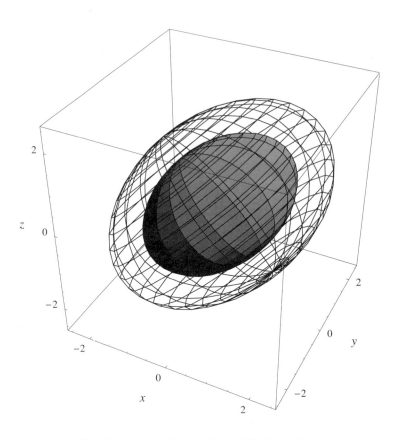

Fig. 19: Quantiles: 60% (solid) and 90% (wireframe)

⊕ *Example 22:* Correlation and Positive Definite Matrix

Let X, Y and Z follow a standardised trivariate Normal distribution. It is known that $\rho_{xy} = 0.9$ and $\rho_{xz} = -0.8$, but ρ_{yz} is not known. What can we say, if anything, about the correlation ρ_{yz}?

Solution: Although there is not enough information to uniquely determine the value of ρ_{yz}, there *is* enough information to specify a range of values for it (of course, $-1 < \rho_{yz} < 1$ must always hold). This is achieved by using the property that Σ must be a positive definite matrix, which implies that the determinant of Σ must be positive:

```
dd = Det[Σ] /. {ρxy → .9, ρxz → -.8}
```

$$-0.45 - 1.44\, \rho_{yz} - \rho_{yz}^2$$

This expression is positive when ρ_{yz} lies in the following interval:

```
<< Algebra`
InequalitySolve[dd > 0, ρyz]
```

$$-0.981534 < \rho_{yz} < -0.458466$$

6.4 C CDF, Probability Calculations and Numerics

While it is generally straightforward to find numerical values for any multivariate Normal pdf, it is not quite as easy to do so for the cdf. To illustrate, we use the trivariate Normal pdf $g(x, y, z)$ = PDF[dist3, {x, y, z}] defined at the start of §6.4 B. We distinguish between two possible scenarios: (i) zero correlation, and (ii) non-zero correlation.

○ *Zero Correlation*

Under zero correlation, it is possible to find an exact *symbolic* solution using **mathStatica** in the usual way.[3,4] Let $G(x, y, z)$ denote the cdf $P(X \le x, Y \le y, Z \le z)$ under zero correlation:

```
Clear[G];    G[x_, y_, z_] = Prob[{x, y, z}, g /. ρ_ → 0]
```

$$\frac{1}{8} \left(1 + \text{Erf}\left[\frac{x}{\sqrt{2}}\right]\right) \left(1 + \text{Erf}\left[\frac{y}{\sqrt{2}}\right]\right) \left(1 + \text{Erf}\left[\frac{z}{\sqrt{2}}\right]\right)$$

This solution is virtuous in two respects: first, it is an exact symbolic expression; second, because the solution is already 'evaluated', it will be computationally efficient in application. Here, for instance, is the exact symbolic solution to $P(X \le -2, Y \le 0, Z \le 2)$:

```
G[-2, 0, 2]
```

$$\frac{1}{8} \left(1 - \text{Erf}\left[\sqrt{2}\right]\right) \left(1 + \text{Erf}\left[\sqrt{2}\right]\right)$$

Because the solution is an exact symbolic expression, we can use *Mathematica*'s arbitrary precision numerical engine to express it as a numerical expression, to any desired number of digits of precision. Here is G[-2,0,2] calculated to 40 digits of precision:

```
N[G[-2, 0, 2], 40]
```

0.01111628172225982147533684086722435761304

If we require the probability content of a region within the domain, we could just type in the whole integral. For instance, the probability of being within the region

$$S = \{(x, y, z): \ 1 < x < 2, \quad 3 < y < 4, \quad 5 < z < 6\}$$

is given by:

```
Integrate[g /. ρ_ → 0,
    {x, 1, 2}, {y, 3, 4}, {z, 5, 6}] // N // Timing
```

{0.27 Second, 5.1178×10^{-11}}

Alternatively, we can use the **mathStatica** function MrSpeedy (*Example 4*). MrSpeedy finds the probability content of a region within the domain just by using the known cdf G[] (which we have already found) and the boundaries of the region, without any need for further integration:

```
S = {{1, 2}, {3, 4}, {5, 6}}; MrSpeedy[G, S] // N // Timing
```

{0. Second, 5.1178×10^{-11}}

`MrSpeedy` often provides enormous speed increases over direct integration.

○ ***Non-Zero Correlation***

In the case of non-zero correlation, a closed form solution to the cdf does not generally exist, so that numerical integration is required. Even if we use the CDF function in *Mathematica*'s `Multinormal` statistics package, ultimately, in the background, we are still resorting to numerical integration. This, in turn, raises the two interrelated motifs of accuracy and computational efficiency, which run throughout this section.

Consider, again, the trivariate Normal pdf $g(x, y, z) = $ `PDF[dist3, {x, y, z}]` defined in §6.4 B. If $\rho_{xy} = \rho_{xz} = \rho_{yz} = \frac{1}{2}$, the cdf is:

```
Clear[G];   G[var__] := CDF[dist3 /. ρ_ → 1/2, {var}]
```

Hence, $P(X \le 1, Y \le -7, Z \le 3)$ evaluates to:[5]

```
G[1, -7, 3]
```

1.27981×10^{-12}

If we require the probability content of a region within the domain, we can again use `MrSpeedy`. The probability of being within the region

$$S = \{(x, y, z): 1 < x < \infty, \quad -3 < y < 4, \quad 5 < z < 6\}$$

is then given by:

```
S = {{1, ∞}, {-3, 4}, {5, 6}}; MrSpeedy[G, S] // Timing
```

{0.55 Second, 2.61015×10^{-7}}

This is a significant improvement over using numerical integration directly, since the latter is both less accurate (at default settings) and *far more resource hungry*.

```
NIntegrate[g /. ρ_ → 1/2,
    {x, 1, ∞}, {y, -3, 4}, {z, 5, 6}] // Timing
```

— NIntegrate::slwcon :
    Numerical integration converging too slowly; suspect one
      of the following: singularity, value of the integration
      being 0, oscillatory integrand, or insufficient
      WorkingPrecision. If your integrand is oscillatory
      try using the option Method->Oscillatory in NIntegrate.

{77.39 Second, 2.61013×10^{-7}}

The direct numerical integration approach can be 'sped up' by sacrificing some accuracy. This can be done by altering the `PrecisionGoal` option; see Rose and Smith (1996a or 1996b). This can be useful when working with a distribution whose cdf is not known (or cannot be derived), such that one has no alternative but to use direct numerical integration.

Finally, it is worth stressing that since the `CDF` function in *Mathematica*'s `Multinormal` statistics package is using numerical integration in the background, the numerical answer that is printed on screen is not exact. Rather, the answer will be correct to several decimal places, and incorrect beyond that; only symbolic entities are exact. To assess the accuracy of the `CDF` function, we can compare the answer it gives with symbolic solutions that are known for special cases. For example, Stuart and Ord (1994, Section 15.10) report symbolic solutions for the standardised bivariate Normal orthant probability $P(X \leq 0, Y \leq 0)$ as:

$$\text{P2} = \frac{1}{4} + \frac{\text{ArcSin}[\rho]}{2\pi};$$

while the standardised trivariate Normal orthant probability $P(X \leq 0, Y \leq 0, Z \leq 0)$ is:

$$\text{P3} = \frac{1}{8} + \frac{1}{4\pi} \left(\text{ArcSin}[\rho_{xy}] + \text{ArcSin}[\rho_{xz}] + \text{ArcSin}[\rho_{yz}] \right);$$

We choose some values for ρ_{xy}, ρ_{xz}, ρ_{yz}:

$$\text{lis} = \left\{ \rho_{xy} \to \frac{1}{17},\ \rho_{xz} \to \frac{1}{12},\ \rho_{yz} \to \frac{2}{5} \right\};$$

Because `P3` is a symbolic entity, we can express it numerically to any desired precision. Here is the correct answer to 30 digits of precision:

N[P3 /. lis, 30]

0.169070356956715121611195785538

By contrast, the `CDF` function yields:

CDF[dist3 /. lis, {0, 0, 0}] // InputForm

0.1690703504574683

In this instance, the `CDF` function has only 8 digits of precision. In other cases, it may offer 12 digits of precision. Even so, 8 digits of precision is better than most competing packages. For more detail on numerical precision in *Mathematica*, see Appendix A.1.

In summary, *Mathematica*'s `CDF` function and **mathStatica**'s `MrSpeedy` function make an excellent team; together, they are more accurate and faster than using numerical integration directly. How then does *Mathematica* compare with highly specialised multivariate Normal computer programs (see Schervish (1984)) such as Bohrer–Schervish, MULNOR, and MVNORM? For zero-correlation, *Mathematica* can easily outperform such programs in both accuracy and speed, due to its symbolic engine. For non-zero correlation, *Mathematica* performs well on accuracy grounds.

6.4 D   Random Number Generation for the Multivariate Normal

○ *Introducing* MVNRandom

The **mathStatica** function MVNRandom[n, $\vec{\mu}$, Σ] generates n pseudo-random m-dimensional drawings from the multivariate Normal distribution with mean vector $\vec{\mu}$, and ($m \times m$) variance-covariance matrix Σ; the function assumes dimension m is an integer larger than 1. Once again, Σ is required to be symmetric and positive definite. The function has been optimised for speed. To demonstrate its application, we generate 6 drawings from a trivariate Normal with mean vector and variance-covariance matrix given by:

$$\vec{\mu} = \{10, 0, -20\}; \quad \Sigma = \begin{pmatrix} 1 & 0.2 & 0.4 \\ 0.2 & 2 & 0.3 \\ 0.4 & 0.3 & 3 \end{pmatrix}; \quad \text{MVNRandom}[6, \vec{\mu}, \Sigma]$$

$$\begin{pmatrix} 10.1802 & 0.792264 & -20.7549 \\ 9.61446 & 0.936577 & -20.3007 \\ 9.00878 & 1.51215 & -17.9076 \\ 10.0042 & -0.749123 & -23.6165 \\ 12.2513 & -1.28886 & -19.8166 \\ 10.7216 & -0.626802 & -15.847 \end{pmatrix}$$

The output from MVNRandom is a set of n lists (here $n = 6$). Each list represents a single pseudo-random drawing from the distribution and so has the dimension of the random variable ($m = 3$). In this way, MVNRandom has recorded 6 pseudo-random drawings from the 3-dimensional $N(\vec{\mu}, \Sigma)$ distribution.

Instead of using **mathStatica**'s MVNRandom function, one can alternatively use the RandomArray function in *Mathematica*'s Multinormal Statistics package. To demonstrate, we generate 20000 drawings using both approaches:

MVNRandom[20000, $\vec{\mu}$, Σ]; // Timing

{0.22 Second, Null}

**RandomArray[
    MultinormalDistribution[$\vec{\mu}$, Σ], 20000] ; // Timing**

{2.53 Second, Null}

In addition to its obvious efficiency, MVNRandom has other advantages. For instance, it advises the user if the variance-covariance matrix is not symmetric and/or if it is not positive definite.

○ *How* MVNRandom *Works*

MVNRandom works by transforming a pseudo-random drawing from an m-dimensional $N(\vec{0}, I_m)$ distribution into a $N(\vec{\mu}, \Sigma)$ drawing: the transformation is essentially the multivariate equivalent of a location shift plus a scale change. The transformation relies upon the spectral decomposition (using Eigensystem) of the variance-covariance matrix; that is, the decomposition of $\Sigma = HDH^T$ into its spectral components H and D. The columns of the $(m \times m)$ matrix H are the eigenvectors of Σ, and the $(m \times m)$ diagonal matrix D contains the eigenvalues of Σ. Then, for a random vector $\vec{Y} \sim N(\vec{0}, I_m)$, a linear transformation from $\vec{Y}$ to a new random vector $\vec{X}$, according to the rule

$$\vec{X} = \vec{\mu} + HD^{1/2} \vec{Y} \qquad (6.31)$$

finds $\vec{X} \sim N(\vec{\mu}, \Sigma)$. By examining the mean vector and variance-covariance matrix, it is easy to see why this transformation works:

$$E[\vec{X}] = E\left[\vec{\mu} + HD^{1/2} \vec{Y}\right] = \vec{\mu}, \quad \text{because } E[\vec{Y}] = \vec{0}$$

and

$$\begin{aligned}
\text{Varcov}(\vec{X}) &= \text{Varcov}\left(\vec{\mu} + HD^{1/2} \vec{Y}\right) \\
&= \text{Varcov}\left(HD^{1/2} \vec{Y}\right) \\
&= HD^{1/2} \text{Varcov}(\vec{Y}) D^{1/2} H^T \\
&= HDH^T \\
&= \Sigma
\end{aligned}$$

because $\text{Varcov}(\vec{Y}) = I_m$. We wish to sample the distribution of $\vec{X}$, which requires that we generate a pseudo-random drawing of $\vec{Y}$ and apply (6.31) to it. So, all that remains is to do the very first step—generate $\vec{Y}$—but that is the easiest bit! Since the components of $\vec{Y}$ are independent, it suffices to combine together m pseudo-random drawings from the univariate standard Normal distribution $N(0, 1)$ into a single column.

○ *Visualising Random Data in 2D and 3D Space*

With *Mathematica*, we can easily visualise random data that has been generated in two or three dimensions. We will use the functions D2 and D3 to plot the data in two-dimensional and three-dimensional space, respectively:

```
D2[x_] := ListPlot[x, PlotStyle → Hue[1],
    AspectRatio → 1, DisplayFunction → Identity];

D3[x_] :=
  Graphics3D[ {Hue[1], Map[Point, x]}, Axes → True]
```

Not only can we plot the data in its appropriate space, but we can also view the data projected onto a hypersphere; for example, two-dimensional data can be projected onto a circle, while three-dimensional data can be projected onto a sphere. This is achieved by normalising the data by using the norm function defined below. Finally, the function MVNPlot provides a neat way of generating our desired diagrams:

```
norm[x_] := Map[ #/√#.# &, x];

MVNPlot[DD_, w_] := Show[GraphicsArray[
    {DD[w], DD[norm[w]]}, GraphicsSpacing → .3]];
```

The Two-Dimensional Case

(i) *Zero correlation:* Fig. 20 shows two plots: the left panel illustrates the generated data in two-dimensional space; the right panel projects this data onto the unit circle. A random vector $\vec{X}$ is said to be *spherically distributed* if its pdf is equivalent to that of $\vec{Y} = H\vec{X}$, for all orthogonal matrices H. Spherically distributed random variables have the property that they are uniformly distributed on the unit circle / sphere / hypersphere. The zero correlation bivariate Normal is a member of the spherical class.[6] This explains why the generated data appears uniform on the circle.

$\vec{\mu} = \{0, 0\};$ $\Sigma = \begin{pmatrix} 1 & 0 \\ 0 & 1 \end{pmatrix};$ w = MVNRandom[1500, $\vec{\mu}$, Σ];
MVNPlot[D2, w];

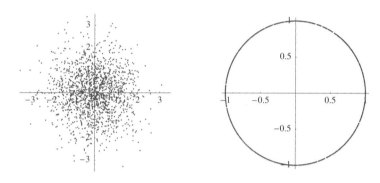

Fig. 20: Zero correlation bivariate Normal: random data

(ii) *Non-zero correlation:* Fig. 21 again shows two plots, but now in the case of non-zero correlation. The left panel shows that the data has high positive correlation. The right panel shows that the distribution is no longer uniform on the unit circle, for there are relatively few points projected onto it in the north-west and south-east quadrants. This is because the correlated bivariate Normal does not belong to the spherical class; instead, it belongs to the elliptical class of distributions. For further details on elliptical distributions, see Muirhead (1982).

$\vec{\mu} = \{0, 0\};$ $\Sigma = \begin{pmatrix} 1 & .95 \\ .95 & 1 \end{pmatrix};$ w = MVNRandom[1500, $\vec{\mu}$, Σ];
MVNPlot[D2, w];

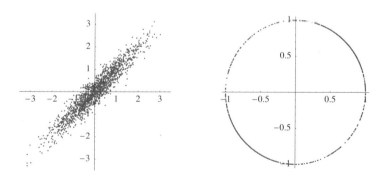

Fig. 21: Correlated bivariate Normal: random data

The Three-Dimensional Case

(i) *Zero correlation:* Fig. 22 again shows two plots. The left panel illustrates the generated data in three-dimensional space. The right panel projects this data onto the unit sphere. The distribution appears uniform on the sphere, as indeed it should, because this particular trivariate Normal is a member of the spherical class.

$$\vec{\mu} = \{0, 0, 0\}; \quad \Sigma = \begin{pmatrix} 1 & 0 & 0 \\ 0 & 1 & 0 \\ 0 & 0 & 1 \end{pmatrix}; \quad w = \text{MVNRandom}[2000, \vec{\mu}, \Sigma];$$

MVNPlot[D3, w];

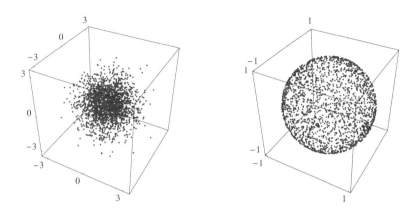

Fig. 22: Zero correlation trivariate Normal: random data

(ii) *Non-zero correlation:* see Fig. 23 below. The three-dimensional plot on the left illustrates that the data is now highly correlated, while the projection onto the unit sphere (on the right) provides ample evidence that this particular trivariate Normal distribution is no longer spherical.

$$\vec{\mu} = \{0, 0, 0\}; \quad \Sigma = \begin{pmatrix} 1 & .95 & .95 \\ .95 & 1 & .95 \\ .95 & .95 & 1 \end{pmatrix}; \quad w = \text{MVNRandom}[2000, \vec{\mu}, \Sigma];$$

MVNPlot[D3, w];

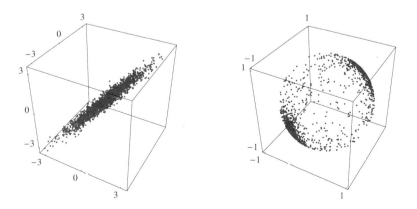

Fig. 23: Correlated trivariate Normal: random data

6.5 The Multivariate *t* and Multivariate Cauchy

Let $(X_1, \ldots, X_m)$ have a joint *standardised* multivariate Normal distribution with correlation matrix R, and let $Y \sim$ Chi-squared(v) be independent of $(X_1, \ldots, X_m)$. Then the joint pdf of

$$T_j = \frac{X_j}{\sqrt{Y/v}}, \qquad (j = 1, \ldots, m) \qquad (6.32)$$

defines the multivariate *t* distribution with v degrees of freedom and correlation matrix R, denoted $t(R, v)$. The multivariate Cauchy distribution is obtained when $R = I_m$ and $v = 1$. The multivariate *t* is included in *Mathematica*'s Multinormal Statistics package, so our discussion here will be brief. First, we ensure the appropriate package is loaded:

```
<< Statistics`
```

Let random variables W_1 and W_2 have joint pdf $t(R, v)$ where $R = \begin{pmatrix} 1 & \rho \\ \rho & 1 \end{pmatrix}$, and ρ denotes the correlation coefficient between W_1 and W_2. So:

```
w⃗ = {w₁, w₂};  R = (1  ρ
                    ρ  1);  cond = {-1 < ρ < 1, v > 0};

dist2 = MultivariateTDistribution[R, v];
```

Then our bivariate *t* pdf $f(w_1, w_2)$ is given by:

```
f = FullSimplify[PDF[dist2, w⃗], cond]
```

$$\frac{v^{\frac{2+v}{2}} (1 - \rho^2)^{\frac{1+v}{2}} (v - v\rho^2 + w_1^2 - 2\rho w_1 w_2 + w_2^2)^{-1-\frac{v}{2}}}{2\pi}$$

with domain of support:

```
domain[f] = Thread[{w⃗, -∞, ∞}] && cond
```

$\{\{w_1, -\infty, \infty\}, \{w_2, -\infty, \infty\}\} \&\& \{-1 < \rho < 1, v > 0\}$

Example 23 below derives this pdf from first principles. The shape of the contours of $f(w_1, w_2)$ depend on ρ. We can plot the specific ellipse that encloses $q\%$ of the distribution by using the function EllipsoidQuantile[*dist, q*]. This is illustrated in Fig. 24 which plots the ellipses that enclose 15% (bold), 90% (dashed) and 99% (plain) of the distribution, respectively, with $\rho = 0.4$ and $v = 2$ degrees of freedom. The long-tailed nature of the *t* distribution is apparent, especially when this diagram is compared with Fig. 13.

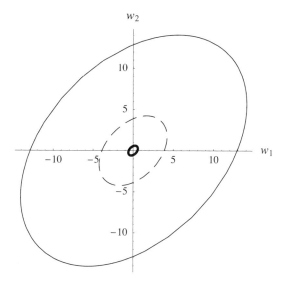

Fig. 24: Quantiles: 15% (bold), 90% (dashed) and 99% (plain)

The bivariate Cauchy distribution is obtained when $R = I_2$ and $v = 1$:

f /. {ρ → 0, v → 1}

$$\frac{1}{2\pi (1 + w_1^2 + w_2^2)^{3/2}}$$

Under these conditions, the marginal distribution of W_1 is the familiar (univariate) Cauchy distribution:

Marginal[w₁, f /. {ρ → 0, v → 1}]

$$\frac{1}{\pi + \pi w_1^2}$$

As in §6.4 C, one can use functions like MrSpeedy in conjunction with *Mathematica*'s CDF function to find probabilities, and RandomArray to generate pseudo-random drawings.

⊕ *Example 23:* Deriving the pdf of the Bivariate *t*

Find the joint pdf of:

$$T_j = \frac{X_j}{\sqrt{Y/v}}, \qquad (j = 1, 2)$$

from first principles, where (X_1, X_2) have a joint *standardised* multivariate Normal distribution, and $Y \sim$ Chi-squared(v) is independent of (X_1, X_2).

Solution: Due to independence, the joint pdf of (X_1, X_2, Y), say $\varphi(x_1, x_2, y)$, is just the pdf of (X_1, X_2) multiplied by the pdf of Y:

$$\varphi = \left(\frac{e^{\frac{x_1^2 - 2\rho x_1 x_2 + x_2^2}{-2 + 2\rho^2}}}{2\pi \sqrt{1-\rho^2}} \right) * \left(\frac{e^{-\frac{y}{2}} y^{\frac{v}{2}-1}}{2^{v/2} \Gamma[\frac{v}{2}]} \right);$$

```
cond = {v > 0, -1 < ρ < 1};
domain[φ] = {{x₁, -∞, ∞}, {x₂, -∞, ∞}, {y, 0, ∞}} && cond;
```

Let $U = Y$. Then, using **mathStatica**'s `Transform` function, the joint pdf of (T_1, T_2, U) is:

$$f = \text{Transform}\left[\left\{ t_1 == \frac{x_1}{\sqrt{y/v}}, \; t_2 == \frac{x_2}{\sqrt{y/v}}, \; u == y \right\}, \varphi \right]$$

$$\frac{2^{-1-\frac{v}{2}} e^{\frac{u(v - v\rho^2 + t_1^2 - 2\rho t_1 t_2 + t_2^2)}{2v(-1+\rho^2)}} u^{v/2}}{\pi v \sqrt{1-\rho^2} \; \Gamma[\frac{v}{2}]}$$

with domain:

```
domain[f] = {{t₁, -∞, ∞}, {t₂, -∞, ∞}, {u, 0, ∞}} && cond;
```

Then, the marginal joint pdf of random variables T_1 and T_2 is:

```
Marginal[{t₁, t₂}, f]
```

$$\frac{v\sqrt{1-\rho^2} \left(\frac{v - v\rho^2 + t_1^2 - 2\rho t_1 t_2 + t_2^2}{v - v\rho^2} \right)^{-1-\frac{v}{2}}}{2\pi (v - v\rho^2)}$$

which is the desired pdf. Note that this output is identical to the answer given to `PDF[dist2, {t₁, t₂}] // FullSimplify`. ∎

6.6 Multinomial and Bivariate Poisson

This section discusses two discrete multivariate distributions, namely the Multinomial and the bivariate Poisson. Both of these distributions are also discussed in *Mathematica*'s `Statistics`MultiDiscreteDistributions`` package.

6.6 A The Multinomial Distribution

The Binomial distribution was discussed in Chapter 3. Here, we present it in its degenerate form: consider an experiment with n independent trials, with two mutually exclusive outcomes per trial ($\mathbb{E}_1$ or $\mathbb{E}_2$). Let p_i ($i = 1, 2$) denote the probability of outcome $\mathbb{E}_i$ (subject to $p_1 + p_2 = 1$, and $0 \le p_i \le 1$), with p_i remaining the same from trial to trial. Let the 'random variables' of interest be X_1 and X_2, where X_i is the number of trials in which outcome $\mathbb{E}_i$ occurs ($x_1 + x_2 = n$). The joint pmf of X_1 and X_2 is

$$f(x_1, x_2) = P(X_1 = x_1, X_2 = x_2) = \frac{n!}{x_1! \, x_2!} \, p_1^{x_1} \, p_2^{x_2}, \qquad x_i \in \{0, 1, \ldots, n\}. \tag{6.33}$$

Since $X_1 + X_2 = n$, one of these 'random variables' is of course degenerate, so that the Binomial is actually a univariate distribution, as in Chapter 3. This framework can easily be generalised into a *Trinomial* distribution, where instead of having just two possible outcomes, we now have three ($\mathbb{E}_1, \mathbb{E}_2$ or $\mathbb{E}_3$), subject to $p_1 + p_2 + p_3 = 1$:

$$f(x_1, x_2, x_3) = P(X_1 = x_1, X_2 = x_2, X_3 = x_3) = \frac{n!}{x_1! \, x_2! \, x_3!} \, p_1^{x_1} \, p_2^{x_2} \, p_3^{x_3}. \tag{6.34}$$

More generally, the *m*-variate *Multinomial* distribution has pmf

$$f(x_1, \ldots, x_m) = P(X_1 = x_1, \ldots, X_m = x_m) = \frac{n!}{x_1! \cdots x_m!} \, p_1^{x_1} \cdots p_m^{x_m} \tag{6.35}$$

$$\text{subject to } \sum_{i=1}^{m} p_i = 1, \text{ and } \sum_{i=1}^{m} x_i = n.$$

Since $\sum_{i=1}^{m} x_i = n$, it follows, for example, that $x_m = n - \sum_{i=1}^{m-1} x_i$. This implies that, given n, the m-variate multinomial can be fully described using only $m - 1$ variables; see also Johnson et al. (1997).[7] We enter (6.35) into *Mathematica* as:

```
Clear[f];    f[X_List, p_List, n_] := n! ∏(i=1 to Length[X]) p[[i]]^X[[i]] / X[[i]]!
```

The multinomial moment generating function is derived in *Example 26* below, where we show that

$$M(\vec{t}) = \left(\sum_{i=1}^{m} p_i \, e^{t_i} \right)^n. \tag{6.36}$$

⊕ **Example 24:** Age Profile

Table 5 gives the age profile of people living in Australia (Australian Bureau of Statistics, 1996 Census). The data is divided into five age classes.

class	age	proportion
I	0–14	21.6 %
II	15–24	14.5 %
III	25–44	30,8 %
IV	45–64	21.0 %
V	65+	12.1 %

Table 5: Age profile of people living in Australia

Let $\vec{p}$ denote the probability vector $(p_1, p_2, p_3, p_4, p_5)$:

$\vec{p}$ = {0.216, 0.145, 0.308, 0.210, 0.121};

(a) If we randomly select 10 people from the population, what is the probability they all come from Class I?

Solution:

$\vec{x}$ = {10, 0, 0, 0, 0};   f[$\vec{x}$, $\vec{p}$, 10]

2.21074×10^{-7}

(b) If we again randomly select 10 people, what is the probability that 3 people will be from Class I, 1 person from Class II, 2 from Class III, 4 from Class IV, and 0 from Class V?

Solution:

$\vec{x}$ = {3, 1, 2, 4, 0};   f[$\vec{x}$, $\vec{p}$, 10]

0.00339687

(c) If we again randomly select 10 people, what is the probability that Class III will contain exactly 1 person?

Solution: If Class III contains 1 person, then the remaining classes must contain 9 people. Thus, we need to calculate every possible way of splitting 9 people over the remaining four classes, then calculate the probability for each case, and then add it all up. The composition of 9 into 4 parts can be obtained using the Compositions function in the DiscreteMath`Combinatorica` package, which we load as follows:

   << DiscreteMath`

Here are the compositions of 9 into 4 parts. The list is very long, so we just display the first few compositions:

   lis = Compositions[9, 4];   lis // Shallow

   {{0, 0, 0, 9}, {0, 0, 1, 8}, {0, 0, 2, 7},
    {0, 0, 3, 6}, {0, 0, 4, 5}, {0, 0, 5, 4}, {0, 0, 6, 3},
    {0, 0, 7, 2}, {0, 0, 8, 1}, {0, 0, 9, 0}, <<210>> }

Since Class III must contain 1 person in our example, we need to insert a '1' at position 3 of each of these lists, so that, for instance, {0, 0, 0, 9} becomes {0, 0, 1, 0, 9}:

   lis2 = Map[Insert[#, 1, 3] &, lis];   lis2 // Shallow

   {{0, 0, 1, 0, 9}, {0, 0, 1, 1, 8},
    {0, 0, 1, 2, 7}, {0, 0, 1, 3, 6}, {0, 0, 1, 4, 5},
    {0, 0, 1, 5, 4}, {0, 0, 1, 6, 3}, {0, 0, 1, 7, 2},
    {0, 0, 1, 8, 1}, {0, 0, 1, 9, 0}, <<210>> }

We can now compute the pmf at each of these cases, and add them all up:

```
Plus @@ Map[f[#, p⃗, 10] &, lis2]
```

0.112074

So, the probability that a random sample of 10 Australians will contain exactly 1 person aged 25–44 is 11.2%. For the 15–24 age group, this probability rises to 35.4%.

An alternative (more automated, but less flexible) approach to solving (c) is to use the summation operator, taking great care to ensure that the summation iterators satisfy the constraint $\sum_{i=1}^{5} x_i = 10$. So, if Class III is fixed at $x_3 = 1$, then x_1 can take values from 0 to 9; x_2 may take values from 0 to $(9-x_1)$; and x_4 may take values from 0 to $(9-x_1-x_2)$. That leaves x_5 which is degenerate: that is, given x_1, x_2, $x_3 = 1$, and x_4, we know that x_5 must equal $9 - x_1 - x_2 - x_4$. Then the required probability is:

```
Sum[f[{x₁, x₂, 1, x₄, x₅}, p⃗, 10],
    {x₁, 0, 9},
    {x₂, 0, 9 - x₁},
    {x₄, 0, 9 - x₁ - x₂},
    {x₅, 9 - x₁ - x₂ - x₄, 9 - x₁ - x₂ - x₄}]
```

0.112074

Example 26 provides another illustration of this summation approach. ∎

⊕ **Example 25:** Working with the mgf

In the case of the Trinomial, the mgf is:

$$\text{mgf} = \left(\sum_{i=1}^{3} p_i\, e^{t_i}\right)^n$$

$(e^{t_1} p_1 + e^{t_2} p_2 + e^{t_3} p_3)^n$

The product raw moments $E[X_1^a X_2^b X_3^c]$ can now be obtained from the mgf in the usual fashion. To keep things neat, we write a little *Mathematica* function `Moment[a, b, c]` function to calculate $E[X_1^a X_2^b X_3^c]$ from the mgf, now noting that $\sum_{i=1}^{m} p_i = 1$:

```
Moment[a_, b_, c_] :=
    D[mgf, {t₁, a}, {t₂, b}, {t₃, c}]  /. t_ → 0  /. ∑ᵢ₌₁³ pᵢ → 1
```

The moments are now easy to obtain. Here is the first moment of X_2, namely $\mu'_{0,1,0}$.

```
Moment[0, 1, 0]
```

n p₂

Here is the second moment of X_2, namely $\acute{\mu}_{0,2,0}$:

Moment[0, 2, 0]

$n\, p_2 + (-1+n)\, n\, p_2^2$

By symmetry, we then have the more general result that $E[X_i] = n\, p_i$ and $E[X_i^2] = n\, p_i + (n-1)\, n\, p_i^2$. Here is the product raw moment $E[X_1^2\, X_2\, X_3] = \acute{\mu}_{2,1,1}$:

Moment[2, 1, 1] // Simplify

$(-2+n)\,(-1+n)\, n\, p_1\,(1 + (-3+n)\, p_1)\, p_2\, p_3$

The covariance between X_1 and X_3 is given by $\mu_{1,0,1}$, which can be expressed in raw moments as:

cov = CentralToRaw[{1, 0, 1}]

$\mu_{1,0,1} \rightarrow -\acute{\mu}_{0,0,1}\, \acute{\mu}_{1,0,0} + \acute{\mu}_{1,0,1}$

Evaluating each $\acute{\mu}$ term with the Moment function then yields this covariance:

cov /. $\acute{\mu}_{r\_}$:→ Moment[r] // Simplify

$\mu_{1,0,1} \rightarrow -n\, p_1\, p_3$

Similarly, the product cumulant $\kappa_{3,1,2}$ is given by:

CumulantToRaw[{3, 1, 2}] /. $\acute{\mu}_{x\_}$:→ Moment[x] // Simplify

$\kappa_{3,1,2} \rightarrow 2\, n\, p_1\, p_2\, p_3\,(1 + p_1^2\,(12 - 60\, p_3) - 3\, p_3 + 9\, p_1\,(-1 + 4\, p_3))$

⊕ *Example 26:* Deriving the Multinomial mgf

Consider a model with $m = 4$ classes. The pmf is:

$\vec{x} = \{x_1,\, x_2,\, x_3,\, x_4\};$
$\vec{p} = \{p_1,\, p_2,\, p_3,\, p_4\};$
$\text{pmf} = f[\vec{x},\, \vec{p},\, n]$

$$\frac{n!\, p_1^{x_1}\, p_2^{x_2}\, p_3^{x_3}\, p_4^{x_4}}{x_1!\, x_2!\, x_3!\, x_4!}$$

Recall that the moment generating function for a discrete distribution is:

$$E\left[e^{\vec{t}\cdot\vec{x}}\right] = \sum_{x_1}\cdots\sum_{x_m} \exp\left(\sum_{i=1}^{m} t_i\, x_i\right) f(x_1, \ldots, x_m).$$

Some care must be taken here to ensure the summation iterators satisfy the constraint $\sum_{i=1}^{m} x_i = n$; thus, if we let x_1 take values from 0 to n, then x_2 may take values from 0 to $n - x_1$, and then x_3 may take values from 0 to $n - x_1 - x_2$. That leaves x_4 which is degenerate; that is, given x_1, x_2 and x_3, we know that x_4 must be equal to $n - x_1 - x_2 - x_3$. Then the mgf is:

```
t⃗ = {t₁, t₂, t₃, t₄};

mgf = FullSimplify[
         n   n-x₁  n-x₁-x₂    n-x₁-x₂-x₃
         ∑    ∑      ∑           ∑          Evaluate[eᵗ⃗·x⃗ pmf],
        x₁=0 x₂=0  x₃=0     x₄=n-x₁-x₂-x₃

         n ∈ Integers] // PowerExpand

(eᵗ¹ p₁ + eᵗ² p₂ + eᵗ³ p₃ + eᵗ⁴ p₄)ⁿ
```

It follows by symmetry that the general solution is $M(\vec{t}) = (\sum_{i=1}^{m} p_i e^{t_i})^n$, where $\sum_{i=1}^{m} p_i = 1$. ∎

6.6 B The Bivariate Poisson

```
Clear[g]
```

Let Y_0, Y_1 and Y_2 be mutually stochastically independent Poisson random variables, with non-negative parameters λ_0, λ_1 and λ_2, respectively, and pmf's $g_i(y_i)$ for $i \in \{0, 1, 2\}$:

$$g_{i\_} = \frac{e^{-\lambda_i} \lambda_i^{y_i}}{y_i!};$$

defined on $y_i \in \{0, 1, 2, ...\}$. Due to independence, the joint pmf of (Y_0, Y_1, Y_2) is:

```
g = g₀ g₁ g₂
```

$$\frac{e^{-\lambda_0 - \lambda_1 - \lambda_2} \lambda_0^{y_0} \lambda_1^{y_1} \lambda_2^{y_2}}{y_0! \, y_1! \, y_2!}$$

with domain:

```
domain[g] = {{y₀, 0, ∞}, {y₁, 0, ∞}, {y₂, 0, ∞}}
             && {λ₀ > 0, λ₁ > 0, λ₂ > 0} && {Discrete};
```

A non-trivial *bivariate Poisson* distribution is the joint distribution of X_1 and X_2 where

$$X_1 = Y_1 + Y_0 \quad \text{and} \quad X_2 = Y_2 + Y_0. \tag{6.37}$$

○ **Probability Mass Function**

We shall consider four approaches for deriving the joint pmf of X_1 and X_2, namely: (i) the transformation method, (ii) the probability generating function (pgf) approach, (iii) limiting forms, and (iv) *Mathematica*'s Statistics package.

(i) *Transformation method*
We wish to find the joint pmf of X_1 and X_2, as defined in (6.37). Let $X_0 = Y_0$ so that the number of new variables X_i is equal to the number of old variables Y_i. Then, the desired transformation here is:

```
eqn = {x₁ == y₁ + y₀, x₂ == y₂ + y₀, x₀ == y₀};
```

Then, the joint pmf of (X_0, X_1, X_2), say $\psi(x_0, x_1, x_2)$, is:

```
ψ = Transform[eqn, g]
```

$$\frac{e^{-\lambda_0 - \lambda_1 - \lambda_2} \, \lambda_0^{x_0} \, \lambda_1^{-x_0 + x_1} \, \lambda_2^{-x_0 + x_2}}{x_0! \, (-x_0 + x_1)! \, (-x_0 + x_2)!}$$

We desire the joint marginal pmf of X_1 and X_2, so we now need to 'sum out' X_0. Since Y_1 is non-negative, it follows that $X_0 \leq X_1$:

```
pmf = ∑_{x₀=0}^{x₁} Evaluate[ψ]
```

$$\left(e^{-\lambda_0 - \lambda_1 - \lambda_2} \, \text{HypergeometricU}\left[-x_1, \, 1 - x_1 + x_2, \, -\frac{\lambda_1 \lambda_2}{\lambda_0}\right] \lambda_1^{x_1} \lambda_2^{x_2} \left(-\frac{\lambda_1 \lambda_2}{\lambda_0}\right)^{-x_1} \right) \Big/ (\Gamma[1+x_1] \, \Gamma[1+x_2])$$

Mathematica, ever the show-off, has found the pmf in terms of the confluent hypergeometric function. Here, for instance, is $P(X_1 = 3, X_2 = 2)$:

```
pmf /. {x₁ → 3, x₂ → 2} // Simplify
```

$$\frac{1}{12} \, e^{-\lambda_0 - \lambda_1 - \lambda_2} \, \lambda_1 \, (6 \lambda_0^2 + 6 \lambda_0 \lambda_1 \lambda_2 + \lambda_1^2 \lambda_2^2)$$

(ii) *Probability generating function approach*
By (6.21), the joint pgf is $E\left[t_1^{X_1} \, t_2^{X_2} \, \cdots \, t_m^{X_m}\right]$:

```
pgf = ∑_{y₀=0}^{∞} ∑_{y₁=0}^{∞} ∑_{y₂=0}^{∞} Evaluate[t₁^{y₁+y₀} t₂^{y₂+y₀} g]
```

$$e^{-\lambda_0 + t_1 t_2 \lambda_0 - \lambda_1 + t_1 \lambda_1 - \lambda_2 + t_2 \lambda_2}$$

The pgf, in turn, determines the probabilities by (6.22). Then, $P(X_1 = r, X_2 = s)$ is:

```
Clear[P];
P[r_, s_] := D[pgf, {t₁, r}, {t₂, s}] / (r! s!) /. {t_ → 0} // Simplify
```

For instance, $P(X_1 = 3, X_2 = 2)$ is:

```
P[3, 2]
```

$$\frac{1}{12} e^{-\lambda_0 - \lambda_1 - \lambda_2} \lambda_1 (6 \lambda_0^2 + 6 \lambda_0 \lambda_1 \lambda_2 + \lambda_1^2 \lambda_2^2)$$

as per our earlier result.

(iii) *Limiting forms*

Just as the univariate Poisson can be obtained as a limiting form of the Binomial, the bivariate Poisson can similarly be obtained as a limiting form of the Multinomial. Hamdan and Al-Bayyati (1969) discuss this approach, while Johnson *et al.* (1997, p. 125) provide an overview.

(iv) *Mathematica's statistics package*

The bivariate Poisson pmf can also be obtained by using *Mathematica*'s `Statistics`MultiDiscreteDistributions`` package, as follows:

```
<< Statistics`

dist = MultiPoissonDistribution[λ₀, {λ₁, λ₂}];
```

Then, the package gives the joint pmf of (X_1, X_2) as:

```
MmaPMF = PDF[dist, {x₁, x₂}] // Simplify
```

$$e^{-\lambda_0 - \lambda_1 - \lambda_2} \lambda_1^{x_1} \lambda_2^{x_2} \left(-\left(\text{HypergeometricPFQ}\left[\{1, 1 + \text{Min}[x_1, x_2] - x_1, 1 + \text{Min}[x_1, x_2] - x_2\}, \{2 + \text{Min}[x_1, x_2]\}, \frac{\lambda_0}{\lambda_1 \lambda_2}\right] \left(\frac{\lambda_0}{\lambda_1 \lambda_2}\right)^{1+\text{Min}[x_1,x_2]} \right) \Big/ (\Gamma[2 + \text{Min}[x_1, x_2]]) \right.$$
$$\Gamma[-\text{Min}[x_1, x_2] + x_1] \Gamma[-\text{Min}[x_1, x_2] + x_2]) +$$
$$\left. \frac{\text{HypergeometricU}\left[-x_1, 1 - x_1 + x_2, -\frac{\lambda_1 \lambda_2}{\lambda_0}\right] \left(-\frac{\lambda_1 \lambda_2}{\lambda_0}\right)^{-x_1}}{\Gamma[1 + x_1] \Gamma[1 + x_2]} \right)$$

While this is not as neat as the result obtained above via the transformation method (i), it nevertheless gives the same results. Here, again, is $P(X_1 = 3, X_2 = 2)$:

```
MmaPMF /. {x₁ → 3, x₂ → 2} // Simplify
```

$$\frac{1}{12} e^{-\lambda_0 - \lambda_1 - \lambda_2} \lambda_1 (6 \lambda_0^2 + 6 \lambda_0 \lambda_1 \lambda_2 + \lambda_1^2 \lambda_2^2)$$

○ *Moments*

We shall consider three approaches for deriving moments, namely: (i) the direct approach, (ii) the mgf approach, and (iii) moment conversion formulae.

(i) *Direct approach*

Even though we know the joint pmf of X_1 and X_2, it is simpler to work with the underlying Y_i random variables. For instance, suppose we wish to find the product moment $\mu'_{1,1}$ for the bivariate Poisson. This can be expressed as:

$$\mu'_{1,1} = E[X_1 X_2] = E[(Y_1 + Y_0)(Y_2 + Y_0)]$$

which is then evaluated as:

$$\sum_{y_2=0}^{\infty} \sum_{y_1=0}^{\infty} \sum_{y_0=0}^{\infty} \texttt{Evaluate}[\,(y_1 + y_0)(y_2 + y_0)\,g\,]\ \ //\ \texttt{Expand}$$

$$\lambda_0 + \lambda_0^2 + \lambda_0 \lambda_1 + \lambda_0 \lambda_2 + \lambda_1 \lambda_2$$

(ii) *MGF approach*

The joint mgf of X_1 and X_2 is:

$$E\big[\exp(t_1 X_1 + t_2 X_2)\big] = E\big[\exp(t_1 Y_1 + t_2 Y_2 + (t_1 + t_2) Y_0)\big]$$

which is then evaluated as:[8]

$$\texttt{mgf} = \texttt{Simplify}\Big[\sum_{y_1=0}^{\infty} \sum_{y_2=0}^{\infty} \sum_{y_0=0}^{\infty} \texttt{Evaluate}[\,e^{t_1 y_1 + t_2 y_2 + (t_1+t_2) y_0}\,g\,]\Big]$$

$$e^{(-1+e^{t_1+t_2})\lambda_0 + (-1+e^{t_1})\lambda_1 + (-1+e^{t_2})\lambda_2}$$

Differentiating the mgf yields the raw product moments, as per (6.18).

$$\texttt{Moment[r\_, s\_]} := \texttt{D[mgf, \{t}_1\texttt{, r\}, \{t}_2\texttt{, s\}] /. t}\_ \to 0$$

Then, $\mu'_{1,1} = E[X_1 X_2]$ is now obtained by:

$$\texttt{Moment[1, 1]}\ \ //\ \texttt{Expand}$$

$$\lambda_0 + \lambda_0^2 + \lambda_0 \lambda_1 + \lambda_0 \lambda_2 + \lambda_1 \lambda_2$$

which is the same result we obtained using the direct method. Here is $\mu'_{3,1} = E[X_1^3 X_2^1]$:

$$\texttt{Moment[3, 1]}$$

$$\lambda_0 + 6\lambda_0(\lambda_0 + \lambda_1) + 3\lambda_0(\lambda_0 + \lambda_1)^2 + (\lambda_0 + \lambda_1)(\lambda_0 + \lambda_2) +$$
$$3(\lambda_0 + \lambda_1)^2(\lambda_0 + \lambda_2) + (\lambda_0 + \lambda_1)^3(\lambda_0 + \lambda_2)$$

The mean vector $\vec{\mu} = (E[X_1], E[X_2])$ is:

> **$\vec{\mu}$ = {Moment[1, 0], Moment[0, 1]}**
>
> $\{\lambda_0 + \lambda_1, \lambda_0 + \lambda_2\}$

By (6.19), the central mgf is given by:

> **$\vec{t}$ = {t_1, t_2};   mgfc = $e^{-\vec{t}\cdot\vec{\mu}}$ mgf   // Simplify**
>
> $e^{(-1+e^{t_1+t_2}-t_1-t_2)\lambda_0 + (-1+e^{t_1}-t_1)\lambda_1 + (-1+e^{t_2}-t_2)\lambda_2}$

Then, $\mu_{1,1} = \text{Cov}(X_1, X_2)$ is:

> **D[mgfc, {t_1, 1}, {t_2, 1}] /. t_ → 0**
>
> λ_0

while the variances of X_1 and X_2 are, respectively:

> **D[mgfc, {t_1, 2}] /. t_ → 0**
>
> $\lambda_0 + \lambda_1$

> **D[mgfc, {t_2, 2}] /. t_ → 0**
>
> $\lambda_0 + \lambda_2$

(iii) *Conversion formulae*

The pgf (derived above) can be used as a factorial moment generating function, as follows:

> **Fac[r_, s_] := D[pgf, {t_1, r}, {t_2, s}] /. t_ → 1**

Thus, the factorial moment $\mu[1,2] = E[X_1^{[1]} X_2^{[2]}]$ is given by:

> **Fac[1, 2]**
>
> $2\lambda_0(\lambda_0 + \lambda_2) + (\lambda_0 + \lambda_1)(\lambda_0 + \lambda_2)^2$

In part (ii), we found $\mu'_{3,1} = E[X_1^3 X_2^1]$ using the mgf approach. We now find the same expression, but this time do so using factorial moments. The solution, in terms of *factorial moments*, is:

> **sol = RawToFactorial[{3, 1}]**
>
> $\mu'_{3,1} \to \mu[1,1] + 3\mu[2,1] + \mu[3,1]$

so $\mu'_{3,1}$ can be obtained as:

```
sol /. µ́[r__] :→ Fac[r]
```

$$\acute{\mu}_{3,1} \to \lambda_0 + 3\,\lambda_0\,(\lambda_0 + \lambda_1)^2 + (\lambda_0 + \lambda_1)\,(\lambda_0 + \lambda_2) +$$
$$(\lambda_0 + \lambda_1)^3\,(\lambda_0 + \lambda_2) + 3\,(2\,\lambda_0\,(\lambda_0 + \lambda_1) + (\lambda_0 + \lambda_1)^2\,(\lambda_0 + \lambda_2))$$

It is easy to show that this is equal to Moment[3,1], as derived above.

6.7 Exercises

1. Let random variables X and Y have Gumbel's bivariate Logistic distribution with joint pdf

$$f(x, y) = \frac{2\,e^{-y-x}}{(1 + e^{-y} + e^{-x})^3}, \quad (x, y) \in \mathbb{R}^2.$$

   (i) Plot the joint pdf; (ii) plot the contours of the joint pdf; (iii) find the joint cdf; (iv) show that the marginal pdf's are Logistic; (v) find the conditional pdf $f(Y \mid X = x)$.

2. Let random variables X and Y have joint pdf

$$f(x, y) = \frac{1}{\lambda \mu}\,\exp\!\left[-\left(\frac{x}{\lambda} + \frac{y}{\mu}\right)\right], \quad \text{defined on } x > 0,\, y > 0$$

   with parameters $\lambda > 0$ and $\mu > 0$. Find the bivariate mgf. Use the mgf to find (i) $E[X]$, (ii) $E[Y]$, (iii) $\acute{\mu}_{3,4} = E[X^3\,Y^4]$, (iv) $\mu_{3,4}$. Verify by deriving each expectation directly.

3. Let random variables X and Y have McKay's bivariate Gamma distribution, with joint pdf

$$f(x, y) = \frac{c^{a+b}}{\Gamma[a]\,\Gamma[b]}\,x^{a-1}\,(y - x)^{b-1}\,e^{-c\,y}, \quad \text{defined on } 0 < x < y < \infty$$

   with parameters $a, b, c > 0$. Hint: use domain[f] = {{x, 0, y}, {y, x, ∞}} etc.

   (i) Show that the marginal pdf of X is Gamma.
   (ii) Find the correlation between X and Y.
   (iii) Derive the bivariate mgf. Use it to find $\acute{\mu}_{3,2} = E[X^3 Y^2]$.
   (iv) Plot $f(x, y)$ when $a = 3$, $b = 2$ and $c = 2$. Hint: use an If statement, as per Plot3D[If[0 < x < y, f, 0], {x, 0, 4}, {y, 0, 4}, etc.]
   (v) Create an animation showing how the pdf plot changes as parameter a increases from 2 to 5—the animation should look similar to the solution given here.

4. Let random variable $X \sim N(0, 1)$ and let $Y = X^2 - 2$. Show that $\text{Cov}(X, Y) = 0$, even though X and Y are clearly dependent.

5. Let random variables X and Y have a Gumbel (1960) bivariate Exponential distribution (see *Example 12*). Find the regression function $E[Y \mid X = x]$ and the scedastic function $\text{Var}(Y \mid X = x)$. Plot both when $\theta = 0$, $\frac{1}{2}$, 1.

6. Find a Normal–Exponential bivariate distribution (*i.e.* a distribution whose marginal pdf's are standard Normal and standard Exponential) using the Morgenstern copula method. Find the joint cdf and the variance-covariance matrix.

7. Find a bivariate distribution whose marginal distributions are both standard Exponential, using Frank's copula method. Plot the joint pdf $h(x, y)$ when $\alpha = -10$. Find the conditional pdf $h(x \mid Y = y)$.

8. Gumbel's bivariate Logistic distribution (defined in Exercise 1) has no parameters. While this is virtuous in being simple, it can also be restrictive.

   (i) Construct a more general bivariate distribution $h(x, y; \alpha)$ whose marginal distributions are both standard Logistic, using the Ali–Mikhail–Haq copula, with parameter α.

   (ii) Show that Gumbel's bivariate Logistic distribution is obtained as the special case $h(x, y; \alpha = 1)$.

   (iii) Plot the joint pdf $h(x, y)$ when $\alpha = \frac{1}{2}$.

   (iv) Find the conditional pdf $h(x \mid Y = y)$.

9. Let $f(x, y; \vec{\mu}, \Sigma)$ denote the joint pdf of a bivariate Normal distribution $N(\vec{\mu}, \Sigma)$. For $0 < \omega < 1$, define a bivariate Normal component-mixture density by:

$$\tilde{f}(x, y) = \omega f(x, y; \vec{\mu}_1, \Sigma_1) + (1 - \omega) f(x, y; \vec{\mu}_2, \Sigma_2)$$

Let $\vec{\mu}_1 = (2, 2)$, $\Sigma_1 = \begin{pmatrix} 1 & 0 \\ 0 & 1 \end{pmatrix}$, $\vec{\mu}_2 = (0, 0)$ and $\Sigma_2 = \begin{pmatrix} 1 & \frac{3}{4} \\ \frac{3}{4} & 1 \end{pmatrix}$.

   (i) Find the functional form for $\tilde{f}(x, y)$.

   (ii) Plot $\tilde{f}(x, y)$ when $\omega = \frac{7}{10}$. Construct contour plots of $\tilde{f}(x, y)$ when $\omega = 0$ and when $\omega = 1$.

   (iii) Create an animation showing how the contour plot changes as ω increases from 0 to 1 in step sizes of 0.025—the animation should look something like the solution given here:

   (iv) Find the marginal pdf of X, namely $\tilde{f}_x(x)$. Find the mean and variance of the latter.

   (v) Plot the marginal pdf derived in (iv) when $\omega = 0, \frac{1}{2}$ and 1.

10. Let random variables (W, X, Y, Z) have a multivariate Normal distribution $N(\vec{\mu}, \Sigma)$, with:

$$\vec{\mu} = (0, 0, 0, 0), \qquad \Sigma = \begin{pmatrix} 1 & \frac{2}{3} & \frac{3}{4} & \frac{4}{5} \\ \frac{2}{3} & 1 & \frac{1}{2} & \frac{8}{15} \\ \frac{3}{4} & \frac{1}{2} & 1 & \frac{3}{5} \\ \frac{4}{5} & \frac{8}{15} & \frac{3}{5} & 1 \end{pmatrix}$$

   (i) Find the joint pdf $f(w, x, y, z)$.

   (ii) Use the multivariate Normal mgf, $\exp(\vec{t}.\vec{\mu} + \frac{1}{2}\vec{t}.\Sigma.\vec{t})$, to find $E[W X Y Z]$ and $E[W X^2 Y Z^2]$.

   (iii) Find $E[W \exp(X + Y + Z)]$.

   (iv) Use Monte Carlo methods (not numerical integration) to check whether the solution to (iii) seems 'correct'.

   (v) Find $P(-3 < W < 3, -2 < X < \infty, -7 < Y < 2, -1 < Z < 1)$.

Chapter 7

Moments of Sampling Distributions

7.1   Introduction

7.1 A   Overview

Let $(X_1, \ldots, X_n)$ denote a random sample of size n drawn from a population random variable X. We can then distinguish between *population moments*:

$$\mu'_r = E[X^r] \qquad \text{raw moment of the population}$$

$$\mu_r = E[(X - \mu)^r] \qquad \text{central moment of the population, where } \mu = E[X]$$

and *sample moments*:

$$m'_r = \frac{1}{n} \sum_{i=1}^{n} X_i^r \qquad \text{sample raw moment}$$

$$m_r = \frac{1}{n} \sum_{i=1}^{n} (X_i - \overline{X})^r \qquad \text{sample central moment, where } \overline{X} = m'_1$$

where r is a positive integer. A *statistic* is a function of $(X_1, \ldots, X_n)$ that does not depend on any unknown parameter. Given this terminology, this chapter addresses two topics:

(i) *Unbiased estimators of population moments*

Given a random sample $(X_1, \ldots, X_n)$, we want to find a statistic that is an unbiased estimator of an unknown population moment. For instance, we might want to find unbiased estimators of population raw moments μ'_r, or of central moments μ_r, or of cumulants κ_r. We might even want to find unbiased estimators of products of population moments such as $\mu_2 \mu_4$. These problems are discussed in §7.2.

(ii) *Population moments of sample moments*

Because $(X_1, \ldots, X_n)$ is a collection of random variables, it follows that statistics like m'_r and m_r are themselves random variables, having their own distribution, and thus their own population moments. Thus, for instance, we may want to find the expectation of m_2. Since $E[m_2]$ is just the first raw moment of m_2, we can denote this problem by $\mu'_1(m_2)$. Similarly, $\text{Var}(m'_1)$ is just the second central moment of m'_1, so we can denote this problem by $\mu_2(m'_1)$. This is the topic of *moments of moments*, and it is discussed in §7.3.

7.1 B Power Sums and Symmetric Functions

Power sums are the *lingua franca* of this chapter. The rth *power sum* is defined as

$$s_r = \sum_{i=1}^{n} X_i^r, \qquad r = 1, 2, \ldots \qquad (7.1)$$

The sample raw moments can easily be expressed in terms of power sums:

$$m'_1 = \frac{s_1}{n}, \quad m'_2 = \frac{s_2}{n}, \quad \ldots, \quad m'_r = \frac{s_r}{n}. \qquad (7.2)$$

One can also express the sample central moments in terms of power sums, and **mathStatica** automates these conversions.[1] Here, for example, we express the 2nd sample central moment m_2 in terms of power sums:

SampleCentralToPowerSum[2]

$$m_2 \to -\frac{s_1^2}{n^2} + \frac{s_2}{n}$$

Next, we express m'_3 and m_4 in terms of power sums:

SampleRawToPowerSum[3]

$$m'_3 \to \frac{s_3}{n}$$

SampleCentralToPowerSum[4]

$$m_4 \to -\frac{3 s_1^4}{n^4} + \frac{6 s_1^2 s_2}{n^3} - \frac{4 s_1 s_3}{n^2} + \frac{s_4}{n}$$

These functions also handle multivariate conversions. For instance, to express the bivariate sample central moment $m_{3,1} = \frac{1}{n} \sum_{i=1}^{n} \left((X_i - \overline{X})^3 (Y_i - \overline{Y})^1\right)$ into power sums, enter:

SampleCentralToPowerSum[{3, 1}]

$$m_{3,1} \to -\frac{3 s_{0,1} s_{1,0}^3}{n^4} + \frac{3 s_{1,0}^2 s_{1,1}}{n^3} + \frac{3 s_{0,1} s_{1,0} s_{2,0}}{n^3} - \frac{3 s_{1,0} s_{2,1}}{n^2} - \frac{s_{0,1} s_{3,0}}{n^2} + \frac{s_{3,1}}{n}$$

where each bivariate power sum $s_{r,t}$ is defined by

$$s_{r,t} = \sum_{i=1}^{n} X_i^r Y_i^t. \qquad (7.3)$$

For a multivariate application, see *Example 7*. Power sums are also discussed in §7.4.

A function $f(x_1, ..., x_n)$ is said to be *symmetric* if it is unchanged after any permutation of the x's; that is, if say $f(x_1, x_2, x_3) = f(x_2, x_1, x_3)$. Thus,

$$x_1 + x_2 + \cdots + x_n = \sum_{i=1}^{n} x_i$$

is a symmetric function of $x_1, x_2, ..., x_n$. Examples of symmetric statistics include moments, product moments, h-statistics (h_r) and k-statistics (k_r). Symmetry is a most desirable property for an estimator to have: it generally amounts to saying that an estimate should *not* depend on the order in which the observations were made. The tools provided in this chapter apply to any rational, integral, algebraic symmetric function. This includes m_r, $m_r k_r$ or $m_r + h_r$, but not m_r / k_r nor $\sqrt{m_r}$. Symmetric functions are also discussed in more detail in §7.4.

7.2 Unbiased Estimators of Population Moments

On browsing through almost any statistics textbook, one encounters an *estimator of population variance* defined by $\frac{1}{n-1} \sum_{i=1}^{n} (X_i - \overline{X})^2$, where $\overline{X}$ is the sample mean. It is only natural to ponder why the denominator in this expression is $n-1$ rather than n. The answer is that $n-1$ yields an unbiased estimator of the population variance, while n yields a biased estimator. This section provides a toolset to attack such questions, not only for the variance, but for any population moment. We introduce h-statistics which are unbiased estimators of central population moments, and k-statistics which are unbiased estimators of population cumulants, and then generalise these statistics to encompass products of moments as well as multivariate moments. To do so, we couch our language in terms of power sums (see §7.1 B), which are closely related to sample moments. Although we assume an infinite universe, the results do extend to finite populations. For the finite univariate case, see Stuart and Ord (1994, Section 12.20); for the finite multivariate case, see Dwyer and Tracy (1980).

7.2 A Unbiased Estimators of Raw Moments of the Population

By the fundamental expectation result (7.15), it can be shown that sample raw moments m'_r are unbiased estimators of population raw moments μ'_r. That is,

$$E[m'_r] = \mu'_r. \tag{7.4}$$

However, products of sample raw moments are *not* unbiased estimators of products of population raw moments. For instance, $m'_2 m'_3$ is not an unbiased estimator of $\mu'_2 \mu'_3$. Unbiased estimators of products of raw moments are discussed in *Example 6* and in §7.4 A.

7.2 B h-statistics: Unbiased Estimators of Central Moments

The h-statistic h_r is an unbiased estimator of μ_r, defined by

$$E[h_r] = \mu_r. \tag{7.5}$$

That is, h_r is the statistic whose expectation is the central moment μ_r. Of all unbiased estimators of μ_r, the h-statistic is the only one that is symmetric. Halmös (1946) showed that not only is h_r unique, but its variance $\text{Var}(h_r) = E[(h_r - \mu_r)^2]$ is a minimum relative to all other unbiased estimators. We express h-statistics in terms of power sums, following Dwyer (1937) who introduced the term h-statistic. Here are the first four h-statistics:

```
Table[HStatistic[i], {i, 4}] // TableForm
```

$h_1 \to 0$

$h_2 \to \dfrac{-s_1^2 + n\, s_2}{(-1+n)\, n}$

$h_3 \to \dfrac{2\, s_1^3 - 3\, n\, s_1\, s_2 + n^2\, s_3}{(-2+n)\, (-1+n)\, n}$

$h_4 \to \dfrac{-3\, s_1^4 + 6\, n\, s_1^2\, s_2 + (9-6\, n)\, s_2^2 + (-12+8\, n - 4\, n^2)\, s_1\, s_3 + (3\, n - 2\, n^2 + n^3)\, s_4}{(-3+n)\, (-2+n)\, (-1+n)\, n}$

If we express the results in terms of sample central moments m_i, they appear neater:

```
Table[HStatisticToSampleCentral[i], {i, 4}] // TableForm
```

$h_1 \to 0$

$h_2 \to \dfrac{n\, m_2}{-1+n}$

$h_3 \to \dfrac{n^2\, m_3}{(-2+n)\, (-1+n)}$

$h_4 \to \dfrac{(9-6\, n)\, n^2\, m_2^2 + n\, (3\, n - 2\, n^2 + n^3)\, m_4}{(-3+n)\, (-2+n)\, (-1+n)\, n}$

⊕ *Example 1:* Unbiased Estimator of the Population Variance

We wish to find an unbiased estimator of the population variance μ_2. It follows immediately that an unbiased estimator of μ_2 is h_2. Here is h_2 expressed in terms of sample central moments:

```
HStatisticToSampleCentral[2]
```

$h_2 \to \dfrac{n\, m_2}{-1+n}$

which is identical to the standard textbook result $\frac{1}{n-1} \sum_{i=1}^n (X_i - \overline{X})^2$. Given that $\frac{n}{n-1} m_2$ is an unbiased estimator of population variance, it follows that m_2 is a biased estimator of population variance. §7.3 provides a toolset that enables one to calculate $F[m_2]$, and hence measure the bias. ∎

⊕ *Example 2:* Unbiased Estimator of μ_5 when $n = 11$

If the sample size n is known, and $n > r$, the function HStatistic[r, n] returns h_r. When $n = 11$, h_5 is:

```
HStatistic[5, 11]
```

$h_5 \to \dfrac{4\, s_1^5 - 110\, s_1^3\, s_2 + 270\, s_1\, s_2^2 + 850\, s_1^2\, s_3 - 990\, s_2\, s_3 - 4180\, s_1\, s_4 + 9196\, s_5}{55440}$

⊕ *Example 3:* Working with Data

The following data is a random sample of 30 lightbulbs, recording the observed life of each bulb in weeks:

```
data = {16.34, 10.76, 11.84, 13.55, 15.85, 18.20,
        7.51, 10.22, 12.52, 14.68, 16.08, 19.43,
        8.12, 11.20, 12.95, 14.77, 16.83, 19.80,
        8.55, 11.58, 12.10, 15.02, 16.83, 16.98,
        19.92,  9.47, 11.68, 13.41, 15.35, 19.11};
```

We wish to estimate the third central moment μ_3 of the population. If we simply calculated m_3 (a biased estimator), we would get the following estimate:

```
<< Statistics`
```

```
CentralMoment[data, 3]
```

-1.30557

By contrast, h_3 is an *unbiased* estimator. Evaluating the power sums $s_r = \sum_{i=1}^{n} X_i^r$ yields:

```
HStatistic[3, 30] /. s_r_ :> Plus @@ data^r
```

$h_3 \to -1.44706$

mathStatica's UnbiasedCentralMoment function automates this process, making it easier to use. That is, UnbiasedCentralMoment[*data*, *r*] estimates μ_r using the unbiased estimator h_r. Of course, it yields the same result:

```
UnbiasedCentralMoment[data, 3]
```

-1.44706

Chapter 5 makes frequent use of this function. ∎

○ *Polyaches* **(Generalised h-statistics)**

The generalised h-statistic (Tracy and Gupta (1974)) is defined by

$$E[h_{\{r,s,\ldots,t\}}] = \mu_r \mu_s \cdots \mu_t. \tag{7.6}$$

That is, $h_{\{r,s,\ldots,t\}}$ is the statistic whose expectation is the product of the central moments $\mu_r \mu_s \cdots \mu_t$. Just as Tukey (1956) created the onomatopoeic term 'polykay' to denote the generalised k-statistic (discussed below), we neologise 'polyache' to denote the generalised h-statistic. Perhaps, to paraphrase Kendall, there really are limits to linguistic miscegenation that should not be exceeded ☺.[2] Note that the polyache of a single term PolyH[{r}] is identical to HStatistic[r].

⊕ **Example 4:** Find an Unbiased Estimator of $\mu_2^2 \mu_3$

The solution is the polyache $h_{\{2,2,3\}}$:

```
PolyH[{2, 2, 3}]
```

$h_{\{2,2,3\}} \to$
$(2 s_1^7 - 7 n s_1^5 s_2 + (30 - 18 n + 8 n^2) s_1^3 s_2^2 + (60 - 63 n + 21 n^2 - 3 n^3) s_1$
$s_2^3 + (-40 + 24 n + n^2) s_1^4 s_3 + (-120 + 96 n - 24 n^2 - 2 n^3) s_1^2 s_2 s_3 +$
$(-20 n + 21 n^2 - 7 n^3 + n^4) s_2^2 s_3 + (80 - 40 n - 4 n^2 + 4 n^3) s_1 s_3^2 +$
$(60 - 8 n - 12 n^2) s_1^3 s_4 + (-120 + 140 n - 63 n^2 + 13 n^3) s_1 s_2 s_4 +$
$(-20 n + 10 n^2 + n^3 - n^4) s_3 s_4 + (48 - 92 n + 30 n^2 + 2 n^3) s_1^2 s_5 +$
$(36 n - 34 n^2 + 12 n^3 - 2 n^4) s_2 s_5 +$
$(-28 n + 42 n^2 - 14 n^3) s_1 s_6 + (4 n^2 - 6 n^3 + 2 n^4) s_7) /$
$((-6 + n)(-5 + n)(-4 + n)(-3 + n)(-2 + n)(-1 + n) n)$

Because h-statistics are symmetric functions, the ordering of the arguments, $h_{\{2,3,2\}}$ versus $h_{\{2,2,3\}}$, does not matter:

```
PolyH[{2, 3, 2}][[2]] == PolyH[{2, 2, 3}][[2]]
```

```
True
```

When using generalised h-statistics $h_{\{r,s,\ldots,t\}}$, the weight of the statistic can easily become quite large. Here, $h_{\{2,2,3\}}$ has weight $7 = 2 + 2 + 3$, and it contains terms such as s_7. Although $h_{\{2,2,3\}}$ is an unbiased estimator of $\mu_2^2 \mu_3$, some care must be taken in small samples because the variance of the estimator may be large. Intuitively, the effect of an outlier in a small sample is accentuated by terms such as s_7. In this vein, *Example 11* compares the performance of $h_{\{2,2\}}$ to h_2^2. ∎

7.2 C   k-statistics: Unbiased Estimators of Cumulants

The k-statistic k_r is an unbiased estimator of κ_r, defined by

$$E[k_r] = \kappa_r, \quad r = 1, 2, \ldots \quad (7.7)$$

That is, k_r is the (unique) symmetric statistic whose expectation is the rth cumulant κ_r. From Halmös (1946), we again know that, of all unbiased estimators of κ_r, the k-statistic is the only one that is symmetric, and its variance $\text{Var}(k_r) = E[(k_r - \kappa_r)^2]$ is a minimum relative to all other unbiased estimators. Following Fisher (1928), we define k-statistics in terms of power sums. Here, for instance, are the first four k-statistics:

```
Table[KStatistic[i], {i, 4}] // TableForm
```

$k_1 \to \frac{s_1}{n}$

$k_2 \to \frac{-s_1^2 + n s_2}{(-1+n) n}$

$k_3 \to \frac{2 s_1^3 - 3 n s_1 s_2 + n^2 s_3}{(-2+n)(-1+n) n}$

$k_4 \to \frac{-6 s_1^4 + 12 n s_1^2 s_2 + (3 n - 3 n^2) s_2^2 + (-4 n - 4 n^2) s_1 s_3 + (n^2 + n^3) s_4}{(-3+n)(-2+n)(-1+n) n}$

Once again, if we express these results in terms of sample central moments m_i, they appear neater:

Table[KStatisticToSampleCentral[i], {i, 4}] // TableForm

$k_1 \to 0$

$k_2 \to \frac{n\, m_2}{-1+n}$

$k_3 \to \frac{n^2\, m_3}{(-2+n)\,(-1+n)}$

$k_4 \to \frac{n^2\,(3n-3n^2)\,m_2^2 + n\,(n^2+n^3)\,m_4}{(-3+n)\,(-2+n)\,(-1+n)\,n}$

Stuart and Ord (1994) provide tables of k-statistics up to $r = 8$, though published results do exist to $r = 12$. Ziaud-Din (1954) derived k_9, and k_{10} (contains errors), Ziaud-Din (1959) derived k_{11} (contains errors), while Ziaud-Din and Ahmad (1960) derived k_{12}. The KStatistic function makes it simple to derive correct solutions 'on the fly', and it extends the analysis well past k_{12}. For instance, it takes just a few seconds to derive the 15th k-statistic on our reference personal computer:

KStatistic[15]; // Timing

{2.8 Second, Null}

But beware — the printed result will fill many pages!

○ *Polykays* **(Generalised k-statistics)**

Dressel (1940) introduced the generalised k-statistic $k_{\{r,s,\ldots,t\}}$ (now also called polykay) defined by

$$E[k_{\{r,s,\ldots,t\}}] = \kappa_r \kappa_s \cdots \kappa_t. \qquad (7.8)$$

That is, a polykay $k_{\{r,s,\ldots,t\}}$ is the statistic whose expectation is the product of the cumulants $\kappa_r \kappa_s \cdots \kappa_t$. Here is the polykay $k_{\{2,4\}}$ in terms of power sums:

PolyK[{2, 4}]

$k_{\{2,4\}} \to (6\,s_1^6 - 18\,n\,s_1^4\,s_2 + (30 - 27\,n + 15\,n^2)\,s_1^2\,s_2^2 +$
$\quad (60 - 60\,n + 21\,n^2 - 3\,n^3)\,s_2^3 + (-40 + 36\,n + 4\,n^2)\,s_1^3\,s_3 +$
$\quad (-120 + 100\,n - 24\,n^2 - 4\,n^3)\,s_1\,s_2\,s_3 + (40 - 10\,n - 10\,n^2 + 4\,n^3)\,s_3^2 +$
$\quad (60 - 20\,n - 15\,n^2 - n^3)\,s_1^2\,s_4 + (-60 + 45\,n - 10\,n^2 + n^4)\,s_2\,s_4 +$
$\quad (24 - 42\,n + 12\,n^2 + 6\,n^3)\,s_1\,s_5 + (-4\,n + 7\,n^2 - 2\,n^3 - n^4)\,s_6)\,/$
$\quad ((-5+n)\,(-4+n)\,(-3+n)\,(-2+n)\,(-1+n)\,n)$

Finally, note that the polykay of a single term PolyK[{r}] is identical to KStatistic[r]; however, they use different algorithms, and the latter function is more efficient computationally.

⊕ **Example 5:** Find an Unbiased Estimator of κ_2^2

Solution: The required unbiased estimator is the polykay $k_{\{2,2\}}$:

```
PolyK[{2, 2}]
```

$$k_{\{2,2\}} \to \frac{s_1^4 - 2 n s_1^2 s_2 + (3 - 3n + n^2) s_2^2 + (-4 + 4n) s_1 s_3 + (n - n^2) s_4}{(-3 + n)(-2 + n)(-1 + n) n}$$

For the lightbulb data set of *Example 3*, this yields the estimate:

```
PolyK[{2, 2}, 30] /. s_r_ :→ Plus @@ data^r
```

$k_{\{2,2\}} \to 154.118$

By contrast, k_2^2 is a biased estimator of κ_2^2, and yields a different estimate:

```
k2 = KStatistic[2, 30] /. s_r_ :→ Plus @@ data^r
k2[[2]]^2
```

$k_2 \to 12.6501$

160.024

Example 11 compares the performance of the unbiased estimator $h_{\{2,2\}}$ to the biased estimator h_2^2 (note that $k_{\{2,2\}} = h_{\{2,2\}}$, and $k_2 = h_2$). ∎

⊕ **Example 6:** Find an Unbiased Estimator of the Product of Raw Moments $\mu_3' \mu_4'$

Polykays can be used to find unbiased estimators of quite general expressions. For instance, to find an unbiased estimator of the product of raw moments $\mu_3' \mu_4'$, we may proceed as follows:

Step (i): Convert $\mu_3' \mu_4'$ into cumulants:

```
p = μ₃' μ₄' /. Table[RawToCumulant[i], {i, 3, 4}] // Expand
```

$\kappa_1^7 + 9 \kappa_1^5 \kappa_2 + 21 \kappa_1^3 \kappa_2^2 + 9 \kappa_1 \kappa_2^3 + 5 \kappa_1^4 \kappa_3 +$
$18 \kappa_1^2 \kappa_2 \kappa_3 + 3 \kappa_2^2 \kappa_3 + 4 \kappa_1 \kappa_3^2 + \kappa_1^3 \kappa_4 + 3 \kappa_1 \kappa_2 \kappa_4 + \kappa_3 \kappa_4$

Step (ii): Find an unbiased estimator of each term in this expression. Since each term is a product of cumulants, the unbiased estimator of each term is a polykay. The first term κ_1^7 becomes $k_{\{1,1,1,1,1,1,1\}}$, while $9 \kappa_1^5 \kappa_2$ becomes $9 k_{\{1,1,1,1,1,2\}}$, and so on. While we could do all this manually, there is an easier way! If $p(x)$ is a symmetric polynomial in x, the **mathStatica** function ListForm[p, x] will convert p into a 'list form' suitable for use by PolyK and many other functions. Note that ListForm should *only* be called on polynomials that have just been expanded using Expand. The order of the terms is now reversed:

```
p1 = ListForm[p, κ]
```

κ[{3, 4}] + 3 κ[{1, 2, 4}] + 4 κ[{1, 3, 3}] + 3 κ[{2, 2, 3}] +
κ[{1, 1, 1, 4}] + 18 κ[{1, 1, 2, 3}] + 9 κ[{1, 2, 2, 2}] +
5 κ[{1, 1, 1, 1, 3}] + 21 κ[{1, 1, 1, 2, 2}] +
9 κ[{1, 1, 1, 1, 1, 2}] + κ[{1, 1, 1, 1, 1, 1}]

Replacing each κ term by PolyK yields the desired estimator:

```
p1 /. κ[x_] :> PolyK[x][[2]]  // Factor
```

$$\frac{s_3 \, s_4 - s_7}{(-1 + n) \, n}$$

which is surprisingly neat. *Example 15* provides a more direct way of finding unbiased estimators of products of raw moments, but requires some knowledge of augmented symmetrics to do so. ∎

7.2 D Multivariate h- and k-statistics

The multivariate h-statistic $h_{r, s, \ldots, t}$ is defined by

$$E[h_{r, s, \ldots, t}] = \mu_{r, s, \ldots, t}. \tag{7.9}$$

That is, $h_{r, s, \ldots, t}$ is the statistic whose expectation is the q-variate central moment $\mu_{r, s, \ldots, t}$ (see §6.2 B), where

$$\mu_{r, s, \ldots, t} = E\left[(X_1 - E[X_1])^r (X_2 - E[X_2])^s \cdots (X_q - E[X_q])^t\right] \tag{7.10}$$

Some care with notation is required here. We use curly brackets {} to distinguish between the multivariate h-statistics $h_{r, s, \ldots, t}$ of this section and the univariate polyaches $h_{\{r, s, \ldots, t\}}$ (generalised h-statistics) discussed in §7.2 B.

The **mathStatica** function HStatistic[{$r, s, \ldots, t$}] yields the multivariate h-statistic $h_{r, s, \ldots, t}$. Here are two bivariate examples:

```
HStatistic[{1, 1}]
HStatistic[{2, 1}]
```

$$h_{1,1} \to \frac{-s_{0,1} \, s_{1,0} + n \, s_{1,1}}{(-1 + n) \, n}$$

$$h_{2,1} \to \frac{2 \, s_{0,1} \, s_{1,0}^2 - 2 \, n \, s_{1,0} \, s_{1,1} - n \, s_{0,1} \, s_{2,0} + n^2 \, s_{2,1}}{(-2 + n) \, (-1 + n) \, n}$$

where each bivariate power sum $s_{r, t}$ is defined by

$$s_{r, t} = \sum_{i=1}^{n} X_i^r \, Y_i^t.$$

Higher variate examples soon become quite lengthy. Here is a simple trivariate example:

HStatistic[{2, 1, 1}]

$h_{2,1,1} \to (-3\, s_{0,0,1}\, s_{0,1,0}\, s_{1,0,0}^2 + n\, s_{0,1,1}\, s_{1,0,0}^2 + 2\, n\, s_{0,1,0}\, s_{1,0,0}\, s_{1,0,1} +$
$\quad 2\, n\, s_{0,0,1}\, s_{1,0,0}\, s_{1,1,0} - 2\, (-3 + 2\, n)\, s_{1,0,1}\, s_{1,1,0} -$
$\quad 2\, (3 - 2\, n + n^2)\, s_{1,0,0}\, s_{1,1,1} + n\, s_{0,0,1}\, s_{0,1,0}\, s_{2,0,0} -$
$\quad (-3 + 2\, n)\, s_{0,1,1}\, s_{2,0,0} - (3 - 2\, n + n^2)\, s_{0,1,0}\, s_{2,0,1} -$
$\quad (3 - 2\, n + n^2)\, s_{0,0,1}\, s_{2,1,0} + n\, (3 - 2\, n + n^2)\, s_{2,1,1}) /$
$\quad ((-3 + n)\, (-2 + n)\, (-1 + n)\, n)$

In similar fashion, the multivariate k-statistic $k_{r,s,\ldots,t}$ is defined by

$$E[k_{r,s,\ldots,t}] = \kappa_{r,s,\ldots,t}. \qquad (7.11)$$

That is, $k_{r,s,\ldots,t}$ is the statistic whose expectation is the multivariate cumulant $\kappa_{r,s,\ldots,t}$. Multivariate cumulants were briefly discussed in §6.2 C and §6.2 D. Here is a bivariate result originally given by Fisher (1928):

KStatistic[{3, 1}]

$k_{3,1} \to$
$\quad (-6\, s_{0,1}\, s_{1,0}^3 + 6\, n\, s_{1,0}^2\, s_{1,1} + 6\, n\, s_{0,1}\, s_{1,0}\, s_{2,0} - 3\, (-1 + n)\, n\, s_{1,1}\, s_{2,0} -$
$\quad 3\, n\, (1 + n)\, s_{1,0}\, s_{2,1} - n\, (1 + n)\, s_{0,1}\, s_{3,0} + n^2\, (1 + n)\, s_{3,1}) /$
$\quad ((-3 + n)\, (-2 + n)\, (-1 + n)\, n)$

Multivariate polykays and multivariate polyaches are not currently implemented in **mathStatica**.

⊕ *Example 7:* American NFL Matches: Estimating the Central Moment $\mu_{2,1}$

The following data is taken from American National Football League games in 1986; see Csörgö and Welsh (1989). Variable X_1 measures the time from the start of the game until the first points are scored by kicking the ball between the end-posts (a field goal), while X_2 measures the time from the start of the game until the first points are scored by a touchdown. Times are given in minutes and seconds. If $X_1 < X_2$, the first score is a field goal; if $X_1 = X_2$, the first score is a converted touchdown; if $X_1 > X_2$, the first score is an unconverted touchdown:

```
data = {{2.03, 3.59}, {7.47, 7.47}, {7.14, 9.41},
    {31.08, 49.53}, {7.15, 7.15}, {4.13, 9.29}, {6.25, 6.25},
    {10.24, 14.15}, {11.38, 17.22}, {14.35, 14.35},
    {17.5, 17.5}, {9.03, 9.03}, {10.34, 14.17}, {6.51, 34.35},
    {14.35, 20.34}, {4.15, 4.15}, {15.32, 15.32}, {8.59, 8.59},
    {2.59, 2.59}, {1.23, 1.23}, {11.49, 11.49}, {10.51, 38.04},
    {0.51, 0.51}, {7.03, 7.03}, {32.27, 42.21}, {5.47, 25.59},
    {1.39, 1.39}, {2.54, 2.54}, {10.09, 10.09}, {3.53, 6.26},
    {10.21, 10.21}, {5.31, 11.16}, {3.26, 3.26}, {2.35, 2.35},
    {8.32, 14.34}, {13.48, 49.45}, {6.25, 15.05}, {7.01, 7.01},
    {8.52, 8.52}, {0.45, 0.45}, {12.08, 12.08}, {19.39, 10.42}};
```

Then, X_1 and X_2 are given by:

```
{X1, X2} = Transpose[data];
```

There are $n = 42$ pairs. An unbiased estimator of the central moment $\mu_{2,1}$ is given by the h-statistic $h_{2,1}$. Using it yields the following estimate of $\mu_{2,1}$:

```
HStatistic[{2, 1}, 42] /. s_{i_,j_} :→ X1^i.X2^j
```

$h_{2,1} \to 752.787$

An alternative estimator of $\mu_{2,1}$ is the sample central moment $m_{2,1}$:

```
m21 = SampleCentralToPowerSum[{2, 1}]
```

$$m_{2,1} \to \frac{2\, s_{0,1}\, s_{1,0}^2}{n^3} - \frac{2\, s_{1,0}\, s_{1,1}}{n^2} - \frac{s_{0,1}\, s_{2,0}}{n^2} + \frac{s_{2,1}}{n}$$

Unfortunately, $m_{2,1}$ is a biased estimator, and it yields a different estimate here:

```
m21 /. {s_{i_,j_} :→ X1^i.X2^j, n → 42}
```

$m_{2,1} \to 699.87$

The CentralMoment function in *Mathematica*'s Statistics package also implements the biased estimator $m_{2,1}$:

```
<< Statistics`MultiDescriptiveStatistics`
CentralMoment[data, {2, 1}]
```

699.87

7.3 Moments of Moments

7.3 A Getting Started

Let $(X_1, ..., X_n)$ denote a random sample of size n drawn from the population random variable X. Because $(X_1, ..., X_n)$ are random variables, it follows that a statistic like the sample central moment m_r is itself a random variable, with its own distribution and its own population moments. Suppose we want to find the expectation of m_2. Since $E[m_2]$ is just the first raw moment of m_2, we can denote this problem by $\mu'_1(m_2)$. Similarly, we might want to find the population variance of m'_1. Since $\text{Var}(m'_1)$ is just the second central moment of m'_1, we can denote this problem by $\mu_2(m'_1)$. Or, we might want to find the fourth cumulant of m_3, which we denote by $\kappa_4(m_3)$. In each of these cases, we are finding a population moment of a sample moment, or, for short, a *moment of a moment*.

The problem of *moments of moments* has attracted a prolific literature containing many beautiful formulae. Such formulae are listed over pages and pages of tables in reference texts and journals. Sometimes these tables contain errors; sometimes one induces errors oneself by typing them in incorrectly; sometimes the desired formula is simply not available and deriving the solution oneself is cumbersome and tricky. Some authors have devoted years to this task! The tools presented in this chapter change all that: they enable one to generate any desired formula, usually in just a few seconds, without even having to worry about typing it in incorrectly.

Although the problem of *moments of moments* has produced a long and complicated literature, conceptually the problem is rather simple. Let $p(s)$ denote any symmetric rational polynomial expressed in terms of power sums s_r (§7.1 B). Our goal is to find the population moments of p, and to express the answer in terms of the population moments of X. Let $\mu'_r(p)$, $\mu_r(p)$ and $\kappa_r(p)$ denote, respectively, the rth raw moment, central moment and cumulant of p. In each case, we can present the solution in terms of raw moments $\mu'_i(X)$ of the population of X, or central moments $\mu_i(X)$ of the population of X, or cumulants $\kappa_i(X)$ of the population of X. As such, the problem can be expressed in 9 different ways:

$$\left. \begin{array}{c} \mu'_r(p) \\ \mu_r(p) \\ \kappa_r(p) \end{array} \right\} \quad \text{in terms of} \quad \left\{ \begin{array}{c} \mu'_i(X) \\ \mu_i(X) \\ \kappa_i(X) \end{array} \right.$$

Consequently, **mathStatica** offers 9 functions to tackle the problem of *moments of moments*, as shown in Table 1.

function	description
RawMomentToRaw[r, p]	$\mu'_r(p)$ in terms of $\mu'_i(X)$
RawMomentToCentral[r, p]	$\mu'_r(p)$ in terms of $\mu_i(X)$
RawMomentToCumulant[r, p]	$\mu'_r(p)$ in terms of $\kappa_i(X)$
CentralMomentToRaw[r, p]	$\mu_r(p)$ in terms of $\mu'_i(X)$
CentralMomentToCentral[r, p]	$\mu_r(p)$ in terms of $\mu_i(X)$
CentralMomentToCumulant[r, p]	$\mu_r(p)$ in terms of $\kappa_i(X)$
CumulantMomentToRaw[r, p]	$\kappa_r(p)$ in terms of $\mu'_i(X)$
CumulantMomentToCentral[r, p]	$\kappa_r(p)$ in terms of $\mu_i(X)$
CumulantMomentToCumulant[r, p]	$\kappa_r(p)$ in terms of $\kappa_i(X)$

Table 1: Moments of moments functions

For instance, consider the function CentralMomentToRaw[r, p]:

- the term CentralMoment indicates that we wish to find $\mu_r(p)$; *i.e.* the rth central moment of p;
- the term ToRaw indicates that we want the answer expressed in terms of raw moments μ'_i of the population of X.

§7.3 A MOMENTS OF SAMPLING DISTRIBUTIONS

These functions nest common operators such as:

- the expectation operator: $E[p] = \mu'_1(p) =$ RawMomentTo?$[1, p]$
- the variance operator: $\text{Var}(p) = \mu_2(p) =$ CentralMomentTo?$[2, p]$

There is often more than one correct way of thinking about these problems. For example, the expectation $E[p^3]$ can be thought of as either $\mu'_1(p^3)$ or as $\mu'_3(p)$. Endnote 3 provides more detail on the ___ToCumulant functions; it should be carefully read before using them.

⊕ *Example 8:* Checking if the Unbiased Estimators Really Are Unbiased

We are now equipped to test, for instance, whether the unbiased estimators introduced in §7.2 *really are* unbiased. In §7.2 C, we obtained the polykay $k_{\{2,4\}}$ in terms of power sums:

p = PolyK[{2, 4}]

$k_{\{2,4\}} \to (6\, s_1^6 - 18\, n\, s_1^4\, s_2 + (30 - 27\, n + 15\, n^2)\, s_1^2\, s_2^2 +$
$(60 - 60\, n + 21\, n^2 - 3\, n^3)\, s_2^3 + (-40 + 36\, n + 4\, n^2)\, s_1^3\, s_3 +$
$(-120 + 100\, n - 24\, n^2 - 4\, n^3)\, s_1\, s_2\, s_3 + (40 - 10\, n - 10\, n^2 + 4\, n^3)\, s_3^2 +$
$(60 - 20\, n - 15\, n^2 - n^3)\, s_1^2\, s_4 + (-60 + 45\, n - 10\, n^2 + n^4)\, s_2\, s_4 +$
$(24 - 42\, n + 12\, n^2 + 6\, n^3)\, s_1\, s_5 + (-4\, n + 7\, n^2 - 2\, n^3 - n^4)\, s_6) /$
$((-5 + n)\, (-4 + n)\, (-3 + n)\, (-2 + n)\, (-1 + n)\, n)$

This statistic is meant to have the property that $E[p] = \kappa_2\, \kappa_4$. Since $E[p] = \mu'_1(p)$, we will use the RawMomentTo?$[1, p]$ function; moreover, since the answer is desired in terms of cumulants, we use the suffix ToCumulant:

RawMomentToCumulant[1, p[[2]]]

$\kappa_2\, \kappa_4$

... so all is well. Similarly, we can check the h-statistics. Here is the 4th h-statistic in terms of power sums:

p = HStatistic[4]

$h_4 \to \dfrac{-3\, s_1^4 + 6\, n\, s_1^2\, s_2 + (9 - 6\, n)\, s_2^2 + (-12 + 8\, n - 4\, n^2)\, s_1\, s_3 + (3\, n - 2\, n^2 + n^3)\, s_4}{(-3 + n)\, (-2 + n)\, (-1 + n)\, n}$

This is meant to have the property that $E[p] = \mu_4$. And ...

RawMomentToCentral[1, p[[2]]]

μ_4

... all is well. ∎

⊕ **Example 9:** The Variance of the Sample Mean $\acute{m}_1$

Step (i): Express $\acute{m}_1$ in terms of power sums: trivially, we have $\acute{m}_1 = \frac{s_1}{n}$.

Step (ii): Since $\text{Var}(\acute{m}_1) = \mu_2\left(\frac{s_1}{n}\right)$, the desired solution is:

CentralMomentToCentral $\left[2, \dfrac{s_1}{n}\right]$

$\dfrac{\mu_2}{n}$

where μ_2 denotes the population variance. This is just the well-known result that the variance of the sample mean is $\text{Var}(X)/n$. ∎

⊕ **Example 10:** The Variance of m_2

Step (i): Convert m_2 into power sums (§7.1 B):

m2 = SampleCentralToPowerSum[2][[2]]

$-\dfrac{s_1^2}{n^2} + \dfrac{s_2}{n}$

Step (ii): Since $\text{Var}(m_2) = \mu_2(m_2)$, the desired solution is:

CentralMomentToCentral[2, m2]

$-\dfrac{(-3+n)\,(-1+n)\,\mu_2^2}{n^3} + \dfrac{(-1+n)^2\,\mu_4}{n^3}$

⊕ **Example 11:** Mean Square Error of Two Estimators

Which is the better estimator of μ_2^2: (a) the square of the second h-statistic h_2^2, or (b) the polyache $h_{\{2,2\}}$?

Solution: We know that the polyache $h_{\{2,2\}}$ is an unbiased estimator of μ_2^2, while h_2^2 is a biased estimator of μ_2^2. But bias is not everything: variance is also important. The *mean square error* of an estimator is a measure that takes account of both bias and variance, defined by $\text{MSE}(\hat{\theta}) = E\left[(\hat{\theta} - \theta)^2\right]$, where $\hat{\theta}$ denotes the estimator, and θ is the true parameter value (see Chapter 9 for more detail). For this particular problem, the two estimators are $\bar{\theta} = h_2^2$ and $\tilde{\theta} = h_{\{2,2\}}$:

$\bar{\theta}$ = **HStatistic[2][[2]]^2**
$\tilde{\theta}$ = **PolyH[{2, 2}][[2]]**

$\dfrac{(-s_1^2 + n\,s_2)^2}{(-1+n)^2\,n^2}$

$\dfrac{s_1^4 - 2n\,s_1^2\,s_2 + (3 - 3n + n^2)\,s_2^2 + (-4 + 4n)\,s_1\,s_3 + (n - n^2)\,s_4}{(-3+n)\,(-2+n)\,(-1+n)\,n}$

§7.3 A   MOMENTS OF SAMPLING DISTRIBUTIONS   265

If we let $p = (\hat{\theta} - \theta)^2$, then $\text{MSE}(\hat{\theta}) = E[p] = \mu'_1(p)$, so the mean square error of each estimator is (in terms of central moments):

MSE[$\bar{\theta}$] = RawMomentToCentral$\left[1, (\bar{\theta} - \mu_2^2)^2\right]$;

MSE[$\tilde{\theta}$] = RawMomentToCentral$\left[1, (\tilde{\theta} - \mu_2^2)^2\right]$;

Now consider the ratio of the mean square errors of the two estimators. We are interested to see whether this ratio is greater than or smaller than 1. If it is always greater than 1, then the polykay $\tilde{\theta} = h_{\{2,2\}}$ is the strictly preferred estimator:

rat = $\dfrac{\text{MSE}[\bar{\theta}]}{\text{MSE}[\tilde{\theta}]}$ // Factor

$$\frac{((-3 + n)(-2 + n)}{(-630\,\mu_2^4 + 885\,n\,\mu_2^4 - 507\,n^2\,\mu_2^4 + 159\,n^3\,\mu_2^4 - 31\,n^4\,\mu_2^4 + 4\,n^5\,\mu_2^4 +}$$

(continued)
$560\,\mu_2\,\mu_3^2 - 840\,n\,\mu_2\,\mu_3^2 + 520\,n^2\,\mu_2\,\mu_3^2 - 168\,n^3\,\mu_2\,\mu_3^2 + 24\,n^4\,\mu_2\,\mu_3^2 +$
$420\,\mu_2^2\,\mu_4 - 690\,n\,\mu_2^2\,\mu_4 + 430\,n^2\,\mu_2^2\,\mu_4 - 138\,n^3\,\mu_2^2\,\mu_4 +$
$30\,n^4\,\mu_2^2\,\mu_4 - 4\,n^5\,\mu_2^2\,\mu_4 - 35\,\mu_4^2 + 60\,n\,\mu_4^2 - 42\,n^2\,\mu_4^2 +$
$12\,n^3\,\mu_4^2 - 3\,n^4\,\mu_4^2 - 56\,\mu_3\,\mu_5 + 104\,n\,\mu_3\,\mu_5 - 72\,n^2\,\mu_3\,\mu_5 +$
$24\,n^3\,\mu_3\,\mu_5 - 28\,\mu_2\,\mu_6 + 64\,n\,\mu_2\,\mu_6 - 48\,n^2\,\mu_2\,\mu_6 +$
$16\,n^3\,\mu_2\,\mu_6 - 4\,n^4\,\mu_2\,\mu_6 + \mu_8 - 3\,n\,\mu_8 + 3\,n^2\,\mu_8 - n^3\,\mu_8)) /$
$(2\,(-1 + n)^2\,n^2\,(-66\,\mu_2^4 + 51\,n\,\mu_2^4 - 17\,n^2\,\mu_2^4 + 2\,n^3\,\mu_2^4 +$
$48\,\mu_2\,\mu_3^2 - 28\,n\,\mu_2\,\mu_3^2 + 4\,n^2\,\mu_2\,\mu_3^2 + 36\,\mu_2^2\,\mu_4 - 36\,n\,\mu_2^2\,\mu_4 +$
$14\,n^2\,\mu_2^2\,\mu_4 - 2\,n^3\,\mu_2^2\,\mu_4 - 6\,\mu_4^2 + 5\,n\,\mu_4^2 - n^2\,\mu_4^2))$

This expression seems too complicated to immediately say anything useful about it, so let us consider an example. If the population is $N(\mu, \sigma^2)$ with pdf $f(x)$:

f = $\dfrac{1}{\sigma\sqrt{2\pi}}\,\text{Exp}\left[-\dfrac{(x-\mu)^2}{2\sigma^2}\right]$;

domain[f] = $\{x, -\infty, \infty\}$ && $\{\mu \in \text{Reals},\ \sigma > 0\}$;

... then the first 8 central moments of the population are:

mgfc = Expect[$e^{t(x-\mu)}$, f];

cm = Table[$\mu_i \to$ D[mgfc, {t, i}] /. t $\to$ 0, {i, 8}]

$\{\mu_1 \to 0,\ \mu_2 \to \sigma^2,\ \mu_3 \to 0,\ \mu_4 \to 3\,\sigma^4,$
$\mu_5 \to 0,\ \mu_6 \to 15\,\sigma^6,\ \mu_7 \to 0,\ \mu_8 \to 105\,\sigma^8\}$

so the ratio becomes:

rr = rat /. cm // Factor

$$\frac{(-3 + n)(-2 + n)\,n\,(3 + n)(1 + 2n)}{2\,(-1 + n)^2\,(3 + 3n - 4n^2 + n^3)}$$

Figure 1 shows that this ratio is always greater than 1, irrespective of σ, so the polyache is strictly preferred, at least for this distribution.

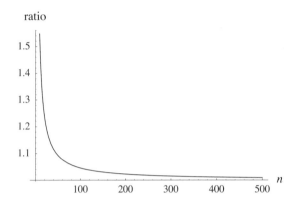

Fig. 1: $\dfrac{\text{MSE}(\bar{\theta})}{\text{MSE}(\tilde{\theta})}$ as a function of n, for the Normal distribution

We plot for $n > 9$ because the *moments of moments* functions are well-defined only for $n > w$, where w is the weight of the statistic. ∎

7.3 B   Product Moments

Product moments (multivariate moments) were introduced in §6.2 B and §6.2 D. We are interested here in expressions such as:

$$\mu'_{r,s}(p_a, p_b) = E[p_a^r \, p_b^s]$$
$$\mu_{r,s}(p_a, p_b) = E\big[(p_a - E[p_a])^r \, (p_b - E[p_b])^s\big]$$
$$\kappa_{r,s}(p_a, p_b)$$

where each p_i is a symmetric polynomial in power sums s_i. All of **mathStatica**'s *moment of moment* functions generalise to neatly handle product moments — given $\mu_r(p)$, simply think of r and p as lists.

⊕ **Example 12:** Find the Covariance Between the Sample Moments m_2 and m_3

Step (i): Express m_2 and m_3 in terms of power sums:

```
m2 = SampleCentralToPowerSum[2][[2]];
m3 = SampleCentralToPowerSum[3][[2]];
```

Step (ii): *Example 13* of Chapter 6 showed that $\text{Cov}(m_2, m_3)$ is just the product moment $\mu_{1,1}(m_2, m_3)$. Thus, the solution is:

CentralMomentToCentral[{1, 1}, {m2, m3}]

$$-\frac{2\,(-2+n)\,(-1+n)\,(-5+2n)\,\mu_2\,\mu_3}{n^4} + \frac{(-2+n)\,(-1+n)^2\,\mu_5}{n^4}$$

7.3 C   Cumulants of k-statistics

Following the work of Fisher (1928), the cumulants of k-statistics have received great attention, for which two reasons are proffered. First, it is often claimed that the cumulants of the k-statistics yield much more compact formulae than other derivations. This is not really true. Experimentation with the *moment of moment* functions shows that $\mu'_r(k_i)$ is just as compact as $\kappa_r(k_i)$, provided both results are expressed *in terms of cumulants*. In this sense, there is nothing special about cumulants of k-statistics per se; the raw moments of the k-statistics are just as compact. Second, Fisher showed how the cumulants of the k-statistics can be derived using a combinatoric method, in contrast to the algebraic method *du jour*. While Fisher's combinatorial approach is less burdensome algebraically, it is tricky and finicky, which can easily lead to errors. Indeed, with **mathStatica**, one can show that even after 70 years, a reference bible such as Stuart and Ord (1994) *still* contains errors in its listings of cumulants of k-statistics; examples are provided below. **mathStatica** uses an internal algebraic approach because (i) this is general, safe and secure, and (ii) the burdensome algebra ceases to be a constraint when you can get a computer to do all the dreary work for you. It is perhaps a little ironic then that modern computing technology has conceptually taken us full circle back to the work of Pearson (1902), Thiele (1903), and 'Student' (1908).

In this section, we will make use of the following k-statistics:

k2 = KStatistic[2][[2]];

k3 = KStatistic[3][[2]];

Here are the first four cumulants of k_2, namely $\kappa_r(k_2)$ for $r = 1, 2, 3, 4$:

CumulantMomentToCumulant[1, k2]

κ_2

CumulantMomentToCumulant[2, k2]

$$\frac{2\kappa_2^2}{-1+n} + \frac{\kappa_4}{n}$$

CumulantMomentToCumulant[3, k2]

$$\frac{8\kappa_2^3}{(-1+n)^2} + \frac{4(-2+n)\kappa_3^2}{(-1+n)^2 n} + \frac{12\kappa_2\kappa_4}{(-1+n)n} + \frac{\kappa_6}{n^2}$$

CumulantMomentToCumulant[4, k2]

$$\frac{48\kappa_2^4}{(-1+n)^3} + \frac{96(-2+n)\kappa_2\kappa_3^2}{(-1+n)^3 n} + \frac{144\kappa_2^2\kappa_4}{(-1+n)^2 n} + \frac{8(6-9n+4n^2)\kappa_4^2}{(-1+n)^3 n^2} + \frac{32(-2+n)\kappa_3\kappa_5}{(-1+n)^2 n^2} + \frac{24\kappa_2\kappa_6}{(-1+n)n^2} + \frac{\kappa_8}{n^3}$$

Next, we derive the product cumulant $\kappa_{3,1}(k_3, k_2)$, expressed in terms of cumulants, as obtained by David and Kendall (1949, p. 433). This takes less than 2 seconds to solve on our reference computer:

`CumulantMomentToCumulant[{3, 1}, {k3, k2}]`

$$\frac{1296\,n\,(-12+5n)\,\kappa_2^4\,\kappa_3}{(-2+n)^2\,(-1+n)^3} + \frac{324\,(164-136n+29n^2)\,\kappa_2\,\kappa_3^3}{(-2+n)^2\,(-1+n)^3} +$$

$$\frac{648\,(137-126n+29n^2)\,\kappa_2^2\,\kappa_3\,\kappa_4}{(-2+n)^2\,(-1+n)^3} +$$

$$\frac{108\,(-390+543n-257n^2+41n^3)\,\kappa_3\,\kappa_4^2}{(-2+n)^2\,(-1+n)^3\,n} +$$

$$\frac{108\,(110-122n+33n^2)\,\kappa_2^3\,\kappa_5}{(-2+n)^2\,(-1+n)^3} +$$

$$\frac{54\,(-564+842n-421n^2+71n^3)\,\kappa_3^2\,\kappa_5}{(-2+n)^2\,(-1+n)^3\,n} +$$

$$\frac{54\,(316-340n+93n^2)\,\kappa_2\,\kappa_4\,\kappa_5}{(-2+n)\,(-1+n)^3\,n} +$$

$$\frac{54\,(178-220n+63n^2)\,\kappa_2\,\kappa_3\,\kappa_6}{(-2+n)\,(-1+n)^3\,n} +$$

$$\frac{9\,(103-134n+49n^2)\,\kappa_5\,\kappa_6}{(-1+n)^3\,n^2} +$$

$$\frac{54\,(-23+12n)\,\kappa_2^2\,\kappa_7}{(-2+n)\,(-1+n)^2\,n} + \frac{27\,(22-31n+11n^2)\,\kappa_4\,\kappa_7}{(-1+n)^3\,n^2} +$$

$$\frac{9\,(-26+17n)\,\kappa_3\,\kappa_8}{(-1+n)^2\,n^2} + \frac{45\,\kappa_2\,\kappa_9}{(-1+n)\,n^2} + \frac{\kappa_{11}}{n^3}$$

⊕ **Example 13:** Find the Correlation Coefficient Between k_2 and k_3

Solution: If ρ_{XY} denotes the correlation coefficient between random variables X and Y, then by definition:

$$\rho_{XY} = \frac{E\big[(X-E[X])(Y-E[Y])\big]}{\sqrt{Var(X)\,Var(Y)}} \quad \text{so that} \quad \rho_{k_2\,k_3} = \frac{E\big[(k_2-\kappa_2)(k_3-\kappa_3)\big]}{\sqrt{\mu_2(k_2)\,\mu_2(k_3)}}$$

The solution (expressed here in terms of cumulants) is thus:

$$\frac{\mathtt{RawMomentToCumulant[1,\,(k2-\kappa_2)\,(k3-\kappa_3)]}}{\sqrt{\mathtt{CentralMomentToCumulant[2,\,k2]\,CentralMomentToCumulant[2,\,k3]}}}$$

$$= \frac{\frac{6\,\kappa_2\,\kappa_3}{-1+n} + \frac{\kappa_5}{n}}{\sqrt{\left(\frac{2\,\kappa_2^2}{-1+n} + \frac{\kappa_4}{n}\right)\left(\frac{6\,n\,\kappa_2^3}{(-2+n)\,(-1+n)} + \frac{9\,\kappa_3^2}{-1+n} + \frac{9\,\kappa_2\,\kappa_4}{-1+n} + \frac{\kappa_6}{n}\right)}}$$

§7.3 C   MOMENTS OF SAMPLING DISTRIBUTIONS   269

Since $E\bigl[(X - E[X])(Y - E[Y])\bigr] = \mu_{1,1}(X, Y)$, we could alternatively derive the numerator as:

CentralMomentToCumulant[{1, 1}, {k2, k3}]

$$\frac{6\,\kappa_2\,\kappa_3}{-1 + n} + \frac{\kappa_5}{n}$$

which gives the same answer. ∎

○ *Product Cumulants*

These tools can be used to check the tables of product cumulants provided in texts such as Stuart and Ord (1994), which in turn are based on Fisher's (1928) results (with corrections). We find full agreement, except for $\kappa_{2,2}(k_3, k_2)$ (Stuart and Ord, equation 12.70) which we correctly obtain as:

CumulantMomentToCumulant[{2, 2}, {k3, k2}]

$$\frac{288\,n\,\kappa_2^5}{(-2+n)\,(-1+n)^3} + \frac{288\,(-23 + 10\,n)\,\kappa_2^2\,\kappa_3^2}{(-2+n)\,(-1+n)^3} +$$

$$\frac{360\,(-7 + 4\,n)\,\kappa_2^3\,\kappa_4}{(-2+n)\,(-1+n)^3} + \frac{36\,(160 - 155\,n + 38\,n^2)\,\kappa_3^2\,\kappa_4}{(-2+n)\,(-1+n)^3\,n} +$$

$$\frac{36\,(93 - 103\,n + 29\,n^2)\,\kappa_2\,\kappa_4^2}{(-2+n)\,(-1+n)^3\,n} +$$

$$\frac{24\,(202 - 246\,n + 71\,n^2)\,\kappa_2\,\kappa_3\,\kappa_5}{(-2+n)\,(-1+n)^3\,n} + \frac{2\,(113 - 154\,n + 59\,n^2)\,\kappa_5^2}{(-1+n)^3\,n^2} +$$

$$\frac{6\,(-131 + 67\,n)\,\kappa_2^2\,\kappa_6}{(-2+n)\,(-1+n)^2\,n} + \frac{3\,(117 - 166\,n + 61\,n^2)\,\kappa_4\,\kappa_6}{(-1+n)^3\,n^2} +$$

$$\frac{6\,(-27 + 17\,n)\,\kappa_3\,\kappa_7}{(-1+n)^2\,n^2} + \frac{37\,\kappa_2\,\kappa_8}{(-1+n)\,n^2} + \frac{\kappa_{10}}{n^3}$$

By contrast, Fisher (1928) and Stuart and Ord (1994) give the coefficient of the $\kappa_2^3\,\kappa_4$ term as $\frac{72\,(-23+14\,n)}{(-2+n)\,(-1+n)^3}$; for the $\kappa_2^2\,\kappa_3^2$ term: $\frac{144\,(-44+19\,n)}{(-2+n)\,(-1+n)^3}$. There is also a small typographic error in Stuart and Ord equation 12.66, $\kappa_{2,1}(k_4, k_2)$, though this is correctly stated in Fisher (1928).

⊕ **Example 14:** Show That Fisher's (1928) Solution for $\kappa_{2,2}(k_3, k_2)$ Is Incorrect

If we can show that Fisher's solution is wrong for one distribution, it must be wrong generally. In this vein, let $X \sim \text{Bernoulli}(\frac{1}{2})$, so that $X^i = X$ for any integer i. Hence, $s_1 = s_2 = s_3 = Y \sim \text{Binomial}(n, \frac{1}{2})$ (cf. *Example 21* of Chapter 4). Recall that the k-statistics k_2 and k_3 were defined above in terms of power sums s_i. We can now replace all power sums s_i in k_2 and k_3 with the random variable Y:

```
K₂ = k2 /. sᵢ_ → y
```

$$\frac{n\,y - y^2}{(-1+n)\,n}$$

```
K₃ = k3 /. sᵢ_ → y
```

$$\frac{n^2\,y - 3\,n\,y^2 + 2\,y^3}{(-2+n)\,(-1+n)\,n}$$

where random variable $Y \sim \text{Binomial}(n, \frac{1}{2})$, with pmf $g(y)$:

```
g = Binomial[n, y] p^y (1 - p)^(n-y) /. p → 1/2;
domain[g] = {y, 0, n} && {n > 0, n ∈ Integers} && {Discrete};
```

We now want to calculate the product cumulant $\kappa_{2,2}(K_3, K_2)$ directly, when $Y \sim \text{Binomial}(n, \frac{1}{2})$. The product cumulant $\kappa_{2,2}$ can be expressed in terms of product raw moments as follows:

```
κ22 = CumulantToRaw[{2, 2}]
```

$$\kappa_{2,2} \to -6\,\mu'^2_{0,1}\,\mu'^2_{1,0} + 2\,\mu'_{0,2}\,\mu'^2_{1,0} + 8\,\mu'_{0,1}\,\mu'_{1,0}\,\mu'_{1,1} - 2\,\mu'^2_{1,1} - 2\,\mu'_{1,0}\,\mu'_{1,2} + 2\,\mu'^2_{0,1}\,\mu'_{2,0} - \mu'_{0,2}\,\mu'_{2,0} - 2\,\mu'_{0,1}\,\mu'_{2,1} + \mu'_{2,2}$$

as given in Cook (1951). Here, each term $\mu'_{r,s}$ denotes $\mu'_{r,s}(K_3, K_2) = E[K_3^r K_2^s]$, and hence can be evaluated with the Expect function. In the next input, we calculate each of the expectations that we require:

```
Ω = κ22[[2]] /. μ'_(r_,s_) :→ Expect[K₃^r K₂^s, g] // Simplify
```

$$\frac{496 - 405\,n + 124\,n^2 - 18\,n^3 + n^4}{32\,(-2+n)\,(-1+n)^3\,n^3}$$

Hence, Ω is the value of $\kappa_{2,2}(k_3, k_2)$ when $X \sim \text{Bernoulli}(\frac{1}{2})$.

Fisher (1928) obtains, for any distribution whose moments exist, that $\kappa_{2,2}(k_3, k_2)$ is:

$$\begin{aligned}
\text{Fisher} = {} & \frac{288\,n\,\kappa_2^5}{(-2+n)\,(-1+n)^3} + \frac{144\,(-44+19\,n)\,\kappa_2^2\,\kappa_3^2}{(-2+n)\,(-1+n)^3} + \\
& \frac{72\,(-23+14\,n)\,\kappa_2^3\,\kappa_4}{(-2+n)\,(-1+n)^3} + \frac{36\,(160-155\,n+38\,n^2)\,\kappa_3^2\,\kappa_4}{(-2+n)\,(-1+n)^3\,n} + \\
& \frac{36\,(93-103\,n+29\,n^2)\,\kappa_2\,\kappa_4^2}{(-2+n)\,(-1+n)^3\,n} + \frac{24\,(202-246\,n+71\,n^2)\,\kappa_2\,\kappa_3\,\kappa_5}{(-2+n)\,(-1+n)^3\,n} + \\
& \frac{2\,(113-154\,n+59\,n^2)\,\kappa_5^2}{(-1+n)^3\,n^2} + \frac{6\,(-131+67\,n)\,\kappa_2^2\,\kappa_6}{(-2+n)\,(-1+n)^2\,n} + \\
& \frac{3\,(117-166\,n+61\,n^2)\,\kappa_4\,\kappa_6}{(-1+n)^3\,n^2} + \frac{6\,(-27+17\,n)\,\kappa_3\,\kappa_7}{(-1+n)^2\,n^2} + \frac{37\,\kappa_2\,\kappa_8}{(-1+n)\,n^2} + \frac{\kappa_{10}}{n^3};
\end{aligned}$$

Now, when $X \sim \text{Bernoulli}(\frac{1}{2})$, with pmf $f(x)$:

```
f = 1/2;  domain[f] = {x, 0, 1} && {Discrete};
```

... the cumulant generating function is:

```
cgf = Log[Expect[e^(t x), f]]
```

$$\text{Log}\left[\frac{1}{2}\,(1 + e^t)\right]$$

and so the first 10 cumulants are:

```
κlis = Table[κ_r → D[cgf, {t, r}] /. t → 0, {r, 10}]
```

$\{\kappa_1 \to \frac{1}{2},\ \kappa_2 \to \frac{1}{4},\ \kappa_3 \to 0,\ \kappa_4 \to -\frac{1}{8},\ \kappa_5 \to 0,$
$\kappa_6 \to \frac{1}{4},\ \kappa_7 \to 0,\ \kappa_8 \to -\frac{17}{16},\ \kappa_9 \to 0,\ \kappa_{10} \to \frac{31}{4}\}$

... so Fisher's solution becomes:

```
Fsol = Fisher /. κlis // Simplify
```

$$\frac{496 - 405\,n + 124\,n^2 - 72\,n^3 + 28\,n^4}{32\,(-2+n)\,(-1+n)^3\,n^3}$$

which is *not* equal to Ω derived above. Hence, Fisher's (1928) solution must be incorrect. How does **mathStatica** fare? When $X \sim \text{Bernoulli}(\frac{1}{2})$, our solution is:

```
CumulantMomentToCumulant[{2, 2}, {k3, k2}]
   /. κlis // Simplify
```

$$\frac{496 - 405\,n + 124\,n^2 - 18\,n^3 + n^4}{32\,(-2+n)\,(-1+n)^3\,n^3}$$

which *is* identical to Ω, as we would expect. How big is the difference between the two solutions? The following output shows that, when $X \sim \text{Bernoulli}(\frac{1}{2})$, Fisher's solution is at least 28 times too large, and as much as 188 times too large:

```
Fsol/Ω /. n → {11, 20, 50, 100, 500, 1000000000} // N
```

{188.172, 68.0391, 38.7601, 32.8029, 28.802, 28.}

This comparison is only valid for n greater than the weight w of $\kappa_{2,2}(k_3, k_2)$, where $w = 10$ here. Weights are defined in the next section. ■

7.4 Augmented Symmetrics and Power Sums

7.4 A Definitions and a Fundamental Expectation Result

This section does not strive to solve new problems; instead, it describes the building blocks upon which unbiased estimators and *moments of moments* are built. Primarily, it deals with converting expressions such as the three-part sum $\sum_{i \neq j \neq k} X_i X_j^2 X_k^2$ into one-part sums such as $\sum_{i=1}^{n} X_i^r$. The former are called *augmented symmetrics functions*, while the latter are one-part symmetrics, more commonly known as *power sums*. Formally, as per §7.1 B, the rth *power sum* is defined as

$$s_r = \sum_{i=1}^{n} X_i^r, \qquad r = 1, 2, \ldots \tag{7.12}$$

Further, let $A_{\{a,b,c,\ldots\}}$ denote an augmented symmetric function of the variates. For example,

$$A_{\{3,2,2,1\}} = \sum_{i \neq j \neq k \neq m} X_i^3 X_j^2 X_k^2 X_m^1 \tag{7.13}$$

where each index in the four-part sum ranges from 1 to n. For any list of positive integers t, the *weight* of A_t is $w = \Sigma t$, while the *order*, or number of parts, is the dimension of t, which we denote by ρ. For instance, $A_{\{3,2,2,1\}}$ has weight 8, and order 4. For convenience, one can notate $A_{\{3,2,2,1,1,1,1\}}$ as $A_{\{3,2^2,1^4\}}$ corresponding to an 'extended form' and 'condensed form' notation, respectively. Many authors would denote $A_{\{3,2^2,1^4\}}$ by the expression [3 2^2 1^4]; unfortunately, this notation is ill-suited to *Mathematica* where [ ] notation is already 'taken'.

This section provides tools that enable one to:

(i) express an augmented symmetric function in terms of power sums; that is, find function f such that $A_t = f(s_1, s_2, \ldots, s_w)$—each term in f will be *isobaric* (have the same weight w);

(ii) express products of power sums (*e.g.* $s_1 s_2 s_3$) in terms of augmented symmetric functions.

Past attempts: Considerable effort has gone into deriving tables to convert between symmetric functions and power sums. This includes the work of O'Toole (1931, weight 6, contains errors), Dwyer (1938, weight 6), Sukhatme (1938, weight 8), and Kerawala and Hanafi (1941, 1942, 1948) for $w = 9$ through 12 (errors). David and Kendall (1949) independently derived a particularly neat set of tables up to weight 12, though this set is also not free of error, though a later version, David *et al.* (1966, weight 12) appears to be correct. With **mathStatica**, we can extend the analysis far beyond weight 12, and derive correct solutions of even weight 20 in just a few seconds.

Augmented Symmetrics to Power Sums

The **mathStatica** function `AugToPowerSum` converts a given augmented symmetric function into power sums. Here we find $[3\, 2^3] = A_{\{3,\,2^3\}}$ in terms of power sums:

AugToPowerSum[{3, 2, 2, 2}]

$A_{\{3,2,2,2\}} \to s_2^3\, s_3 - 3\, s_2\, s_3\, s_4 - 3\, s_2^2\, s_5 + 3\, s_4\, s_5 + 2\, s_3\, s_6 + 6\, s_2\, s_7 - 6\, s_9$

The integers in `AugToPowerSum[{3, 2, 2, 2}]` do not need to be any particular order. In fact, one can even use 'condensed-form' notation:[4]

AugToPowerSum[{3, 2³}]

$A_{\{3,2,2,2\}} \to s_2^3\, s_3 - 3\, s_2\, s_3\, s_4 - 3\, s_2^2\, s_5 + 3\, s_4\, s_5 + 2\, s_3\, s_6 + 6\, s_2\, s_7 - 6\, s_9$

Standard tables also list the related monomial symmetric functions, though these are generally less useful than the augmented symmetrics. Using condensed form notation, the *monomial symmetric* $M_{\{a^\alpha,\, b^\beta,\, c^\chi,\, \ldots\}}$ is defined by:

$$M_{\{a^\alpha,\, b^\beta,\, c^\chi,\, \ldots\}} = \frac{A_{\{a^\alpha,\, b^\beta,\, c^\chi,\, \ldots\}}}{\alpha!\, \beta!\, \chi!\, \ldots}\,. \tag{7.14}$$

mathStatica provides a function to express monomial symmetric functions in terms of power sums. Here is $M_{\{3,\, 2^3\}}$:

MonomialToPowerSum[{3, 2³}]

$M_{\{3,2,2,2\}} \to \frac{1}{6}\, s_2^3\, s_3 - \frac{1}{2}\, s_2\, s_3\, s_4 - \frac{1}{2}\, s_2^2\, s_5 + \frac{s_4\, s_5}{2} + \frac{s_3\, s_6}{3} + s_2\, s_7 - s_9$

Power Sums to Augmented Symmetrics

The **mathStatica** function `PowerSumToAug` converts products of power sums into augmented symmetric functions. For instance, to find $s_1\, s_2^3$ in terms of $A_{\{\}}$:

PowerSumToAug[{1, 2, 2, 2}]

$s_1\, s_2^3 \to A_{\{7\}} + 3\, A_{\{4,3\}} + 3\, A_{\{5,2\}} + A_{\{6,1\}} + 3\, A_{\{3,2,2\}} + 3\, A_{\{4,2,1\}} + A_{\{2,2,2,1\}}$

Here is an example with weight 20 and order 20. It takes less than a second to find the solution, but many pages to display the result:

PowerSumToAug[{1²⁰}]; // Timing

{0.93 Second, Null}

Like most other converter functions, these functions also allow one to specify ones own notation. Here, we keep 's' to denote power sums, but change the $A_{\{\}}$ terms to $\lambda_{\{\}}$:

PowerSumToAug[{3, 2, 3}, s, λ]

$s_2 \, s_3^2 \rightarrow \lambda_{\{8\}} + 2 \, \lambda_{\{5,3\}} + \lambda_{\{6,2\}} + \lambda_{\{3,3,2\}}$

○ *A Fundamental Expectation Result*

A fundamental expectation result (Stuart and Ord (1994), Section (12.5)) is that

$$E[A_{\{a,b,c,\ldots\}}] = \acute{\mu}_a \, \acute{\mu}_b \, \acute{\mu}_c \cdots \times n(n-1) \cdots (n-p+1) \qquad (7.15)$$

where, given A_t, the symbol p denotes the number of elements in the list t. This result is important because it lies at the very heart of both the unbiased estimation of population moments, and the *moments of moments* literature (see §7.4 B and C below). As a simple illustration, suppose we want to prove that $\acute{m}_r$ is an unbiased estimator of $\acute{\mu}_r$ (7.4): to do so, we first express $\acute{m}_r = \frac{s_r}{n} = \frac{A_{\{r\}}}{n}$ so that we have an expression in $A_{\{r\}}$, and then apply (7.15) to yield $E[\acute{m}_r] = \frac{1}{n} E[A_{\{r\}}] = \acute{\mu}_r$.

We can implement (7.15) in *Mathematica* as follows:

ExpectAug[t_] :=

$\left(\text{Thread}[\acute{\mu}_t] \, / . \, \text{List} \rightarrow \text{Times}\right) \prod_{i=0}^{\text{Length}[t]-1} (n - i)$

Thus, the expectation of say $A_{\{2,2,3\}}$ is given by:

ExpectAug[{2, 2, 3}]

$(-2 + n) \, (-1 + n) \, n \, \acute{\mu}_2^2 \, \acute{\mu}_3$

⊕ *Example 15:* An Unbiased Estimator of $\acute{\mu}_3 \, \acute{\mu}_4$

In *Example 6*, we found an unbiased estimator of $\acute{\mu}_3 \, \acute{\mu}_4$ by converting to cumulants, and then finding an unbiased estimator for each cumulant by using polykays. It is much easier to apply the expectation theorem (7.15) directly, from which it follows immediately that an unbiased estimator of $\acute{\mu}_3 \, \acute{\mu}_4$ is $\frac{A_{\{3,4\}}}{n(n-1)}$, where $A_{\{3,4\}}$ is given by:

AugToPowerSum[{3, 4}]

$A_{\{3,4\}} \rightarrow s_3 \, s_4 - s_7$

as we found in *Example 6*.

7.4 B Application 1: Understanding Unbiased Estimation
Augmented Symmetrics → Power Sums

Let us suppose that we wish to find an unbiased estimator of $\kappa_2 \kappa_1 \kappa_1$ from first principles. Now, $\kappa_2 \kappa_1 \kappa_1$ can be written in terms of raw moments:

```
z1 = Times @@
        Map[CumulantToRaw[#][[2]] &, {2, 1, 1}] // Expand
```

$-\mu_1'^4 + \mu_1'^2 \mu_2'$

We have just found the coefficients of the polykay $k_{(2,1,1)}$ in terms of so-called *Wishart Tables* (see Table 1 of Wishart (1952) or Appendix 11 of Stuart and Ord (1994)). To obtain the inverse relation in such tables, use `RawToCumulant` instead of `CumulantToRaw`. In `ListForm` notation (noting that the order of the terms is now reversed), we have:

```
z2 = ListForm[z1, μ́]
```

$\acute{\mu}[\{1, 1, 2\}] - \acute{\mu}[\{1, 1, 1, 1\}]$

By the fundamental expectation result (7.15), an unbiased estimator of z1 (or z2) is:

```
z3 = z2 /. μ́[x_] :> AugToPowerSum[x][[2]] / ∏_{i=0}^{Length[x]-1} (n - i)   // Factor
```

$$\frac{-s_1^4 + 3 s_1^2 s_2 + n s_1^2 s_2 - n s_2^2 - 2 s_1 s_3 - 2 n s_1 s_3 + 2 n s_4}{(-3 + n)(-2 + n)(-1 + n) n}$$

This result is identical to `PolyK[{2,1,1}]`, other than the ordering of the terms.

7.4 C Application 2: Understanding Moments of Moments
Products of Power Sums → Augmented Symmetrics

We wish to find an exact method for finding moments of sampling distributions in terms of population moments, which is what the *moments of moments* functions do, but now from first principles. Equation (7.15) enables one to find the expectation of a moment, by implementing the following three steps:

(i)   convert that moment into power sums,
(ii) convert the power sums into augmented symmetrics, and
(iii) then apply the fundamental expectation result (7.15) using `ExpectAug`.

For example, to find $E[m_4]$, we first convert m_4 into power sums s_i:

```
m4 = SampleCentralToPowerSum[4][[2]]
```

$$-\frac{3 s_1^4}{n^4} + \frac{6 s_1^2 s_2}{n^3} - \frac{4 s_1 s_3}{n^2} + \frac{s_4}{n}$$

Then, after converting into `ListForm`, convert into augmented symmetrics:

```
z1 = ListForm[m4, s] /. s[x_] :> PowerSumToAug[x][[2]]
```

$$\frac{A_{\{4\}}}{n} - \frac{4\,(A_{\{4\}} + A_{\{3,1\}})}{n^2} + \frac{6\,(A_{\{4\}} + A_{\{2,2\}} + 2\,A_{\{3,1\}} + A_{\{2,1,1\}})}{n^3} - \frac{3\,(A_{\{4\}} + 3\,A_{\{2,2\}} + 4\,A_{\{3,1\}} + 6\,A_{\{2,1,1\}} + A_{\{1,1,1,1\}})}{n^4}$$

We can now apply the fundamental expectation result (7.15):

```
z2 = z1 /. A_t_ :> ExpectAug[t]  // Simplify
```

$$-\frac{1}{n^3}\left((-1+n)\left(3\,(6-5n+n^2)\,\mu_1'^4 - 6\,(6-5n+n^2)\,\mu_1'^2\,\mu_2' + (9-6n)\,\mu_2'^2 + 4\,(3-3n+n^2)\,\mu_1'\,\mu_3' - (3-3n+n^2)\,\mu_4'\right)\right)$$

This output is identical to that given by `RawMomentToRaw[1, m4]`, except that the latter does a better job of ordering the terms of the resulting polynomial.

7.5 Exercises

1. Which of the following are rational, integral, algebraic symmetric functions?

   (i) $\sum_{i=1}^{n} X_i^2$ (ii) $\left(\sum_{i=1}^{n} X_i\right)^2$ (iii) $\frac{1}{n}\sum_{i=1}^{n}(X_i - \bar{X})^2$ (iv) $\sqrt{\sum_{i=1}^{n}(X_i - \bar{X})^2}$

   (v) $h_2\,m_3^2$ (vi) h_2/m_3^2 (vii) $h_2 + m_3^2$ (viii) $\sqrt{h_2\,m_3^2}$

2. Express each of the following in terms of power sums:

   (i) $\sum_{i=1}^{n} X_i^4$ (ii) $\left(\sum_{i=1}^{n} X_i\right)^2$ (iii) $m_3 = \frac{1}{n}\sum_{i=1}^{n}(X_i - \bar{X})^3$

   (iv) $k_4\,m_2^3$ (v) $(h_3 - 5)^2$ (vi) $\sum_{i=1}^{n}\left((X_i - \bar{X})^3\,(Y_i - \bar{Y})^2\right)$

3. Find an unbiased estimator of: (i) μ_3 (ii) $\mu_3^2\,\mu_2$ (iii) κ_{13} ☺   (iv) the sixth factorial moment. Verify that each solution is, in fact, an unbiased estimator.

4. Solve the following: (i) $\text{Var}(m_4)$ (ii) $E\left[\sum_{i=1}^{n} X_i^2\right]$ (iii) $E\left[\left(\sum_{i=1}^{n} X_i\right)^2\right]$

   (iv) $\kappa_4(k_2)$ (v) $\mu_{3,2}(h_2, h_3)$.

5. Let $(X_1, \ldots, X_n)$ denote a random sample of size n drawn from $X \sim \text{Lognormal}(\mu, \sigma)$. Let $Y = \sum_{i=1}^{n} X_i$. Find the first 4 raw moments of Y.

6. Find the covariance between $\frac{1}{n-1}\sum_{i=1}^{n}(X_i - \bar{X})^2$ and $\frac{1}{n}\sum_{i=1}^{n} X_i$. What can be said about the covariance if the population is symmetric?

Chapter 8

Asymptotic Theory

8.1 Introduction

Asymptotic theory is often used to justify the selection of particular estimators. Indeed, it is commonplace in modern statistical practice to base inference upon a suitable asymptotic theory. Desirable asymptotic properties — *consistency* and *limiting Normality* — can sometimes be ascribed to an estimator, even when there is relatively little known, or assumed known, about the population in question. In this chapter, we focus on both of these properties in the context of the sample mean,

$$\overline{X}_n = \frac{1}{n} \sum_{i=1}^{n} X_i$$

and the sample sum,

$$S_n = \sum_{i=1}^{n} X_i$$

where symbol n denotes the sample size. We have especially attached n as a subscript to emphasise that $\overline{X}_n$ and S_n are random variables that depend on n. In subsequent chapters, we shall examine the asymptotic properties of estimators with more complicated structures than $\overline{X}_n$ and S_n. Our discussion of asymptotic theory centres on asking: What happens to an estimator (such as the sample mean) as n becomes large (in fact, as $n \to \infty$)? Thus, our presentation of asymptotic theory can be viewed as a theory relevant to increasing sample sizes. Of course, we require that the random variables used to form $\overline{X}_n$ and S_n must exist at each and every value of n. Accordingly, for an asymptotic theory to make sense, infinite-length sequences of random variables must be allowed to exist. For example, for $\overline{X}_n$ and S_n, the sequence of underlying random variables would be

$$(X_1, X_2, \ldots, X_i, X_{i+1}, \ldots) = \{X_n\}_{n=1}^{\infty}.$$

Throughout this chapter, apart from one or two exceptions, we shall work with examples dealing with the simplest of cases; namely, when all variables in the sequence are independent and identically distributed. Our treatment is therefore pitched at an elementary level.

The asymptotic properties of consistency and asymptotic normality are due to, respectively, the concepts of *convergence in probability* (§8.5) and *convergence in distribution* (§8.2). Moreover, these properties can often be established in a variety of situations through application of two fundamental theorems of asymptotic theory: *Khinchine's Weak Law of Large Numbers* and *Lindeberg–Lévy's Central Limit Theorem*.

The *Mathematica* tools needed in a chapter on asymptotic theory depend, not surprisingly, in large part on the built-in `Limit` function; however, we will also use the add-on package `Calculus`Limit``. The add-on *removes* and *replaces* the built-in `Limit` function with an alternate algorithm for computing limits. As its development ceased a few years ago, we would ideally prefer to ignore this package altogether and use only the built-in `Limit` function, for the latter *is* subject to ongoing research and development.[1] Unfortunately, the world is not ideal! The built-in `Limit` function in Version 4 of *Mathematica* is unable to perform some limits that are commonplace in statistics, whereas if `Calculus`Limit`` is implemented, a number of these limits can be computed correctly. The solution that we adopt is to load and unload the add-on as needed. To illustrate our approach, consider the following limit (see *Example 2*) which cannot be solved by built-in `Limit` (try it and see!):

$$\lim_{n \to \infty} \text{Binomial}[n, x] \left(\frac{\theta}{n}\right)^x \left(1 - \frac{\theta}{n}\right)^{n-x}.$$

With `Calculus`Limit`` loaded, a solution to the limit is reported. Enter the following:

```
<< Calculus`Limit`
Limit[Binomial[n, x] (θ/n)^x (1 - θ/n)^(n-x), n → ∞];
Unprotect[Limit]; Clear[Limit];
```

The limit is computed correctly — we suppress the output here — what is important to see is the procedure for loading and unloading the `Calculus`Limit`` add-on.

Asymptotic theory is so widespread in its application that there is already an extensive field of literature in probability and statistics that contributes to its development. Accordingly, we shall cite only a select collection of works that we have found to be of particular use in preparing this chapter: Amemiya (1985), Bhattacharya and Rao (1976), Billingsley (1995), Chow and Teicher (1978), Hogg and Craig (1995), McCabe and Tremayne (1993) and Mittelhammer (1996).

8.2   Convergence in Distribution

The cumulative distribution function (cdf) has three attractive properties associated with it, namely (i) all random variables possess a cdf, (ii) the cdf has a range that is bounded within the closed unit interval [0,1], and (iii) the cdf is monotonic increasing. So when studying the behaviour of a sequence of random variables, we may, possibly just as easily, consider the behaviour of the infinite sequence of associated cdf's. This leads to the concept of convergence in distribution, a definition of which follows.

Let the random variable X_n have cdf F_n at each value of $n = 1, 2, \ldots$. Also, let the random variable X have cdf F, where X and F do not depend upon n. If it can be shown that

$$\lim_{n \to \infty} F_n(x) = F(x) \tag{8.1}$$

for all points x at which $F(x)$ is continuous, then X_n is said to *converge in distribution* to X.[2] A common notation to denote convergence in distribution is

$$X_n \xrightarrow{d} X. \tag{8.2}$$

F is termed the *limit distribution* of X_n.

⊕ **Example 1:** The Limit Distribution of a Sample Mean

In this example, the limiting distribution of the sample mean is derived, assuming that the population from which random samples are drawn is $N(0, 1)$. For a random sample of size n, the sample mean $\overline{X}_n \sim N(0, \frac{1}{n})$ (established in *Example 24* of Chapter 4). Therefore, the pdf and support of $\overline{X}_n$ are:

```
f = e^(-x̄²/(2/n)) / √(2π/n) ;     domain[f] = {x̄, -∞, ∞} && {n > 0};
```

while the cdf (evaluated at a point x) is:

```
Fₙ = Prob[x, f]
```

$$\frac{1}{2}\left(1 + \text{Erf}\left[\frac{\sqrt{n}\, x}{\sqrt{2}}\right]\right)$$

The limiting behaviour of the cdf depends on the sign of x. Here, we evaluate $\lim_{n \to \infty} F_n(x)$ when x is negative (say $x = -1$), zero, and positive (say $x = 1$):

```
<< Calculus`Limit`

Limit[Fₙ /. x → {-1, 0, 1}, n → ∞]

Unprotect[Limit]; Clear[Limit];
```

$$\left\{0, \frac{1}{2}, 1\right\}$$

The left-hand side of (8.1) is, in this case, a step function with a discontinuity at the origin, as the left panel of Fig. 1 shows.

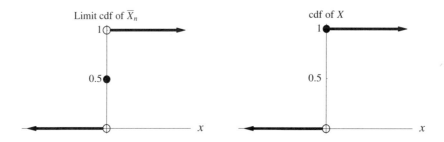

Fig. 1: Limit cdf of $\overline{X}_n$, and cdf of X

Now consider a random variable X whose cdf evaluated at a point x is given by

$$F(x) = \begin{cases} 0 & \text{if } x < 0 \\ 1 & \text{if } x \geq 0 \end{cases}$$

Comparing the graph of the cdf of X (given in the right panel of Fig. 1) to the graph of the limit of the cdf of $\overline{X}_n$, we see that both are identical at all points apart from when $x = 0$. However, because both graphs are discontinuous at $x = 0$, it follows that definition (8.1) holds, and so

$$\overline{X}_n \xrightarrow{d} X.$$

F is the limiting distribution function of $\overline{X}_n$. Now, focusing upon the random variable X and its cdf F, notice that F assigns all probability to a single point at the origin. Since X takes only one value, 0, with probability one, then X is a degenerate random variable, and F is termed a *degenerate distribution*. This is one instance where the limiting distribution provides information about the probability space of the underlying random variable. ∎

⊕ *Example 2:* The Poisson as the Limit Distribution of a Binomial

It is sometimes possible to show convergence in distribution by deriving the limiting behaviour of functions other than the cdf, such as the pdf/pmf, the mgf, or the cf. This means that convergence in distribution becomes an issue of convergence of an infinite-length sequence of pdf/pmf, mgf, or cf.

In this example, convergence in distribution is illustrated by deriving the limit of a sequence of pmf. Recall that the Binomial(n, p) distribution has mean np. Suppose that $X_n \sim$ Binomial$(n, \theta/n)$ (then $0 < \theta < n$); furthermore, assume that θ remains finite as n increases. To interpret the assumption on θ, note that $E[X_n] = n\theta/n = \theta$; thus, for every sample size n, the mean remains fixed and finite at the value of θ. Let f denote the pmf of X_n. Then:

```
f = Binomial[n, x] (θ/n)^x (1 - θ/n)^(n-x);

domain[f] =
   {x, 0, n} && {0 < θ < n, n > 0, n ∈ Integers} && {Discrete};
```

§8.2   ASYMPTOTIC THEORY   281

```
<< Calculus`Limit`
Limit[f, n → ∞]
Unprotect[Limit]; Clear[Limit];
```

$$\frac{e^{-\theta}\,\theta^x}{\Gamma[1+x]}$$

Because $\Gamma[1+x] = x!$ for integer $x \geq 0$, this expression is equivalent to the pmf of a variable which is Poisson distributed with parameter θ. Therefore, under our assumptions,

$$X_n \xrightarrow{d} X \sim \text{Poisson}(\theta).$$

The limiting distribution of the Binomial random variable X_n is thus Poisson(θ).

⊕ **Example 3:** The Normal as the Limit Distribution of a Binomial

In the previous example, both the limit distribution and the random variables in the sequence were defined over a discrete sample space. However, this equivalence need not always occur: the limit distribution of a discrete variable may be continuous, or a continuous random variable may have a discrete limit distribution, as seen in *Example 1* (albeit that it was a degenerate limit distribution).

In this example, convergence in distribution is illustrated by deriving the limit of a sequence of moment generating functions (mgf). Suppose that $X_n \sim \text{Binomial}(n, \theta)$, where $0 < \theta < 1$. Unlike the previous example where the probability of a 'success' diminished with n, in this example the probability stays fixed at θ for all n. Let f once again denote the pmf of X_n:

```
    f = Binomial[n, x] θ^x (1 - θ)^(n-x);
domain[f] =
    {x, 0, n} && {0 < θ < 1, n > 0, n ∈ Integers} && {Discrete};
```

Then, the mgf of X_n is derived as:

```
mgf_X = Expect[e^(t x), f]
```

$$(1 + (-1 + e^t)\,\theta)^n$$

Now consider the standardised random variable Y_n defined as

$$Y_n = \frac{X_n - E[X_n]}{\sqrt{\text{Var}(X_n)}} = \frac{X_n - n\theta}{\sqrt{n\,\theta(1-\theta)}}.$$

Y_n necessarily has a mean of 0 and a variance of 1. The mgf of Y_n can be obtained using the MGF Theorem (§2.4 D), setting a and b in that theorem equal to:

$$a = \frac{-n\theta}{\sqrt{n\theta(1-\theta)}}; \quad b = \frac{1}{\sqrt{n\theta(1-\theta)}};$$

to find:

$$\mathtt{mgf_Y = e^{at}\ (mgf_X\ /.\ t \to b\,t)}$$

$$e^{-\frac{n t \theta}{\sqrt{n(1-\theta)\theta}}} \left(1 + \left(-1 + e^{\frac{t}{\sqrt{n(1-\theta)\theta}}}\right)\theta\right)^n$$

Executing built-in `Limit`, we find the limit mgf of the infinite sequence of mgf's equal to:

$$\mathtt{Limit[mgf_Y,\ n \to \infty]}$$

$$e^{\frac{t^2}{2}}$$

As this last expression is equivalent to the mgf of a $N(0, 1)$ variable, it follows that

$$Y_n \xrightarrow{d} Z \sim N(0, 1).$$

Thus, the limiting distribution of a standardised Binomial random variable is the standard Normal distribution. ∎

8.3 Asymptotic Distribution

Suppose, for example, that we have established the following limiting distribution for a random variable X_n:

$$X_n \xrightarrow{d} Z \sim N(0, 1). \tag{8.3}$$

Let n_* denote a fixed and finite sample size; for example, n_* might correspond to the sample size of the data set with which we are working. In the absence of any knowledge about the exact distribution of X_n, it makes sense to use the limiting distribution of X_n as an approximation to the distribution of X_{n_*}, for if n_* is *sufficiently large*, the discrepancy between the exact distribution and the posited approximation must be small due to (8.3). This approximation is referred to as the *asymptotic distribution*. A commonly used notation for the asymptotic distribution is

$$X_{n_*} \stackrel{a}{\sim} N(0, 1) \tag{8.4}$$

which reads literally as 'the asymptotic distribution of X_{n_*} is $N(0, 1)$', or 'the approximate distribution of X_{n_*} is $N(0, 1)$'.

Of course, the variable that is of interest to us need not necessarily be X_{n_*}. However, if we know the relationship between X_{n_*} and the variable of interest, Y_{n_*} say, it is often possible to derive the asymptotic distribution for the latter. For example, if

$Y_{n_*} = \mu + \sigma X_{n_*}$, where $\mu \in \mathbb{R}$ and $\sigma \in \mathbb{R}_+$, then the asymptotic distribution of Y_{n_*} may be obtained directly from (8.4) using the properties of the Normal distribution:

$$Y_{n_*} \stackrel{a}{\sim} N(\mu, \sigma^2).$$

As a second example, suppose that W_{n_*} is related to X_{n_*} by the transformation $W_{n_*} = X_{n_*}^2$. Once again, the asymptotic distribution of W_{n_*} may be deduced by using the properties of the Normal distribution:

$$W_{n_*} \stackrel{a}{\sim} \text{Chi-squared}(1).$$

Typically, the distinction between arbitrary n and a specific value n_* is made implicit by dropping the * subscript. We too shall adopt this convention from now on.

⊕ ***Example 4:*** The Asymptotic Distribution of a Method of Moments Estimator

Let $X \sim \text{Chi-squared}(\theta)$, where $\theta \in \mathbb{R}_+$ is unknown. Let $(X_1, X_2, ..., X_n)$ denote a random sample of size n drawn on X. The *method of moments* (§5.6) estimator of θ is the sample mean $\bar{X}_n$. Further, let Z_n be related to $\bar{X}_n$ by the following location shift and scale change,

$$Z_n = \frac{\bar{X}_n - \theta}{\sqrt{2\theta/n}} \tag{8.5}$$

Since it can be shown that $Z_n \stackrel{d}{\longrightarrow} Z \sim N(0, 1)$, it follows that the asymptotic distribution of the estimator is

$$\bar{X}_n \stackrel{a}{\sim} N\left(\theta, \frac{2\theta}{n}\right).$$ ■

○ **van Beek Bound**

One way to assess the accuracy of the asymptotic distribution is to calculate an upper bound on the approximation error of its cdf. Such a bound has been derived by van Beek,[3] and generally applies when the limiting distribution is the standard Normal. The relevant result is typically expressed in the form of an inequality.

Let $(W_1, ..., W_n)$ be a set of n independent variables, each with *zero mean* and finite third absolute moment. Define

$$\mu_2 = \frac{1}{n} \sum_{i=1}^{n} E[W_i^2]$$

$$\mu_3^+ = \frac{1}{n} \sum_{i=1}^{n} E\big[|W_i|^3\big] \tag{8.6}$$

$$B = \frac{1}{\sqrt{n}} 0.7975 \, \mu_3^+ \, \mu_2^{-3/2}$$

and let

$$W_* = \frac{\overline{W}}{\sqrt{\mu_2/n}}$$

where $\overline{W}$ denotes the sample mean $\frac{1}{n}\sum_{i=1}^{n} W_i$. Then van Beek's inequality holds for all w_* in the support of the variable W_*, namely,

$$\left| F_n(w_*) - \Phi(w_*) \right| \le B \tag{8.7}$$

where $F_n(w_*)$ is the cdf of W_* evaluated at w_*, and $\Phi(w_*)$ is the cdf of a $N(0, 1)$ variable evaluated at the same point.[4] Some features of this result that are worth noting are: (i) the variables $(W_1, \ldots, W_n)$ need not be identically distributed, nor does their distribution need to be specified; (ii) van Beek's bound B decreases as the sample size increases, eventually reaching zero in the limit; and (iii) if $(W_1, \ldots, W_n)$ are identical in distribution to a random variable W, then $\mu_2 = E[W^2] = \text{Var}(W)$ and $\mu_3^+ = E\left[|W|^3\right]$. These simplifications will be useful in the next example.

⊕ *Example 5:* van Beek's Bound for the Method of Moments Estimator

We shall derive van Beek's bound B on the error induced by using the $N(0, 1)$ distribution to approximate the distribution of Z_n, where Z_n is the scaled method of moments estimator given in (8.5) in *Example 4*. Recall that $Z_n = (\overline{X}_n - \theta)/\sqrt{2\theta/n}$, where $\overline{X}_n$ is the sample mean of n independent and identically distributed Chi-squared(θ) random variables, each with pdf $f(x)$:

```
f = x^(θ/2-1) e^(-x/2) / (Γ[θ/2] 2^(θ/2));    domain[f] = {x, 0, ∞} && {θ > 0};
```

Note that van Beek's bound assumes a zero mean, whereas X has mean θ. To resolve this difference, we shall work *about the mean* and take $W = X - \theta$. We now derive $\mu_2 = E[W^2]$:

```
w = x - θ;    μ₂ = Expect[w², f]
```

2θ

To derive $\mu_3^+ = E\left[|W|^3\right] = E\left[|X - \theta|^3\right]$, note that *Mathematica* has difficulty integrating expressions with absolute values. Fortunately, **mathStatica** allows us to replace $|W|$ with an If[] statement. The calculation takes about 30 seconds on our reference machine:[5]

```
μ₃⁺ = Expect[ If[x < θ, -w, w]³, f]
```

$$\theta\left(-8 + \frac{1}{\Gamma[4 + \frac{\theta}{2}]} \left((2e)^{-\theta/2} \theta^{\frac{4+\theta}{2}} (6 + \theta)\right.\right.$$
$$\left.\left.\left(2 + \theta + e^{\theta/2} \theta \, \text{ExpIntegralE}\left[-2 - \frac{\theta}{2}, \frac{\theta}{2}\right]\right)\right)\right)$$

Since $\mu_2 = 2\theta$, we have

$$Z_n = \frac{\overline{X}_n - \theta}{\sqrt{2\theta/n}} = \frac{\overline{W}}{\sqrt{\mu_2/n}} = W_*$$

allowing us to apply van Beek's bound (8.7):

$$\mathbf{B} = \frac{0.7975}{\sqrt{\mathbf{n}}} \frac{\mu_3^+}{\mu_2^{3/2}};$$

which depends on θ and n. To illustrate, we select a sample size of $n = 20$ and set $\theta = 1$, to find:

$$\mathbf{B}\ /.\ \{\mathbf{n} \to 20,\ \theta \to 1\}\ //\ \mathbf{N}$$

0.547985

At our chosen point, van Beek's bound is particularly large, and so will not be of any real use in judging the effectiveness of the asymptotic distribution in this case. Fortunately, with **mathStatica**, it is reasonably straightforward to evaluate the exact value of the approximation error by computing the left-hand side of (8.7). Recalling that $S_n = \sum_{i=1}^{n} X_i$, we have

$$\begin{aligned}
F_n(w_*) &= P(Z_n \leq w_*) \\
&= P\left(\frac{n^{-1} S_n - \theta}{\sqrt{2\theta/n}} \leq w_*\right) \\
&= P(S_n \leq w_* \sqrt{2\theta n} + n\theta).
\end{aligned}$$

Example 23 of Chapter 4 shows that the random variable $S_n \sim$ Chi-squared($n\theta$). Its pdf $g(s_n)$ is thus:

$$\mathbf{g} = \frac{\mathbf{s_n}^{\frac{n\theta}{2}-1} e^{-\frac{\mathbf{s_n}}{2}}}{2^{\frac{n\theta}{2}} \Gamma[\frac{n\theta}{2}]}; \quad \mathbf{domain[g]} = \{\mathbf{s_n},\ 0,\ \infty\}\ \&\&\ \{\theta > 0,\ \mathbf{n} > 0\};$$

Then, $F_n(w_*)$ is:

$$\mathbf{F_n} = \mathbf{Prob}\left[w_* \sqrt{2\theta n} + n\theta,\ g\right]$$

$$1 - \frac{\text{Gamma}\left[\frac{n\theta}{2},\ \frac{n\theta}{2} + \frac{\sqrt{n\theta}\ w_*}{\sqrt{2}}\right]}{\Gamma[\frac{n\theta}{2}]}$$

After evaluating $\Phi(w_*)$, we can plot the *actual* error caused by approximating F_n with a Normal distribution, as shown in Fig. 2.

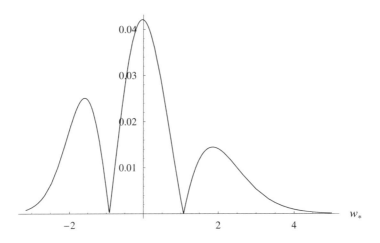

Fig. 2: Actual approximation error ($n = 20, \theta = 1$) in absolute value

It is easy to see from this diagram that at our selected values of n and θ, the discrepancy (in absolute value) between the exact cdf and the cdf of the asymptotic distribution is no larger than approximately 0.042. This is considerably lower than the reported van Beek bound of approximately 0.548. The error the asymptotic distribution induces is nevertheless fairly substantial in this case. Of course, as sample size increases, the size of the error must decline. ∎

8.4 Central Limit Theorem

§8.2 discussed the convergence in distribution of a sequence of random variables whose distribution was known. In practice, such information is often not available, thus jeopardising the derivation of the limiting distribution. In such cases, if the variables in the sequence are used to form sums and averages, such as S_n and $\overline{X}_n$, the limiting distribution can often be derived by applying the famous *Central Limit Theorem*. Since many estimators *are* functions of sums of random variables, the Central Limit Theorem is of considerable importance in statistics. See Le Cam (1986) for an interesting discussion of the history of the Central Limit Theorem.

We consider random variables constructed in the following manner,

$$\frac{S_n - a_n}{b_n} \tag{8.8}$$

where $\{a_n\}_{n=1}^{\infty}$ and $\{b_n\}_{n=1}^{\infty}$ represent sequences of real numbers. The random variables appearing in the sum S_n, namely, $\{X_i\}_{i=1}^{n}$, are the first n elements of the infinite-length sequence $\{X_n\}_{n=1}^{\infty}$. If we set

$$a_n = \sum_{i=1}^{n} E[X_i] \quad \text{and} \quad b_n^2 = \sum_{i=1}^{n} \text{Var}(X_i) \tag{8.9}$$

then (8.8) would be a standardised random variable — it has mean 0 and variance 1. Notice that this construction necessarily requires that the mean and variance of every random variable in the sequence $\{X_n\}_{n=1}^{\infty}$ exists. The Central Limit Theorem states the conditions for $\{X_n\}$, $\{a_n\}$ and $\{b_n\}$ in order that

$$\frac{S_n - a_n}{b_n} \xrightarrow{d} Z \qquad (8.10)$$

for some random variable Z. We shall only consider cases for which $Z \sim N(0, 1)$.

We present the *Lindeberg–Lévy* version of the Central Limit Theorem, which applies when the variables $\{X_n\}_{n=1}^{\infty}$ are mutually independent and identically distributed (iid). The Lindeberg–Lévy version is particularly relevant for determining asymptotic properties of estimators such as $\overline{X}_n$, where $\overline{X}_n$ is constructed from size n random samples collected on some variable which we may label X. Assuming that $E[X] = \mu$ and $\text{Var}(X) = \sigma^2$, under the iid assumption, each variable in $\{X_n\}_{n=1}^{\infty}$ may be viewed as a copy of X. Hence $E[X_i] = \mu$ and $\text{Var}(X_i) = \sigma^2$. The constants in (8.9) therefore become

$$a_n = n\mu \quad \text{and} \quad b_n^2 = n\sigma^2$$

and the theorem states the conditions that μ and σ^2 must satisfy in order that the limiting distribution of $(S_n - n\mu)/\sqrt{n\sigma^2}$ is $Z \sim N(0, 1)$.

Theorem (Lindeberg–Lévy): Let the random variables in the sequence $\{X_n\}_{n=1}^{\infty}$ be independent and identically distributed, each with finite mean μ and finite variance σ^2. Then the random variable

$$\frac{S_n - n\mu}{\sigma\sqrt{n}} \qquad (8.11)$$

converges in distribution to a random variable $Z \sim N(0, 1)$.

Proof: See, for example, Mittelhammer (1996, p. 270).

The strength of the Central Limit Theorem is that the distribution of X need not be known. Of course, if $X \sim N(\mu, \sigma^2)$, then the theorem holds trivially, since the sampling distribution of the sample sum is also Normal. On the other hand, for any non-Normal random variable X that possesses a finite mean and variance, the theorem permits us to construct an approximation to the sampling distribution of the sample sum which will become increasingly accurate with sample size. Thus, for the sample sum,

$$S_n \stackrel{a}{\sim} N(n\mu, n\sigma^2) \qquad (8.12)$$

and, for the sample mean,

$$\overline{X}_n \stackrel{a}{\sim} N\!\left(\mu, \frac{\sigma^2}{n}\right). \qquad (8.13)$$

⊕ *Example 6:* The Sample Mean and the Uniform Distribution

Let $X \sim$ Uniform(0, 1), the Uniform distribution on the interval (0, 1). Enter its pdf $f(x)$ as:

f = 1;   domain[f] = {x, 0, 1};

The mean μ and the variance σ^2 of X are, respectively:

Expect[x, f]

$\dfrac{1}{2}$

Var[x, f]

$\dfrac{1}{12}$

Let $\overline{X}_3$ denote the sample mean of a random sample of size $n = 3$ collected on X. Now suppose, for some reason, that we wish to obtain the probability:

$$p = P\left(\tfrac{1}{6} < \overline{X}_3 < \tfrac{5}{6}\right).$$

As the conditions of the Central Limit Theorem are satisfied, it follows from (8.13) that the asymptotic distribution of $\overline{X}_3$ is:

$$\overline{X}_3 \stackrel{a}{\sim} N\left(\tfrac{1}{2}, \tfrac{1}{36}\right).$$

We may therefore use this asymptotic distribution to find an approximate solution for p. Let $g(\overline{x})$ denote the pdf of the asymptotic distribution:

g = $\dfrac{1}{\sigma \sqrt{2\pi}}$ e$^{-\frac{(\bar{x}-\mu)^2}{2\sigma^2}}$ /. $\left\{\mu \to \dfrac{1}{2}, \sigma \to \dfrac{1}{6}\right\}$;

domain[g] = {$\bar{x}$, -∞, ∞};

Then p is approximated by:

Prob$\left[\dfrac{5}{6}, g\right]$ - Prob$\left[\dfrac{1}{6}, g\right]$ // N

0.9545

Just as we were concerned about the accuracy of the asymptotic distribution in *Example 5*, it is quite reasonable to be concerned about the accuracy of the asymptotic approximation for the probability that we seek; after all, a sample size of $n = 3$ is far from large! Generally speaking, the answer to 'How large does n need to be?' is context dependent. Thus, our answer when $X \sim$ Uniform(0, 1) may be quite inadequate under different distributional assumptions for X.

○ Small Sample Accuracy

In this subsection, we wish to compare the exact solution for p, with our asymptotic approximation 0.9545. For the exact solution, we require the sampling distribution of $\bar{X}_3$. More generally, if $X \sim \text{Uniform}(0, 1)$, the sampling distribution of $\bar{X}_n$ is known as Bates's distribution; for a derivation, see Bates (1955) or Stuart and Ord (1994, Example 11.9). The Bates(n) distribution has an n-part piecewise structure:

```
Bates[x_, n_] := Table[{ k/n ≤ x < (k+1)/n ,
            Expand[ (n^n ∑_{i=0}^{k} (-1)^i Binomial[n, i] (x - i/n)^(n-1)) / (n-1)! ]},
  {k, 0, n-1}]
```

For instance, when $n = 3$, the pdf of $Y = \bar{X}_3$ has the 3-part form:

Bates[y, 3]

$$\begin{pmatrix} 0 \le y < \frac{1}{3} & \frac{27 y^2}{2} \\ \frac{1}{3} \le y < \frac{2}{3} & -\frac{9}{2} + 27 y - 27 y^2 \\ \frac{2}{3} \le y < 1 & \frac{27}{2} - 27 y + \frac{27 y^2}{2} \end{pmatrix}$$

This means if $0 \le y < \frac{1}{3}$, the pdf of Y is given by $h(y) = \frac{27 y^2}{2}$, and so on. In the past, we have used If statements to represent 2-part piecewise functions. However, for functions with at least three parts, a Which statement is required. Given $Y = \bar{X}_n \sim \text{Bates}(n)$ with pdf $h(y)$, we may create the Which structure as follows:

```
h[y_] = Which @@ Flatten[Bates[y, 3]]
domain[h[y]] = {y, 0, 1};
```

$$\text{Which}\Big[0 \le y < \frac{1}{3},\ \frac{27 y^2}{2},\ \frac{1}{3} \le y < \frac{2}{3},\\ -\frac{9}{2} + 27 y - 27 y^2,\ \frac{2}{3} \le y < 1,\ \frac{27}{2} - 27 y + \frac{27 y^2}{2}\Big]$$

Then, the natural way to find p with **mathStatica** would be to evaluate Prob[$\frac{5}{6}$, h[y]] − Prob[$\frac{1}{6}$, h[y]]. Unfortunately, at present, neither *Mathematica* nor **mathStatica** can perform integration on Which statements. However, implementation of this important feature is already being planned for version 2 of **mathStatica**. Nevertheless, we can still compute the exact value of p manually, as follows:

$$\int_{\frac{1}{6}}^{\frac{1}{3}} \frac{27 y^2}{2}\, dy + \int_{\frac{1}{3}}^{\frac{2}{3}} \left(-\frac{9}{2} + 27 y - 27 y^2\right) dy + \int_{\frac{2}{3}}^{\frac{5}{6}} \left(\frac{27}{2} - 27 y + \frac{27 y^2}{2}\right) dy$$

$$\frac{23}{24}$$

where $23/24 \approx 0.958333$. By contrast, the approximation based on the asymptotic distribution was 0.9545. Thus, asymptotic theory is doing fairly well here — especially when we remind ourselves that the sample size is only three! Figure 3 illustrates the pdf of $\overline{X}_3$, which certainly has that nice 'bell-shaped' look associated with the Normal distribution.

```
PlotDensity[h[y]];
```

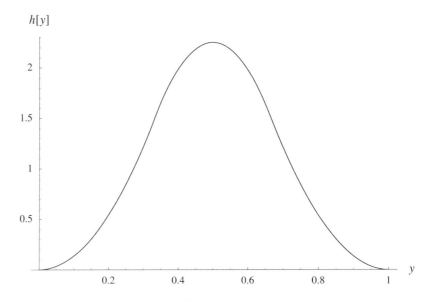

Fig. 3: Density of $\overline{X}_3$ — the Bates(3) distribution

Next, we examine the approximation provided by the cdf of the asymptotic $N(0, 1)$ distribution. In *Example 5*, a similar exercise was undertaken using the van Beek bound, as well as plotting the absolute difference of the exact to the asymptotic distribution. This time, however, we shall take a different route. We now conduct a *Monte Carlo* exercise to compare an artificially generated distribution with the asymptotic distribution. To do so, we generate a pseudo-random sample of size $n = 3$ from the Uniform(0, 1) distribution using *Mathematica*'s internal pseudo-random number generator: `Random[]`. The sample mean $\overline{X}_3$ is then computed. This exercise is repeated $T = 2000$ times. Here then are T realisations of the random variable $\overline{X}_3$:

```
realisations =
    Table[ Plus @@ Table[Random[], {3}]
           ─────────────────────────────── , {2000}];
                        3
```

We now standardise these realisations using the true mean ($\frac{1}{2}$) and the true standard deviation ($\frac{1}{6}$):

```
            realisations - 1/2
Sdata = ─────────────────────;
                  1/6
```

We may use a *quantile–quantile plot* to examine the closeness of the realised standardised sample means to the $N(0, 1)$ distribution. If the plot lies close to the 45° line, it suggests that the distribution of the standardised realisations is close to the $N(0, 1)$. The **mathStatica** function QQPlot constructs this quantile–quantile plot.

```
QQPlot[Sdata];
```

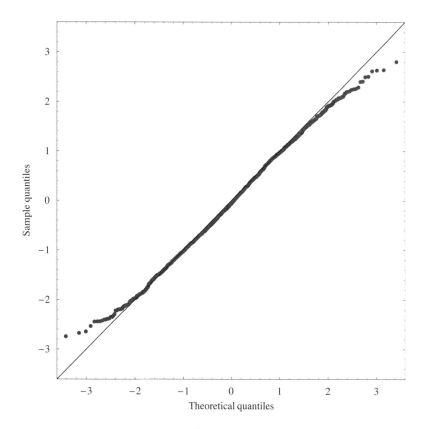

Fig. 4: Quantiles of $\overline{X}_3$ against the quantiles of $N(0, 1)$

The plotted points appear slightly S-shaped, with the elongated centre portion (from values of about -2 to $+2$ along the horizontal axis) closely hugging the 45° line. However, in the tails of the distribution (values below -2, and above $+2$), the accuracy of the Normal approximation to the true cdf weakens. The main reason for this is that the standardised statistic $6(\overline{X}_3 - \frac{1}{2})$ is bounded between -3 and $+3$ (notice that the plot stays within this interval of the vertical axis), whereas the Normal is unbounded. Evidently, the asymptotic distribution provides an accurate approximation except in the tails.

These ideas have practical value: they can be used to construct a pseudo-random number generator for standard Normal random variables. The Normal pseudo-random number generators considered previously were based on the inverse cdf method (see §2.6 B and §2.6 C) and the rejection method (see §2.6 D). By appealing to the Central Limit Theorem, a third possibility arises. We have seen that the cdf of $6(\overline{X}_3 - \frac{1}{2})$ performs fairly well in mimicking the cdf of the $N(0, 1)$ distribution, apart from in the tails. This suggests, due to the Central Limit Theorem, that an increase in sample size might improve tail behaviour; in this respect, using a sample size of $n = 12$ is a common choice. When $n = 12$, the statistic with a limiting $N(0, 1)$ distribution is

$$12(\overline{X}_{12} - \tfrac{1}{2}) = S_{12} - 6$$

which is now bounded between -6 and $+6$. The generator works by taking 12 pseudo-random drawings from the Uniform(0, 1) distribution, and then subtracts 6 from their sum — easy!

```
N01RNG := Plus @@ Table[Random[], {12}] - 6
```

The function N01RNG returns a single number each time it is executed. For example:

N01RNG

-0.185085

The suitability of this generator can be investigated by using QQPlot.[6] ∎

8.5 Convergence in Probability

8.5 A Introduction

For a sequence of random variables $\{X_n\}_{n=1}^{\infty}$, *convergence in probability* is concerned with establishing whether or not the outcomes of those variables become increasingly close to the outcomes of another random variable X with high probability. A formal definition follows:

Let the sequence of random variables $\{X_n\}_{n=1}^{\infty}$ and the random variable X be defined on the same probability space. If for every $\varepsilon > 0$,

$$\lim_{n \to \infty} P(|X_n - X| \geq \varepsilon) = 0 \tag{8.14}$$

then X_n is said to converge in probability to X, written $X_n \xrightarrow{p} X$.

The implication of the definition is that, if indeed $\{X_n\}_{n=1}^{\infty}$ is converging in probability to X, then for a fixed and finite value of n, say n_*, the outcomes of X can be used to approximate the outcomes of X_{n_*}. As we are now referring to outcomes of random variables, it is necessary to insist that all random variables in $\{X_n\}_{n=1}^{\infty}$ be measured in the same sample space as X.[7] This was not the case when we considered convergence in distribution, for this property concerned only the cdf function, and variables measured in different sample spaces are not generally restricted from having equivalent cdf's. Accordingly, convergence in probability is a stronger concept than convergence in distribution.

The following rule establishes the relationship between convergence in probability and convergence in distribution. If $X_n \xrightarrow{p} X$, then it follows that the limiting cdf of X_n must be identical to that of X, and hence,

$$X_n \xrightarrow{p} X \text{ implies } X_n \xrightarrow{d} X. \qquad (8.15)$$

On the other hand, by the argument of the preceding paragraph, the converse is not generally true. The situation when the converse is true occurs only when X is a degenerate random variable, for then convergence in distribution specifies exactly what that value must be. For a fixed constant c,

$$X_n \xrightarrow{p} X = c \text{ implies and is implied by } X_n \xrightarrow{d} X = c. \qquad (8.16)$$

The following two examples show the use of **mathStatica** in establishing convergence in probability.

⊕ **Example 7:** Convergence in Probability to a Normal Random Variable

Suppose that the random variable $X_n = (1 + \frac{1}{n}) X$, where $n = 1, 2, \ldots$. Clearly, X_n and X must lie within the same sample space for all n, as they are related by a simple scaling transformation. Moreover, it is easy to see that $|X_n - X| = \frac{1}{n} |X|$. Therefore,

$$P(|X_n - X| \geq \varepsilon) = P(|X| \geq n\varepsilon). \qquad (8.17)$$

For any random variable X, and any scalar $\alpha > 0$, we may express the event $\{|X| \geq \alpha\}$ as the union of two disjoint events, $\{X \geq \alpha\} \cup \{X \leq -\alpha\}$. Therefore, the occurrence probability can be written as

$$P(|X| \geq \alpha) = P(X \geq \alpha) + P(X \leq -\alpha). \qquad (8.18)$$

Now if we suppose that $X \sim N(0, 1)$, and take $\alpha = n\varepsilon$, the right-hand side of (8.17) becomes

$$(1 - \Phi(n\varepsilon)) + \Phi(-n\varepsilon) = 2(1 - \Phi(n\varepsilon))$$

where Φ denotes the cdf of X, and the symmetry of the pdf of X about zero has been exploited. This can be entered into *Mathematica* as:

```
f = e^(-x²/2) / √(2π);  domain[f] = {x, -∞, ∞};

sol = 2 (1 - Prob[n ε, f]) // Simplify

1 - Erf[n ε / √2]
```

In light of definition (8.14), we now show that X_n converges in probability to X because the following limit is equal to zero:

```
<< Calculus`Limit`

Limit[sol, n → ∞]

Unprotect[Limit]; Clear[Limit];
```

0

As the limit of (8.17) is zero, $X_n \xrightarrow{P} X$. Of course, this outcome should be immediately obvious by inspection of the relationship between X_n and X; the transforming scalar $(1 + \frac{1}{n}) \to 1$ as $n \to \infty$. ∎

Showing convergence in probability often entails complicated calculations, for as definition (8.14) shows, the joint distribution of the random variables X_n and X must typically be known for all n. This, fortunately, was not necessary in the previous example because the relation $X_n = (1 + \frac{1}{n})X$ was known. In any case, from now on, our concern lies predominantly with convergence in probability to a *constant*. Although this type of convergence is easier to deal with, this does not mean that it is less important. In fact, when it comes to determining properties of estimators, it is of vital importance to establish whether or not the estimator converges in probability to the (constant) parameter for which it is proposed. Under this scenario, we take X to be constant in (8.14). Then X can be thought of as representing a parameter θ, while X_n may be viewed as the estimator proposed to estimate it. Under these conditions, if (8.14) holds, X_n is said to be *consistent* for θ, or X_n is a *consistent estimator* of θ.

⊕ ***Example 8:*** Convergence in Probability to a Constant

For a random sample of size n from a $N(\theta, \sigma^2)$ population, the sample mean $\overline{X}_n$ is proposed as an estimator of θ. We shall show, using definition (8.14), that $\overline{X}_n$ is a consistent estimator of θ; that is, we shall show, for every $\varepsilon > 0$,

$$\lim_{n \to \infty} P(|\overline{X}_n - \theta| \geq \varepsilon) = 0.$$

Input into *Mathematica* the pdf of $\overline{X}_n$, which we know to be exactly $N(\theta, \frac{\sigma^2}{n})$:

```
f = 1/(σ √(2 π)) e^(-(x̄ - μ)²/(2 σ²)) /. {μ → θ, σ → σ/√n};

domain[f] =
    {x̄, -∞, ∞} && {θ ∈ Reals, σ > 0, n > 0, n ∈ Integers};
```

Now, by (8.18),

$$P(|\overline{X}_n - \theta| \geq \varepsilon) = P(\overline{X}_n - \theta \geq \varepsilon) + P(\overline{X}_n - \theta \leq -\varepsilon)$$

$$= P(\overline{X}_n \geq \varepsilon + \theta) + P(\overline{X}_n \leq -\varepsilon + \theta)$$

which is equal to:

```
sol = 1 - Prob[ε + θ, f] + Prob[-ε + θ, f] // FullSimplify
```

$$\text{Erfc}\left[\frac{\sqrt{n}}{\sqrt{2}}\frac{\varepsilon}{\sigma}\right]$$

Taking the limit, we find:

```
<< Calculus`Limit`

lsol = Limit[sol, n → ∞]

Unprotect[Limit]; Clear[Limit];
```

$$\frac{e^{-\infty \frac{\text{Sign}[\varepsilon]^2}{\text{Sign}[\sigma]^2}} \sigma}{\varepsilon}$$

The output is not zero as we had hoped for, but if we apply Simplify along with the conditions on ε and σ:

```
Simplify[lsol, {ε > 0, σ > 0}]

0
```

Thus, $\overline{X}_n \xrightarrow{p} \theta$; that is, $\overline{X}_n$ is a consistent estimator of θ. ∎

8.5 B Markov and Chebyshev Inequalities

In the previous example, the sample mean was shown to be a consistent estimator of the population mean (under Normality) by applying the definition of convergence in probability (8.14). Essentially, this requires deriving the cdf of the estimator, followed by taking a limit. This procedure may become less feasible in more complicated settings. Fortunately, it is often possible to establish consistency (or otherwise) of an estimator by only knowing its first two moments. This is done using probability inequalities. Consider, initially, *Markov's Inequality*

$$P(|X| \geq \alpha) \leq \alpha^{-k} E[|X|^k] \tag{8.19}$$

valid for $\alpha > 0$ and provided the kth moment of X exists. Notice that the inequality holds for X having any distribution. For a proof of Markov's Inequality, see Billingsley (1995). A special case of Markov's Inequality is obtained by replacing $|X|$ with $|X - \mu|$, where $\mu = E[X]$, and setting $k = 2$. Doing so yields

$$P(|X - \mu| \geq \alpha) \leq \alpha^{-2} E[(X - \mu)^2] = \alpha^{-2} \text{Var}(X) \tag{8.20}$$

which is usually termed *Chebyshev's Inequality*.

⊕ **Example 9:** Applying the Inequalities

Let X denote the number of customers using a particular gas pump on any given day. What can be said about $P(150 < X < 250)$ when it is known that:

(i) $E[X] = 200$ and $E[(X - 200)^2] = 400$, and

(ii) $E[X] = 200$ and $E[(X - 200)^4] = 10^6$?

Solution (i): We have $\mu = 200$ and $\text{Var}(X) = 400$. Note that

$$P(150 < X < 250) = P(|X - 200| < 50) = 1 - P(|X - 200| \geq 50).$$

By Chebyshev's Inequality (8.20), with $\alpha = 50$,

$$P(|X - 200| \geq 50) \leq \frac{400}{2500} = 0.16.$$

Thus, $P(150 < X < 250) \geq 0.84$. The probability that the gas pump will be used by between 150 and 250 customers each day is at least 84%.

Solution (ii): Applying Markov's Inequality (8.19) with X replaced by $X - 200$, with α set to 50 and k set to 4, finds

$$P(|X - 200| \geq 50) \leq \frac{10^6}{50^4} = 0.16.$$

In this case, the results from (i) and (ii) are equivalent. ∎

8.5 C Weak Law of Large Numbers

There exist general conditions under which estimators such as $\overline{X}_n$ converge in probability, as sample size n increases. Inequalities such as Chebyshev's play a vital role in this respect, as we now show.

In Chebyshev's Inequality (8.20), replace X, μ and α with the symbols $\overline{X}_n$, θ and ε, respectively. That is,

$$P(|\overline{X}_n - \theta| \geq \varepsilon) \leq \varepsilon^{-2} E\left[(\overline{X}_n - \theta)^2\right] \tag{8.21}$$

where we interpret θ to be a parameter, and given constant $\varepsilon > 0$. Let MSE denote the expectation on the right-hand side of (8.21). Under the assumption that $(X_1, \ldots, X_n)$ is a random sample of size n drawn on a random variable X, it can be shown that:[8]

$$\text{MSE} = E\left[(\overline{X}_n - \theta)^2\right] = \frac{1}{n} E[(X - \theta)^2] + \frac{n-1}{n}(E[X] - \theta)^2. \tag{8.22}$$

In the following example, MSE is used to show that the sample mean $\overline{X}_n$ is a consistent estimator of the population mean.

⊕ Example 10: Consistent Estimation

Let $X \sim \text{Uniform}(0, 1)$ with pdf:

```
f = 1;   domain[f] = {x, 0, 1};
```

Let parameter $\theta \in (0, 1)$. We may evaluate MSE (8.22) as follows:

```
MSE = 1/n Expect[(x - θ)², f] +
      (n - 1)/n (Expect[x, f] - θ)²  // Simplify
```

$$\frac{1}{4} + \frac{1}{12n} - \theta + \theta^2$$

Accordingly, the right-hand side of (8.21) is given simply by $\varepsilon^{-2} (\frac{1}{4} + \frac{1}{12n} - \theta + \theta^2)$, when $X \sim \text{Uniform}(0, 1)$.

Taking limits of both sides of (8.21) yields

$$\lim_{n\to\infty} P\big(\big|\overline{X}_n - \theta\big| \geq \varepsilon\big) \leq \varepsilon^{-2}(\tfrac{1}{4} - \theta + \theta^2).$$

Figure 5 plots the limit of MSE across the parameter space of θ:

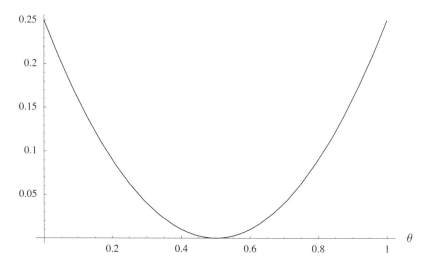

Fig. 5: Limit of MSE against θ

Since the plot is precisely 0 when $\theta = \theta_0 = \frac{1}{2}$, for every $\varepsilon > 0$, it follows from the definition of convergence in probability (8.14) that $\overline{X}_n \xrightarrow{P} \frac{1}{2}$, and ensures, due to uniqueness, that $\overline{X}_n$ cannot converge in probability to any other point in the parameter space. $\overline{X}_n$ is a consistent estimator of $\theta_0 = \frac{1}{2}$. What, if anything, is special about $\frac{1}{2}$ here? Put simply, $E[X] = \frac{1}{2}$. Thus, the sample mean $\overline{X}_n$ is a consistent estimator of the population mean. ∎

Example 10 is suggestive of a more general result encapsulated in a set of theorems known as *Laws of Large Numbers*. These laws are especially relevant when trying to establish consistency of parameter estimators. We shall present just one — *Khinchine's Weak Law of Large Numbers*:

Theorem (Khinchine): Let $\{X_n\}_{n=1}^{\infty}$ be a sequence of mutually independent and identically distributed random variables with finite mean μ. The sample mean:

$$\overline{X}_n \xrightarrow{p} \mu. \tag{8.23}$$

Proof: See, for example, Mittelhammer (1996, pp. 259–260).

In Khinchine's theorem, existence of a finite variance σ^2 for the random variables in the sequence is not required. If σ^2 is known to exist, a simple proof of (8.23) is to use Chebyshev's Inequality, because $E[(\overline{X}_n - \mu)^2] = \text{Var}(\overline{X}_n) = \frac{\sigma^2}{n}$.

8.6 Exercises

1. Let $X_n \sim \text{Bernoulli}(\frac{1}{2} + \frac{1}{2n})$, for $n \in \{1, 2, \ldots\}$. Show that $X_n \xrightarrow{d} X \sim \text{Bernoulli}(\frac{1}{2})$ using (i) the pmf of X_n, (ii) the mgf of X_n, and (iii) the cdf of X_n.

2. Let $X \sim \text{Poisson}(\lambda)$. Derive the cf of $X_\lambda = (X - \lambda)/\sqrt{\lambda}$. Then, use it to show that $X_\lambda \xrightarrow{d} Z \sim N(0, 1)$ as $\lambda \to \infty$.

3. Let $X \sim \text{Uniform}(0, \theta)$, where $\theta > 0$. Define $X_{(j)}$ as the jth order statistic from a random sample of size n drawn on X, for $j \in \{1, \ldots, n\}$; see §9.4 for details on order statistics. Consider the transformation of $X_{(j)}$ to Y_j such that $Y_j = n(\theta - X_{(j)})$. By making use of **mathStatica**'s OrderStat, OrderStatDomain, Transform, TransformExtremum and Prob functions, derive the limit distribution of Y_j when (i) $j = n$, (ii) $j = n - 1$, and (iii) $j = n - 2$. From this pattern, can you deduce the limit distribution of Y_{n-k}, where constant k is a non-negative integer k?

4. Let $X \sim \text{Cauchy}$, and let $(X_1, \ldots, X_n)$ denote a random sample of size n drawn on X. Derive the cf of X. Then, use it to show that the sample mean $\overline{X}_n = \frac{1}{n} \sum_{i=1}^{n} X_i$ cannot converge in probability to a constant.

5. Let $X \sim \text{Uniform}(0, \pi)$, and let $(X_1, X_2, \ldots, X_n)$ denote a random sample of size n drawn on X. Determine a_n and b_n such that $\frac{S_n - a_n}{b_n} \xrightarrow{d} Z \sim N(0, 1)$, where $S_n = \sum_{i=1}^{n} \cos(X_i)$. Then evaluate van Beek's bound.

6. **Simulation I:** At the conclusion of *Example 6*, the function

    ```
    N01RNG := Plus @@ Table[Random[], {12}] - 6
    ```

    was proposed as an approximate pseudo-random number generator for a random variable $X \sim N(0, 1)$. Using QQPlot, investigate the performance of N01RNG.

7. **Simulation II:** Let $X \sim N(0, 1)$, and let $Y = X^2 \sim$ Chi-squared(1). From the relation between X and Y, it follows that NO1RNG^2 is an approximate pseudo-random number generator for Y. That is, if

$$\text{NO1RNG} \xrightarrow{d} X, \text{ then } \text{NO1RNG}^2 \xrightarrow{d} Y.$$

   (i) Noting that the sum of m independent Chi-squared(1) random variables is distributed Chi-squared(m), propose an approximate pseudo-random number generator for $Z \sim$ Chi-squared(m) based on NO1RNG.

   (ii) Provided that X and Z are independent, $T = X / \sqrt{Z/m} \sim$ Student's $t(m)$. Hence, propose an approximate pseudo-random number generator for T based on NO1RNG, and investigate its performance when $m = 1$ and 10.

8. **Simulation III:** Let $(W_1, W_2, ..., W_m)$ be mutually independent random variables such that $W_i \sim N(\mu_i, 1)$. Define $V = \sum_{i=1}^{m} W_i^2 \sim$ Noncentral Chi-squared (m, λ), where $\lambda = \sum_{i=1}^{m} \mu_i^2$.

   (i) Use the relationship between V and $\{W_i\}$ to propose an approximate pseudo-random number generator for V based on NO1RNG, as a *Mathematica* function of m and λ.

   (ii) Use NO1RNG and DiscreteRNG to construct an approximate pseudo-random number generator for V based on the parameter-mix

$$\text{Noncentral Chi-squared}(m, \lambda) = \text{Chi-squared}(m + 2K) \bigwedge_{K} \text{Poisson}\left(\frac{\lambda}{2}\right)$$

   as a *Mathematica* function of m and λ.

9. For a random variable X with mean $\mu \ne 0$ and variance σ^2, reformulate the Chebyshev Inequality (8.20) in terms of the *relative mean deviation* $\frac{|X - \mu|}{|\mu|} = \left|\frac{X - \mu}{\mu}\right|$. That is, using pen and paper, show that

$$P\left(\left|\frac{X - \mu}{\mu}\right| \ge \beta\right) \le (r\beta)^{-2}$$

   where $\beta > 0$, and r denotes the signal-to-noise ratio $|\mu|/\sigma$. Then evaluate r^2 for the Binomial(n, p), Uniform(a, b), Exponential(λ) and Fisher $F(a, b)$ distributions.

10. Let X denote a random variable with mean μ and variance σ^2. In Chebyshev's Inequality, show (you need only use pencil and paper) that if $\alpha \ge 10\sigma$, then $P(|X - \mu| \ge \alpha) \le 0.01$. Next, suppose there is more known about X; namely, $X \sim N(\mu, \sigma^2)$. By evaluating $P(|X - \mu| \ge \alpha)$, show that the assumption of Normality has the effect of allowing the inequality to hold over a larger range of values for α.

11. Let $X \sim$ Binomial(n, p), and let $a \le b$ be non-negative integers. The Normal approximation to the Binomial is given by

$$P(a \le X \le b) \simeq \Phi(d) - \Phi(c)$$

   where Φ denotes the cdf of a standard Normal distribution, and

$$c = \frac{a - np - \frac{1}{2}}{\sqrt{np(1-p)}} \quad \text{and} \quad d = \frac{b - np + \frac{1}{2}}{\sqrt{np(1-p)}}.$$

Investigate the accuracy of the approximation by plotting the error of the approximation when $a = 20$, $b = 80$ and $p = 0.1$, against values of n from 100 to 500 in increments of 10.

12. Let $X \sim \text{Binomial}(n, p)$ and $Y \sim \text{Poisson}(np)$, and let $a \leq b$ be non-negative integers. The Poisson approximation to the Binomial is given by

$$P(a \leq X \leq b) \simeq P(a \leq Y \leq b).$$

Investigate the accuracy of the approximation.

Chapter 9

Statistical Decision Theory

9.1 Introduction

Statistical decision theory is an approach to decision making for problems involving random variables. For any given problem, we use the notation D to denote the set that contains all the different decisions that can be made. There could be as few as two elements in D, or even an uncountably large number of possibilities. The aim is to select a particular decision from D that is, in some sense, optimal. A wide range of statistical problems can be tackled using the tools of decision theory, including estimator selection, hypothesis testing, forecasting, and prediction. For discussion ranging across different types of problems, see, amongst others, Silvey (1975), Gourieroux and Monfort (1995), and Judge *et al.* (1985). In this chapter, emphasis focuses on using decision theory for estimator selection.

Because the decision problem involves random variables, the impact of any particular choice will be uncertain. We represent uncertainty by assuming the existence of a parameter $\theta \in \Theta$ whose true value θ_0 is unknown. The decision problem is then to select an estimator from the set D whose elements are the estimators proposed for a given problem. Our goal, therefore, is to select an estimator from D in an optimal fashion.

9.2 Loss and Risk

Optimality in decision theory is defined according to a *loss structure*, the latter being a function that applies without variation to all estimators in the decision set D. The *loss function*, denoted by $L = L(\hat{\theta}, \theta)$, measures the disadvantages of selecting an estimator $\hat{\theta} \in D$. Loss takes a single, non-negative value for each and every combination of values of $\hat{\theta} \in D$ and $\theta \in \Theta$, but, apart from that, its mathematical form is *discretionary*. This, for example, means that two individuals tackling the same decision problem, can reach different least-loss outcomes, the most likely reason being that their chosen loss functions differ. Moreover, since L is a function of the random variable $\hat{\theta}$, L is itself a random variable, so that the criterion of minimisation of loss is not meaningful. Although we cannot minimise loss L, we can minimise the *expected loss*. The expected loss of $\hat{\theta}$ is also known as the *risk* of $\hat{\theta}$, where the risk function is defined as

$$R_{\hat{\theta}}(\theta) = E[L(\hat{\theta}, \theta)] \tag{9.1}$$

where $\hat{\theta}$ is a random variable with density $g(\hat{\theta}; \theta)$. Because the expectation is with respect to the density of $\hat{\theta}$, risk is a non-random function of θ. Notice that because the loss L is non-negative, risk must also be non-negative. Given a particular estimator chosen from D, we solve (9.1) to obtain its risk. As its name would suggest, the smaller the value of risk, the better off we are—the decision criterion is to *minimise risk*.[1]

With the aid of risk, we now return to the basic question of how to choose amongst the estimators in the decision set. Consider two estimators of θ_0, namely, $\hat{\theta}$ and $\tilde{\theta}$, both of which are members of a decision set D. We say that $\hat{\theta}$ *dominates* $\tilde{\theta}$ if the risk of the former is no greater than the risk of the latter throughout the entire parameter space, with the added proviso that the risk of $\hat{\theta}$ be strictly smaller in some part of the parameter space; that is, $\hat{\theta}$ dominates $\tilde{\theta}$ if

$$R_{\hat{\theta}}(\theta) \leq R_{\tilde{\theta}}(\theta), \quad \text{for all } \theta \in \Theta$$

along with

$$R_{\hat{\theta}}(\theta) < R_{\tilde{\theta}}(\theta), \quad \text{for some } \theta \in \Theta^* \subset \Theta$$

where Θ^* is a non-null set. Notice that dominance is a *binary* relationship between estimators in D. This means that if there are d estimators in D, then there are $d(d-1)/2$ dominance relations that can be tested. Once an estimator is shown to be dominated, then we may rule it out of our decision process, for we can always do better by using the estimator(s) that dominate it; a dominated estimator is termed *inadmissible*. Finally, if an estimator is not dominated by any of the other estimators in D, then it is deemed to be *admissible*; an admissible estimator is eligible to be selected to estimate θ_0. Figure 1 illustrates these concepts.

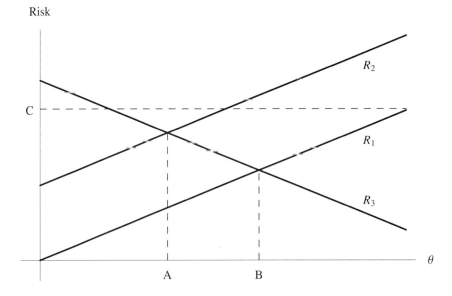

Fig. 1: Risk comparison

The decision set here is $D = \{\hat{\theta}_1, \hat{\theta}_2, \hat{\theta}_3\}$, and the risk of each estimator is plotted as a function of θ, and is labelled on each (continuous) line; the diagram is plotted over the entire parameter space Θ. The first feature to observe is that the risk of estimator $\hat{\theta}_1$ (denoted by R_1) is everywhere below the risk of $\hat{\theta}_2$ (denoted by R_2). Thus, $\hat{\theta}_1$ dominates $\hat{\theta}_2$ (therefore $\hat{\theta}_2$ is inadmissible). The next feature concerns the risk functions of $\hat{\theta}_1$ and $\hat{\theta}_3$ (denoted by R_3); they cross at B, and therefore neither estimator dominates the other. To the left of B, $\hat{\theta}_1$ has smaller risk and is preferred to $\hat{\theta}_3$, whereas to the right of B the ranking is reversed. It follows that both $\hat{\theta}_1$ and $\hat{\theta}_3$ are admissible estimators. Of course, if we knew (for example) that the true parameter value θ_0 lay in the region to the left of B, then $\hat{\theta}_1$ is dominant in D and therefore preferred. However, generally this type of knowledge is not available. The following example, based on Silvey (1975, Example 11.2), illustrates some of these ideas.

⊕ **Example 1:** The Risk of a Normally Distributed Estimator

Suppose that a random variable $X \sim N(\theta, 1)$, where $\theta \in \mathbb{R}$ is an unknown parameter. The random variable $\hat{\theta} = X + k$ is proposed as an estimator of θ, where constant $k \in \mathbb{R}$. Thus, the estimate of θ is formed by adding k to a single realisation of the random variable X. The decision set, in this case, consists of all possible choices for k. Thus, $D = \{k : k \in \mathbb{R}\}$ is a set with an uncountably infinite number of elements. By the linearity property of the Normal distribution, it follows that estimator $\hat{\theta}$ is Normally distributed; that is, $\hat{\theta} \sim N(\theta + k, 1)$ with pdf $f(\hat{\theta}; \theta)$:

```
f    =   ───────  e^(-½ (θ̂ - (θ + k))²) ;
         √(2 π)

domain[f] = {θ̂, -∞, ∞} && {-∞ < θ < ∞, -∞ < k < ∞};
```

Let $c_1 \in \mathbb{R}_+$ and $c_2 \in \mathbb{R}_+$ be two constants chosen by an individual, and suppose that the loss structure specified for this problem is

$$L(\hat{\theta}, \theta) = \begin{cases} c_1(\hat{\theta} - \theta) & \text{if } \hat{\theta} \geq \theta \\ c_2(\theta - \hat{\theta}) & \text{if } \hat{\theta} < \theta. \end{cases} \quad (9.2)$$

In **mathStatica**, we enter this as:

```
L = If[θ̂ ≥ θ, c₁ (θ̂ - θ), c₂ (θ - θ̂)];
```

Figure 2 plots the loss function when $c_1 = 2$, $c_2 = 1$ and $\theta = 0$. The asymmetry in the loss function leads to differing magnitudes of loss depending on whether the estimate is larger or smaller than $\theta = 0$. In Fig. 2, an over-estimate of θ causes a greater loss than an under-estimate of the same size. In this case, intuition suggests that we search for a $k < 0$, for if we are to err, we will do better if the error results from an under-estimate. In a similar vein, if $c_1 < c_2$, then over-estimates are preferred to under-estimates, so we would expect to choose a $k > 0$; and when the loss is symmetric $c_1 = c_2$, no correction would be necessary and we would choose $k = 0$. We now show that it is possible to identify the unique value of k that minimises risk. Naturally, it will depend on the values of c_1 and c_2.

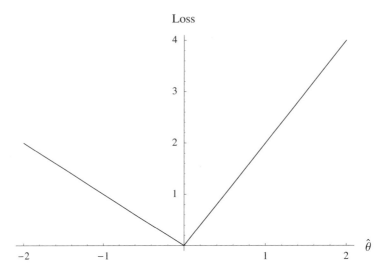

Fig. 2: An asymmetric loss function ($c_1 = 2$ and $c_2 = 1$)

Risk is expected loss:

Risk = Expect[L, f]

$$\frac{e^{-\frac{k^2}{2}}(c_1 + c_2)}{\sqrt{2\pi}} +$$
$$\frac{1}{2} k \left(\left(1 + \mathrm{Erf}\left[\frac{k}{\sqrt{2}}\right]\right) c_1 + \left(-1 + \mathrm{Erf}\left[\frac{k}{\sqrt{2}}\right]\right) c_2 \right)$$

from which we see that risk is dependent on factors that are under our control; that is, it does not depend on values of θ. For given values of c_1 and c_2, the value of k which minimises risk can be found in the usual way. Here is the first derivative of risk with respect to k:

d1 = D[Risk, k] // Simplify

$$\frac{1}{2} \left(\left(1 + \mathrm{Erf}\left[\frac{k}{\sqrt{2}}\right]\right) c_1 + \left(-1 + \mathrm{Erf}\left[\frac{k}{\sqrt{2}}\right]\right) c_2 \right)$$

and here is the second derivative:

d2 = D[Risk, {k, 2}] // Simplify

$$\frac{e^{-\frac{k^2}{2}}(c_1 + c_2)}{\sqrt{2\pi}}$$

Notice that the second derivative d2 is positive for all k.

Next, set the first derivative to zero and solve for k:

```
solk = Solve[d1 == 0, k]
```

$$\left\{\left\{k \to \sqrt{2}\ \text{InverseErf}\left[0, \frac{-c_1 + c_2}{c_1 + c_2}\right]\right\}\right\}$$

This value for k must globally minimise risk, because the second derivative d2 is positive for all k. Let us calculate some optimal values for k for differing choices of c_1 and c_2:

```
solk /. {c₁ → 2, c₂ → 1} // N
```

$\{\{k \to -0.430727\}\}$

```
solk /. {c₁ → 2, c₂ → 3} // N
```

$\{\{k \to 0.253347\}\}$

```
solk /. {c₁ → 1, c₂ → 1} // N
```

$\{\{k \to 0.\}\}$

For example, the first output shows that the estimator that minimises risk when $c_1 = 2$ and $c_2 = 1$ is

$$\hat{\theta} = X - 0.430727.$$

This is the only admissible estimator in D for it dominates all others with respect to the loss structure (9.2). In each of the three previous outputs, notice that the optimal value of k depends on the asymmetry of the loss function as induced by the values of c_1 and c_2, and that its sign varies in accord with the intuition given earlier. ∎

Of course, all decision theory outcomes are conditional upon the assumed loss structure, and as such may alter if a different loss structure is specified. Consider, for example, the *minimax* decision rule: the particular estimator $\hat{\theta}$ is preferred if

$$\hat{\theta} = \arg\min\,(\max_{\theta \in \Theta}\ R_{\hat{\Theta}}(\theta)), \quad \text{for all } \hat{\Theta} \in D.$$

In other words, $\hat{\theta}$ is preferred over all other estimators in the decision set if its maximum risk is no greater than the maximum risk of all other estimators. If two estimators have the same maximum risk (which may not necessarily occur at the same points in the parameter space), then we would be indifferent between them under this criterion. The minimax criterion is conservative in the sense that it selects the estimator with the least worst risk. To illustrate, consider Fig. 1 once again. We see that for the admissible estimators, $\hat{\theta}_1$ and $\hat{\theta}_3$, maximum risk occurs at the extremes of the parameter space. The value C corresponds to the maximum risk of $\hat{\theta}_1$. Since the maximum risk of $\hat{\theta}_3$ is greater than C, it follows that $\hat{\theta}_1$ is the minimax estimator.

9.3 Mean Square Error as Risk

The *Mean Square Error* (MSE) of an estimator $\hat{\theta}$ is defined as

$$\text{MSE}(\hat{\theta}) = E\left[\left(\hat{\theta} - \theta\right)^2\right].$$

Thus, if a quadratic loss function $L(\hat{\theta}, \theta) = (\hat{\theta} - \theta)^2$ is specified, $\text{MSE}(\hat{\theta})$ is equivalent to risk; that is, MSE is risk under quadratic loss. MSE can be expressed in terms of the first two moments of $\hat{\theta}$. To see this, let $\overline{\theta} = E[\hat{\theta}]$ for notational convenience, and write

$$\begin{aligned} \text{MSE}(\hat{\theta}) &= E\left[\left((\hat{\theta} - \overline{\theta}) - (\theta - \overline{\theta})\right)^2\right] \\ &= E\left[(\hat{\theta} - \overline{\theta})^2\right] + E\left[(\theta - \overline{\theta})^2\right] - 2 E\left[(\hat{\theta} - \overline{\theta})(\theta - \overline{\theta})\right] \quad (9.3) \\ &= \text{Var}(\hat{\theta}) + \left(\text{Bias}(\hat{\theta})\right)^2. \end{aligned}$$

Bias is defined as $E[\hat{\theta}] - \theta = \overline{\theta} - \theta$. Thus, the first term in the second line defines the variance of $\hat{\theta}$; the second term is the squared bias of $\hat{\theta}$, and as it is non-stochastic, the outer expectation is superfluous; the third term is zero because

$$E\left[(\hat{\theta} - \overline{\theta})(\theta - \overline{\theta})\right] = (\theta - \overline{\theta}) E\left[\hat{\theta} - \overline{\theta}\right] = (\theta - \overline{\theta})\left(E[\hat{\theta}] - \overline{\theta}\right) = 0.$$

As the last line of (9.3) shows, estimator choice under quadratic loss depends on both variance and bias. If the decision set D consists only of unbiased estimators, then choosing the estimator with the smallest risk coincides with choosing the estimator with least variance. But should the decision set also include biased estimators, then choice based on risk is no longer as straightforward, as there is now potential to trade off variance against bias. The following diagram illustrates.

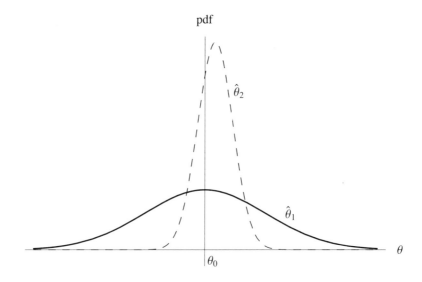

Fig. 3: Estimator densities: $\hat{\theta}_1$ has large variance (——), $\hat{\theta}_2$ is biased (– – –)

Figure 3 depicts the (scaled) pdf of two estimators of θ_0, labelled $\hat{\theta}_1$ and $\hat{\theta}_2$. Here, $\hat{\theta}_1$ is unbiased for θ_0, whereas $\hat{\theta}_2$ has a slight bias. On the other hand, the variance, or spread, of $\hat{\theta}_1$ is far greater than it is for $\hat{\theta}_2$. On computing MSE for each estimator, it would not be surprising to find $\text{MSE}(\hat{\theta}_1) > \text{MSE}(\hat{\theta}_2)$, meaning that $\hat{\theta}_2$ is preferred to $\hat{\theta}_1$ under quadratic loss. The trade-off between bias and variance favours the biased estimator in this case. However, if we envisage the pdf of $\hat{\theta}_2$ (the dashed curve) shifting further and further to the right, the cost of increasing bias would soon become overwhelming, until eventually $\text{MSE}(\hat{\theta}_2)$ would exceed $\text{MSE}(\hat{\theta}_1)$.

⊕ *Example 2:* Estimators for the Normal Variance

Consider a random variable $X \sim N(\mu, \theta)$, and let $(X_1, \ldots, X_n)$ denote a random sample of size n drawn on X. Asymptotic arguments may be used to justify estimating the variance parameter θ using the statistic $T = \sum_{i=1}^{n}(X_i - \overline{X})^2$, because, for example, the estimator

$$\hat{\theta} = \frac{T}{n} \xrightarrow{p} \theta.$$

That is, $\hat{\theta}$ is a consistent estimator of θ. However, the estimator remains consistent if the denominator n is replaced by, for example, $n-1$. Doing so yields the estimator $\tilde{\theta} = T/(n-1)$, for which $\tilde{\theta} \xrightarrow{p} \theta$ as the subtraction of 1 from n in the denominator becomes insignificant as n becomes larger. We therefore cannot distinguish between $\hat{\theta}$ and $\tilde{\theta}$ using asymptotic theory. As we have seen in *Example 1* of Chapter 7, estimator $\tilde{\theta}$ is an unbiased estimator of θ; consequently, given that $\hat{\theta} < \tilde{\theta}$, it follows that $\hat{\theta}$ must be biased downward for θ (i.e. $E[\hat{\theta}] < \theta$). On the other hand, the variance of $\tilde{\theta}$ is larger than that of $\hat{\theta}$. To summarise the situation: both estimators are asymptotically equivalent, but in finite samples there is a bias–variance trade-off between them. Proceeding along decision theoretic lines, we impose a quadratic loss structure $L(\Theta, \theta) = (\Theta - \theta)^2$ on the estimators in the decision set $D = \{\Theta : \Theta = \hat{\theta} \text{ or } \tilde{\theta}\}$.

From *Example 27* of Chapter 4, we know that $T/\theta \sim \text{Chi-squared}(n-1)$. Therefore, the pdf of T, say $f(t)$, is:

```
           t^(n-1/2 - 1)  e^-t/(2 θ)
    f  =  ─────────────────────────── ;
              (2 θ)^(n-1/2)  Γ[n-1/2]

    domain[f] = {t, 0, ∞} && {n > 0, θ > 0};
```

The MSE of each estimator can be derived by:

```
    MŜE = Expect[(t/n - θ)^2, f]
```

— This further assumes that: {n > 1}

$$\frac{(-1 + 2n)\,\theta^2}{n^2}$$

$$\text{MSE} = \text{Expect}\left[\left(\frac{t}{n-1} - \theta\right)^2, f\right]$$

— This further assumes that: {n > 1}

$$\frac{2\,\theta^2}{-1+n}$$

Both MSEs depend upon θ and sample size n. However, in this example, it is easy to rank the two estimators, because $\hat{\text{MSE}}$ is strictly smaller than $\tilde{\text{MSE}}$ for any value of θ:

$\hat{\text{MSE}} - \tilde{\text{MSE}}$ // Simplify

$$\frac{(1-3n)\,\theta^2}{(-1+n)\,n^2}$$

Therefore, $\hat{\theta}$ dominates $\tilde{\theta}$ given quadratic loss, so $\tilde{\theta}$ is inadmissible given quadratic loss.

In this problem, it is possible to broaden the decision set from two estimators to an uncountably infinite number of estimators (all of which retain the asymptotic property of consistency) and then determine the (unique) dominant estimator; that is, the estimator that minimises MSE. To do so, we need to suppose that all estimators in the decision set have general form $\hat{\Theta} = T/(n+k)$, for some real value of k that is independent of n. The estimators that we have already examined are special cases of $\hat{\Theta}$, corresponding to $k = -1$ (for $\tilde{\theta}$) and 0 (for $\hat{\theta}$). For arbitrary k, the MSE is:

$$\text{MSEk} = \text{Expect}\left[\left(\frac{t}{n+k} - \theta\right)^2, f\right]$$

— This further assumes that: {n > 1}

$$\frac{(-1 + k\,(2+k) + 2\,n)\,\theta^2}{(k+n)^2}$$

The minimum MSE can be obtained in the usual way by solving the first-order condition:

Solve[D[MSEk, k] == 0, k]

{{k → 1}}

... because the sign of the second derivative when evaluated at the solution is positive:

D[MSEk, {k, 2}] /. k → 1 // Simplify

$$\frac{2\,(-1+n)\,\theta^2}{(1+n)^3}$$

We conclude that θ^* dominates all other estimators with respect to quadratic loss, where

$$\theta^* = \frac{T}{n+1} = \frac{1}{n+1}\sum_{i=1}^{n}(X_i - \overline{X})^2.$$

∎

⊕ Example 3: Sample Mean Versus Sample Median for Bernoulli Trials

Suppose that $Y \sim \text{Bernoulli}(\theta)$; that is, Y is a Bernoulli random variable such that $P(Y = 1) = \theta$ and $P(Y = 0) = 1 - \theta$, where θ is an unknown parameter taking real values within the unit interval, $0 < \theta < 1$. Suppose that a random sample of size n is drawn on Y, denoted by $(Y_1, \ldots, Y_n)$. We shall consider two estimators, namely, the sample mean $\hat{\theta}$, and the sample median $\tilde{\theta}$, and attempt to decide between them on the basis of quadratic loss. The decision set is $D = \{\hat{\theta}, \tilde{\theta}\}$.

The sample mean

$$\hat{\theta} = \frac{1}{n} \sum_{i=1}^{n} Y_i$$

is clearly a function of the sample sum S. In *Example 21* of Chapter 4, we established $S = \sum_{i=1}^{n} Y_i \sim \text{Binomial}(n, \theta)$, the Binomial distribution with index n and parameter θ. Therefore, $\hat{\theta}$ is a discrete random variable that may take values in the sample space $\hat{\Omega} = \{0, n^{-1}, 2n^{-1}, \ldots, 1\}$. Let $f(s)$ denote the pmf of S:

```
f = Binomial[n, s] θ^s (1 - θ)^(n-s);

domain[f] = {s, 0, n} &&
            {0 < θ < 1, n > 0, n ∈ Integers} && {Discrete};
```

The MSE of $\hat{\theta}$, the sample mean, is given by

$$\widehat{\text{MSE}} = \text{Expect}\left[\left(\frac{s}{n} - \theta\right)^2, f\right]$$

$$-\frac{(-1 + \theta)\theta}{n}$$

The sample space of the sample median also depends upon the sample sum, but it is also important to identify whether the sample size is odd- or even-valued. To see this, consider first when n is odd: $\tilde{\theta}$, the sample median, will take values from $\tilde{\Omega}_{\text{odd}} = \{0, 1\}$. If the estimate is zero, then there have to be more zeroes than ones in the observed sample; this occurs when $S \leq (n-1)/2$. The reverse occurs if $S \geq (n+1)/2$, for then there must be more ones than zeroes: hence the sample median must be 1. The next case is when n is even: now $\tilde{\theta}$ can take values from $\tilde{\Omega}_{\text{even}} = \{0, \frac{1}{2}, 1\}$. The outcome of $\frac{1}{2}$ exists (by convention) in even-sized samples because the number of zeroes can match exactly the number of ones.

Let us assume the sample size n is even. Then

$P(\tilde{\Theta} = \tilde{\theta})$:	$P(S \leq \frac{n}{2} - 1)$	$P(S = \frac{n}{2})$	$P(S \geq \frac{n}{2} + 1)$
$\tilde{\theta}$:	0	$\frac{1}{2}$	1

Table 1: The pmf of $\tilde{\theta}$ when n is even

Let $g(\tilde{\theta})$ denote the pmf of $\tilde{\theta}$. We enter this using List Form:

```
g = {Prob[n/2 - 1, f],   f /. s → n/2,   1 - Prob[n/2, f]};
domain[g] =
   {θ̃, {0, 1/2, 1}} && {n > 0, n/2 ∈ Integers} && {Discrete};
```

Then, the MSE of $\tilde{\theta}$, the sample median, is:

```
MSẼ = Expect[(θ̃ - θ)^2, g]
```

$$\theta^2 + \left(-\frac{1}{2} + \theta\right)^2 (-(-1+\theta)\,\theta)^{n/2} \text{Binomial}\left[n, \frac{n}{2}\right] -$$

$$\frac{1}{\Gamma[2+\tfrac{n}{2}]\,\Gamma[\tfrac{n}{2}]} \left((-1+\theta)^{1+n}\,\theta\left(\frac{\theta}{1-\theta}\right)^{n/2} \Gamma[1+n]\right.$$

$$\text{Hypergeometric2F1}\left[1,\, 1-\frac{n}{2},\, 2+\frac{n}{2},\, \frac{\theta}{-1+\theta}\right]\Bigg) -$$

$$\frac{1}{\Gamma[1+\tfrac{n}{2}]^2} \left((1-\theta)^{-n/2}(-1+\theta)^n\,\theta^{\tfrac{4+n}{2}}\,\Gamma[1+n]\right.$$

$$\text{Hypergeometric2F1}\left[1,\, -\frac{n}{2},\, 1+\frac{n}{2},\, \frac{\theta}{-1+\theta}\right]\Bigg)$$

The complicated nature of this expression rules out the possibility of a simple analytical procedure to compare MSEs for arbitrary n. However, by selecting a specific value of n, say $n = 4$, we can compare the estimators by plotting their MSEs, as illustrated in Fig. 4.

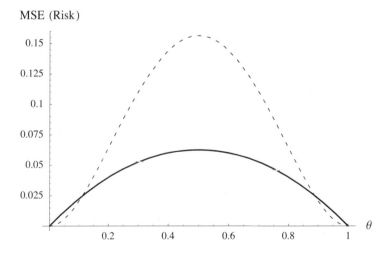

Fig. 4: MSÊ (——) and MSẼ (– – –) when $n = 4$

Evidently, the risk of the sample mean (MSÊ) is *nearly* everywhere below the risk of the sample median (MSẼ). Nevertheless, the plot shows that there exist values towards the edges of the parameter space where the sample median has lower risk than the sample mean; consequently, when $n = 4$, both estimators are admissible with respect to quadratic loss. Thus, to make a decision, we need to know θ_0, the true value of θ. Of course, it is precisely because θ_0 is unknown that we started this investigation. So, our decision-theoretic approach has left us in a situation of needing to know θ_0 in order to decide how to estimate it! Clearly, further information is required in order to reach a decision. To progress, we might do the following:

(i) Experiment with increasing n: that is, replace '4' with larger even numbers in the above analysis. On doing so, we find that the parameter region for which the sample mean is preferred also increases. This procedure motivates a formal asymptotic analysis of each MSE for $n \to \infty$, returning us to the type of analysis developed in Chapter 8.

(ii) Alter the decision criterion: for example, suppose the decision criterion was to select the minimax estimator — the estimator that has the smaller maximum risk. From the diagram, we see that the maximum risk of $\hat{\theta}$ (occurring at $\theta = \frac{1}{2}$) is smaller than the maximum risk of $\tilde{\theta}$ (also occurring at $\theta = \frac{1}{2}$). Hence, the sample mean is the minimax estimator.

(iii) Finally, comparing the number of points in $\tilde{\Omega}_{\text{odd}}$ or $\tilde{\Omega}_{\text{even}}$ (the sample space of the sample median), relative to the number of points in $\hat{\Omega}$ (the sample space of the sample mean) is probably sufficient argument to motivate selecting the sample mean over the sample median for Bernoulli trials, because the parameter space is the (0, 1) interval of the real line. ∎

9.4 Order Statistics

9.4 A Definition and OrderStat

Let X denote a continuous random variable with pdf $f(x)$ and cdf $F(x)$, and let $(X_1, X_2, \ldots, X_n)$ denote a random sample of size n drawn on X. Suppose that we place the variables in the random sample in ascending order. The re-ordered variables, which we shall label $(X_{(1)}, X_{(2)}, \ldots, X_{(n)})$, are known as *order statistics*. By construction, the order statistics are such that $X_{(1)} < X_{(2)} < \cdots < X_{(n)}$; for example, $X_{(1)} = \min(X_1, \ldots, X_n)$ is the smallest order statistic and corresponds to the sample minimum, and $X_{(n)}$ is the largest order statistic and corresponds to the sample maximum. Each order statistic is a continuous random variable (this is inherited from X), and each has domain of support equivalent to that of X. For example, the pdf of $X_{(r)}$, the rth order statistic ($r \in \{1, \ldots, n\}$), is given by[2]

$$\frac{n!}{(r-1)!\,(n-r)!}\, F(x)^{r-1} \left(1 - F(x)\right)^{n-r} f(x) \qquad (9.4)$$

where x represents values assigned to $X_{(r)}$. Finally, because X is continuous, any ties (*i.e.* two identical outcomes) between the order statistics can be disregarded as ties occur with probability zero. For further discussion of order statistics, see David (1981), Balakrishnan and Rao (1998a, 1998b) and Hogg and Craig (1995).

mathStatica's `OrderStat` function automates the construction of the pdf of order statistics for a size n random sample drawn on a random variable X with pdf f. In particular, `OrderStat[r, f]` finds the pdf of the rth order statistic, while `OrderStat[{r, s, ..., t}, f]` finds the joint pdf of the order statistics indicated in the list. An optional third argument, `OrderStat[r, f, m]`, sets the sample size to m.

⊕ *Example 4:* Order Statistics for the Uniform Distribution

Let $X \sim$ Uniform(0, 1) with pdf $f(x)$:

f = 1;   **domain[f] = {x, 0, 1};**

Let $(X_1, ..., X_n)$ denote a random sample of size n drawn on X, and let $(X_{(1)}, ..., X_{(n)})$ denote the corresponding order statistics. Then, the pdf of the smallest order statistic $X_{(1)}$ is given by:

OrderStat[1, f]

$n (1 - x)^{-1+n}$

The pdf of the largest order statistic $X_{(n)}$ is given by:

OrderStat[n, f]

$n x^{-1+n}$

and the pdf of the rth order statistic is given by:

OrderStat[r, f]

$$\frac{(1 - x)^{n-r} x^{-1+r} n!}{(n - r)! (-1 + r)!}$$

Note that `OrderStat` assumes an arbitrary sample size n. If a specific value for sample size is required, or if you wish to use your own notation for 'n', then this may be conveyed using a third argument to `OrderStat`. For example, if $n = 5$, the pdf of the rth order statistic is:

OrderStat[r, f, 5]

$$\frac{120 (1 - x)^{5-r} x^{-1+r}}{(5 - r)! (-1 + r)!}$$

In each case, the domain of support of $X_{(r)} = x \in (0, 1)$. ∎

⊕ **Example 5:** Operating on Order Statistics

Let $X \sim$ Exponential(λ) with pdf $f(x)$:

$$\mathtt{f = \frac{1}{\lambda}\ e^{-x/\lambda}\ ; \qquad domain[f] = \{x,\ 0,\ \infty\}\ \&\&\ \{\lambda > 0\};}$$

Let $(X_1, \ldots, X_n)$ denote a random sample of size n drawn on X, and let $(X_{(1)}, \ldots, X_{(n)})$ denote the corresponding order statistics. Here is $g(x)$, the pdf of the rth order statistic:

$$\mathtt{g = OrderStat[r,\ f]}$$

$$\frac{e^{-\frac{(1+n-r)x}{\lambda}}\ (1 - e^{-\frac{x}{\lambda}})^{-1+r}\ n!}{\lambda\ (n-r)!\ (-1+r)!}$$

The `domain` statement to accompany `g` may be found using **mathStatica**'s `OrderStatDomain` function:

$$\mathtt{domain[g] = OrderStatDomain[r,\ f]}$$

$$\{x,\ 0,\ \infty\}\ \&\&\ \{n \in \text{Integers},\ r \in \text{Integers},\ \lambda > 0,\ 1 \le r \le n\}$$

Figure 5 plots the pdf of $X_{(r)}$ as r increases.

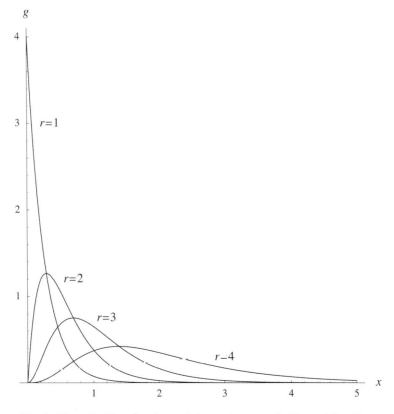

Fig. 5: The pdf of the rth order statistic, as r increases (with $n = 4$, $\lambda = 1$)

We can now operate on $X_{(r)}$ using **mathStatica** functions. For example, here is the mean of $X_{(r)}$ as a function of n, r and λ:

Expect[x, g]

λ (HarmonicNumber[n] - HarmonicNumber[n - r])

and here is the variance:

Var[x, g]

λ^2 (-PolyGamma[1, 1 + n] + PolyGamma[1, 1 + n - r])

⊕ *Example 6:* Joint Distributions of Order Statistics

Once again, let $X \sim$ Exponential(λ) with pdf:

$$f = \frac{1}{\lambda} e^{-x/\lambda}; \qquad \text{domain}[f] = \{x, 0, \infty\} \, \&\& \, \{\lambda > 0\};$$

For a size n random sample drawn on X, the joint pdf of the order statistics $(X_{(1)}, X_{(2)})$ is:

g = OrderStat[{1, 2}, f]

$$\frac{e^{-\frac{x_1 + (-1+n) x_2}{\lambda}} \Gamma[1 + n]}{\lambda^2 \, \Gamma[-1 + n]}$$

domain[g] = OrderStatDomain[{1, 2}, f]

— The domain is: $\{0 < x_1 < x_2 < \infty\}$, which we enter into mathStatica as:

$\{\{x_1, 0, x_2\}, \{x_2, x_1, \infty\}\} \, \&\& \, \{n \in \text{Integers}, \lambda > 0, 2 \leq n\}$

where x_1 denotes values assigned to $X_{(1)}$, and x_2 denotes values assigned to $X_{(2)}$. In this bivariate case, the domain of support of $(X_{(1)}, X_{(2)})$ is given by the non-rectangular region $\Omega = \{(x_1, x_2): 0 < x_1 < x_2 < \infty\}$. At present, **mathStatica** does not support non-rectangular regions (see §6.1 B). However, **mathStatica** functions such as Expect, Var, Cov and Corr do know how to operate on triangular regions which have general form $a < x < y < z < \cdots < b$, where a and b are constants. Here, for example, is the correlation coefficient between $X_{(1)}$ and $X_{(2)}$:

Corr[{x₁, x₂}, g]

$$\frac{-1 + n}{\sqrt{1 - 2n + 2n^2}}$$

The non-zero correlation coefficient and (especially) the non-rectangular domain of support of $(X_{(1)}, X_{(2)})$ illustrate a general property of order statistics—they are mutually dependent. ∎

⊕ **Example 7:** Order Statistics for the Laplace Distribution

The `OrderStat` function also supports pdf's which take a piecewise form. For example, let random variable $X \sim \text{Laplace}(\mu, \sigma)$ with piecewise pdf:

```
f = If[x < μ, e^((x-μ)/σ)/(2σ), e^(-(x-μ)/σ)/(2σ)];
domain[f] = {x, -∞, ∞} && {μ ∈ Reals, σ > 0};
```

The pdf of the rth order statistic, $X_{(r)}$, is given by:[3]

```
OrderStat[r, f]
```

$$\text{If}\left[x < \mu, \; \frac{2^{-r} \, e^{\frac{r(x-\mu)}{\sigma}} \left(1 - \frac{1}{2} e^{\frac{x-\mu}{\sigma}}\right)^{n-r} n!}{\sigma \, (n-r)! \, (-1+r)!}, \; \frac{2^{-1-n+r} \, e^{\frac{(1-n-r)(-x+\mu)}{\sigma}} \left(1 - \frac{1}{2} e^{-\frac{x-\mu}{\sigma}}\right)^{-1+r} n!}{\sigma \, (n-r)! \, (-1+r)!}\right]$$

Notice that **mathStatica**'s output is in piecewise form too.

As a special case, let $X_{(1)}$ denote the smallest order statistic from a random sample of size n drawn on the standardised Laplace distribution (*i.e.* the Laplace(0, 1) distribution). The pdf of $X_{(1)}$ is given by:

```
g₁ = OrderStat[1, f /. {μ → 0, σ → 1}]
```

$$\text{If}\left[x < 0, \; \frac{1}{2} e^x \left(1 - \frac{e^x}{2}\right)^{-1+n} n, \; 2^{-n} \, e^{-nx} \, n\right]$$

```
domain[g₁] = OrderStatDomain[1, f /. {μ → 0, σ → 1}]
```

$\{x, -\infty, \infty\}$ && $\{n \in \text{Integers}, 1 \le n\}$

Figure 6 shows how the pdf of $X_{(1)}$ varies as n increases. It is evident that the bulk of the mass of the pdf of $X_{(1)}$ shifts to the left, as n increases.

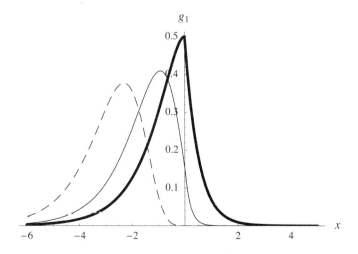

Fig. 6: pdf of $X_{(1)}$: $n = 2$ (——), $n = 5$ (——), $n = 20$ (– – –)

As a final illustration, consider the joint pdf of the two smallest order statistics, $X_{(1)}$ and $X_{(2)}$, when the sample size is $n = 5$:

$$\mathbf{g_{12} = OrderStat[\{1, 2\}, f\ /.\ \{\mu \to 0,\ \sigma \to 1\},\ 5]}$$

$$20\ \text{If}\left[x_1 < 0,\ \frac{e^{x_1}}{2},\ \frac{e^{-x_1}}{2}\right]\ \text{If}\left[x_2 < 0,\ -\frac{1}{16}\ e^{x_2}\ (-2 + e^{x_2})^3,\ \frac{1}{16}\ e^{-4 x_2}\right]$$

The joint pdf is illustrated from differing perspectives in Fig. 7 and Fig. 8.

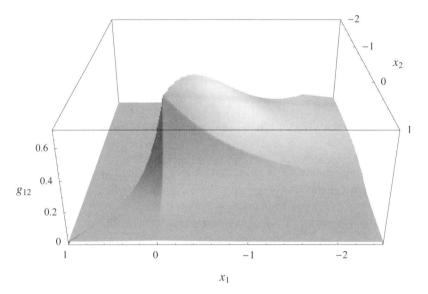

Fig. 7: pdf of g_{12} ('front' view)

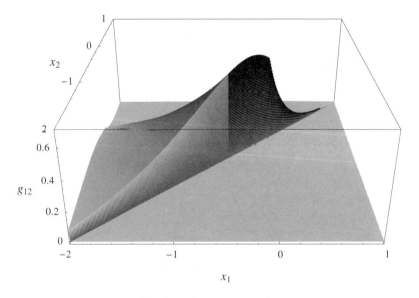

Fig. 8: pdf of g_{12} ('rear' view)

§9.4 A   STATISTICAL DECISION THEORY   317

In Fig. 7, ridges are evident along the lines $x_1 = 0$ and $x_2 = 0$; this is consistent with the piecewise nature of g_{12}. In Fig. 8, the face of the plane $x_1 = x_2$ is prominent; this neatly illustrates the domain of support of $X_{(1)}$ and $X_{(2)}$ (viz. the triangular region $\{(x_1, x_2) : -\infty < x_1 < x_2 < \infty\}$). The domain of support can also be illustrated as the shaded region in Fig. 9.

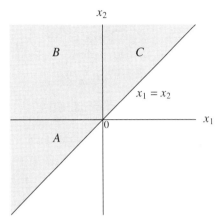

Fig. 9: Domain of support of $X_{(1)}$ and $X_{(2)}$ (shaded)

mathStatica cannot operate on the pdf, because the pdf has a multiple If structure. However, we may proceed by separating the domain of support into three distinct regions, labelled A, B and C in Fig. 9. In the triangular region A, the pdf of $X_{(1)}$ and $X_{(2)}$ is given by:

Ag$_{12}$ = Simplify[g$_{12}$, {x$_1$ < 0, x$_2$ < 0}]

$$-\frac{5}{8} e^{x_1 + x_2} (-2 + e^{x_2})^3$$

... while in the rectangular region B, the pdf is given by:

Bg$_{12}$ = Simplify[g$_{12}$, {x$_1$ < 0, x$_2$ > 0}]

$$\frac{5}{8} e^{x_1 - 4 x_2}$$

... and finally, in region C, the pdf is:

Cg$_{12}$ = Simplify[g$_{12}$, {x$_1$ > 0, x$_2$ > 0}]

$$\frac{5}{8} e^{-x_1 - 4 x_2}$$

In this way, we can verify that the pdf integrates to unity over its domain of support:

$$\int_{-\infty}^{0} \int_{-\infty}^{x_1} Ag_{12} \, dx_1 \, dx_2 + \int_{0}^{\infty} \int_{-\infty}^{0} Bg_{12} \, dx_1 \, dx_2 + \int_{0}^{\infty} \int_{0}^{x_2} Cg_{12} \, dx_1 \, dx_2$$

1

9.4 B Applications

Estimators such as the sample median (used to estimate location) and the sample interquartile range (to estimate scale) may be constructed from the order statistics of a random sample. In *Example 8*, we derive the MSE of the sample median, while in *Example 9* we derive the MSE of the sample range (a function of two order statistics).

⊕ *Example 8:* Sample Median versus Sample Mean

Two estimators of location are the sample median and the sample mean. In this example, we compare the MSE performance of each estimator when $X \sim \text{Logistic}(\theta)$, the location-shifted Logistic distribution with pdf $f(x)$:

```
f = e^-(x-θ) / (1 + e^-(x-θ))^2 ;    domain[f] = {x, -∞, ∞} && {θ ∈ Reals};
```

where $\theta \in \mathbb{R}$ is the location parameter (the mean of X). For simplicity, we assume that a random sample of size n drawn on X is odd-sized (*i.e.* n is odd), and so we shall write $n = 2r + 1$, for $r \in \{1, 2, ...\}$. Therefore, the sample median, which we denote by M, corresponds to the middle order statistic $X_{(r+1)}$. Thus, the pdf of M is given by:

```
g = OrderStat[r + 1, f, 2 r + 1] /. x → m
```

$$\frac{e^{(1+r)(m+\theta)} (e^m + e^\theta)^{-2(1+r)} (1 + 2r)!}{r!^2}$$

Here is the domain of support of M:

```
domain[g] = {m, -∞, ∞} && {θ ∈ Reals, r > 0};
```

The MSE of the sample median is given by $E[(M - \theta)^2]$. Unfortunately, if we evaluate `Expect[(m - θ)^2, g]`, an unsolved integral is returned. There are two possible reasons for this: (i) either *Mathematica* does not know how to solve this integral, or (ii) *Mathematica* can solve the integral, but needs a bit of help![4] In this case, we can help out by expressing the integrand in a simpler form. Since we want $E[(M - \theta)^2] = E[U^2]$, consider transforming M to the new variable $U = M - \theta$. The pdf of U, say g_u, is obtained using **mathStatica**'s `Transform` function:

```
g_u = Transform[u == m - θ, g]
```

$$\frac{e^{(1+r)u} (1 + e^u)^{-2(1+r)} (1 + 2r)!}{r!^2}$$

```
domain[g_u] = TransformExtremum[u == m - θ, g]
```

$\{u, -\infty, \infty\} \&\& \{r > 0\}$

Since the functional form of the pdf of U does not depend upon θ, it follows that the MSE cannot depend on the value of θ. To make things even simpler, we make the further transformation $V = e^U$. Then, the pdf of V, denoted g_v, is:

```
g_v = Transform[v == e^u, g_u]
```

$$\frac{v^r (1 + v)^{-2 (1+r)} (1 + 2 r)!}{r!^2}$$

```
domain[g_v] = TransformExtremum[v == e^u, g_u]
```

$$\{v, 0, \infty\} \;\&\&\; \{r > 0\}$$

Since $V = \exp(U)$, it follows that $E[U^2] = E[(\log V)^2]$. Therefore, the MSE of the sample median is:

```
MSE_med = Expect[Log[v]^2, g_v]
```

$$2 \, \text{PolyGamma}[1, 1 + r]$$

Our other estimator of location is the sample mean $\overline{X}$. To obtain its MSE, we must evaluate $E[(\overline{X} - \theta)^2]$. Because $\overline{X} = s_1/n$, where $s_1 = \sum_{i=1}^{n} X_i$ is the sample sum, the MSE is an expression involving power sums, and we can therefore use **mathStatica**'s Moments of Moments toolset (see §7.3) to solve the expectation. The MSE corresponds to the 1st raw moment of $(\frac{1}{n} s_1 - \theta)^2$, and so we shall present the answer in terms of raw population moments of X (hence ToRaw):

```
sol = RawMomentToRaw[1, (s_1/n - θ)^2]
```

$$-2\, \theta\, \acute{\mu}_1 + \frac{(-1 + n)\, \acute{\mu}_1^{\,2}}{n} + \frac{n\, \theta^2 + \acute{\mu}_2}{n}$$

We now find $\acute{\mu}_1$ and $\acute{\mu}_2$, and substitute these values into the solution:

```
MSE_mean =
  sol /. Table[μ_i → Expect[x^i, f], {i, 2}] // Simplify
```

$$\frac{\pi^2}{3\,n}$$

where $n = 2r + 1$.

Both MSE$_{med}$ and MSE$_{mean}$ are independent of θ, but vary with sample size. We can compare the performance of each estimator by plotting their respective MSE for various values of r, see Fig. 10.

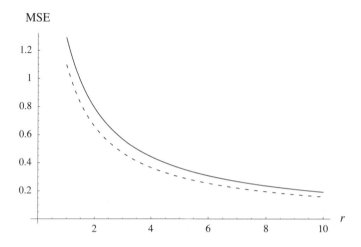

Fig. 10: MSE of sample mean (– – –) and sample median (——)

We see that the MSE of the sample mean (the dashed line) is everywhere below the MSE of the sample median (the unbroken line), and that this persists for all r. Hence, the sample mean dominates the sample median in mean square error (risk under quadratic loss) when estimating θ. We conclude that the sample median is inadmissible in this situation. However, this does not imply that the sample mean is admissible, for there may exist another estimator that dominates the sample mean under quadratic loss. ∎

⊕ *Example 9:* Sample Range versus Largest Order Statistic

Let $X \sim$ Uniform$(0, \theta)$, where $\theta \in \mathbb{R}_+$ is an unknown parameter, with pdf:

```
f = 1/θ ;     domain[f] = {x, 0, θ} && {θ > 0};
```

The sample range R is defined as the distance between the smallest and largest order statistics; that is, $R = X_{(n)} - X_{(1)}$. It may be used to estimate θ. Another estimator is the sample maximum, corresponding to the largest order statistic $X_{(n)}$. In this example, we compare the performance of both estimators on the basis of their respective MSE.

To derive the distribution of R, we first obtain the joint pdf of $X_{(1)}$ and $X_{(n)}$:

```
g = OrderStat[{1, n}, f] // FunctionExpand
```

$$\frac{(-1+n)\, n\, \left(\frac{-x_1+x_n}{\theta}\right)^n}{(-x_1+x_n)^2}$$

with non-rectangular domain of support:

```
domain[g] = OrderStatDomain[{1, n}, f]
```

— The domain is: $\{0 < x_1 < x_n < \theta\}$, which we enter into mathStatica as:

$\{\{x_1,\, 0,\, x_n\},\, \{x_n,\, x_1,\, \theta\}\}$ && $\{n \in$ Integers, $\theta > 0,\, 1 < n\}$

We use **mathStatica**'s Transform function to perform the transformation from $(X_{(1)}, X_{(n)})$ to (R, S), where $S = X_{(1)}$. Here is the joint pdf of (R, S):

$$g_{rs} = \text{Transform}[\{r == x_n - x_1, s == x_1\}, g]$$

$$\frac{(-1+n) \, n \, \left(\frac{r}{\theta}\right)^n}{r^2}$$

with non-rectangular support $\{(r, s) : 0 < r < \theta, \; 0 < s < \theta - r\}$. Integrating out S yields the pdf of R:

$$g_r = \int_0^{\theta-r} g_{rs} \, ds$$

$$\frac{(-1+n) \, n \, \left(\frac{r}{\theta}\right)^n (-r + \theta)}{r^2}$$

$$\text{domain}[g_r] = \{r, 0, \theta\} \, \&\& \, \{\theta > 0, n > 1, n \in \text{Integers}\};$$

The MSE for the sample range is:

$$\text{MSE}_{\text{range}} = \text{Expect}[(r - \theta)^2, g_r]$$

$$\frac{6 \, \theta^2}{2 + 3 \, n + n^2}$$

Our other estimator of θ is the sample maximum $X_{(n)}$. The pdf of $X_{(n)}$ is:

$$g_n = \text{OrderStat}[n, f]$$

$$\frac{n \, \left(\frac{x}{\theta}\right)^n}{x}$$

$$\text{domain}[g_n] = \text{OrderStatDomain}[n, f]$$

$$\{x, 0, \theta\} \, \&\& \, \{n \in \text{Integers}, \theta > 0, 1 \le n\}$$

The MSE of $X_{(n)}$ is:

$$\text{MSE}_{\text{max}} = \text{Expect}[(x - \theta)^2, g_n]$$

$$\frac{2 \, \theta^2}{2 + 3 \, n + n^2}$$

so $\text{MSE}_{\text{range}} = 3 \, \text{MSE}_{\text{max}}$ for all permissible values of θ and n. Therefore, the sample range is inadmissible.

Inadmissibility of the sample range does not imply that the sample maximum is admissible. Indeed, consider the following estimator that scales the sample maximum:

$$X^*_{(n)} = \frac{n+1}{n} X_{(n)}.$$

The MSE of the scaled estimator is:

`MSE`$_{\text{scaled}}$ `= Expect`$\left[\left(\frac{n+1}{n} x - \theta \right)^2 , g_n \right]$

$$\frac{\theta^2}{n(2+n)}$$

Dividing by the MSE of $X_{(n)}$ finds:

$\frac{\text{MSE}_{\text{scaled}}}{\text{MSE}_{\text{max}}}$ `// Simplify`

$$\frac{1+n}{2n}$$

which is strictly less than unity for all $n > 1$, implying that the sample maximum $X_{(n)}$ is inadmissible too! ■

9.5 Exercises

1. Let $\hat{\theta}$ denote an estimator of an unknown parameter θ, and let $a > 0$ and $b > 0$ ($a \neq b$) denote constants. Consider the asymmetric quadratic loss function

$$L(\hat{\theta}, \theta) = \begin{cases} a(\hat{\theta} - \theta)^2 & \text{if } \hat{\theta} > \theta \\ b(\hat{\theta} - \theta)^2 & \text{if } \hat{\theta} \leq \theta. \end{cases}$$

   Plot the loss function against values of $(\hat{\theta} - \theta)$, when $a = 1$ and $b = 2$.

2. Varian (1975) introduced the linex (linear–exponential) loss function

$$L(\hat{\theta}, \theta) = e^{c(\hat{\theta}-\theta)} - c(\hat{\theta} - \theta) - 1$$

   where $\hat{\theta}$ denotes an estimator of an unknown parameter θ, and constant $c \neq 0$.

   (i) Investigate the loss function by plotting L against $(\hat{\theta} - \theta)$ for various values of c.

   (ii) Using linear–exponential loss in the context of *Example 1* (i.e. $X \sim N(\theta, 1)$ and $\hat{\theta} = X + k$), determine the value of k which minimises risk.

3. Suppose that $X \sim \text{Exponential}(\theta)$, where $\theta > 0$ is an unknown parameter. The random variable $\hat{\theta} = X/k$ is proposed as an estimator of θ, where constant $k > 0$. Obtain the risk, and the value of k which minimises risk, when the loss function is:

   (i) symmetric quadratic $L_1(\hat{\theta}, \theta) = (\hat{\theta} - \theta)^2$.

   (ii) linear–exponential $L_2(\hat{\theta}, \theta) = e^{\hat{\theta}-\theta} - (\hat{\theta} - \theta) - 1$.

4. Let random variable T have the same pdf $f(t)$ as used in *Example 2*. For estimators of θ of general form $\hat{\Theta} = T/(n+k)$, where real $k > -n$, consider the asymmetric quadratic loss function

$$L(\hat{\Theta}, \theta) = \begin{cases} (\hat{\Theta} - \theta)^2 & \text{if } \hat{\Theta} > \theta \\ b(\hat{\Theta} - \theta)^2 & \text{if } \hat{\Theta} \leq \theta. \end{cases}$$

   (i) After transforming from T to $\hat{\Theta}$, derive the risk of $\hat{\Theta}$ as a function of θ, n, k and b (the solution takes about 140 seconds to compute on our reference machine).

   (ii) Explain why the minimum risk estimator does not depend on θ.

   (iii) Setting $n = 10$, use numerical methods to determine the value of k which yields the minimum risk estimator when (a) $b = \frac{1}{2}$ and (b) $b = 2$. Do your results make sense?

5. Let $X_{(n)}$ denote the largest order statistic of a random sample of size n from $X \sim \text{Beta}(a, b)$.

   (i) Derive the pdf of $X_{(n)}$.

   (ii) Use `PlotDensity` to plot (on a single diagram) the pdf of $X_{(n)}$ when $a = 2$, $b = 3$ and $n = 2, 4$ and 6.

6. Let $X_{(1)}$, $X_{(2)}$ and $X_{(3)}$ denote the order statistics of a random sample of size $n = 3$ from $X \sim N(0, 1)$.

   (i) Derive the pdf and cdf of each order statistic.

   (ii) Use `PlotDensity` to plot (on a single diagram) the pdf of each order statistic (use the interval $(-3, 3)$).

   (iii) Determine $E[X_{(r)}]$ for $r = 1, 2, 3$.

   (iv) The pdf of $X_{(1)}$ and the pdf of $X_{(3)}$ appear to be similar—perhaps they differ by a simple mean shift? Test this assertion by plotting (on a single diagram) the pdf of $X_{(3)}$ and Y, where the random variable $Y = X_{(1)} + 3/\sqrt{\pi}$.

7. Apply the loss function $L_k(\hat{\Theta}, \theta) = |\hat{\Theta} - \theta|^k$ in the context of *Example 9* (note: symmetric quadratic loss corresponds to the special case $k = 2$). Find the values of k for which the sample maximum dominates the sample range.

Chapter 10

Unbiased Parameter Estimation

10.1 Introduction

10.1 A Overview

For any given statistical model, there are any number of estimators that can be constructed in order to estimate unknown population parameters. In the previous chapter, we attempted to distinguish between estimators by specifying a loss structure, from which we hoped to identify the least risk estimator. Unfortunately, this process rarely presents a suitable overall winner. However, two important factors emerged from that discussion (especially for risk computed under quadratic loss), namely, the extent of bias, and the extent of variance inflation. Accounting for these factors yields a search for a preferred estimator from amongst classes of estimators, where the class members are forced to have a specific statistical property. This is precisely the approach taken in this chapter. Attention is restricted to the *class of unbiased estimators*, from which we wish to select the estimator that has least variance. We have already encountered the same type of idea in Chapter 7, where concern lay with unbiased estimation of population moments. In this chapter, on the other hand, we focus on unbiased estimation of the parameters of statistical models.

The chapter begins by measuring the statistical information that is present on a parameter in a given statistical model. This is done using Fisher Information and Sample Information (§10.2). This then leads to the so-called Cramer–Rao Lower Bound (a lower bound on the variance of any unbiased estimator), and to Best Unbiased Estimators, which are the rare breed of estimator whose variance achieves the lower bound (§10.3). The remaining two sections (§10.4 and §10.5) provide for the theoretical development of Minimum Variance Unbiased Estimators (MVUE). Vital to this is the notion of a sufficient statistic, its completeness, and its relation to the MVUE via a famous theorem due to Rao and Blackwell.

The statistical literature on MVUE estimation is extensive. The reference list that follows offers a sample of a range of treatments. In rough order of decreasing technical difficulty are Lehmann (1983), Silvey (1975), Cox and Hinkley (1974), Stuart and Ord (1991), Gourieroux and Monfort (1995), Mittelhammer (1996) and Hogg and Craig (1995).

10.1 B SuperD

In this chapter, it is necessary to activate the **mathStatica** function SuperD. This tool enhances *Mathematica*'s differentiator D (or, equivalently, ∂), allowing differentiation with respect to powers of variables. To illustrate, consider the derivative of $\sigma^{3/2}$ with respect to σ^2:

    `D[σ^(3/2), σ^2]`

    — General::ivar : σ^2 is not a valid variable.

    $\partial_{\sigma^2}\, \sigma^{3/2}$

Mathematica does not allow this operation because σ^2 is not a Symbol variable; in fact, it is stored as Power (*i.e.* Head[σ^2] = Power). However, by turning On the **mathStatica** function SuperD:

    `SuperD[On]`

    — SuperD is now On.

derivatives, such as the former, can now be performed:

    `D[σ^(3/2), σ^2]`

    $\dfrac{3}{4\sqrt{\sigma}}$

At any stage, this enhancement to D may be removed by entering SuperD[Off].

10.2 Fisher Information

10.2 A Fisher Information

Let a random variable X have density $f(x; \theta)$, where θ is an unknown parameter which, for the moment, we assume is a scalar. The amount of statistical information about θ that is contributed per observation on X is defined to be

$$i_\theta = E\left[\left(\frac{\partial \log f(X; \theta)}{\partial \theta}\right)^2\right] \tag{10.1}$$

and is termed *Fisher's Information* on θ, after R. A. Fisher who first formulated it.

⊕ **Example 1:** Fisher's Information on the Lindley Parameter

Let $X \sim \text{Lindley}(\delta)$, the Lindley distribution with parameter $\delta \in \mathbb{R}_+$, with pdf $f(x; \delta)$. Then, from **mathStatica**'s *Continuous* palette, the pdf of X is:

$$f = \frac{\delta^2}{\delta+1}(x+1)\,e^{-\delta x};$$
$$\text{domain}[f] = \{x, 0, \infty\}\ \&\&\ \{\delta > 0\};$$

Then i_δ, the Fisher Information on δ, is given by (10.1) as:

Expect[D[Log[f], δ]², f]

$$\frac{2}{\delta^2} - \frac{1}{(1+\delta)^2}$$

⊕ *Example 2:* An Imprecise Survey: Censoring a Poisson Variable

Over a 1-week period, assume that the number of over-the-counter banking transactions by individuals is described by a discrete random variable $X \sim \text{Poisson}(\lambda)$, where $\lambda \in \mathbb{R}_+$ is an unknown parameter. Suppose, when collecting data from individuals, a market research company adopts the following survey policy: four or fewer transactions are recorded correctly, whereas five or more are recorded simply as five. Study the loss of statistical information on λ that is incurred by this data recording method.

Solution: Let $f(x; \lambda)$ denote the pmf of X:

$$f = \frac{e^{-\lambda}\lambda^x}{x!};$$
$$\text{domain}[f] = \{x, 0, \infty\}\ \&\&\ \{\lambda > 0\}\ \&\&\ \{\text{Discrete}\};$$

Now define a discrete random variable Y, related to X as follows:

$$Y = \begin{cases} X & \text{if } X \leq 4 \\ 5 & \text{if } X \geq 5. \end{cases}$$

Notice that the survey method samples Y, not X. Random variable X is said to be *right-censored* at 5. The pmf of Y is given by

$$P(Y = y) = \begin{cases} P(X = y) & \text{if } y \leq 4 \\ P(X \geq 5) & \text{if } y = 5. \end{cases}$$

Let $g(y; \lambda)$ denote the pmf of Y in List Form, as shown in Table 1.

$P(Y = y):$	$f(0; \lambda)$	$f(1; \lambda)$	$f(2; \lambda)$	$f(3; \lambda)$	$f(4; \lambda)$	$P(X \geq 5)$
$y:$	0	1	2	3	4	5

Table 1: List Form pmf of Y

We enter this into *Mathematica* as follows:

```
            g = Append[Table[f, {x, 0, 4}], 1 - Prob[4, f]];
    domain[g] = {y, {0, 1, 2, 3, 4, 5}} && {λ > 0} && {Discrete};
```

where $P(Y = 5) = P(X \geq 5) = 1 - P(X \leq 4)$ is used. If an observation on X is recorded correctly, the Fisher Information on λ per observation, denoted by $i_{\lambda,X}$, is equal to:

```
i_{λ,x} = Expect[D[Log[f], λ]², f]
```

$$\frac{1}{\lambda}$$

On the other hand, the Fisher Information on λ per observation collected in the actual survey, denoted by $i_{\lambda,Y}$, is:

```
i_{λ,Y} = Expect[D[Log[g], λ]², g]
```

$$-(e^{-\lambda}(-144 - 288\lambda - 288\lambda^2 - 192\lambda^3 - 66\lambda^4 - 12\lambda^5 - \lambda^6 + 6e^{\lambda}(24 + 24\lambda + 12\lambda^2 + 4\lambda^3 - 4\lambda^4 + \lambda^5)))/(6\lambda(24 - 24e^{\lambda} + 24\lambda + 12\lambda^2 + 4\lambda^3 + \lambda^4))$$

Figure 1 plots relative information $i_{\lambda,Y}/i_{\lambda,X}$ against values of λ.

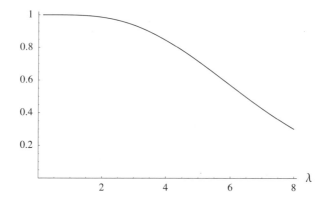

Fig. 1: Relative Fisher Information on λ

The figure shows that as λ increases, relative information declines. When, say, $\lambda = 5$, the relative information is:

```
i_{λ,Y}
------ /. λ → 5 // N
i_{λ,x}

0.715636
```

which means that about 28.5% of relative information on λ per observation has been lost by using this survey methodology. This would mean that to obtain the same amount of statistical information on λ as would be observed in a correctly recorded sample of say 100 individuals, the market research company would need to record data from about 140 (= 100/0.716) individuals. ■

10.2 B Alternate Form

Subject to some regularity conditions (*e.g.* Silvey (1975, p. 37) or Gourieroux and Monfort (1995, pp. 81–82)), an alternative expression for Fisher's Information to that given in (10.1) is

$$i_\theta = -E\left[\frac{\partial^2 \log f(X;\theta)}{\partial \theta^2}\right]. \tag{10.2}$$

For a proof of (10.2), see Silvey (1975, p. 40). When it is valid, this form of Fisher's Information can often be more convenient to compute, especially if the second derivative is not stochastic.

⊕ *Example 3:* First Derivative Form versus Second Derivative Form

Suppose the discrete random variable $X \sim$ RiemannZeta(ρ). Then, from **mathStatica**'s *Discrete* palette, the pmf $f(x; \rho)$ of X is given by:

```
f = x^-(ρ+1) / Zeta[1 + ρ];
domain[f] = {x, 1, ∞} && {ρ > 0} && {Discrete};
```

Following (10.1), $\left(\frac{\partial \log f(x;\rho)}{\partial \rho}\right)^2$ is given by:

```
d = D[Log[f], ρ]^2 // Simplify
```

$$\frac{(\text{Log}[x]\, \text{Zeta}[1+\rho] + \text{Zeta}'[1+\rho])^2}{\text{Zeta}[1+\rho]^2}$$

This is a stochastic expression for it depends on x, the values of X. Applying `Expect` yields the Fisher Information on ρ:

```
Expect[d, f]
```

$$\frac{-\text{Zeta}'[1+\rho]^2 + \text{Zeta}[1+\rho]\, \text{Zeta}''[1+\rho]}{\text{Zeta}[1+\rho]^2}$$

Alternately, following (10.2), we find:

```
-D[Log[f], {ρ, 2}] // Simplify
```

$$\frac{-\text{Zeta}'[1+\rho]^2 + \text{Zeta}[1+\rho]\, \text{Zeta}''[1+\rho]}{\text{Zeta}[1+\rho]^2}$$

This output is non-stochastic, and is clearly equivalent to the previous output. In this case, (10.2) yields Fisher's Information on ρ, without the need to even apply `Expect`. ∎

⊕ **Example 4:** Regularity Conditions

Suppose $X \sim \text{Uniform}(\theta)$, where parameter $\theta \in \mathbb{R}_+$. The pdf of X is:

$$f = \frac{1}{\theta}; \quad \text{domain}[f] = \{x, 0, \theta\} \,\&\&\, \{\theta > 0\};$$

According to the definition (10.1), the Fisher Information on θ is:

`Expect[D[Log[f], `θ`]`^{2}`, f]`

$$\frac{1}{\theta^2}$$

Next, consider the following output calculated according to (10.2):

`-Expect[D[Log[f], {`θ`, 2}], f]`

$$-\frac{1}{\theta^2}$$

Clearly, this expression cannot be correct, because Fisher Information cannot be negative. The reason why our second computation is incorrect is because a regularity condition is violated—the condition that permits interchangeability between the differential and integral operators. In general, it can be shown (see Silvey (1975, p. 40)) that (10.2) is equivalent to (10.1) if

$$\frac{\partial^2}{\partial \theta^2} \int_0^\theta f\, dx = \int_0^\theta \frac{\partial^2 f}{\partial \theta^2}\, dx \qquad (10.3)$$

where $f = 1/\theta$ is the pdf of X. In this case, (10.3) is not true as the value of the pdf at $x = \theta$ is strictly positive. Indeed, as a general rule, the regularity conditions permitting computation of Fisher Information according to (10.2) are violated whenever the domain of support of a random variable depends on unknown parameters, when the density at those points is strictly positive. ∎

10.2 C Automating Computation: `FisherInformation`

In light of (10.1) and (10.2), **mathStatica**'s `FisherInformation` function automates the computation of Fisher Information. In an obvious notation, the function's syntax is `FisherInformation[`θ`, f]`, with options `Method → 1` (default) for computation according to (10.1), or `Method → 2` for computation according to (10.2).

⊕ **Example 5:** `FisherInformation`

Suppose that $X \sim N(\mu, 1)$. Then, its pdf is given by:

$$f = \frac{1}{\sqrt{2\pi}}\, e^{-\frac{1}{2}(x-\mu)^2}; \quad \text{domain}[f] = \{x, -\infty, \infty\} \,\&\&\, \{\mu \in \text{Reals}\};$$

The Fisher Information on μ may be derived using the one-line command:

```
FisherInformation[μ, f]
```

```
1
```

It is well worth contrasting the computational efficiency of the two methods of calculation, (10.1) and (10.2):

```
FisherInformation[μ, f, Method → 1] // Timing
```

```
{0.72 Second, 1}
```

```
FisherInformation[μ, f, Method → 2] // Timing
```

```
{0.11 Second, 1}
```

Generally, the second method is more efficient; however, the second method is only valid under regularity conditions. In this example, the regularity conditions are satisfied. ∎

10.2 D Multiple Parameters

The discussion so far has been concerned with statistical information on a single parameter. Of course, many statistical models have multiple parameters. Accordingly, we now broaden the definition of Fisher Information (10.1) to the case when θ is a $(k \times 1)$ vector of unknown parameters. Fisher's Information on θ is now a square, symmetric matrix of dimension $(k \times k)$. The $(i, j)^{\text{th}}$ element of the Fisher Information matrix i_θ is

$$E\left[\left(\frac{\partial \log f(X;\theta)}{\partial \theta_i}\right)\left(\frac{\partial \log f(X;\theta)}{\partial \theta_j}\right)\right] \quad (10.4)$$

for $i, j \in \{1, \ldots, k\}$. Notice that when $i = j$, (10.4) becomes (10.1), and is equivalent to the Fisher Information on θ_i. The multi-parameter analogue of (10.2) is given by

$$-E\left[\frac{\partial^2 \log f(X;\theta)}{\partial \theta_i \, \partial \theta_j}\right] \quad (10.5)$$

which corresponds to the $(i, j)^{\text{th}}$ element of i_θ, provided the regularity conditions hold. **mathStatica**'s `FisherInformation` function extends to the multi-parameter setting.

⊕ *Example 6:* Fisher Information Matrix for Gamma Parameters

Suppose that $X \sim \text{Gamma}(a, b)$, where $\theta = \binom{a}{b}$ is a (2×1) vector of unknown parameters. Let $f(x; \theta)$ denote the pdf of X:

```
      x^(a-1) e^(-x/b)
f = ─────────────────;    domain[f] = {x, 0, ∞} && {a > 0, b > 0};
         Γ[a] b^a
```

The elements of Fisher's Information on θ, a (2×2) matrix, are:

FisherInformation[{a, b}, f]

$$\begin{pmatrix} \text{PolyGamma}[1, a] & \frac{1}{b} \\ \frac{1}{b} & \frac{a}{b^2} \end{pmatrix}$$

where the placement of the elements in the matrix is important; for example, the top-left element corresponds to Fisher's Information on a. ∎

10.2 E Sample Information

As estimation of parameters is typically based on a sample of data drawn from a population, it is important to contemplate the amount of information that is contained by a sample about any parameters. Once again, Fisher's formulation may be used to measure statistical information. However, this time we focus upon the joint distribution of the random sample, as opposed to the distribution of the population from which the sample is drawn. We use the symbol I_θ to denote the statistical information contained by a sample, terming this *Sample Information*, as distinct from i_θ for Fisher Information.[1]

Let $\vec{X} = (X_1, \ldots, X_n)$ denote a random sample of size n drawn on a random variable X. Denote the joint density of $\vec{X}$ by $f(\vec{x}; \theta)$, where scalar θ is an unknown parameter. The Sample Information on θ is defined as

$$I_\theta = E\left[\left(\frac{\partial \log f(\vec{X}; \theta)}{\partial \theta}\right)^2\right]. \tag{10.6}$$

If $\vec{X}$ is a collection of n independent and identically distributed (iid) random variables, each with density $f(x_i; \theta)$ ($i = 1, \ldots, n$), equivalent in functional form, then the joint density of the collection $\vec{X}$ is given by $f(\vec{x}; \theta) = \prod_{i=1}^n f(x_i; \theta)$. Furthermore, if the regularity condition $E\left[\left(\frac{\partial \log f(X; \theta)}{\partial \theta}\right)\right] = 0$ is satisfied, then

$$\begin{aligned} I_\theta &= E\left[\left(\frac{\partial}{\partial \theta} \log \prod_{i=1}^n f(X_i; \theta)\right)^2\right] \\ &= \sum_{i=1}^n E\left[\left(\frac{\partial \log f(X_i; \theta)}{\partial \theta}\right)^2\right] \\ &= n\, i_\theta. \end{aligned} \tag{10.7}$$

If it is valid to do so, it is well worth exploiting (10.7), as the derivation of I_θ through the multivariate expectation (10.6) can be difficult. For example, for n observations collected on $X \sim \text{Lindley}(\delta)$, the Sample Information is simply $n\, i_\delta$, where i_δ was derived in *Example 1*. On the other hand, for models that generate observations according to underlying regimes (*e.g.* the censoring model discussed in *Example 2* is of this type), the relationship between Fisher Information and Sample Information is generally more complicated than that described by (10.7), even if the random sample consists of a collection of iid random variables.

10.3 Best Unbiased Estimators

10.3 A The Cramér–Rao Lower Bound

Let θ denote the parameter of a statistical model, and let $g(\theta)$ be some differentiable function of θ that we are interested in estimating. The *Cramér–Rao Lower Bound* (CRLB) establishes a lower bound below which the variance of an *unbiased estimator* of $g(\theta)$ cannot go. Often the CRLB is written in the form of an inequality—the *Cramér–Rao Inequality*. Let $\hat{g}$ denote an unbiased estimator of $g(\theta)$ constructed from a random sample of n observations. Then, subject to some regularity conditions, the Cramér–Rao Inequality is given by

$$\text{Var}(\hat{g}) \geq \left(\frac{\partial g(\theta)}{\partial \theta}\right)^2 \bigg/ I_\theta \qquad (10.8)$$

where I_θ denotes Sample Information (§10.2 E). If we are interested in estimating θ, then set $g(\theta) = \theta$, in which case (10.8) simplifies to

$$\text{Var}(\hat{\theta}) \geq 1/I_\theta \qquad (10.9)$$

where $\hat{\theta}$ is an unbiased estimator of θ. When estimating $g(\theta)$, the CRLB is the quantity on the right-hand side of (10.8); similarly, when estimating θ, the CRLB is the right-hand side of (10.9). The inverse relationship between the CRLB and Sample Information is intuitive. After all, the more statistical information that a sample contains on θ, the better should an (unbiased) estimator of θ (or $g(\theta)$) perform. In our present context, 'better' refers to smaller variance.

If θ, or $g(\theta)$, represent vectors of parameters, say θ is $(k \times 1)$ and $g(\theta)$ is $(m \times 1)$ with $m \leq k$, then the CRLB expresses a lower bound on the variance-covariance matrix of unbiased estimators. In this instance, (10.8) becomes

$$\text{Varcov}(\hat{g}) \geq G \times I_\theta^{-1} \times G^\text{T} \qquad (10.10)$$

where the $(m \times k)$ matrix of derivatives

$$G = \frac{\partial g(\theta)}{\partial \theta^\text{T}}.$$

Equation (10.9) becomes

$$\text{Varcov}(\hat{\theta}) \geq I_\theta^{-1} \qquad (10.11)$$

where the notation $A \geq B$ indicates that $A - B$ is a positive semi-definite matrix, and I_θ^{-1} denotes the inverse of the Sample Information matrix. For proofs of the Cramér–Rao Inequality for both scalar and vector cases, plus discussion on the regularity conditions, see Silvey (1975), Mittelhammer (1996), or Gourieroux and Monfort (1995).

⊕ **Example 7:** The CRLB for the Poisson Parameter

Suppose that $X \sim \text{Poisson}(\lambda)$. Derive the CRLB for all unbiased estimators of λ.

Solution: Let $f(x; \lambda)$ denote the pmf of X:

```
        e^-λ λ^x
f  =  ──────── ;
          x !
domain[f] = {x, 0, ∞} && {λ > 0} && {Discrete};
```

The right-hand side of (10.9) gives the general formula for the CRLB for unbiased estimators. Thus, for random samples of size n drawn on X, the CRLB for the Poisson parameter λ is:

```
            1
───────────────────────
 n FisherInformation[λ, f]
```

$$\frac{\lambda}{n}$$

where we have exploited the relationship between Sample Information and Fisher Information given in (10.7). ∎

⊕ **Example 8:** The CRLB for the Inverse Gaussian Mean and Variance

Let $X \sim \text{InverseGaussian}(\mu, \lambda)$, and let $\theta = \begin{pmatrix} \mu \\ \lambda \end{pmatrix}$. Derive the CRLB for unbiased estimators of $g(\theta)$, where

$$g(\theta) = g(\mu, \lambda) = \begin{pmatrix} \mu \\ \mu^3/\lambda \end{pmatrix}.$$

Solution: Enter the pdf of X:

```
         ┌─────           (x - μ)²
f  =    │  λ      Exp[-λ ────────── ];
        √ ─────           2 μ² x
          2 π x³
domain[f] = {x, 0, ∞} && {μ > 0, λ > 0};
```

The CRLB for θ is equal to the (2×2) matrix:

```
CRLB = Inverse[n FisherInformation[{μ, λ}, f]]
```

$$\begin{pmatrix} \frac{\mu^3}{n\lambda} & 0 \\ 0 & \frac{2\lambda^2}{n} \end{pmatrix}$$

To find the CRLB for $g(\mu, \lambda) = (\mu, \mu^3/\lambda)^T$, a (2×1) vector, the right-hand side of (10.10) must be evaluated. First, we derive the (2×2) matrix of derivatives $G = \partial g(\theta) / \partial \theta^T$ using the **mathStatica** function Grad:

G = Grad[{μ, $\frac{\mu^3}{\lambda}$}, {μ, λ}]

$$\begin{pmatrix} 1 & 0 \\ \frac{3\mu^2}{\lambda} & -\frac{\mu^3}{\lambda^2} \end{pmatrix}$$

Then, the CRLB is given by the (2×2) matrix:

G.CRLB.Transpose[G] // Simplify

$$\begin{pmatrix} \frac{\mu^3}{n\lambda} & \frac{3\mu^5}{n\lambda^2} \\ \frac{3\mu^5}{n\lambda^2} & \frac{\mu^6(2\lambda+9\mu)}{n\lambda^3} \end{pmatrix}$$

10.3 B Best Unbiased Estimators

Suppose that $\hat{g}$ is an unbiased estimator of $g(\theta)$ that satisfies all regularity conditions, and that Var($\hat{g}$) attains the CRLB. In this event, we can do no better (in terms of variance minimisation) by adopting another unbiased estimator of $g(\theta)$; consequently, $\hat{g}$ is preferred over all other unbiased estimators. Because Var($\hat{g}$) is equivalent to the CRLB, $\hat{g}$ is referred to as the *Best Unbiased Estimator* (BUE) of $g(\theta)$.

⊕ **Example 9:** The BUE of the Poisson Parameter

Suppose that $X \sim \text{Poisson}(\lambda)$, with pmf:

$$f = \frac{e^{-\lambda}\lambda^x}{x!}; \quad \text{domain}[f] = \{x, 0, \infty\} \&\& \{\lambda > 0\} \&\& \{\text{Discrete}\};$$

Let $(X_1, \ldots, X_n)$ denote a random sample of size n drawn on X. We have already seen that the CRLB for unbiased estimators of λ is given by λ/n (see *Example* 7). Consider then the estimator $\hat{\lambda} = \frac{1}{n}\sum_{i=1}^{n} X_i = \overline{X}$, the sample mean. Whatever the value of index i, X_i is a copy of X, so the mean of $\hat{\lambda}$ is given by:

$$\frac{1}{n}\sum_{i=1}^{n} \text{Expect}[x, f]$$

λ

In addition, because X_i is independent of X_j for all $i \neq j$, the variance of $\hat{\lambda}$ is given by:

$$\frac{1}{n^2}\sum_{i=1}^{n} \text{Var}[x, f]$$

$\frac{\lambda}{n}$

From these results, we see that $\hat{\lambda}$ is an unbiased estimator of λ, and its variance corresponds to the CRLB. Thus, $\hat{\lambda}$ is the BUE of λ. ■

⊕ **Example 10:** Estimation of the Extreme Value Scale Parameter

Let the continuous random variable X have the following pdf:

```
f = 1/σ Exp[-x/σ - e^(-x/σ)];
domain[f] = {x, -∞, ∞} && {σ > 0};
```

Thus, $X \sim$ ExtremeValue, with unknown scale parameter $\sigma \in \mathbb{R}_+$. The CRLB for unbiased estimators of σ is given by:

```
CRLB = 1 / (n FisherInformation[σ, f])
```

$$\frac{6\sigma^2}{n(6(-1+\text{EulerGamma})^2 + \pi^2)}$$

where n denotes the size of the random sample drawn on X. In numeric terms:

```
CRLB // N
```

$$\frac{0.548342\,\sigma^2}{n}$$

Now consider the expectation $E[|X|]$:

```
Expect[If[x < 0, -x, x], f]
```

$$\sigma\,(\text{EulerGamma} - 2\,\text{ExpIntegralEi}[-1])$$

Let γ denote EulerGamma, and let Ei(−1) denote ExpIntegralEi[−1]. Knowing $E[|X|]$, it is easy to construct an unbiased estimator of the scale parameter σ, namely

$$\hat{\sigma} = \frac{1}{n(\gamma - 2\,\text{Ei}(-1))} \sum_{i=1}^{n} |X_i|$$

$$= \frac{0.984268}{n} \sum_{i=1}^{n} |X_i|$$

where γ and Ei(−1) have been assigned their respective numeric value. Following the method of *Example 9*, the variance of $\hat{\sigma}$ is:

$$\frac{\sum_{i=1}^{n} \text{Var}[\text{If}[x < 0, -x, x], f]}{(n\,(\text{EulerGamma} - 2\,\text{ExpIntegralEi}[-1]))^2} \text{ // N}$$

$$\frac{0.916362\,\sigma^2}{n}$$

Clearly, Var($\hat{\sigma}$) > CRLB, in which case $\hat{\sigma}$ is *not* the BUE of σ. ∎

10.4 Sufficient Statistics

10.4 A Introduction

Unfortunately, there are many statistical models for which the BUE of a given parameter does not exist.[2] In this case, even if it is straightforward to construct unbiased estimators, how can we be sure that the particular estimator we select has least variance? After all, unless we inspect the variance of every unbiased estimator — keep in mind that this class of estimator may well have an infinite number of members — the least variance unbiased estimator may simply not happen to be amongst those we examined. Nevertheless, if our proposed estimator has *used all available statistical information on the parameter of interest*, then intuition suggests that our selection may have least variance. A statistic that retains all information about a parameter is said to be *sufficient* for that parameter.

Let X denote the population of interest, dependent on some unknown parameter θ (which may be a vector). Then, the 'information' referred to above is that which is derived from a size n random sample drawn on X, the latter denoted by $\vec{X} = (X_1, ..., X_n)$. A sufficient statistic S is a function of the random sample; that is, $S = S(\vec{X})$. Obviously $S(\vec{X})$ is a random variable, but for a particular set of observed data, $\vec{x} = (x_1, ..., x_n)$, $S(\vec{x})$ must be numeric.

A statistic S, whose values we shall denote by s, is sufficient for a parameter θ if the conditional distribution of $\vec{X}$ given $S = s$ does not depend on θ. Immediately, then, the identity statistic $S = \vec{X}$ must be sufficient; however, it is of no use as it has dimension n. This is because the key idea behind sufficiency is to reduce the dimensionality of $\vec{X}$, without losing information. Finally, if another statistic $T = T(\vec{X})$ is such that it *loses* all information about a parameter, then it is termed *ancillary* for that parameter. It is also possible that a statistic $U = U(\vec{X})$ can be neither sufficient nor ancillary for a parameter.

⊕ ***Example 11:*** Sufficiency in Bernoulli Trials

Let $X \sim \text{Bernoulli}(p)$, where $p = P(X = 1)$ denotes the success probability. Given a random sample $\vec{X}$, we would expect the number of successes $S = \sum_{i=1}^{n} X_i \sim \text{Binomial}(n, p)$ to be influential when estimating the success probability p. In fact, for values $x_i \in \{0, 1\}$, and value $s \in \{0, 1, ..., n\}$ such that $s = \sum_{i=1}^{n} x_i$, the conditional distribution of $\vec{X}$ given $S = s$ is

$$P(\vec{X} \mid S = s) = \frac{P(X_1 = x_1, ..., X_n = x_n)}{P(S = s)} = \frac{p^n(1-p)^{n-s}}{\binom{n}{s} p^n (1-p)^{n-s}} = \frac{1}{\binom{n}{s}}.$$

As the conditional distribution does not depend on p, the one dimensional statistic $S = \sum_{i=1}^{n} X_i$ is sufficient for p. On the other hand, the statistic T, defined here as the chronological order in which observations occur, contributes nothing to our knowledge of the success probability: T is ancillary for p. A third statistic, the sample median M, is neither sufficient for p, nor is it ancillary for p.

It is interesting to examine the loss in Sample Information incurred as a result of using M to estimate p. For simplicity, set $n = 4$. Then, the sample sum $S \sim \text{Binomial}(4, p)$, with pmf $f(s; p)$:

```
f = Binomial[4, s] p^s (1 - p)^(4-s);

domain[f] = {s, 0, 4} && {0 < p < 1} && {Discrete};
```

From *Example 3* of Chapter 9, when $n = 4$, the sample median M has pmf $g(m, p)$, as given in Table 2.

$P(M = m)$:	$P(S \leq 1)$	$P(S = 2)$	$P(S \geq 3)$
m:	0	$\frac{1}{2}$	1

Table 2: The pmf of M when $n = 4$

We enter the pmf of M in List Form:

```
g = {Prob[1, f],   f /. s → 2,   1 - Prob[2, f]}
```

$\{-(-1 + p)^3 (1 + 3p),\ 6(1 - p)^2 p^2,\ 4p^3 - 3p^4\}$

with domain of support:

```
domain[g] = {m, {0, 1/2, 1}} && {Discrete};
```

To compute the Sample Information on p, we use the fact that it is equivalent to the Fisher Information on p per observation on the sufficient statistic S:

```
FisherInformation[p, f]
```

$$\frac{4}{p - p^2}$$

Similarly, the amount of Sample Information on p that is captured by statistic M is equivalent to the Fisher Information on p per observation on M:

```
FisherInformation[p, g]
```

$$-\frac{24(4 - p + p^2)}{(-4 + 3p)(1 + 3p)}$$

Figure 2 plots the amount of Sample Information captured by each statistic against values of p. Evidently, the farther the true value of p lies from $\frac{1}{2}$, the greater is the loss of information about p incurred by the sample median M.

§10.4 A   UNBIASED PARAMETER ESTIMATION   339

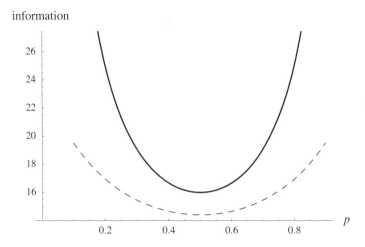

Fig. 2: Information on p due to statistics S (———) and M (– – –) when $n=4$

10.4 B The Factorisation Criterion

The *Factorisation Criterion* provides a way to identify sufficient statistics. Once again, let X denote the population of interest, dependent on some unknown parameter θ, and let $\vec{X}$ denote a size n random sample drawn on X with joint density $f_*(\vec{x}; \theta)$. A necessary and sufficient condition for a statistic $S = S(\vec{X})$ to be sufficient for θ is that the density of $\vec{X}$ can be factored into the product,

$$f_*(\vec{x}; \theta) = g_*(s; \theta)\, h_*(\vec{x}) \tag{10.12}$$

where $g_*(s; \theta)$ denotes the density of S, and $h_*(\vec{x})$ is a non-negative function that does not involve θ; for discussion of the proof of this result, see Stuart and Ord (1991, Chapter 17). The factorisation (10.12) requires knowledge of the density of S which can, on occasion, add unnecessary difficulties. Fortunately, (10.12) can be weakened to

$$f_*(\vec{x}; \theta) = g(s; \theta)\, h(\vec{x}) \tag{10.13}$$

where $g(s; \theta)$ is a non-negative function (not necessarily a density function), and $h(\vec{x})$ is a non-negative function that does not involve θ. From now on, we shall adopt (10.13) to identify sufficient statistics.[3]

The **mathStatica** function `Sufficient[f]` constructs the joint density $f_*(\vec{x}; \theta)$ of a size n random sample $\vec{X} = (X_1, \ldots, X_n)$ drawn on a random variable X, and then simplifies it. The output from `Sufficient` can be useful when attempting to identify sufficient statistics for a parameter.

Finally, sufficient statistics are not unique; indeed, if a statistic S is sufficient for a parameter θ, then so too is a one-to-one function of S. To illustrate, suppose that statistic $S = (\sum_{i=1}^{n} X_i, \sum_{i=1}^{n} X_i^2)$ is sufficient for a parameter θ. Then, $T = \left(\overline{X}, \frac{1}{n-1} \sum_{i=1}^{n} (X_i - \overline{X})^2\right)$ is also sufficient for θ, as T and S are related by a one-to-one transformation.

⊕ **Example 12:** A Sufficient Statistic for the Poisson Parameter

Let $X \sim \text{Poisson}(\lambda)$ with pmf $f(x; \lambda)$:

$$\texttt{f} = \frac{e^{-\lambda} \lambda^x}{x!}; \quad \texttt{domain[f]} = \{\texttt{x}, 0, \infty\} \,\&\&\, \{\lambda > 0\} \,\&\&\, \{\texttt{Discrete}\};$$

The joint density of $\vec{X}$, a random sample of size n drawn on X, is given by $f_*(\vec{x}; \lambda) = \prod_{i=1}^n f(x_i; \lambda)$. This is derived by `Sufficient` as follows:

Sufficient[f]

$$e^{-n\lambda} \, \lambda^{\sum_{i=1}^n x_i} \prod_{i=1}^n \frac{1}{x_i!}$$

If we define $S = \sum_{i=1}^n X_i$, and let $g(s; \lambda) = e^{-n\lambda} \lambda^s$ and $h(\vec{x}) = \prod_{i=1}^n \frac{1}{x_i!}$, then, in view of (10.13), it follows that S is sufficient for λ. ■

⊕ **Example 13:** Sufficient Statistics for the Normal Parameters

Let $X \sim N(\mu, \sigma^2)$ with pdf $f(x; \mu, \sigma^2)$:

$$\texttt{f} = \frac{1}{\sigma \sqrt{2\pi}} \,\texttt{Exp}\!\left[-\frac{(x-\mu)^2}{2\sigma^2}\right];$$
$$\texttt{domain[f]} = \{\texttt{x}, -\infty, \infty\} \,\&\&\, \{\mu \in \texttt{Reals}, \sigma > 0\};$$

Let $\vec{X}$ denote a random sample of size n drawn on X. Identify sufficient statistics when: (i) μ is unknown and σ^2 is known, (ii) μ is known and σ^2 unknown, (iii) both μ and σ^2 are unknown, and (iv) $\mu = \sigma = \theta$ is unknown.

Solution: In each case we must inspect the joint density of $\vec{X}$ produced by:

Sufficient[f]

$$e^{-\frac{n\mu^2 - 2\mu \sum_{i=1}^n x_i + \sum_{i=1}^n x_i^2}{2\sigma^2}} \, (2\pi)^{-n/2} \, \sigma^{-n}$$

(i) Define $S_1 = \sum_{i=1}^n X_i$. Because the value of σ^2 is known, let

$$g(s_1; \mu) = \exp\!\left(-\frac{n\mu^2 - 2\mu s_1}{2\sigma^2}\right)$$

$$h(\vec{x}) = \exp\!\left(-\frac{1}{2\sigma^2} \sum_{i=1}^n x_i^2\right) (2\pi)^{-n/2} \, \sigma^{-n}.$$

Then, by (10.13), it follows that S_1 is sufficient for μ.

(ii) Define $S_2 = n\mu^2 - 2\mu \sum_{i=1}^{n} X_i + \sum_{i=1}^{n} X_i^2 = \sum_{i=1}^{n} (X_i - \mu)^2$. As μ is known, let

$$g(s_2; \sigma^2) = \exp\left(-\frac{s_2}{2\sigma^2}\right)\sigma^{-n}$$

$$h(\vec{x}) = (2\pi)^{-n/2}.$$

Since $g(s_2; \sigma^2)h(\vec{x})$ is equivalent to the joint density of $\vec{X}$, it follows that S_2 is sufficient for σ^2.

(iii) Define $S_3 = (S_{31}, S_{32}) = (\sum_{i=1}^{n} X_i, \sum_{i=1}^{n} X_i^2)$. Setting

$$g(s_3; \mu, \sigma^2) = \exp\left(-\frac{n\mu^2 - 2\mu s_{31} + s_{32}}{2\sigma^2}\right)\sigma^{-n}$$

$$h(\vec{x}) = (2\pi)^{-n/2}$$

it follows that the two-dimensional statistic S_3 is sufficient for (μ, σ^2).

(iv) For $\mu = \sigma = \theta$, and S_3 as defined in part (iii), set

$$g(s_3; \theta) = \exp\left(\frac{2\theta s_{31} - s_{32}}{2\theta^2}\right)\theta^{-n}$$

$$h(\vec{x}) = e^{-n/2}(2\pi)^{-n/2}.$$

Then, the two-dimensional statistic S_3 is sufficient for the scalar parameter θ. This last example serves to illustrate a more general point: the number of sufficient statistics need not match the number of unknown parameters. ■

10.5 Minimum Variance Unbiased Estimation

10.5 A Introduction

So far, we have armed ourselves with a sufficient statistic that captures all the statistical information that exists about a parameter. The next question is then how to use that statistic to construct an unbiased estimator of the unknown parameter. Intuition suggests that such an estimator should distinguish itself by having least variance. In other words, the estimator should be a *minimum variance unbiased estimator* (MVUE). This section focuses on the search for the MVUE of a parameter. Important to this development are theorems due to Rao and Blackwell (§10.5 B) and Lehmann and Scheffé (§10.5 D), and the notion of a complete sufficient statistic (§10.5 C).

10.5 B The Rao–Blackwell Theorem

The following theorem, due to Rao and Blackwell, is critical in the search for a MVUE:

Theorem (Rao–Blackwell): Let $S = S(\vec{X})$ be a sufficient statistic for a parameter θ, and let another statistic $T = T(\vec{X})$ be an unbiased estimator of $g(\theta)$ with finite variance. Define the function $\hat{g}(s) = E[T \mid S = s]$. Then:

(i) $E[\hat{g}(S)] = g(\theta)$; that is, $\hat{g}(S)$ is an unbiased estimator of $g(\theta)$.

(ii) $\text{Var}(\hat{g}(S)) \leq \text{Var}(T)$.

Proof: See, for example, Silvey (1975, pp. 28–29). For discussion, see Hogg and Craig (1995, p. 326).

⊕ **Example 14:** A Conditional Expectation

Let $X \sim N(\mu, 1)$, and let $\overline{X}$ denote the sample mean from a random sample of size $n = 2r + 1$ drawn on X (for integer $r \geq 1$). Derive $E[T \mid \overline{X} = \overline{x}]$, where $T = T(\vec{X})$ denotes the sample median.

Solution (partial): We know from *Example 13*(i) that $S = \sum_{i=1}^{n} X_i$ is sufficient for μ. Thus, $\overline{X}$ will also be sufficient for μ as it is a one-to-one function of S. It follows that $E[T \mid \overline{X} = \overline{x}]$ can only be some function of $\overline{x}$, say $\hat{g}(\overline{x})$; that is, $E[T \mid \overline{X} = \overline{x}] = \hat{g}(\overline{x})$. The next step is to try and narrow down the possibilities for $\hat{g}(\overline{x})$. This is where part (i) of the Rao–Blackwell Theorem is used, for after deriving $E[T] = g(\mu)$, we may then be able to deduce those functions $\hat{g}(\overline{x})$ satisfying $E[\hat{g}(\overline{X})] = g(\mu)$, as we know $\overline{X} \sim N(\mu, \frac{1}{n})$.

Our strategy requires that we determine $E[T]$. Enter f, the pdf of X:

```
f = 1/√(2π) Exp[- (x - μ)²/2 ];

domain[f] = {x, -∞, ∞} && {μ ∈ Reals};
```

In a sample of size $n = 2r + 1$, the sample median T corresponds to the $(r+1)^{\text{th}}$ order statistic. We can use OrderStat to determine the pdf of T:

```
g = OrderStat[r + 1, f, 2 r + 1]
```

$$\frac{2^{-\frac{1}{2} - 2r} \, e^{-\frac{1}{2}(x-\mu)^2} \left(1 - \text{Erf}\left[\frac{x-\mu}{\sqrt{2}}\right]^2\right)^r (1 + 2r)!}{\sqrt{\pi} \, r!^2}$$

```
domain[g] = {x, -∞, ∞} && {μ ∈ Reals};
```

Transforming $T \to Q$, such that $Q = T - \mu$, yields the pdf of Q:

```
h = Transform[q == x - μ, g]
```

$$\frac{2^{-\frac{1}{2}-2r} e^{-\frac{q^2}{2}} \left(1 - \text{Erf}\left[\frac{q}{\sqrt{2}}\right]^2\right)^r (1 + 2r)!}{\sqrt{\pi} \, r!^2}$$

```
domain[h] = {q, -∞, ∞};
```

From this, we find $E[Q]$:

```
Expect[q, h]

0
```

Thus, $E[Q] = E[T - \mu] = 0$; that is, $E[T] = g(\mu) = \mu$. Substituting into part (i) of Rao–Blackwell's Theorem finds $E[\hat{g}(\overline{X})] = \mu$.

Now it is also true that $E[\overline{X}] = \mu$, as $\overline{X} \sim N(\mu, \frac{1}{n})$. Therefore, one solution for $\hat{g}(\overline{x})$ is the identity function $\hat{g}(\overline{x}) = \overline{x}$; that is,

$$E[T \mid \overline{X} = \overline{x}] = \overline{x}.$$

However, we cannot at this stage eliminate the possibility of other solutions to the conditional expectation (at least not under the Rao–Blackwell Theorem). In fact, for our solution to be unique, the concept of a *complete sufficient statistic* is required. We turn to this next. ∎

10.5 C Completeness and MVUE

Suppose that a statistic S is sufficient for a parameter θ. Let $h(S)$ denote any function of S such that $E[h(S)] = 0$; note that the expectation is taken with respect to distributions of S. If this expectation only holds in the degenerate case when $h(S) = 0$, for all θ, then *the family of distributions of S is complete*.[4] A slightly different nomenclature is to refer to S as a *complete sufficient statistic*. We will not concern ourselves with establishing the completeness of a sufficient statistic; in fact, with the exception of the sufficient statistic derived in *Example 13*(iv), every other sufficient statistic we have encountered has been complete.

Completeness is important because of the uniqueness it confers on expectations of a sufficient statistic. In particular, if S is a complete sufficient statistic such that $E[S] = g(\theta)$, then there can be no other function of S that is unbiased for $g(\theta)$. In other words, completeness ensures that S is the *unique unbiased estimator* of $g(\theta)$. We may now finish *Example 14*. Since the sufficient statistic $S = \sum_{i=1}^{n} X_i$ is complete, our tentative solution is, in fact, the only solution. Thus, $E[T \mid \overline{X} = \overline{x}] = \overline{x}$.

The presence of a complete sufficient statistic in the Rao–Blackwell Theorem yields a MVUE. To see this, let S be a complete sufficient statistic for θ. Now, for any other statistic T that is unbiased for $g(\theta)$, the Rao–Blackwell Theorem yields, without exception, the function $\hat{g}(S)$, which is *unbiased* for $g(\theta)$; that is, $E[\hat{g}(S)] = g(\theta)$. By completeness, $\hat{g}(S)$

is the *unique* unbiased estimator of $g(\theta)$ amongst all functions of S. Furthermore, by the Rao–Blackwell Theorem, $\hat{g}(S)$ has variance *no larger* than that of any other unbiased estimator of $g(\theta)$. In combination, these facts ensure that $\hat{g}(S)$ is the MVUE of $g(\theta)$.

⊕ ***Example 15:*** Estimation of Probabilities

Let random variable $X \sim$ Exponential(λ), with pdf $f(x; \lambda)$:

```
f = 1/λ e^-x/λ ;       domain[f] = {x, 0, ∞} && {λ > 0};
```

and let $\vec{X} = (X_1, \ldots, X_n)$ denote a random sample of size n drawn on X. In this example, we shall derive the MVUE of the survival function $g(\lambda) = P(X > k)$, namely:

```
g = 1 - Prob[k, f]
```

$$e^{-\frac{k}{\lambda}}$$

where k is a known positive constant. Estimation of probabilistic quantities, such as $g(\lambda)$, play a prominent role in many continuous time statistical models, especially duration models (*e.g.* see Lancaster (1992)).

The first thing we must do is to identify a complete sufficient statistic for λ. This is quite straightforward after we apply `Sufficient`:

```
Sufficient[f]
```

$$e^{-\frac{\sum_{i=1}^n x_i}{\lambda}} \lambda^{-n}$$

Here, $S = \sum_{i=1}^n X_i$ fills our requirements (we state completeness of S without proof). Next, consider statistics $T = T(\vec{X})$ that are unbiased for $g(\lambda)$. One such statistic is the Bernoulli random variable defined as[5]

$$T = \begin{cases} 0 & \text{if } X_n \le k \\ 1 & \text{if } X_n > k. \end{cases}$$

Then, let

$$\hat{g}(s) = E[T \mid S = s] = P(T = 1 \mid S = s) = P(X_n > k \mid S = s). \tag{10.14}$$

By the Rao–Blackwell Theorem, $\hat{g}(S)$ is the MVUE of $g(\lambda)$. The next step is therefore clear. We must find $P(X_n > k \mid S = s)$.

To derive the distribution of $X_n \mid (S = s)$, we first require the bivariate distribution of (S, X_n). Now this bivariate distribution is found from the joint density of the n random variables in the random sample $\vec{X}$. Superficially the problem appears complicated: we must transform $\vec{X}$ to $(S, X_2, \ldots, X_n)$, followed by $n - 2$ integrations to remove the unwanted variables $(X_2, \ldots, X_{n-1})$. However, if we define $S_{(n)} = \sum_{i=1}^{n-1} X_i$ (the sum of the first $n - 1$ components of $\vec{X}$), with density $f_{(n)}(s_{(n)}; \lambda)$, then, by independence, the joint

§10.5 C   UNBIASED PARAMETER ESTIMATION   345

density of $(S_{(n)}, X_n)$ is equal to the product $f_{(n)}(s_{(n)}; \lambda) f(x_n; \lambda)$. The joint density of (S, X_n) is then found by a simple transformation, because $S = S_{(n)} + X_n$. Determining $f_{(n)}(s_{(n)}; \lambda)$ is the key; fortunately, §4.5 contains a number of useful results concerning the density of sums of random variables. For our particular case, from *Example 22* of Chapter 4, we know that $S_{(n)} \sim \text{Gamma}(n-1, \lambda)$. Thus, the joint density of $(S_{(n)}, X_n)$ is given by:

```
h1 = (s_n^(a-1) e^(-s_n/b) / Γ[a] b^a  /. {a → n - 1, b → λ}) * (f /. x → x_n);
domain[h1] =
    {{s_n, 0, ∞}, {x_n, 0, ∞}} && {λ > 0, n > 1, n ∈ Integers};
```

Transforming $(S_{(n)}, X_n)$ to (S, X_n), where $S = S_{(n)} + X_n$, gives the pdf of (S, X_n):

```
h2 = Transform[{s == s_n + x_n, y == x_n}, h1] /. y → x_n
```

$$\frac{e^{-\frac{s}{\lambda}} \lambda^{-n} (s - x_n)^{-2+n}}{\Gamma[-1+n]}$$

The domain of support for (S, X_n) is all points in $\mathbb{R}_+^2$ such that $0 < x_n < s < \infty$. Thus:

```
domain[h2] =
    {{s, x_n, ∞}, {x_n, 0, s}} && {λ > 0, n > 1, n ∈ Integers};
```

The conditional distribution $X_n \mid (S = s)$ is given by:

```
h3 = Conditional[x_n, h2]
domain[h3] = {x_n, 0, s} && {n > 1, n ∈ Integers};
```

— Here is the conditional pdf h2 ($x_n \mid s$):

$$\frac{s^{1-n} \Gamma[n] (s - x_n)^{-2+n}}{\Gamma[-1+n]}$$

We now have all the ingredients in place ready to evaluate $\hat{g}(s) = P(X_n > k \mid S = s)$ and so determine the functional form of the MVUE:

```
Simplify[1 - Prob[k, h3], s > 0]
```

$$\left(1 - \frac{k}{s}\right)^{-1+n}$$

We conclude that $\hat{g}(S)$, the MVUE of $g(\lambda) = e^{-k/\lambda}$, is given by

$$\hat{g} = \begin{cases} 0 & \text{if } \sum_{i=1}^{n} X_i \le k \\ \left(1 - \dfrac{k}{\sum_{i=1}^{n} X_i}\right)^{n-1} & \text{if } \sum_{i=1}^{n} X_i > k. \end{cases}$$

Notice that $\hat{g}$ is a function of the complete sufficient statistic $S = \sum_{i=1}^{n} X_i$. ■

10.5 D Conclusion

In the previous example, the fact that the sufficient statistic was complete enabled us to construct the MVUE of $g(\lambda)$ by direct use of the Rao–Blackwell Theorem. Now, if in a given problem there exists a complete sufficient statistic, the key feature to notice from the Rao–Blackwell Theorem is that the MVUE will be *a function of the complete sufficient statistic*. We can, therefore, confine ourselves to examining the expectation of functions of complete sufficient statistics in order to derive minimum variance unbiased estimators. The following theorem summarises:

Theorem (Lehmann–Scheffé): Let S be a complete sufficient statistic for a parameter θ. If there is a function of S that has expectation $g(\theta)$, then this function is the MVUE of $g(\theta)$.

Proof: See, for example, Silvey (1995, p. 33). Also, Hogg and Craig (1995, p. 332).

⊕ **Example 16:** MVUE of the Normal Parameters

Let $X \sim N(\mu, \sigma^2)$ and define (see *Example 13*(iii)),

$$S = \begin{pmatrix} \sum_{i=1}^{n} X_i \\ \sum_{i=1}^{n} X_i^2 \end{pmatrix}$$

which is a complete sufficient statistic for (μ, σ^2). Let

$$T = \begin{pmatrix} \overline{X} \\ \hat{\sigma}^2 \end{pmatrix}$$

where $\overline{X} = \frac{1}{n} \sum_{i=1}^{n} X_i$ denotes the sample mean, and $\hat{\sigma}^2 = \frac{1}{n-1} \sum_{i=1}^{n} (X_i - \overline{X})^2$ is the sample variance. T is related one-to-one with S, and therefore it too is complete and sufficient for (μ, σ^2). Now we know that

$$E[T] = \begin{pmatrix} E[\overline{X}] \\ E[\hat{\sigma}^2] \end{pmatrix} = \begin{pmatrix} \mu \\ \sigma^2 \end{pmatrix}.$$

Therefore, by the Rao–Blackwell and Lehmann–Scheffé theorems, $\overline{X}$ is the MVUE of μ, and $\hat{\sigma}^2$ is the MVUE of σ^2. ∎

MVUE estimation relies on the existence of a complete sufficient statistic (whose variance exists). Without such a statistic, the rather elegant theory encapsulated in the Rao–Blackwell and Lehmann–Scheffé theorems cannot be applied. If it so happens that MVUE estimation is ruled out, how then do we proceed to estimate unknown parameters? We can return to considerations based on asymptotically desirable properties (Chapter 8), or choice based on decision loss criteria (Chapter 9), or choice based on maximising the content of statistical information (§10.2). Fortunately, there is another estimation technique—maximum likelihood estimation—which combines together features of each of these methods; the last two chapters of this book address aspects of this topic.

10.6 Exercises

1. Let the random variable $X \sim \text{Rayleigh}(\sigma)$, where parameter $\sigma > 0$. Derive Fisher's Information on σ.

2. Let the random variable $X \sim \text{Laplace}(\mu, \sigma)$. Obtain the CRLB for (μ, σ^2).

3. Let the random variable $X \sim \text{Lindley}(\delta)$. The sample mean $\overline{X}$ is the BUE of
$$g(\delta) = \frac{2+\delta}{\delta + \delta^2}.$$
Using *Mathematica*'s SolveAlways function, show that
$$h(\delta) = \frac{(3\delta+2)(2\delta+1)}{2\delta(\delta+1)}$$
is a linear function of $g(\delta)$. Hence, obtain the BUE of $h(\delta)$.

4. Let the random variable $X \sim \text{Laplace}(0, \sigma)$, and $(X_1, \ldots, X_n)$ denote a random sample of size n collected on X. Show that $\hat{\sigma} = \frac{1}{n}\sum_{i=1}^{n} |X_i|$ is the BUE of σ.

5. Referring to *Example 10*, show that the estimator $\tilde{\sigma} = \frac{1}{n\gamma}\sum_{i=1}^{n} X_i$ is unbiased for σ. Give reasons as to why $\hat{\sigma}$, given in *Example 10*, is preferred to $\tilde{\sigma}$ as an estimator of σ.

6. Let $X \sim \text{RiemannZeta}(\rho)$, and let $(X_1, \ldots, X_n)$ denote a random sample of size n drawn on X. Use the Factorisation Criterion to identify a sufficient statistic for ρ.

7. Let the pair (X, Y) be bivariate Normal with $E[X] = E[Y] = 0$, $\text{Var}(X) = \text{Var}(Y) = 1$ and correlation coefficient ρ. Use the Factorisation Criterion to identify a sufficient statistic for ρ.

8. Let $X \sim \text{Gamma}(a, b)$, and let $(X_1, \ldots, X_n)$ denote a random sample of size n drawn on X. Use the Factorisation Criterion to identify a sufficient statistic for (a, b).

9. Using the technique of *Example 15*, obtain the MVUE of $P(X = 0) = e^{-\lambda}$, where $X \sim \text{Poisson}(\lambda)$.

Chapter 11

Principles of Maximum Likelihood Estimation

11.1   Introduction

11.1 A Review

The previous chapter concentrated on obtaining unbiased estimators for parameters. The existence of unbiased estimators with minimum variance—the so-called MVUE class of estimators—required the sufficient statistics of the statistical model to be complete. Unfortunately, in practice, statistical models often falter in this respect. Therefore, parameter estimators must be found from other sources. The suitability of estimators based on large sample considerations such as consistency and limiting Normal distribution has already been addressed, as has the selection of estimators based on small sample properties dependent upon assumed loss structures. However, in both cases, the estimators that arose did so in an ad-hoc fashion. Fortunately, in the absence of complete sufficient statistics, there are other possibilities available. Of particular interest, here and in the following chapter, is the method of Maximum Likelihood (ML). ML techniques provide a way to generate parameter estimators that share some of the optimality properties, principally asymptotic ones.

§11.2 introduces the likelihood function. §11.3 defines the Maximum Likelihood Estimator (MLE) and shows how *Mathematica* can be used to determine its functional form. §11.4 discusses the statistical properties of the estimator. From the viewpoint of small sample sizes, the properties of the MLE depend very much on the particular statistical model in question. However, from a large sample perspective, the properties of the MLE are widely applicable and desirable: consistency, limiting Normal distribution and asymptotic efficiency. Desirable asymptotic properties and functional invariance (the Invariance Property) help to explain the popularity of ML in practice. §11.5 examines further the asymptotic properties of the MLE, using regularity conditions to establish these.

The statistical literature on ML methods is extensive with many texts devoting at least a chapter to the topic. The list of references that follow offers at least a sample of a range of treatments. In rough order of decreasing technical difficulty are Lehmann (1983), Amemiya (1985), Dhrymes (1970), Silvey (1975), Cox and Hinkley (1974), Stuart and Ord (1991), Gourieroux and Monfort (1995), Cramer (1986), McCabe and Tremayne (1993), Nerlove (2002), Mittelhammer (1996) and Hogg and Craig (1995). Currie (1995) gives numerical examples of computation of ML estimates using Version 2 of *Mathematica*, while Rose and Smith (2000) discuss computation under Version 4.

11.1 B SuperLog

Before embarking, we need to activate the **mathStatica** function `SuperLog`. This tool enhances *Mathematica*'s ability to simplify `Log[Product[]]` expressions. For instance, consider the following expression:

$$\mathtt{f} = \prod_{i=1}^{n} (1-\theta)^{1-x_i}\, \theta^{x_i}\,; \qquad \mathtt{Log[f]}$$

$$\mathrm{Log}\left[\prod_{i=1}^{n} (1-\theta)^{1-x_i}\, \theta^{x_i}\right]$$

Mathematica has not simplified `Log[f]` at all. However, if we turn `SuperLog` on:

`SuperLog[On]`

— SuperLog is now On.

and try again:

`Log[f]`

$$n\, \mathrm{Log}[1-\theta] + (-\mathrm{Log}[1-\theta] + \mathrm{Log}[\theta])\sum_{i=1}^{n} x_i$$

we obtain a significant improvement on *Mathematica*'s previous effort. `SuperLog` is part of the **mathStatica** suite. It modifies *Mathematica*'s `Log` function so that `Log[Product[]]` 'objects' or 'terms' get converted into sums of logarithms. At any stage, this enhancement may be removed by entering `SuperLog[Off]`.

11.2 The Likelihood Function

In this section, we define the likelihood function and illustrate its construction in a variety of settings. To establish notation, let X denote the variable(s) of interest that has (or is assumed to have) a pdf $f(x; \theta)$ dependent upon a $(k \times 1)$ parameter $\theta \in \Theta \subset \mathbb{R}^k$ whose true value θ_0 is unknown; we assume that the functional form of f is known. Next, we let $(X_1, \ldots, X_n)$ denote a random sample of size n drawn on X. It is assumed that the pdf of the random sample $f_{1,\ldots,n}(x_1, \ldots, x_n; \theta)$ can be derived from the knowledge we have about f, and hence that the joint density depends on the unknown parameter θ. A key point is that the likelihood function is mathematically equivalent to the joint distribution of the sample. Instead of regarding it as a function of the X_i, the likelihood is interpreted as a function of θ defined over the parameter space Θ for fixed values of each $X_i = x_i$. The *likelihood* for θ is thus

$$L(\theta \mid x_1, \ldots, x_n) \equiv f_{1,\ldots,n}(x_1, \ldots, x_n; \theta). \qquad (11.1)$$

Often, we will shorten the notation for the likelihood to just $L(\theta)$. Construction of the joint pdf may at first sight seem a daunting task. However, if the variables in $(X_1, \ldots, X_n)$ are

§11.2   PRINCIPLES OF MAXIMUM LIKELIHOOD ESTIMATION   351

mutually independent, then the joint pdf is given by the product of the marginals,

$$f_{1,\ldots,n}(x_1, \ldots, x_n; \theta) = \prod_{i=1}^{n} f(x_i; \theta) \tag{11.2}$$

which usually makes it easy to construct the joint pdf and hence the likelihood for θ.

We often need to distinguish between two forms of the likelihood for θ, namely, the likelihood function, and the observed likelihood. The *likelihood function* is defined as the likelihood for θ given the random sample prior to observation; it is given by $L(\theta | X_1, \ldots, X_n)$, and is a random variable. Where there is no possibility of confusion, we use 'likelihood' and 'likelihood function' interchangeably. The second form, the *observed likelihood*, is defined as the likelihood for θ evaluated for a given sample of observed data, and it is *not* random. The following examples illustrate the construction of the likelihood, and its observed counterpart.

⊕ **Example 1:** The Likelihood and Observed Likelihood for an Exponential Model

Let random variable $X \sim$ Exponential(θ), with pdf:

```
f = 1/θ e^-x/θ;     domain[f] = {x, 0, ∞} && {θ > 0};
```

Let $(X_1, \ldots, X_n)$ denote a random sample of size n collected on X. Then, the *likelihood* for θ is equivalent to the joint pdf of the random sample (11.1), and as $(X_1, \ldots, X_n)$ are mutually independent, then it can be constructed as per (11.2):

```
Lθ = ∏_{i=1}^{n} (f /. x → x_i)
```

$$\prod_{i=1}^{n} \frac{e^{-\frac{x_i}{\theta}}}{\theta}$$

Given a random sample of size $n = 4$ on X, let us suppose that the observed data are:

```
data = {1, 2, 1, 4};
```

There are two main methods to construct the *observed likelihood* for θ:

Method 1: Substitute the data into the likelihood:

```
Lθ /. {n → Length[data], x_i_ :→ data[[i]]}
```

$$\frac{e^{-8/\theta}}{\theta^4}$$

Note the use of *delayed* replacement :→ (which is entered as :>). By contrast, *immediate* replacement → (which is entered as −>) would fail.

Method 2: Substitute the data into the density:

> **`Times @@ (f /. x → data)`**

$$\frac{e^{-8/\theta}}{\theta^4}$$

Here, the immediate replacement `f /. x → data` yields a list of empirical densities $\{f(1;\theta), f(2;\theta), f(1;\theta), f(4;\theta)\}$. The observed likelihood for θ is obtained by multiplying the elements of the list together using `Times` (the `@@` is 'shorthand' for the `Apply` function). ∎

⊕ *Example 2:* The Likelihood and Observed Likelihood for a Bernoulli Model

Now suppose that X is discrete, and, in particular, that $X \sim$ Bernoulli(θ):

> **`f = θ^x (1 - θ)^(1-x);`**
> **`domain[f] = {x, 0, 1} && {0 < θ < 1} && {Discrete};`**

where $0 < \theta < 1$. For $(X_1, ..., X_n)$, a random sample of size n drawn on X, the likelihood for θ is equivalent to the joint pmf of the random sample (11.1), and as $(X_1, ..., X_n)$ are mutually independent, it can be constructed as per (11.2):

$$L\theta = \prod_{i=1}^{n} (f /. x \to x_i)$$

$$\prod_{i=1}^{n} (1-\theta)^{1-x_i} \theta^{x_i}$$

Suppose that observations were recorded as follows:

> **`data = {1, 1, 0, 1, 0, 0, 1, 1, 0};`**

We again construct the observed likelihood using our two methods:

Method 1: Substitute the data into the likelihood:

> **`∏_{i=1}^{n} (f /. x → x_i) /. {n → Length[data], x_i_ ⧴ data[[i]]}`**

$$(1-\theta)^4 \theta^5$$

Method 2: Substitute the data into the pmf:

> **`Times @@ (f /. x → data)`**

$$(1-\theta)^4 \theta^5$$

§11.2 PRINCIPLES OF MAXIMUM LIKELIHOOD ESTIMATION

⊕ *Example 3:* The Likelihood and Observed Likelihood for a Latent Variable Model

There are many instances where care is needed in deriving the likelihood. One important situation is when the variable of interest is latent (meaning that it cannot be observed), but a variable that is functionally related to it can be observed. To construct the likelihood for the parameters in a statistical model for a latent variable, we need to know the function (or the sampling scheme) that relates the observable variable to the latent variable.

Let X be the examination mark of a student in percent; thus $X = x \in [0, 100]$. Suppose that the mark is only revealed to us if the exam is passed; that is, X is disclosed provided $X \geq 50$. On the other hand, if the student fails the exam, then we receive a datum of 0 (say) and know only that $X < 50$. Thus, X is only partially observed by us and therefore it is latent. Let Y denote the observed variable, which is related to X by

$$Y = \begin{cases} X & \text{if } X \in [50, 100] \\ 0 & \text{if } X \in [0, 50). \end{cases} \tag{11.3}$$

We propose to model X with the (scaled) Beta distribution, $X \sim 100 \times \text{Beta}(a, b)$. Let $f(x; \theta)$ denote the statistical model for X:

```
f = (x/100)^(a-1) (1 - x/100)^(b-1)
    ─────────────────────────────────── ;
              100 Beta[a, b]

domain[f] = {x, 0, 100} && {a > 0, b > 0};
```

Although we cannot fully observe X, it is still possible to elicit information about the parameter $\theta = (a, b)$, as the relationship linking X to Y is known. Thus, given the distribution of X, we can derive the distribution of Y. The density of Y is non-standard in the sense that it has both discrete and continuous components. The discrete component of the density is a mass measured at the origin, while the continuous component of the density is equivalent to the pdf of X for values of 50 or more. By (11.3), the value of the mass at the origin is $P(Y = 0) = P(X < 50)$, which equals:

```
P₀ = Prob[50, f]

Γ[a] Hypergeometric2F1Regularized[a, a + b, 1 + a, -1]
──────────────────────────────────────────────────────
                    Beta[a, b]
```

Let $(Y_1, \ldots, Y_n)$ denote a random sample of size n collected on Y (remember it is Y that is observed, not X). The likelihood for θ is, by (11.1), equivalent to the joint density of the random sample. Because of the component structure of the distribution of Y, it is convenient to introduce a quantity, n_0, defined to be the number of zeroes observed in the random sample—clearly $0 \leq n_0 \leq n$. Now, for a particular random sample $(y_1, \ldots, y_n)$, the likelihood is made up of contributions from both types of observations. For the n_0 zero observations it is

$$\prod_0 P(Y_i = 0) = (P(Y = 0))^{n_0}$$

where the product is taken over the n_0 zero observations. The contribution of the non-zero observations to the likelihood is

$$\prod_+ f(y_i; \theta)$$

where the product is taken over the $(n - n_0)$ observations in the sample which are at least equal to 50, and f denotes the scaled Beta pdf. The likelihood is therefore

$$L(\theta) = (P(Y = 0))^{n_0} \prod_+ f(y_i; \theta). \tag{11.4}$$

To illustrate construction of the observed likelihood, we load the CensoredMarks data set into *Mathematica*:

```
data = ReadList["CensoredMarks.dat"];
```

There are a total of $n = 264$ observations in this data set:

```
n = Length[data]
```

264

Next, we select the marks of only those students that passed, storing them in the PassMark list:

```
PassMark = Select[data, (# ≥ 50) &];

n₀ = n - Length[PassMark]
```

40

Calculation reveals that 40 of the 264 students must have received marks below 50, which implies a censoring (failure) rate of around 15%. As per (11.4), the observed likelihood for θ, given this data, is:

```
P₀^n₀ * Times @@ (f /. x → PassMark)
```

$$\frac{1}{\text{Beta}[a, b]^{264}}$$
$(2^{-40-202\,a-206\,b}\ 3^{-186+100\,a+86\,b}\ 5^{304-376\,a-376\,b}\ 7^{-65+31\,a+34\,b}$
$11^{-40+20\,a+20\,b}\ 13^{-25+14\,a+11\,b}\ 17^{-23+10\,a+13\,b}$
$19^{-31+17\,a+14\,b}\ 23^{-13+6\,a+7\,b}\ 29^{-20+13\,a+7\,b}\ 31^{-18+13\,a+5\,b}$
$37^{-15+4\,a+11\,b}\ 47^{-8+8\,b}\ 53^{-8+8\,a}\ 59^{-9+9\,a}\ 71^{-7+7\,a}$
$79^{-1+a}\ 1763^{-10+a+9\,b}\ 4087^{-6+6\,a}\ 6059^{-3+3\,a}\ \Gamma[a]^{40}$
$\text{Hypergeometric2F1Regularized}[a, a + b, 1 + a, -1]^{40})$

```
ClearAll[data, n, PassMark]; Unset[n₀]; Unset[P₀];
```

⊕ *Example 4:* The Likelihood and Observed Likelihood for a Time Series Model

In the previous examples, the likelihood function was easily constructed, since due to mutual independence, the joint distribution of the random sample was simply the product of the marginal distributions. In some situations, however, mutual independence amongst the sampling variables does not occur, and so the derivation of the likelihood function requires more effort. Examples include time series models, pertaining to variables collected through time that depend on their past.

Consider a random walk with drift model

$$X_t = \mu + X_{t-1} + U_t$$

with initial condition $X_0 = 0$. The drift is given by the constant $\mu \in \mathbb{R}$, while the disturbances U_t are assumed to be independently Normally distributed with zero mean and common variance $\sigma^2 \in \mathbb{R}_+$; that is, $U_t \sim N(0, \sigma^2)$, for all $t = 1, \ldots, T$, and $E[U_t U_s] = 0$ for all $t \neq s$.

We wish to construct the likelihood for parameter $\theta = (\mu, \sigma^2)$. One approach is to use conditioning arguments. We begin by considering the joint distribution of the sample $(X_1, \ldots, X_T)$. This cannot be written as the product of the marginals (*cf.* (11.2)) as X_t depends on $X_{t-1}, \ldots, X_0$, for all $t = 1, \ldots, T$. However, in light of this dependence, suppose instead that we decompose the joint distribution of the entire sample into the distribution of X_T conditional on all previous variables, multiplied by the joint distribution of all the conditioning variables:

$$f_{1,\ldots,T}(x_1, \ldots, x_T; \theta) = f_{T|1,\ldots,T-1}(x_T \mid x_1, \ldots, x_{T-1}; \theta) \\ \times f_{1,\ldots,T-1}(x_1, \ldots, x_{T-1}; \theta) \quad (11.5)$$

where $f_{T|1,\ldots,T-1}$ denotes the distribution of X_T conditional on $X_1 = x_1, \ldots, X_{T-1} = x_{T-1}$, and $f_{1,\ldots,T-1}$ denotes the joint distribution of $(X_1, \ldots, X_{T-1})$. From the form of the random walk model, it is clear that when fixing any X_t, all previous X_s ($s < t$) must also be fixed. This enables us to simplify the notation, for the conditional pdf on the right-hand side of (11.5) may be written as

$$f_{T|1,\ldots,T-1}(x_T \mid x_1, \ldots, x_{T-1}; \theta) = f_{T|T-1}(x_T \mid x_{T-1}; \theta). \quad (11.6)$$

From the assumptions on the disturbances, it follows that

$$X_T \mid (X_{T-1} = x_{T-1}) \sim N(\mu + x_{T-1}, \sigma^2) \quad (11.7)$$

which makes it is easy to write down the conditional density given in (11.6). Consider now the joint distribution of $(X_1, \ldots, X_{T-1})$ on the right-hand side of (11.5). Here, again, the same idea is used to decompose the joint distribution of the remaining variables: the appropriate equations are (11.5) and (11.6) but with T replaced by $T-1$. By recursion,

$$f_{1,\ldots,T}(x_1, \ldots, x_T; \theta) = f_{T|T-1}(x_T \mid x_{T-1}; \theta) \times f_{T-1|T-2}(x_{T-1} \mid x_{T-2}; \theta) \times \cdots \\ \times f_{2|1}(x_2 \mid x_1; \theta) \times f_{1|0}(x_1 \mid (X_0 = 0); \theta)$$

$$= \prod_{t=1}^{T} f_{t|t-1}(x_t \mid x_{t-1}; \theta) \tag{11.8}$$

where each of the conditional densities in (11.8) is equivalent to (11.6) for $t = 2, \ldots, T$, and $f_{1|0}$ is the pdf of a $N(\mu, \sigma^2)$ distribution because of the assumption $X_0 = 0$. By (11.1), (11.8) is equivalent to the likelihood for θ.

To enter this likelihood into *Mathematica*, we begin by entering the time t conditional pdf given in (11.7):

```
f = 1/(σ √(2 π)) Exp[- (xₜ - μ - xₜ₋₁)²/(2 σ²)];
```

Let us suppose we have data $\{x_1, \ldots, x_6\} = \{1, 2, 4, 2, -3, -2\}$:

```
xdata = {1, 2, 4, 2, -3, -2};
```

To obtain the observed likelihood, we use a modified form of *Method 1* that accounts for the initial condition $x_0 = 0$:

```
xlis = Thread[x_Range[Length[xdata]]];
xrules = Join[{x₀ → 0}, Thread[xlis → xdata]]
```

$\{x_0 \to 0, x_1 \to 1, x_2 \to 2, x_3 \to 4, x_4 \to 2, x_5 \to -3, x_6 \to -2\}$

Then, the observed likelihood for $\theta = (\mu, \sigma^2)$ is obtained by substituting in the observational rules:

```
obsLθ = ∏ₜ₌₁⁶ f /. xrules // Simplify
```

$$\frac{e^{-\frac{18+2\mu+3\mu^2}{\sigma^2}}}{8 \pi^3 \sigma^6}$$

Figure 1 plots the observed likelihood against values of μ and σ^2. Evidently, obsLθ is maximised in the neighbourhood of $(\mu, \sigma^2) = (0, 6)$.

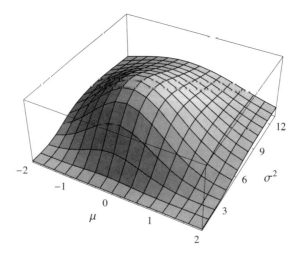

Fig. 1: Observed likelihood for μ and σ^2

11.3 Maximum Likelihood Estimation

Maximum likelihood parameter estimation is based on choosing values for θ so as to maximise the likelihood function. That is, the MLE of θ, denoted $\hat{\theta}$, is the solution to the optimisation problem:

$$\hat{\theta} = \arg\max_{\theta \in \Theta} L(\theta \mid X_1 = x_1, \ldots, X_n = x_n). \tag{11.9}$$

Thus, $\hat{\theta}$ is the value of the argument of the likelihood, selected from anywhere in the parameter space, that maximises the value of the likelihood after we have been given the sample. In other words, we seek the particular value of θ, namely, $\hat{\theta}$, which makes it most likely to have observed the sample that we actually have. We may view the solution to (11.9) in two ways depending on whether the objective function is the *likelihood function* or the *observed likelihood function*. If the objective is the likelihood, then (11.9) defines the ML *estimator*, $\hat{\theta} = \hat{\theta}(X_1, \ldots, X_n)$; since this is a function of the random sample, $\hat{\theta}$ is a random variable. If the objective is the observed likelihood, then (11.9) defines the ML *estimate*, $\hat{\theta} = \hat{\theta}(x_1, \ldots, x_n)$, where $(x_1, \ldots, x_n)$ denotes observed data; in this case $\hat{\theta}$ is a point estimate.

The solution to (11.9) is invariant to any monotonic increasing transformation of the objective. Since the natural logarithm is a monotonic transformation, it follows that

$$\hat{\theta} = \arg\max_{\theta \in \Theta} \log L(\theta) \tag{11.10}$$

which we shall use, from now on, as the definition of the estimator (estimate). The natural logarithm of the likelihood, $\log L(\theta)$, is called the *log-likelihood function*. A weaker definition of the MLE, but one that, in practice, is often equivalent to (11.10) is

$$\hat{\theta} = \arg\max_{\tilde{\theta} \in \tilde{\Theta}} \log L(\tilde{\theta}) \tag{11.11}$$

where $\tilde{\Theta}$ denotes a finite, non-null set whose elements $\tilde{\theta}$ satisfy the conditions

$$\frac{\partial}{\partial \theta} \log L(\tilde{\theta}) = 0 \quad \text{and} \quad \frac{\partial^2}{\partial \theta^2} \log L(\tilde{\theta}) < 0. \tag{11.12}$$

The two parts of (11.12) express, respectively, the *first-* and *second-order conditions* familiar from basic calculus for determining local maxima of a function.[1] Generally speaking, we shall determine MLE through (11.12), although *Example 7* below relies on (11.10) alone. One further piece of notation is the so-called *score* (or 'efficient score' in some texts), defined as the gradient of the log-likelihood,

$$S(\theta) = \frac{\partial}{\partial \theta} \log L(\theta).$$

For example, the first-order condition is simply $S(\tilde{\theta}) = 0$.

```
Clear[n];
```

⊕ **Example 5:** The MLE for the Exponential Parameter

Let $X \sim \text{Exponential}(\theta)$, where parameter $\theta \in \mathbb{R}_+$. Here is its pdf:

```
f = 1/θ e^-x/θ ;     domain[f] = {x, 0, ∞} && {θ > 0};
```

For a random sample of size n drawn on X, the log-likelihood function is:

```
logLθ = Log[∏_{i=1}^{n} (f /. x → x_i)]
```

$$-\frac{n\theta \, \text{Log}[\theta] + \sum_{i=1}^{n} x_i}{\theta}$$

Of course, this will only work if `SuperLog` has been activated (see §11.1 B). The score function is the gradient of the log-likelihood with respect to θ:

```
score = Grad[logLθ, θ]
```

$$\frac{-n\theta + \sum_{i=1}^{n} x_i}{\theta^2}$$

where we have applied **mathStatica**'s `Grad` function. Setting the score to zero and solving for θ corresponds to the first-order condition given in (11.12). We find:

```
solθ = Solve[score == 0, θ]
```

$$\left\{\left\{\theta \to \frac{\sum_{i=1}^{n} x_i}{n}\right\}\right\}$$

The unique solution, `solθ`, appears in the form of a replacement rule and corresponds to the sample mean. The nature of the solution is not yet clear; that is, does the sample mean correspond to a local minimum, local maximum, or saddle point of the log-likelihood? A check of the second-order condition, evaluated at `solθ`:

```
Hessian[logLθ, θ] /. Flatten[solθ]
```

$$-\frac{n^3}{\left(\sum_{i=1}^{n} x_i\right)^2}$$

... reveals that the Hessian is strictly negative at the sample mean and therefore the log-likelihood is maximised at the sample mean. Hence, the MLE of θ is

$$\hat{\theta} = \frac{1}{n}\sum_{i=1}^{n} X_i.$$

Note that `Hessian[f, x]` is a **mathStatica** function. ∎

⊕ **Example 6:** The MLE for the Normal Parameters

Let $X \sim N(\mu, \sigma^2)$, where $\mu \in \mathbb{R}$ and $\sigma^2 \in \mathbb{R}_+$, with pdf $f(x; \mu, \sigma^2)$:

```
f = ─────── Exp[- (x - μ)² ];        domain[f] = {x, -∞, ∞};
    σ √(2π)       2 σ²
```

For a random sample of size n drawn on X, the log-likelihood for parameter $\theta = (\mu, \sigma)$ is:[2]

```
logLθ = Log[ ∏ᵢ₌₁ⁿ (f /. x → xᵢ) ]
```

$$-\frac{n(\mu^2 + \sigma^2 \log[2\pi] + 2\sigma^2 \log[\sigma]) - 2\mu \sum_{i=1}^n x_i + \sum_{i=1}^n x_i^2}{2\sigma^2}$$

The score vector $S(\theta) = S(\mu, \sigma)$ is given by:

```
score = Grad[logLθ, {μ, σ}]
```

$$\left\{ \frac{-n\mu + \sum_{i=1}^n x_i}{\sigma^2}, \frac{n\mu^2 - n\sigma^2 - 2\mu\sum_{i=1}^n x_i + \sum_{i=1}^n x_i^2}{\sigma^3} \right\}$$

Mathematica's Solve command is quite flexible in allowing various forms of the first-order conditions to be entered; for example, {score[[1]] == 0, score[[2]] == 0} or score == {0, 0}, or score == 0. Setting the score to zero and solving yields:

```
solθ = Solve[score == 0, {μ, σ}]
```

$$\left\{ \left\{ \sigma \to -\frac{\sqrt{-\frac{(\sum_{i=1}^n x_i)^2}{n} + \sum_{i=1}^n x_i^2}}{\sqrt{n}}, \mu \to \frac{\sum_{i=1}^n x_i}{n} \right\}, \right.$$

$$\left. \left\{ \sigma \to \frac{\sqrt{-\frac{(\sum_{i=1}^n x_i)^2}{n} + \sum_{i=1}^n x_i^2}}{\sqrt{n}}, \mu \to \frac{\sum_{i=1}^n x_i}{n} \right\} \right\}$$

Clearly, the negative-valued solution for σ lies outside the parameter space and is therefore invalid; thus, the only permissible solution to the first-order conditions is:

```
solθ = solθ[[2]]
```

$$\left\{ \sigma \to \frac{\sqrt{-\frac{(\sum_{i=1}^n x_i)^2}{n} + \sum_{i=1}^n x_i^2}}{\sqrt{n}}, \mu \to \frac{\sum_{i=1}^n x_i}{n} \right\}$$

Then $\hat{\theta} = (\hat{\mu}, \hat{\sigma})$ is the MLE of θ, where $\hat{\mu}$ and $\hat{\sigma}$ are the formulae given in solθ (we check second-order conditions below). The functional form given by *Mathematica* for $\hat{\sigma}$

may appear unfamiliar. However, if we utilise the following identity for the sum of squared deviations about the sample mean,

$$\sum_{i=1}^{n} (X_i - \overline{X})^2 = \sum_{i=1}^{n} X_i^2 - n\overline{X}^2$$

where $\overline{X} = \frac{1}{n} \sum_{i=1}^{n} X_i$, then

$$\hat{\sigma} = \sqrt{\frac{1}{n} \sum_{i=1}^{n} (X_i - \overline{X})^2}.$$

By the Invariance Property (see §11.4 E), the MLE of σ^2 is

$$(\hat{\sigma})^2 = \frac{1}{n} \sum_{i=1}^{n} (X_i - \overline{X})^2$$

which is the 2nd sample central moment.

The second-order conditions may, for example, be checked by examining the eigenvalues of the Hessian matrix evaluated at $\hat{\theta}$:

Eigenvalues[Hessian[logLθ, {μ, σ}] /. solθ] // Simplify

$$\left\{ \frac{n^3}{\left(\sum_{i=1}^{n} x_i\right)^2 - n \sum_{i=1}^{n} x_i^2}, \frac{2 n^3}{\left(\sum_{i=1}^{n} x_i\right)^2 - n \sum_{i=1}^{n} x_i^2} \right\}$$

Given the identity for the sum of squared deviations, the eigenvalues of the Hessian are $-n\hat{\sigma}^{-2}$ and $-2n\hat{\sigma}^{-2}$, which clearly are negative. Thus, the Hessian is negative definite at $\hat{\theta}$ and therefore the log-likelihood is maximised at $\hat{\theta}$. ∎

⊕ ***Example 7:*** The MLE for the Pareto Parameters

Let $X \sim \text{Pareto}(\alpha, \beta)$, where parameters $\alpha \in \mathbb{R}_+$ and $\beta \in \mathbb{R}_+$. The pdf of X is given by:

f = α β^α x^{-(α + 1)}; domain[f] = {x, β, ∞} && {α > 0, β > 0};

Since $X \geq \beta$, there exists dependence between the parameter and sample spaces. Given a random sample of size n collected on X, the log-likelihood for $\theta = (\alpha, \beta)$ is:

logLθ = Log$\left[\prod_{i=1}^{n} (f /. x \to x_i)\right]$

$$n (\text{Log}[\alpha] + \alpha \text{Log}[\beta]) - (1 + \alpha) \sum_{i=1}^{n} \text{Log}[x_i]$$

The score vector is given by:

```
score = Grad[logLθ, {α, β}]
```

$$\left\{ n\left(\frac{1}{\alpha} + \text{Log}[\beta]\right) - \sum_{i=1}^{n} \text{Log}[x_i], \frac{n\alpha}{\beta} \right\}$$

If we attempt to solve the first-order conditions in the usual way:

```
Solve[score == 0, {α, β}]
```

{}

... we see that Solve cannot find a solution to the equations. However, if we focus on solving just the first of the first-order conditions, we find:[3]

```
solα = Solve[score[[1]] == 0, α]
```

$$\left\{\left\{\alpha \to -\frac{n}{n\,\text{Log}[\beta] - \sum_{i=1}^{n}\text{Log}[x_i]}\right\}\right\}$$

This time a solution is provided, albeit in terms of β; that is, $\hat{\alpha} = \hat{\alpha}(\beta)$. We now take this solution and substitute it back into the log-likelihood:

```
logLθ /. Flatten[solα] // Simplify
```

$$n\left(-1 + \text{Log}\left[\frac{n}{-n\,\text{Log}[\beta] + \sum_{i=1}^{n}\text{Log}[x_i]}\right]\right) - \sum_{i=1}^{n}\text{Log}[x_i]$$

This function is known as the *concentrated log-likelihood*. It corresponds to $\log L(\hat{\alpha}(\beta), \beta)$. Since it no longer involves α, we can maximise it with respect to β. Let $\hat{\beta}$ denote the solution to this optimisation problem. This solution can then be substituted back to recover $\hat{\alpha} = \hat{\alpha}(\hat{\beta})$; then $\hat{\theta} = (\hat{\alpha}, \hat{\beta})$ would be the MLE of θ. In general, when the first-order conditions can be solved uniquely for some subset of parameters in θ, then those solutions can be substituted back into the log-likelihood to yield the concentrated log-likelihood. The concentrated log-likelihood is then maximised with respect to the remaining parameters, usually using numerical techniques.

For our example, maximising the concentrated log-likelihood using standard calculus will not work. This is because the parameter space depends on the sample space. However, by inspection, it is apparent that the concentrated log-likelihood is increasing in β. Therefore, we should select β as large as possible. Now, since each $X_i \geq \beta$, we can choose β no larger than the smallest observation. Hence, the MLE for β is

$$\hat{\beta} = \min(X_1, X_2, \ldots, X_n)$$

which is the smallest order statistic. Replacing β in $\hat{\alpha}(\beta)$ with $\hat{\beta}$ yields the MLE for α,

$$\hat{\alpha} = n \Bigg/ \sum_{i=1}^{n} \log\left(\frac{X_i}{\min(X_1, X_2, \ldots, X_n)}\right).$$

■

11.4 Properties of the ML Estimator

11.4 A Introduction

This section considers the small and large sample statistical properties of the MLE. Typically, small sample properties of a MLE are determined on a case-by-case basis. Finding the distribution of the estimator is the most important — its pdf and/or cdf, mgf or cf — for from this we can determine the moments of the estimator and construct confidence intervals about point estimates, and so on. Unlike, say, the MVUE class of estimator, whose properties are supported by a set of elegant theorems, the MLE has only limited small sample properties. Generally though, the MLE has the 'property' of being biased. The MLE properties are listed in Table 1.

Sufficiency	The MLE is a function of sufficient statistics.
Efficiency	If an estimator is BUE, then it is equivalent to the MLE, provided that the MLE is the unique solution to the first-order condition that maximises the log-likelihood function.
Asymptotic	Under certain regularity conditions, the MLE is *consistent*; it has a *limiting Normal distribution* when suitably scaled; and it is *asymptotically efficient*.
Invariance	If $\hat{\theta}$ is the MLE of θ, then $g(\hat{\theta})$ is the MLE of $g(\theta)$.

Table 1: General properties of ML estimators

For proofs of these properties see, amongst others, Stuart and Ord (1991). The *Invariance* property is particularly important for estimation and it will be extensively exploited in the following chapter. Under fairly general conditions, the *Asymptotic* properties of the MLE are quite desirable; it is the attractiveness of its large sample properties which has contributed to the popularity of this estimator in practice. Even if the functional form of the MLE is not known (*i.e.* the solution to (11.12) can only be obtained by numerical methods), one can assert asymptotic properties by checking regularity conditions; in such situations, it is popular to use simulation techniques to determine small sample properties.

In §11.4 B, we examine the small sample properties of the MLE. Then, in §11.4 C, some of the estimators asymptotic properties are derived. In §11.4 D, further asymptotic properties of the MLE are revealed as a result of the model being shown to satisfy certain regularity conditions. Finally, in §11.4 E, the invariance property is illustrated. We begin with *Example 8*, which describes the model and derives the MLE.

⊕ *Example 8:* The MLE of θ

Let the continuous random variable X have pdf $f(x; \theta)$:

```
f = θ x^(θ-1);    domain[f] = {x, 0, 1} && {θ > 0};
```

where parameter $\theta \in \mathbb{R}_+$. The distribution of X can be viewed as either a special case of the Beta distribution (*i.e.* Beta(θ, 1)), or as a special case of the Power Function distribution (*i.e.* PowerFunction(θ, 1)). Assuming `SuperLog` has been activated (see §11.1 B), the log-likelihood for θ is derived with:

```
logLθ = Log[∏(f /. x → xᵢ)]
         i=1
```

$$n \, \text{Log}[\theta] + (-1 + \theta) \sum_{i=1}^{n} \text{Log}[x_i]$$

In this example, the MLE of θ is the unique solution to the first-order condition:

```
solθ = Solve[Grad[logLθ, θ] == 0, θ]
```

$$\left\{\left\{\theta \to -\frac{n}{\sum_{i=1}^{n} \text{Log}[x_i]}\right\}\right\}$$

... because the log-likelihood is globally concave with respect to θ; that is, the Hessian is negative-valued at all points in the parameter space:

```
Hessian[logLθ, θ]
```

$$-\frac{n}{\theta^2}$$

Thus, the MLE of θ is

$$\hat{\theta} = -\frac{n}{\sum_{i=1}^{n} \log(X_i)} \cdot \blacksquare \tag{11.13}$$

11.4 B Small Sample Properties

The sufficiency and efficiency properties listed in Table 1 pertain to the small sample performance of the MLE. The first property (sufficiency; see §10.4), is desirable because sufficient statistics retain all statistical information about parameters, and therefore so too must the MLE. Despite this, the MLE does not always use this information in an optimal fashion, for generally the MLE is a biased estimator.[4] Consequently, the second property (efficiency; see §10.3), should be seen as a special situation in which the MLE is unbiased and its variance attains the Cramér–Rao Lower Bound.

⊕ *Example 9:* Sufficiency, Efficiency and $\hat{\theta}$

Consider again the model given in *Example 8*, with pdf $f(x; \theta)$:

```
f = θ x^(θ-1);  domain[f] = {x, 0, 1} && {θ > 0};
```

The first property claims that there should exist a functional relationship between a sufficient statistic for θ and the MLE $\hat{\theta}$, given in (11.13). This can be shown by identifying a sufficient statistic for θ. Following the procedure given in §10.4, we apply **mathStatica**'s Sufficient function to find:

Sufficient[f]

$$\theta^n \prod_{i=1}^{n} x_i^{-1+\theta}$$

Then, by the Factorisation Criterion, the statistic $S = \prod_{i=1}^{n} X_i$ is sufficient for θ. We therefore have

$$\hat{\theta} = -\frac{n}{\log(S)}$$

and so the MLE is indeed a function of a sufficient statistic for θ.

The second property states that the MLE is the BUE provided the latter exists, and provided the MLE is the unique solution to the first-order conditions. Unfortunately, even though it was demonstrated in *Example 8* that $\hat{\theta}$ uniquely solved the first-order conditions, there is no BUE in this case. Nevertheless, the MVUE of θ does exist (since S is a complete sufficient statistic for θ) and it is given by

$$\tilde{\theta} = -\frac{n-1}{\log(S)}.$$

It is easy to see that the MLE $\hat{\theta}$ and the MVUE $\tilde{\theta}$ are related by a simple scaling transformation, $\tilde{\theta} = \frac{n-1}{n} \hat{\theta}$. In light of this, it follows immediately that the MLE must be biased upwards. ∎

⊕ ***Example 10:*** The Distribution of $\hat{\theta}$

Consider again the model given in *Example 8*, with pdf $f(x; \theta)$:

f = θ x^(θ-1);   domain[f] = {x, 0, 1} && {θ > 0};

In this example, we derive the (small sample) distribution of the MLE

$$\hat{\theta} = -\frac{n}{\sum_{i=1}^{n} \log(X_i)}$$

by applying the MGF Theorem (see §2.4 D). We begin by deriving the mgf of

$$\overline{\log X} = -\frac{1}{n} \sum_{i=1}^{n} \log(X_i)$$

and then matching it to the mgf of a known distribution. In this way, we obtain the distribution of $\overline{\log X}$. The final step involves transforming from $\overline{\log X}$ to $\hat{\theta}$.

By the MGF Theorem, the mgf of $\overline{\log X}$ is:

Expect[e^(t Log[x]), f]^n /. t → -t/n

— This further assumes that: $\{t + \Theta > 0\}$

$$\left(\frac{\Theta}{-\frac{t}{n} + \Theta}\right)^n$$

This expression matches the mgf of a Gamma$(n, \frac{1}{n\theta})$ distribution.[5] Hence, $\overline{\log X} \sim$ Gamma$(n, \frac{1}{n\theta})$. Then, since $\hat{\theta} = 1/\overline{\log X}$, it follows that $\hat{\theta}$ has an Inverse Gamma distribution with parameters n and $\frac{1}{n\theta}$. That is,

$$\hat{\theta} \sim \text{InverseGamma}(n, \tfrac{1}{n\theta}).$$

The pdf of $\hat{\theta}$, say $f_{\hat{\theta}}$, can be entered from **mathStatica**'s *Continuous* palette:

$$f_{\hat{\theta}} = \frac{\hat{\theta}^{-(a+1)} e^{-\frac{1}{b\hat{\theta}}}}{\Gamma[a]\, b^a} \,/.\, \left\{a \to n,\ b \to \frac{1}{n\Theta}\right\};$$

domain[$f_{\hat{\theta}}$] = $\{\hat{\theta}, 0, \infty\}$ && $\{n > 0,\ n \in \text{Integers},\ \Theta > 0\}$;

We now determine the mean (although we have already deduced its nature through the relation between $\hat{\theta}$ and $\tilde{\theta}$ given in *Example 9*) and the variance of the MLE:

Expect[$\hat{\theta}$, $f_{\hat{\theta}}$]

— This further assumes that: $\{n > 1\}$

$$\frac{n\Theta}{-1+n}$$

Var[$\hat{\theta}$, $f_{\hat{\theta}}$] // FullSimplify

— This further assumes that: $\{n > 2\}$

$$\frac{n^2 \Theta^2}{(-2+n)(-1+n)^2}$$

11.4 C Asymptotic Properties

Recall that estimators may possess large sample properties such as asymptotic unbiasedness, consistency, asymptotic efficiency, be limit Normally distributed when suitably scaled, and so on. These properties are also relevant to ML estimators. Like the small sample properties, large sample properties can be examined on a case-by-case basis. Analysis might proceed by applying the appropriate Central Limit Theorem and Law of Large Numbers.

⊕ **Example 11:** Asymptotic Unbiasedness and Consistency of $\hat{\theta}$

Consider the model of *Example 8*, with pdf $f(x; \theta)$:

```
f = θ x^(θ-1);   domain[f] = {x, 0, 1} && {θ > 0};
```

Since we have already shown $E[\hat{\theta}] = \frac{n\theta}{n-1}$ in *Example 10*, it is particularly easy to establish whether or not $\hat{\theta}$ is asymptotically unbiased for θ:

```
Limit[ nθ/(n-1), n → ∞ ]
```

θ

As the mean of $\hat{\theta}$ tends to θ as n increases, we say that $\hat{\theta}$ is asymptotically unbiased for θ. Here we have defined asymptotic unbiasedness such that $\lim_{n \to \infty} E[\hat{\theta}] = \theta$. Note that there are other definitions of asymptotic unbiasedness in use in the literature. For example, an estimator may be termed asymptotically unbiased if the mean of its asymptotic distribution is θ. In most cases, such as the present one, this second definition will coincide with the first so that there is no ambiguity.

We can also establish whether or not $\hat{\theta}$ is a consistent estimator of θ by using Khinchine's Weak Law of Large Numbers (see §8.5 C), and the Continuous Mapping Theorem. Consider

$$\overline{\log X} = \frac{1}{n} \sum_{i=1}^{n} (-\log(X_i))$$

which is in the form of a sample mean. Each variable in the sum is mutually independent, identically distributed, with mean

```
Expect[-Log[x], f]
```

$\frac{1}{\theta}$

Therefore, by Khinchine's Theorem, $\overline{\log X} \xrightarrow{p} \theta^{-1}$. As $\hat{\theta} = 1/(\overline{\log X})$, $\hat{\theta} \xrightarrow{p} \theta$ by the Continuous Mapping Theorem.[6] Therefore, the MLE $\hat{\theta}$ is a consistent estimator of θ.

The next asymptotic property concerns the limiting distribution of $\sqrt{n}(\hat{\theta} - \theta)$. Unfortunately, in this case, it is *not* possible to derive the limiting distribution using the asymptotic theory presented so far. If we apply Lindeberg–Lévy's version of the Central Limit Theorem (see §8.4) to $-\sum_{i=1}^{n} \log(X_i)$, we can only get as far as stating,

$$\frac{\sum_{i=1}^{n}(-\log(X_i)) - n\theta^{-1}}{\theta^{-1}\sqrt{n}} = \sqrt{n}\left(\frac{\theta}{\hat{\theta}} - 1\right) \xrightarrow{d} Z \sim N(0, 1).$$

To proceed any further, we must establish whether or not certain regularity conditions are satisfied by the distribution of X.[7] ■

11.4 D Regularity Conditions

To derive (some of) the asymptotic properties of $\hat{\theta}$, we used the fact that we knew the estimator's functional form, just as we did when determining its small sample properties. Alas, the functional form of the MLE is often unknown; how then are we to determine the asymptotic properties of the MLE? Fortunately, there exist sets of regularity conditions that, if satisfied, permit us to make relatively straightforward statements about the asymptotic properties of the MLE. Those stated here apply if the random sample is a collection of mutually independent, identically distributed random variables, if the parameter θ is a scalar, and if there is a unique solution to the first-order condition that globally maximises the log-likelihood function. This ideal setting fits our particular case.

Let θ_0 denote the 'true value' of θ, let i_0 denote the Fisher Information on θ evaluated at $\theta = \theta_0$, and let n denote the sample size. Under the previously mentioned conditions, the MLE has the following asymptotic properties,

consistency	$\hat{\theta} \xrightarrow{p} \theta_0$
limit Normal distribution	$\sqrt{n} \, (\hat{\theta} - \theta_0) \xrightarrow{d} N(0, i_0^{-1})$
asymptotic efficiency	relative to all other consistent uniformly limiting Normal estimators

Table 2: Asymptotic properties of the MLE, given regularity conditions

under the following *regularity conditions*:

1. The parameter space Θ is an open interval of the real line within which θ_0 lies.

2. The probability distributions defined by any two different values of θ are distinct.

3. For any finite n, the first three derivatives of the log-likelihood function with respect to θ exist in an open neighbourhood of θ_0.

4. In an open neighbourhood of θ_0, the information identity for Fisher Information holds:

$$i_0 = E\left[\left(\frac{\partial}{\partial \theta} \log f(X; \theta_0)\right)^2\right] = -E\left[\frac{\partial^2}{\partial \theta^2} \log f(X; \theta_0)\right].$$

Moreover, i_0 is finite and positive.

5. In an open neighbourhood of θ_0:

   (i) $\quad \frac{1}{\sqrt{n}} \frac{\partial}{\partial \theta} \log L(\theta_0) \xrightarrow{d} N(0, i_0)$

   (ii) $\quad -\frac{1}{n} \frac{\partial^2}{\partial \theta^2} \log L(\theta_0) \xrightarrow{p} i_0$

   (iii) For some constant $M < \infty$, $\frac{1}{n} \left| \frac{\partial^3}{\partial \theta^3} \log L(\theta_0) \right| \xrightarrow{p} M$.

For discussion about the role of regularity conditions in determining asymptotic properties of estimators such as the MLE, see, for example, Cox and Hinkley (1974), Amemiya (1985) and McCabe and Tremayne (1993).

⊕ **Example 12:** Satisfying Regularity Conditions

The model of *Example 8*, with pdf $f(x; \theta_0)$, is given by:

```
f = θ₀ x^(θ₀-1);   domain[f] = {x, 0, 1} && {θ₀ > 0};
```

Note that the parameter of the distribution is given at its true value θ_0.

The first regularity condition is satisfied as the parameter space $\Theta = \{\theta : \theta \in \mathbb{R}_+\}$ is an open interval of the real line, within which we assume θ_0 lies. The second condition pertains to parameter identification and is satisfied in our single-parameter case. For the third condition, the first three derivatives of the log-likelihood function evaluated at θ_0 are:

```
Table[D[Log[∏_{i=1}^{n} (f /. x → xᵢ)], {θ₀, j}], {j, 3}]
```

$$\left\{ \frac{n}{\theta_0} + \sum_{i=1}^{n} \log[x_i], \ -\frac{n}{\theta_0^2}, \ \frac{2n}{\theta_0^3} \right\}$$

and each exists within a neighbourhood about θ_0 (wherever that might be). Next, the information identity is satisfied:

```
FisherInformation[θ₀, f, Method → 1] ==
  FisherInformation[θ₀, f, Method → 2]
```

True

Moreover, the Fisher Information i_0 is equal to:

```
FisherInformation[θ₀, f]
```

$$\frac{1}{\theta_0^2}$$

which is finite, so the fourth condition is satisfied. From the derivatives of the log-likelihood function, we can establish that the fifth condition is satisfied. For 5(i),

$$\frac{1}{\sqrt{n}} \frac{\partial}{\partial \theta} \log L(\theta_0) = \frac{1}{\sqrt{n}} \sum_{i=1}^{n} (\log X_i + \theta_0^{-1})$$

which, by the Lindeberg–Lévy version of the Central Limit Theorem, is $N(0, i_0)$ in the limit, as each term in the summand has mean and variance:

```
Expect[Log[x] + 1/θ₀, f]
```

0

$$\text{Var}\left[\text{Log}[x] + \frac{1}{\theta_0}, \text{f}\right]$$

$$\frac{1}{\theta_0^2}$$

For 5(ii),

$$-\frac{1}{n}\frac{\partial^2}{\partial \theta^2} \log L(\theta_0) = \theta_0^{-2} = i_0$$

for every n, including in the limit. For 5(iii),

$$\frac{1}{n}\left|\frac{\partial^3}{\partial \theta^3} \log L(\theta_0)\right| = 2\,\theta_0^{-3}$$

is non-stochastic and finite for every n, including in the limit. In conclusion, each regularity condition is satisfied. Thus, $\hat{\theta}$ is consistent for θ_0, $\sqrt{n}\,\hat{\theta}$ has a limit Normal distribution, in particular, $\sqrt{n}\,(\hat{\theta} - \theta_0) \xrightarrow{d} N(0, \theta_0^2)$, and $\hat{\theta}$ is asymptotically efficient. These results enable us, for example, to construct the estimator's asymptotic distribution: $\hat{\theta} \stackrel{a}{\sim} N(\theta_0, \theta_0^2/n)$, which may be contrasted against the estimator's exact distribution $\hat{\theta} \sim \text{InverseGamma}\left(n, \frac{1}{n\theta_0}\right)$ found in §11.4 B. ∎

11.4 E Invariance Property

Throughout this section, our example has concentrated on estimation of θ. But suppose another parameter λ, related functionally to θ, is also of interest. Given what we already know about $\hat{\theta}$, it is usually possible to obtain the MLE of λ and to establish its statistical properties by the Invariance Property (see Table 1), provided we know the functional form that links λ to θ.

Consider a multi-parameter setting in which θ is a $(k \times 1)$ vector and λ is a $(j \times 1)$ vector, where $j \leq k$. The link from θ to λ is through a vector function g; that is, $\lambda = g(\theta)$, where g is assumed known. The parameters are such that $\theta \in \Theta$ and $\lambda \in \Lambda$, with the particular true values once again indicated by a 0 subscript. The parameter spaces are $\Theta \subset \mathbb{R}^k$ and $\Lambda \subset \mathbb{R}^j$, so that $g : \Theta \to \Lambda$. Moreover, we assume that g is a continuous function of θ, and that the $(j \times k)$ matrix of partial derivatives

$$G(\theta) = \frac{\partial g(\theta)}{\partial \theta^\mathrm{T}}$$

has finite elements and is of full row rank; that is, $\text{rank}(G(\theta)) = j$, for all $\theta \in \Theta$.

Of particular use is the case when $j = k$, for then the dimensions of θ and λ are the same and $G(\theta)$ becomes a square matrix having full rank (which means that the inverse function g^{-1} must exist). In this case, the parameter λ is said to represent a *re-parameterisation* of θ. There are a number of examples of re-parameterisation in the next chapter, the idea there being to transform a constrained optimisation problem in θ (occurring when Θ is a proper subset of $\mathbb{R}^k$) into an unconstrained optimisation problem in λ (re-parameterisation achieves $\Lambda = \mathbb{R}^k$).

The key results of the Invariance Property apply to the MLE of $g(\theta)$ and to its asymptotic properties. First, if $\hat{\theta}$ is the MLE of θ, then $g(\hat{\theta})$ is the MLE of $\lambda = g(\theta)$. This is an extremely useful property for it means that if we already know $\hat{\theta}$, then we do *not* need to find the MLE of λ_0 by maximising the log-likelihood $\log L(\lambda)$. Second, if $\hat{\theta}$ is consistent, and has a limiting Normal distribution when suitably scaled, and is asymptotically efficient, then so too is $\hat{\lambda} = g(\hat{\theta})$. That is, if

$$\hat{\theta} \xrightarrow{p} \theta_0 \qquad (11.14)$$

$$\sqrt{n}\left(\hat{\theta} - \theta_0\right) \xrightarrow{d} N(\vec{0}, i_0^{-1}) \qquad (11.15)$$

then

$$g(\hat{\theta}) \xrightarrow{p} g(\theta_0) \qquad (11.16)$$

$$\sqrt{n}\left(g(\hat{\theta}) - g(\theta_0)\right) \xrightarrow{d} N(\vec{0}, G(\theta_0) \times i_0^{-1} \times G(\theta_0)^{\mathrm{T}}). \qquad (11.17)$$

The small sample properties of $\hat{\lambda}$ generally cannot be deduced from those of $\hat{\theta}$, but must be examined on a case-by-case basis. To see this, a simple example suffices. Let $\lambda = g(\theta) = \theta^2$, and suppose that the MLE $\hat{\theta}$ is unbiased. By the Invariance Property, the MLE of λ is $\hat{\lambda} = \hat{\theta}^2$; however, it is *not* necessarily true that $\hat{\lambda}$ is unbiased for λ, for in general $E[\hat{\theta}^2] \neq \left(E[\hat{\theta}]\right)^2$.

⊕ ***Example 13:*** The Invariance Property

The model of *Example 8*, with pdf $f(x; \theta)$, is given by:

```
f = Θ x^(Θ-1);   domain[f] = {x, 0, 1} && {Θ > 0};
```

Consider the parameter $\lambda = E[X]$:

λ = **Expect[x, f]**

$$\frac{\theta}{1+\theta}$$

Clearly, parameter $\lambda \in \Lambda = (0, 1)$ is a function of θ; $\lambda = g(\theta) = \theta/(1+\theta)$, with true value $\lambda_0 = g(\theta_0)$. To estimate λ_0, one possibility is to re-parameterise the pdf of X from θ to λ and repeat the same ML estimation procedures from the very beginning. But we can do better by applying the Invariance Property, for we already have the functional form of $\hat{\theta}$ (see (11.13)) as well as its asymptotic properties. The MLE of λ_0 is given by

$$\hat{\lambda} = \frac{\hat{\theta}}{1+\hat{\theta}} = \frac{n}{n - \sum_{i=1}^{n} \log(X_i)}.$$

Since g is continuously differentiable with respect to θ, it follows from (11.17) that the limiting distribution of $\hat{\lambda}$ is

$$\sqrt{n}\left(\hat{\lambda}-\lambda_0\right) \xrightarrow{d} N\left(0, \left(\tfrac{\partial}{\partial\theta} g(\theta_0)\right)^2 \Big/ i_0\right).$$

In particular, the variance of the limiting distribution of $\sqrt{n}\left(\hat{\lambda}-\lambda_0\right)$ in terms of θ_0, is given by:

```
    Grad[λ, θ]²
 ────────────────────── /. θ → θ₀
 FisherInformation[θ, f]
```

$$\frac{\theta_0^2}{(1+\theta_0)^4}$$

The asymptotic distribution of the MLE of λ_0 is therefore

$$\hat{\lambda} \stackrel{a}{\sim} N\left(\lambda_0, \frac{\theta_0^2}{n(1+\theta_0)^4}\right). \qquad\blacksquare$$

11.5 Asymptotic Properties: Extensions

The asymptotic properties of the MLE — consistency, a limiting Normal distribution when suitably scaled, and asymptotic efficiency — generally hold in a variety of circumstances far weaker than those considered in §11.4. In fact, there exists a range of regularity conditions designed to cater for a variety of settings involving various combinations of non-independent and/or non-identically distributed samples, parameter θ a vector, multiple local optima, and so on. In this section, we consider two departures from the setup in §11.4 D. Texts that discuss proofs of asymptotic properties of the MLE and regularity conditions include Amemiya (1985), Cox and Hinkley (1974), Dhrymes (1970), Lehmann (1983), McCabe and Tremayne (1993) and Mittelhammer (1996).

11.5 A More Than One Parameter

Suppose we now allow parameter θ to be k-dimensional, but otherwise keep the statisticial setup described in §11.4 unaltered; namely, the random sample consists of mutually independent and identically distributed random variables, and there is a unique solution to the first-order condition — a system of k equations — that maximises the log-likelihood function. Then, it seems reasonable to expect that regularity conditions 1, 4 and 5 given in §11.4 D need only be extended to account for the higher dimensionality of θ:

1a. The k-dimensional parameter space Θ must be of finite dimension as sample size n increases, it must be an open subset of $\mathbb{R}^k$, and it must contain the true value θ_0 within its interior.

4a. In an open neighbourhood of θ_0, the information identity for Fisher Information (a $(k \times k)$ symmetric matrix) holds. That is:

$$i_0 = E\left[\left(\frac{\partial}{\partial \theta} \log f(X; \theta_0)\right)\left(\frac{\partial}{\partial \theta} \log f(X; \theta_0)\right)^T\right]$$

$$= -E\left[\frac{\partial^2}{\partial \theta\, \partial \theta^T} \log f(X; \theta_0)\right].$$

Moreover, every element of i_0 is finite, and i_0 is positive definite.

5a. In an open neighbourhood of θ_0:

(i) $\dfrac{1}{\sqrt{n}} \dfrac{\partial}{\partial \theta} \log L(\theta_0) \xrightarrow{d} N(\vec{0}, i_0)$

(ii) $-\dfrac{1}{n} \dfrac{\partial^2}{\partial \theta\, \partial \theta^T} \log L(\theta_0) \xrightarrow{p} i_0$

(iii) Let indexes $u, v, w \in \{1, \ldots, k\}$ pick out elements of θ. For constants $M_{u,v,w} < \infty$,

$$\frac{1}{n}\left|\frac{\partial^3}{\partial \theta_u\, \partial \theta_v\, \partial \theta_w} \log L(\theta_0)\right| \xrightarrow{p} M_{u,v,w}.$$

If these conditions hold, as well as conditions 2 and 3, then the MLE $\hat\theta$ is a consistent estimator of θ, $\sqrt{n}\,(\hat\theta - \theta_0) \xrightarrow{d} N(\vec{0}, i_0^{-1})$, and $\hat\theta$ is asymptotically efficient (*cf.* Table 2).

⊕ **Example 14:** The Asymptotic Distribution of $\hat\theta$: $X \sim$ Normal

Let $X \sim N(\mu_0, \sigma_0^2)$, with pdf $f(x; \mu_0, \sigma_0^2)$:

```
f  =     1      Exp[- (x - μ₀)² ];
      ─────────         ───────
      σ₀ √2 π            2 σ₀²

domain[f] = {x, -∞, ∞} && {μ₀ ∈ Reals, σ₀ > 0};
```

In this case, the parameter $\theta = (\mu, \sigma^2)$ is two-dimensional ($k = 2$), with true value $\theta_0 = (\mu_0, \sigma_0^2)$. In *Example 6*, where $(X_1, \ldots, X_n)$ denoted a size n random sample drawn on X, the MLE of θ was derived as

$$\hat\theta = \begin{pmatrix}\hat\mu \\ \hat\sigma^2\end{pmatrix} = \begin{pmatrix}\dfrac{1}{n}\sum_{i=1}^{n} X_i \\ \dfrac{1}{n}\sum_{i=1}^{n} (X_i - \hat\mu)^2\end{pmatrix}.$$

The regularity conditions 1a, 2, 3, 4a, 5a hold in this case. The dimension k is fixed at 2 for all n, the parameter space $\Theta = \{\theta = (\mu, \sigma^2) : \mu \in \mathbb{R}, \sigma^2 \in \mathbb{R}_+\}$ is an open subset of $\mathbb{R}^2$ within which we assume θ_0 lies, and the information identity holds:

```
FisherInformation[{μ₀, σ₀²}, f, Method → 1] ==
   FisherInformation[{μ₀, σ₀²}, f, Method → 2]

True
```

§11.5 A PRINCIPLES OF MAXIMUM LIKELIHOOD ESTIMATION

The Fisher Information matrix i_0 is equal to:

$\mathtt{i_0 = FisherInformation[\{\mu_0, \sigma_0^2\}, f]}$

$$\begin{pmatrix} \frac{1}{\sigma_0^2} & 0 \\ 0 & \frac{1}{2\sigma_0^4} \end{pmatrix}$$

and it has finite elements and is positive definite. The asymptotic conditions 5a are satisfied too. We demonstrate 5a(i), leaving verification of 5a(ii) and 5a(iii) to the reader. For 5a(i), we require the derivatives of the log-likelihood function with respect to the elements of θ. Here is the log-likelihood:

$\mathtt{logL\theta = Log\left[\prod_{i=1}^{n} (f \; /. \; x \to x_i)\right]}$

$$-\frac{n\mu_0^2 + n(\text{Log}[2\pi] + 2\text{Log}[\sigma_0])\sigma_0^2 - 2\mu_0 \sum_{i=1}^{n} x_i + \sum_{i=1}^{n} x_i^2}{2\sigma_0^2}$$

and here are the derivatives:

$\mathtt{Grad[logL\theta, \{\mu_0, \sigma_0^2\}]}$

$$\left\{ \frac{-n\mu_0 + \sum_{i=1}^{n} x_i}{\sigma_0^2}, \; \frac{n\mu_0^2 - n\sigma_0^2 - 2\mu_0 \sum_{i=1}^{n} x_i + \sum_{i=1}^{n} x_i^2}{2\sigma_0^4} \right\}$$

For the first element, we have for 5a(i),

$$\frac{1}{\sqrt{n}} \frac{\partial}{\partial \mu} \log L(\theta_0) = \frac{1}{\sqrt{n}} \sum_{i=1}^{n} \frac{X_i - \mu_0}{\sigma_0^2}$$

which, by the Lindeberg–Lévy version of the Central Limit Theorem, is $N(0, \sigma_0^{-2})$ in the limit, as each term in the summand has mean and variance:

$\mathtt{Expect\left[\frac{x - \mu_0}{\sigma_0^2}, f\right]}$

0

$\mathtt{Var\left[\frac{x - \mu_0}{\sigma_0^2}, f\right]}$

$\frac{1}{\sigma_0^2}$

Similarly, for the derivative with respect to σ^2,

$$\frac{1}{\sqrt{n}} \frac{\partial}{\partial \sigma^2} \log L(\theta_0) = \frac{1}{\sqrt{n}} \sum_{i=1}^{n} \frac{1}{2\sigma_0^2}\left(\left(\frac{X_i - \mu_0}{\sigma_0}\right)^2 - 1\right)$$

which is $N(0, \frac{1}{2}\sigma_0^{-4})$ in the limit, as each term in the summand has mean and variance:

Expect$\left[\dfrac{1}{2\sigma_0^2}\left(\left(\dfrac{\mathbf{x}-\mu_0}{\sigma_0}\right)^2 - 1\right), \mathbf{f}\right]$

0

Var$\left[\dfrac{1}{2\sigma_0^2}\left(\left(\dfrac{\mathbf{x}-\mu_0}{\sigma_0}\right)^2 - 1\right), \mathbf{f}\right]$

$\dfrac{1}{2\sigma_0^4}$

Finally then, as $\sum_{i=1}^n X_i$ and $\sum_{i=1}^n X_i^2$ are independent (see *Example 27* of Chapter 4):

$$\frac{1}{\sqrt{n}}\frac{\partial}{\partial \theta}\log L(\theta_0) \xrightarrow{d} N(\vec{0}, i_0).$$

As all regularity conditions hold, $\sqrt{n}(\hat{\theta}-\theta_0) \xrightarrow{d} N(\vec{0}, i_0^{-1})$, with the variance-covariance matrix of the limiting distribution given by:

Inverse$[\,i_0\,]$

$\begin{pmatrix} \sigma_0^2 & 0 \\ 0 & 2\sigma_0^4 \end{pmatrix}$

From this result we can find, for example, the asymptotic distribution of the MLE

$$\hat{\theta} \stackrel{a}{\sim} N\left(\begin{pmatrix}\mu_0 \\ \sigma_0^2\end{pmatrix}, \begin{pmatrix}\sigma_0^2/n & 0 \\ 0 & 2\sigma_0^4/n\end{pmatrix}\right).$$

This can be contrasted against the small sample distributions: $\hat{\mu} \sim N(\mu_0, \sigma_0^2/n)$ independent of $n\hat{\sigma}^2/\sigma_0^2 \sim$ Chi-squared$(n-1)$. ∎

11.5 B Non-identically Distributed Samples

Suppose that the statistical setup described in §11.5 A is further extended such that ML estimation is based on a random sample which does *not* consist of identically distributed random variables. Despite the loss of identicality, mutual independence between the variables $(X_1, \ldots, X_n)$ in the sample ensures that the log-likelihood remains a sum:

$$\log L(\theta) = \sum_{i=1}^n \log f_i(x_i; \theta)$$

where $f_i(x_i; \theta)$ is the pdf of X_i. Accordingly, for the MLE to have the usual trio of asymptotic properties (see Table 1), the regularity conditions will need to be weakened even further in order that certain forms of the Central Limit Theorem and Law of Large Numbers relevant to sums of non-identically distributed random variables remain valid. The conditions requiring weakening are 4a, 5a(i) and 5a(ii):

§11.5 B   PRINCIPLES OF MAXIMUM LIKELIHOOD ESTIMATION

4b. In an open neighbourhood of θ_0, the information identity for *asymptotic* Fisher Information (a $(k \times k)$ symmetric matrix) holds. That is:

$$i_0^{(\infty)} = \lim_{n \to \infty} E\left[\frac{1}{n} \sum_{i=1}^{n} \left(\frac{\partial}{\partial \theta} \log f(X_i; \theta_0)\right)\left(\frac{\partial}{\partial \theta} \log f(X_i; \theta_0)\right)^{\mathrm{T}}\right]$$

$$= \lim_{n \to \infty} E\left[-\frac{1}{n} \frac{\partial^2}{\partial \theta \, \partial \theta^{\mathrm{T}}} \log L(\theta_0)\right].$$

Moreover, every element of $i_0^{(\infty)}$ is finite, and $i_0^{(\infty)}$ is positive definite.

5b. In an open neighbourhood of θ_0:

(i) $\dfrac{1}{\sqrt{n}} \dfrac{\partial}{\partial \theta} \log L(\theta_0) \xrightarrow{d} N(\vec{0}, i_0^{(\infty)})$

(ii) $-\dfrac{1}{n} \dfrac{\partial^2}{\partial \theta \, \partial \theta^{\mathrm{T}}} \log L(\theta_0) \xrightarrow{p} i_0^{(\infty)}$.

Should these conditions hold, as well as 1a, 2, 3 and 5a(iii), then the MLE $\hat{\theta}$ is a consistent estimator of θ, $\sqrt{n}\,(\hat{\theta} - \theta_0) \xrightarrow{d} N(\vec{0}, (i_0^{(\infty)})^{-1})$, and $\hat{\theta}$ is asymptotically efficient.

⊕ **Example 15:** The Asymptotic Distribution of $\hat{\theta}$: Exponential Regression

Suppose that a positive-valued random variable Y depends on another random variable X, both of which are observed in pairs $((Y_1, X_1), (Y_2, X_2), \ldots)$. For example, Y may represent sales of a firm, and X may represent the firms advertising expenditure. We may represent this dependence by specifying a *conditional* statistical model for Y; that is, by specifying a pdf for Y, given that a value $x \in \mathbb{R}$ is assigned to X. One such model is the *Exponential Regression*, where $Y \mid (X = x) \sim \text{Exponential}(\exp(\alpha_0 + \beta_0 x))$, with pdf $f(y \mid X = x; \theta_0)$:

```
f = ──────────────── Exp[- ──────────────── ];
     Exp[α₀ + β₀ x]         Exp[α₀ + β₀ x]
        1                         y

domain[f] = {y, 0, ∞} && {α₀ ∈ Reals, β₀ ∈ Reals, x ∈ Reals};
```

The parameter $\theta = (\alpha, \beta) \in \mathbb{R}^2$, and its true value $\theta_0 = (\alpha_0, \beta_0)$ is unknown. The regression function is given by the conditional mean $E[Y \mid (X = x)]$, and this is equal to:

```
Expect[y, f]
```

$e^{\alpha_0 + x\,\beta_0}$

Despite the fact that the functional form of the MLE $\hat{\theta}$ cannot be derived in this case,[8] we can still obtain the asymptotic properties of the MLE by determining if the regularity conditions 1a, 2, 3, 4b, 5b(i), 5b(ii) and 5a(iii) are satisfied. In this example, we shall focus on obtaining the asymptotic Fisher Information matrix $i_0^{(\infty)}$ given in 4b. We begin by deriving the Fisher Information:

```
FisherInformation[{α₀, β₀}, f]
```

$\begin{pmatrix} 1 & x \\ x & x^2 \end{pmatrix}$

This output reflects the non-identicality of the distribution of $Y|(X=x)$, for Fisher Information quite clearly depends on the value assigned to X. Let $((Y_1, X_1), \ldots, (Y_n, X_n))$ denote a random sample of size n on the pair (Y, X). Because the distribution of $Y_i | (X_i = x_i)$ need not be identical to the distribution of $Y_j | (X_j = x_j)$ (for x_i need not equal x_j), then the Sample Information matrix is no longer given by Fisher Information multiplied by sample size; rather, Sample Information is given by the sample sum:

$$I_0 = \begin{pmatrix} n & \sum_{i=1}^n x_i \\ \sum_{i=1}^n x_i & \sum_{i=1}^n x_i^2 \end{pmatrix}.$$

Under independence, the log-likelihood is made up of a sum of contributions,

$$\log L(\theta) = \sum_{i=1}^n \log f(y_i | (X_i = x_i); \theta)$$

implying that $\frac{1}{n} I_0$ is exactly the expectation given in regularity condition 4b, when computed either way because

```
FisherInformation[{α₀, β₀}, f, Method → 1] ==
FisherInformation[{α₀, β₀}, f, Method → 2]

True
```

To obtain the asymptotic Fisher Information matrix, we must examine the limiting behaviour of the elements of $\frac{1}{n} I_0$. This will require further assumptions about the marginal distribution of X. If the random variable X has finite mean μ, finite variance σ^2, with neither moment depending on n, then by Khinchine's Weak Law of Large Numbers,

$$\frac{1}{n} \sum_{i=1}^n X_i \xrightarrow{p} \mu \quad \text{and} \quad \frac{1}{n} \sum_{i=1}^n X_i^2 \xrightarrow{p} \sigma^2 + \mu^2.$$

Under these further assumptions, we obtain the asymptotic Fisher Information matrix as

$$i_0^{(\infty)} = \begin{pmatrix} 1 & \mu \\ \mu & \sigma^2 + \mu^2 \end{pmatrix}$$

which is positive definite. Establishing conditions 5b(i), 5b(ii) and 5a(iii) involves similar manipulations, and in this case can be shown to hold under the assumptions concerning the behaviour of X. In conclusion, the asymptotic distribution of the MLE $\hat{\theta}$ of $\theta_0 = (\alpha_0, \beta_0)$ is, under the assumptions placed on X, given by

$$\hat{\theta} \stackrel{a}{\sim} N\left(\begin{pmatrix} \alpha_0 \\ \beta_0 \end{pmatrix}, \frac{1}{n\sigma^2} \begin{pmatrix} \sigma^2 + \mu^2 & -\mu \\ -\mu & 1 \end{pmatrix}\right).$$

∎

11.6 Exercises

1. Let $X \sim \text{Poisson}(\lambda)$, where parameter $\lambda \in \mathbb{R}_+$. Let $(X_1, X_2, \ldots, X_n)$ denote a size n random sample drawn on X. (i) Derive $\hat{\lambda}$, the ML estimator of λ. (ii) Obtain the exact distribution of $\hat{\lambda}$. (iii) Obtain the asymptotic distribution of $\hat{\lambda}$ (check regularity conditions).

2. Let $X \sim \text{Geometric}(p)$, where parameter p is such that $0 < p < 1$. Let $(X_1, X_2, \ldots, X_n)$ denote a size n random sample drawn on X. Derive $\hat{p}$, the ML estimator of p, and obtain its asymptotic distribution.

3. Let $X \sim N(\mu, 1)$, where parameter $\mu \in \mathbb{R}$. Let $(X_1, X_2, \ldots, X_n)$ denote a size n random sample drawn on X. (i) Derive $\hat{\mu}$, the ML estimator of μ. (ii) Obtain the exact distribution of $\hat{\mu}$. (iii) Obtain the asymptotic distribution of $\hat{\mu}$ (check regularity conditions).

4. Let $X \sim \text{ExtremeValue}(\theta)$, with pdf $f(x; \theta) = \exp(-(x - \theta) - e^{-(x-\theta)})$, where $\theta \in \mathbb{R}$ is an unknown parameter. Let $(X_1, X_2, \ldots, X_n)$ denote a size n random sample drawn on X. (i) Obtain $\hat{\theta}$, the ML estimator of θ. (ii) Obtain the asymptotic distribution of $\hat{\theta}$ (check regularity conditions).

5. For the pdf of the $N(0, \sigma^2)$ distribution, specify a *replacement rule* that serves to replace σ and its powers in the pdf. In particular, the rule you construct should act to convert the pdf from an input of

$$\frac{1}{\sigma \sqrt{2\pi}} \text{Exp}\left[-\frac{x^2}{2\sigma^2}\right] \quad \text{to an output of} \quad \frac{1}{\sqrt{\theta}\sqrt{2\pi}} \text{Exp}\left[-\frac{x^2}{2\theta}\right]$$

6. Let $X \sim N(0, \sigma^2)$, where parameter $\sigma^2 \in \mathbb{R}_+$. Let $(X_1, X_2, \ldots, X_n)$ denote a size n random sample drawn on X.
   (i) Derive $\hat{\sigma}^2$, the ML estimator of σ^2.
   (ii) Obtain the exact distribution of $\hat{\sigma}^2$.
   (iii) Obtain the asymptotic distribution of $\hat{\sigma}^2$ (check regularity conditions).
   Hint: use your solution to Exercise 5.

7. Let $X \sim \text{Rayleigh}(\sigma^2)$, where parameter $\sigma^2 \in \mathbb{R}_+$. Let $(X_1, X_2, \ldots, X_n)$ denote a size n random sample drawn on X.
   (i) Derive $\hat{\sigma}^2$, the ML estimator of σ^2.
   (ii) Obtain the exact distribution of $\hat{\sigma}^2$.
   (iii) Obtain the asymptotic distribution of $\hat{\sigma}^2$ (check regularity conditions).

8. Let $X \sim \text{Uniform}(0, \theta)$, where parameter $\theta \in \mathbb{R}_+$ is unknown, and, of course, $X < \theta$. Let $(X_1, X_2, \ldots, X_n)$ denote a size n random sample drawn on X. Show that the largest order statistic $\hat{\theta} = X_{(n)} = \max(X_1, X_2, \ldots, X_n)$ is the ML estimator of θ. Using **mathStatica**'s OrderStat function, obtain the exact distribution of $\hat{\theta}$. Transform $\hat{\theta} \to Y$ such that $Y = n(\theta - \hat{\theta})$. Then derive the limiting distribution of $n(\theta - \hat{\theta})$. Propose an asymptotic approximation to the exact distribution of $\hat{\theta}$.

Chapter 12

Maximum Likelihood Estimation in Practice

12.1 Introduction

The previous chapter focused on the theory of maximum likelihood (ML) estimation, using examples for which analytic closed form solutions were possible. In practice, however, ML problems rarely yield closed form solutions. Consequently, ML estimation generally requires numerical methods that iterate progressively from one potential solution to the next, designed to terminate (at some pre-specified tolerance) at the point that maximises the likelihood.

This chapter emphasises the numerical aspects of ML estimation, using illustrations that have appeared in statistical practice. In §12.2, ML estimation is tackled using **mathStatica**'s FindMaximum function; this function is the mirror image of *Mathematica*'s built-in minimiser FindMinimum. Following this, §12.3 examines the performance of FindMinimum / FindMaximum as both a constrained and an unconstrained optimiser. We come away from this with the firm opinion that FindMinimum / FindMaximum should only be used for unconstrained optimisation. §12.4 discusses statistical inference applied to an estimated statistical model. We emphasise asymptotic methods, mainly because the asymptotic distribution of the ML estimator, being Normal, is simple to use. We then encounter a significant weakness in FindMinimum / FindMaximum, in that it only yields ML estimates. Further effort is required to estimate the (asymptotic) variance-covariance matrix of the ML estimator, which is required for inference. The remaining three sections focus on details of optimisation algorithms, especially the so-called gradient-method algorithms implemented in FindMinimum / FindMaximum. §12.5 describes how these algorithms are built, while §12.6 and §12.7 give code for the more popular algorithms of this family, namely the BFGS algorithm and the Newton–Raphson algorithm.

This chapter requires that we activate the **mathStatica** function SuperLog:

    SuperLog[On]

      — SuperLog is now On.

SuperLog modifies *Mathematica*'s Log function so that Log[Product[]] 'objects' or 'terms' get converted into sums of logarithms; see §11.1 B for more detail on SuperLog.

12.2 FindMaximum

Optimisation plays an important role throughout statistics, just as it does across a broad spectrum of sciences. When analytic solutions for ML estimators are not possible, as is typically the case in statistical practice, we must resort to numerical methods. There are numerous optimisation algorithms, a number of which are implemented in *Mathematica*'s FindMinimum function. However, we want to maximise an observed log-likelihood, not minimise it, so **mathStatica**'s FindMaximum function is designed for this purpose. FindMaximum is a simple mirror image of FindMinimum:

? FindMaximum

> FindMaximum is identical to
>   the built-in function FindMinimum, except
>   that it finds a Max rather than a Min.

To illustrate usage of FindMaximum, we use a random sample of biometric data attributed to Fatt and Katz by Cox and Lewis (1966):

xdata = ReadList["nerve.dat", Number];

The data represents a random sample of size $n = 799$ observations on a continuous random variable X, where X is defined as the time interval (measured in units of one second) between successive pulses along a nerve fibre. We term this the 'Nerve data'. A frequency polygon of the data is drawn in Fig. 1 using **mathStatica**'s FrequencyPlot function. The statistical model for X that generated the data is unknown; however, its appearance resembles an Exponential distribution (*Example 1*), or a generalisation of it to the Gamma distribution (*Example 2*).

FrequencyPlot[xdata];

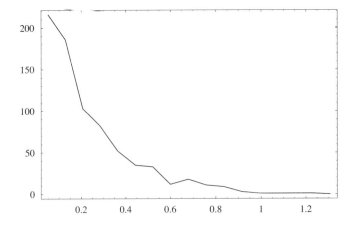

Fig. 1: The Nerve data

§12.2 MAXIMUM LIKELIHOOD ESTIMATION IN PRACTICE

⊕ **Example 1:** `FindMaximum` — Part I

Assume $X \sim$ Exponential(λ), with pdf $f(x; \lambda)$, where $\lambda \in \mathbb{R}_+$:

```
f = 1/λ e^-x/λ;        domain[f] = {x, 0, ∞} && {λ > 0};
```

For $(X_1, \ldots, X_n)$, a size n random sample drawn on X, the log-likelihood is given by:

```
logLλ = Log[∏_{i=1}^{n} (f /. x → x_i)]
```

$$-\frac{n \lambda \, \text{Log}[\lambda] + \sum_{i=1}^{n} x_i}{\lambda}$$

For the Nerve data, the observed log-likelihood is given by:

```
obslogLλ = logLλ /. {n → Length[xdata], x_i_ :→ xdata〚i〛}
```

$$-\frac{174.64 + 799 \, \lambda \, \text{Log}[\lambda]}{\lambda}$$

To obtain the MLE of λ, we use `FindMaximum` to numerically maximise `obslogLλ`.[1] For example:

```
solλ = FindMaximum[obslogLλ, {λ, {0.1, 1}}]
```

{415.987, {λ → 0.218573}}

The output states that the ML estimate of λ is 0.218573, and that the maximised value of the observed log-likelihood is 415.987. Here is a plot of the data overlaid with the fitted model:

```
FrequencyPlot[xdata, f /. solλ〚2〛];
```

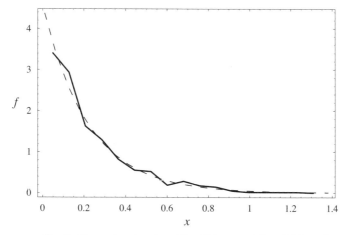

Fig. 2: Nerve data (——) and fitted Exponential model (– – –)

The Exponential model yields a close fit to the data, except in the neighbourhood of zero where the fit over-predicts. In the next example, we specify a more general model in an attempt to overcome this weakness. ∎

⊕ **Example 2:** FindMaximum — Part II

Assume that $X \sim \text{Gamma}(\alpha, \beta)$, with pdf $f(x; \alpha, \beta)$:

$$f = \frac{x^{\alpha-1} e^{-x/\beta}}{\Gamma[\alpha] \beta^{\alpha}}; \quad \text{domain}[f] = \{x, 0, \infty\} \text{ \&\& } \{\alpha > 0, \beta > 0\};$$

where $\alpha \in \mathbb{R}_+$ and $\beta \in \mathbb{R}_+$. ML estimation of the parameter $\theta = (\alpha, \beta)$ proceeds in two steps. First, we obtain a closed form solution for either $\hat{\alpha}$ or $\hat{\beta}$ in terms of the other parameter (*i.e.* we can obtain either $\hat{\alpha}(\beta)$ or $\hat{\beta}(\alpha)$). We then estimate the remaining parameter using the appropriate concentrated log-likelihood via numerical methods (FindMaximum).

The log-likelihood $\log L(\alpha, \beta)$ is:

$$\text{logL}\theta = \text{Log}\left[\prod_{i=1}^{n} (f /. x \to x_i)\right]$$

$$-\frac{1}{\beta}\left(n\beta(\alpha \text{Log}[\beta] + \text{Log}[\Gamma[\alpha]]) + (\beta - \alpha\beta)\sum_{i=1}^{n}\text{Log}[x_i] + \sum_{i=1}^{n}x_i\right)$$

The score vector $\frac{\partial}{\partial \theta} \log L(\theta)$ is derived using **mathStatica**'s Grad function:

$$\text{score} = \text{Grad}[\text{logL}\theta, \{\alpha, \beta\}]$$

$$\left\{-n(\text{Log}[\beta] + \text{PolyGamma}[0, \alpha]) + \sum_{i=1}^{n}\text{Log}[x_i], \frac{-n\alpha\beta + \sum_{i=1}^{n}x_i}{\beta^2}\right\}$$

The ML estimator of α in terms of β is obtained as:

$$\text{sol}\alpha = \text{Solve}[\text{score}[\![2]\!] == 0, \alpha] \text{ // Flatten}$$

$$\left\{\alpha \to \frac{\sum_{i=1}^{n} x_i}{n\beta}\right\}$$

That is,

$$\hat{\alpha}(\beta) = \frac{1}{n\beta} \sum_{i=1}^{n} X_i.$$

Substituting this solution into the log-likelihood yields the concentrated log-likelihood $\log L(\hat{\alpha}(\beta), \beta)$, which we denote logLc:

§12.2 MAXIMUM LIKELIHOOD ESTIMATION IN PRACTICE

logLc = logLθ /. solα

$$-\frac{1}{\beta}\left(\sum_{i=1}^{n} x_i + \left(\sum_{i=1}^{n} \text{Log}[x_i]\right)\left(\beta - \frac{\sum_{i=1}^{n} x_i}{n}\right) + n\beta\left(\text{Log}\left[\Gamma\left[\frac{\sum_{i=1}^{n} x_i}{n\beta}\right]\right] + \frac{\text{Log}[\beta]\sum_{i=1}^{n} x_i}{n\beta}\right)\right)$$

Next, we substitute the data into the concentrated log-likelihood:

obslogLc = logLc /. {n → Length[xdata], x$_{i\_}$:→ xdata〚i〛};

Then, we estimate β using `FindMaximum`:

solβ = FindMaximum[obslogLc, {β, {0.1, 1}}]〚2〛

{β → 0.186206}

For the Nerve data, and assuming $X \sim \text{Gamma}(\alpha, \beta)$, the ML estimate of β is $\hat{\beta} = 0.186206$. Therefore, the ML estimate of α, $\hat{\alpha}(\hat{\beta})$, is:

**solα /.
   Flatten[{solβ, n → Length[xdata], x$_{i\_}$:→ xdata〚i〛}]**

{α → 1.17382}

Here is a plot of the data overlaid by the fitted model:

FrequencyPlot[xdata, f /. {α → 1.17382, β → 0.186206}];

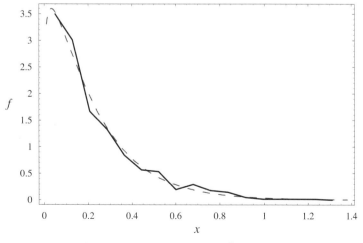

Fig. 3: Nerve data (——) and fitted Gamma model (– – –)

The Gamma model (see Fig. 3) achieves a better fit than the Exponential model (see Fig. 2), especially in the neighbourhood of zero. ∎

12.3 A Journey with `FindMaximum`

In this section, we take a closer look at the performance of `FindMaximum`. This is done in the context of a statistical model that has become popular amongst analysts of financial time series data—the so-called autoregressive conditional heteroscedasticity model (ARCH model). Originally proposed by Engle (1982), the ARCH model is designed for situations in which the variance of a random variable seems to alternate between periods of relative stability and periods of pronounced volatility. We will consider only the simplest member of the ARCH suite, known as the ARCH(1) model.

Load the following data:

pdata = ReadList["BML.dat"];

The data lists the daily closing price (in Australian dollars) of Bank of Melbourne shares on the Australian Stock Exchange from October 30, 1996, until October 10, 1997 (non-trading days have been removed). Figure 4 illustrates the data.

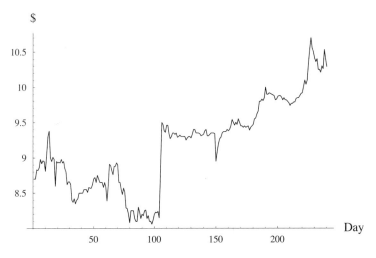

Fig. 4: The Bank of Melbourne data

Evidently, there are two dramatic increases in price: +$0.65 on day 105, and +$0.70 on day 106. These movements were caused by a takeover rumour that swept the market on those days, which was officially confirmed by the bank during day 106. Further important dates in the takeover process included: day 185, when approval was granted by the government regulator; day 226, when complete details of the financial offer were announced to shareholders; day 231, when shareholders voted to accept the offer; and day 240, the bank's final trading day.

Our analysis begins by specifying a variant of the random walk with drift model (see *Example 4* in Chapter 11) which, as we shall see upon examining the estimated residuals, leads us to specify an ARCH model later to improve fit. Let variable P_t denote the closing price on day t, and let $\bar{x}_t$ denote a vector of regressors. Then, the random walk model we consider is

§12.3 MAXIMUM LIKELIHOOD ESTIMATION IN PRACTICE

$$\Delta P_t = \vec{x}_t . \beta + U_t \tag{12.1}$$

where $\Delta P_t = P_t - P_{t-1}$, and the notation $\vec{x}_t . \beta$ indicates the dot product between the vectors $\vec{x}_t$ and β. We assume $U_t \sim N(0, \sigma^2)$; thus, $\Delta P_t \sim N(\vec{x}_t . \beta, \sigma^2)$. For this example, we specify a model with five regressors for vector $\vec{x}_t$, all of which are dummy variables: they consist of a constant intercept (the drift term), and day-specific intercept dummies corresponding to the takeover, the regulator, the disclosure and the vote. We denote the regression function by

$$\vec{x}_t . \beta = \beta_1 + x_2 \beta_2 + x_3 \beta_3 + x_4 \beta_4 + x_5 \beta_5.$$

For all n observations, we enter the price change:

```
Δp = Drop[pdata, 1] - Drop[pdata, -1];
```

and then the regressors: x2 for the takeover, x3 for the regulator, x4 for the disclosure and x5 for the vote:

```
x2 = x3 = x4 = x5 = Table[0, {239}];
x2[[104]] = x2[[105]] = x3[[184]] = x4[[225]] = x5[[230]] = 1;
```

Note that the estimation period is from day 2 to day 240; hence, the reduction of 1 in the day-specific dummies. The statistical model (12.1) is in the form of a *Normal linear regression model*. To estimate the parameters of our model, we apply the `Regress` function given in *Mathematica*'s `Statistics`LinearRegression`` package. The `Regress` function is built using an ordinary least squares (OLS) estimator. The differences between OLS and ML estimates of the parameters of our model are minimal.[2] To use `Regress`, we must first load the Statistics add-on:

```
<< Statistics`
```

and then manipulate the data to the required format:

```
rdata = Transpose[{x2, x3, x4, x5, Δp}];
```

The estimation results are collected in `olsθ`:

```
olsθ = Regress[rdata,
    {takeover, regulator, disclosure, vote},
    {takeover, regulator, disclosure, vote},
    RegressionReport → {ParameterTable,
       EstimatedVariance, FitResiduals}];
```

Table 1 lists the OLS estimates of the parameters $(\beta_1, \beta_2, \beta_3, \beta_4, \beta_5)$; the estimates correspond to the coefficients of drift (labelled 1) and the day-specific dummies (labelled takeover, regulator, disclosure and vote).

	Estimate	SE	TStat
1	−0.000171	0.0059496	−0.02873
takeover	0.675171	0.0646293	10.44680
regulator	0.140171	0.0912057	1.53687
disclosure	0.190171	0.0912057	2.08508
vote	−0.049829	0.0912057	−0.54633
σ^2	0.008283		

Table 1: OLS estimates of the Random Walk with Drift model

Notice that the only regressors to have t-statistics that exceed 2 in absolute value are the takeover and disclosure day-specific dummies. Figure 5 plots the time series of fitted residuals.

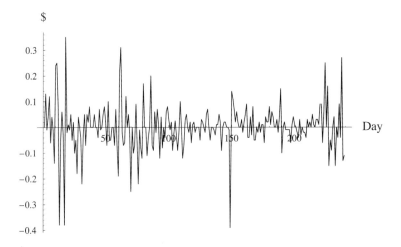

Fig. 5: Fitted OLS residuals

The residuals exhibit clusters of variability (approximately, days 2–30, 60–100, 220–240) interspersed with periods of stability (day 150 providing an exception to this). This suggests that an ARCH specification for the model disturbance U_t may improve the fit of (12.1); for details on formal statistical testing procedures for ARCH disturbances, see Engle (1982).

To specify an ARCH(1) model for the disturbances, we extend (12.1) to

$$\Delta P_t = \tilde{x}_t . \beta + U_t \qquad (12.2)$$

$$U_t = V_t \sqrt{\alpha_1 + \alpha_2 U_{t-1}^2} \qquad (12.3)$$

where $V_t \sim N(0, 1)$. We now deduce conditional moments of the disturbance U_t holding U_{t-1} fixed at a specific value u_{t-1}. The conditional mean and variance of U_t are $E[U_t \mid U_{t-1} = u_{t-1}] = 0$ and $\text{Var}(U_t \mid U_{t-1} = u_{t-1}) = \alpha_1 + \alpha_2 u_{t-1}^2$, respectively. These results imply that $\Delta P_t \mid (U_{t-1} = u_{t-1}) \sim N(\tilde{x}_t . \beta, \alpha_1 + \alpha_2 u_{t-1}^2)$. The likelihood function is

§12.3 MAXIMUM LIKELIHOOD ESTIMATION IN PRACTICE

the product of the distribution of the initial condition and the conditional distributions; the theory behind this construction is similar to that discussed in *Example 4* of Chapter 11. Given the initial condition $U_0 = 0$, the likelihood is

$$L(\theta) = \frac{1}{\sqrt{2\pi\alpha_1}} \exp\left(\frac{-(\Delta p_1 - \vec{x}_1 \cdot \beta)^2}{2\alpha_1}\right)$$
$$\times \prod_{t=2}^{n} \frac{1}{\sqrt{2\pi(\alpha_1 + \alpha_2 u_{t-1}^2)}} \exp\left(\frac{-(\Delta p_t - \vec{x}_t \cdot \beta)^2}{2(\alpha_1 + \alpha_2 u_{t-1}^2)}\right) \quad (12.4)$$

where Δp_t is the datum observed on ΔP_t and $u_t = \Delta p_t - \vec{x}_t \cdot \beta$, for $t = 1, \ldots, n$. We now enter the log-likelihood into *Mathematica*. It is convenient to express the log-likelihood in terms of u_t:

```
Clear[n];
logLθ =
    FullSimplify[ Log[ Exp[-u₁²/(2 α1)] / √(2 π α1) ∏(t=2 to n) Exp[-u_t²/(2(α1 + α2 u²_{t-1}))] / √(2 π (α1 + α2 u²_{t-1}))],
    {u₁ ∈ Reals, α1 > 0}] // Expand
```

$$-\frac{1}{2} n \operatorname{Log}[2\pi] - \frac{\operatorname{Log}[\alpha 1]}{2} - \frac{u_1^2}{2\alpha 1} - \frac{1}{2}\sum_{t=2}^{n}\operatorname{Log}[\alpha 1 + \alpha 2\, u_{-1+t}^2] - \frac{1}{2}\sum_{t=2}^{n}\frac{u_t^2}{\alpha 1 + \alpha 2\, u_{-1+t}^2}$$

To obtain the observed log-likelihood, we first enter in the value of n; we then redefine the regressors in $\vec{x}$, reducing the number of regressors from five down to just the two significant regressors (takeover and disclosure) from the random walk model fitted previously:

```
n = 239;    xdata = Transpose[{x2, x4}];
```

Next, we enter the disturbances u_t defined, via (12.2), as $U_t = \Delta P_t - \vec{x}_t \cdot \beta$:

```
uvec = Δp - xdata.{β2, β4};
```

Finally, we create a set of replacement rules called `urules`:[3]

```
urules = Table[u_i → uvec[[i]], {i, n}];    Short[urules]
```

$\{u_1 \to 0., u_2 \to 0.13, \ll 236 \gg, u_{239} \to -0.11\}$

Substituting `urules` into the log-likelihood yields the observed log-likelihood:

```
obslogLθ = logLθ /. urules;
```

Note that our *Mathematica* inputs use the parameter notation (β2, β4, α1, α2) rather than the neater subscript form $(\beta_2, \beta_4, \alpha_1, \alpha_2)$; this is because FindMinimum / FindMaximum does not handle subscripts well.[4]

We undertake maximum likelihood estimation of the parameters $(\beta_2, \beta_4, \alpha_1, \alpha_2)$ with FindMaximum. To begin, we apply it blindly, selecting as initial values for (β_2, β_4) the estimates from the random walk model, and choosing arbitrary initial values for α_1 and α_2:

```
sol = FindMaximum[obslogLθ,
    {β2, .675171}, {β4, .190171}, {α1, 1}, {α2, 1}]
```

- FindMinimum::fmnum :
    Objective function 122.878 + 375.42 i is not real at
    {β2, β4, α1, α2} = {0.675165, 0.190167, -≪19≫, 0.988813}.
- FindMinimum::fmnum :
    Objective function 122.878 + 375.42 i is not real at
    {β2, β4, α1, α2} = {0.675165, 0.190167, -≪19≫, 0.988813}.
- FindMinimum::fmnum :
    Objective function 122.878 + 375.42 i is not real at
    {β2, β4, α1, α2} = {0.675165, 0.190167, -≪19≫, 0.988813}.
- General::stop : Further output of FindMinimum::fmnum will
    be suppressed during this calculation.

Why has it crashed? Our first clue comes from the error message, which tells us that the observed log-likelihood 'is not real' for some set of values assigned to the parameters. Of course, all log-likelihoods *must* be real-valued at all points in the *parameter space*, so the problem must be that FindMaximum has drifted outside the parameter space. Indeed, from the error message we see that α_1 has become negative, which may in turn cause the conditional variance, $\text{Var}(\Delta P_t \mid (U_{t-1} = u_{t-1})) = \alpha_1 + \alpha_2 u_{t-1}^2$ to become negative, causing *Mathematica* to report a complex value for $\log(\alpha_1 + \alpha_2 u_{t-1}^2)$. It is easy to see that if $\alpha_2 = 0$, the ARCH model, (12.2) and (12.3), reduces to the random walk model (12.1) in which case $\alpha_1 = \sigma^2$, so we require $\alpha_1 > 0$. Similarly, we must insist on $\alpha_2 \geq 0$. Finally, Engle (1982, Theorem 1) derives an upper bound for α_2 which must hold if higher order even moments of the ARCH(1) process are to exist. Imposing $\alpha_2 < 1$ ensures that the unconditional variance, $\text{Var}(U_t)$, exists.

In order to obtain the ML estimates, we need to incorporate the parameter restrictions $\alpha_1 > 0$ and $0 \leq \alpha_2 < 1$ into *Mathematica*. There are two possibilities open to us:

(i) to use FindMaximum as a constrained optimiser, or

(ii) to re-parameterise the observed log-likelihood function so that the constraints are not needed.

For approach (i), we implement FindMaximum with the constraints entered at the command line; for example, we might enter:

```
sol1 = FindMaximum[obslogLθ, {β2, .675171}, {β4, .190171},
    {α1, 1, 0.00001, 100}, {α2, 0.5, 0, 1}, MaxIterations → 40]
```

{243.226, {β2 → 0.693842,
    β4 → 0.191731, α1 → 0.00651728, α2 → 0.192958}}

§12.3 MAXIMUM LIKELIHOOD ESTIMATION IN PRACTICE

In this way, FindMinimum / FindMaximum is being used as a *constrained* optimiser. The constraints entered above correspond to $0.00001 \leq \alpha_1 \leq 100$ and $0 \leq \alpha_2 \leq 1$. Also, note that we had to increase the maximum possible number of iterations to 40 (10 more than the default) to enable FindMinimum / FindMaximum to report convergence. Unfortunately, FindMinimum / FindMaximum often encounters difficulties when parameter constraints are entered at the command line.

Approach (ii) improves on the previous method by re-parameterising the observed log-likelihood in such a way that the constraints are eliminated. In doing so, FindMinimum / FindMaximum is implemented as an *unconstrained* optimiser, which is a task it can cope with. Firstly, the constraint $\alpha_1 > 0$ is satisfied for all real γ_1 provided $\alpha_1 = e^{\gamma_1}$. Secondly, the constraint $0 \leq \alpha_2 < 1$ is (almost) satisfied for all real γ_2 provided $\alpha_2 = (1 + \exp(\gamma_2))^{-1}$. A replacement rule is all that is needed to re-parameterise the observed log-likelihood:

obslogLλ = obslogLθ /. $\left\{\alpha 1 \to e^{\gamma 1},\ \alpha 2 \to \dfrac{1}{1 + e^{\gamma 2}}\right\}$;

We now attempt:

**sol2 = FindMaximum[obslogLλ,
    {β2, .675171}, {β4, .190171}, {γ1, 0}, {γ2, 0}]**

{243.534, {$\beta 2 \to 0.677367$,
    $\beta 4 \to 0.305868,\ \gamma 1 \to -5.07541,\ \gamma 2 \to 1.02915$}}

The striking feature of this result is that even though the starting points of this and our earlier effort are effectively the same, the maximised value of the observed log-likelihood yielded by the current solution sol2 is strictly superior to that of the former sol1:

sol2[[1]] > sol1[[1]]

True

It would, however, be unwise to state unreservedly that sol2 represents the ML estimates! In practice, it is advisable to experiment with different starting values. Suppose, for example, that the algorithm is started from a different location in the parameter space:

**sol3 = FindMaximum[obslogLλ,
    {β2, .675171}, {β4, .190171}, {γ1, -5}, {γ2, 0}]**

{243.534, {$\beta 2 \to 0.677263$,
    $\beta 4 \to 0.305021,\ \gamma 1 \to -5.07498,\ \gamma 2 \to 1.03053$}}

This solution is slightly better than the former one, the difference being detectable at the 5th decimal place:

**NumberForm[sol2[[1]], 9]
NumberForm[sol3[[1]], 9]**

243.53372

243.533752

Nevertheless, we observe that the parameter estimates output from both runs are fairly close, so it seems reasonable enough to expect that sol3 is in the *neighbourhood* of the solution.[5]

There are still two features of the proposed solution that need to be checked, and these concern the gradient:

```
g = Grad[obslogLλ, {β2, β4, γ1, γ2}];
g /. sol3[[2]]
```

{0.0553552, 0.000195139, 0.0116302, -0.000123497}

and the Hessian:

```
h = Hessian[obslogLλ, {β2, β4, γ1, γ2}];
Eigenvalues[h /. sol3[[2]]]
```

{-359.682, -96.3175, -79.1461, -2.60905}

The gradient at the maximum (or minimum or saddle point) should disappear—but this is far from true here. It would therefore be a mistake to claim that sol3 is the ML estimate! On the other hand, all eigenvalues at sol3 are negative, so the observed log-likelihood is concave in this neighbourhood. This is useful information, as we shall see later on. For now, let us return to the puzzle of the non-zero gradient!

Why does FindMinimum / FindMaximum fail to detect a non-zero gradient at what it claims is the optimum? The answer lies with the algorithm's stopping rule. Quite clearly, FindMinimum / FindMaximum does not check the magnitude of the gradient, for if it did, further iterations would be performed. So what criterion does FindMinimum use in deciding whether to stop or proceed to a new iteration? After searching the documentation on FindMinimum, the criterion is not revealed. So, at this stage, our answer is incomplete; we can only say for certain what criterion is *not* used. Perhaps, like many optimisers, FindMinimum iterates until the improvement in the objective function is smaller than some critical number? Alternatively, perhaps FindMinimum iterates until the absolute change in the choice variables is smaller than some critical value? Further discussion of stopping rule criteria appears in §12.5.

Our final optimisation assault utilises the fact that, at sol3 (our current best 'solution'), we have reached a neighbourhood of the parameter space in which the observed log-likelihood is concave, since the eigenvalues of the Hessian matrix are negative at sol3. In practice, it is nearly always advisable to 'finish off' an optimisation with iterations of the Newton–Raphson algorithm, provided it is computationally feasible to do so. This algorithm can often be costly to perform, for it requires computation of the Hessian matrix at each iteration, but this is exactly where *Mathematica* comes into its own because it is a wonderful differentiator! And for our particular problem, provided that we do not print it to screen, the Hessian matrix takes less than no time for *Mathematica* to compute—as we have already witnessed when it was computed for the re-parameterised observed log-likelihood and stored as h. The Newton–Raphson algorithm can be run by supplying an option to FindMaximum. Starting our search at sol3, we find:

```
sol4 = FindMaximum[obslogLλ,
                   {β2, 0.677263}, {β4, 0.305021},
                   {γ1, -5.07498}, {γ2, 1.03053},
                                   Method → Newton]
```

{243.534, {β2 → 0.677416,
  β4 → 0.304999, γ1 → -5.07483, γ2 → 1.03084}}

Not much appears to have changed in going from sol3 to sol4. The value of the observed log-likelihood increases slightly at the 6th decimal place:

```
NumberForm[sol3[[1]], 10]
NumberForm[sol4[[1]], 10]
```

243.5337516

243.5337567

which necessarily forces us to replace sol3 with sol4, the latter now being a possible contender for the maximum. The parameter estimates alter slightly too:

```
sol3[[2]]
```

{β2 → 0.677263, β4 → 0.305021,
  γ1 → -5.07498, γ2 → 1.03053}

```
sol4[[2]]
```

{β2 → 0.677416, β4 → 0.304999,
  γ1 → -5.07483, γ2 → 1.03084}

But what about our concerns over the gradient and the Hessian?

```
g /. sol4[[2]]
```

{0.0000131549, -6.90917×10^{-7},
  4.09054×10^{-6}, 3.40381×10^{-7}}

```
Eigenvalues[h /. sol4[[2]]]
```

{-359.569, -96.3068, -79.1165, -2.60887}

Wonderful! All elements of the gradient are numerically much closer to zero, and the eigenvalues of the Hessian matrix are all negative, indicating that it is negative definite. FindMinimum / FindMaximum has, with some effort on our part, successfully navigated its way through the numerical optimisation maze and presented to us the point estimates that maximise the re-parameterised observed log-likelihood. However, our work is not yet finished! The ML estimates of the parameters of the original ARCH(1) model must be determined:

```
{β2, β4, e^γ1, 1/(1 + e^γ2)} /. sol4[[2]]

{0.677416, 0.304999, 0.00625216, 0.262921}
```

We conclude our 'journey' by presenting the ML estimates in Table 2.

	Estimate
takeover	0.677416
disclosure	0.304999
α_1	0.006252
α_2	0.262921

Table 2: ML estimates of the ARCH(1) model

12.4 Asymptotic Inference

Inference refers to topics such as hypothesis testing and diagnostic checking of fitted models, confidence interval construction, within-sample prediction, and out-of-sample forecasting. For statistical models fitted using ML methods, inference is often based on large sample results, as ML estimators (suitably standardised) have a limiting Normal distribution.

12.4 A Hypothesis Testing

Asymptotic inference is operationalised by replacing unknowns with consistent estimates. To illustrate, consider the Gamma(α, β) model, with mean $\mu = \alpha\beta$. Suppose we want to test

$$H_0 : \mu = \mu_0 \quad \text{against} \quad H_1 : \mu \neq \mu_0$$

where $\mu_0 \in \mathbb{R}_+$ is known. Letting $\hat{\mu}$ denote the ML estimator of μ, we find (see *Example 4* for the derivation):

$$\sqrt{n} \, (\hat{\mu} - \mu) \xrightarrow{d} N(0, \alpha \beta^2).$$

Assuming H_0 to be true, we can (to give just two possibilities) base our hypothesis test on either of the following asymptotic distributions for $\hat{\mu}$:

$$\hat{\mu} \stackrel{a}{\sim} N\left(\mu_0, \frac{1}{n} \hat{\alpha}\hat{\beta}^2\right) \quad \text{or} \quad \hat{\mu} \stackrel{a}{\sim} N\left(\mu_0, \frac{1}{n} \mu_0 \hat{\beta}\right).$$

Depending on which distribution is used, it is quite possible to obtain conflicting outcomes to the tests. The potential for arbitrary outcomes in asymptotic inference has, on occasion, 'ruffled the feathers' of those advocating that inference should be based on small sample performance!

§12.4 A MAXIMUM LIKELIHOOD ESTIMATION IN PRACTICE

⊕ *Example 3:* The Gamma or the Exponential?

In this example, we consider whether there is a statistically significant improvement in using the Gamma(α, β) model to fit the Nerve data (*Example 2*) when compared to the Exponential(λ) model (*Example 1*). In a Gamma distribution, restricting the shape parameter α to unity yields an Exponential distribution; that is, Gamma(1, β) = Exponential(β). Hence, we shall conduct a hypothesis test of

$$H_0 : \alpha = 1 \quad \text{against} \quad H_1 : \alpha \neq 1.$$

We use the asymptotic theory of ML estimators to perform the test of H_0 against H_1. Here is the pdf of $X \sim$ Gamma(α, β):

```
f = (x^(α-1) e^(-x/β)) / (Γ[α] β^α);    domain[f] = {x, 0, ∞} && {α > 0, β > 0};
```

Since the MLE is regular (conditions 1a, 2, 3, 4a, and 5a are satisfied; see §11.4 D and §11.5 A),

$$\sqrt{n}\left(\hat{\theta} - \theta_0\right) \xrightarrow{d} N(0, i_0^{-1})$$

where $\hat{\theta}$ denotes the MLE of $\theta_0 = (\alpha, \beta)$. We can evaluate i_0^{-1}:

```
Inverse[FisherInformation[{α, β}, f]] // Simplify
```

$$\begin{pmatrix} \dfrac{\alpha}{-1+\alpha \, \text{PolyGamma}[1,\alpha]} & \dfrac{\beta}{1-\alpha \, \text{PolyGamma}[1,\alpha]} \\ \dfrac{\beta}{1-\alpha \, \text{PolyGamma}[1,\alpha]} & \dfrac{\beta^2 \, \text{PolyGamma}[1,\alpha]}{-1+\alpha \, \text{PolyGamma}[1,\alpha]} \end{pmatrix}$$

Let σ^2 denote the top left element of i_0^{-1}; note that σ^2 depends only on α. From the (joint) asymptotic distribution of $\hat{\theta}$, we find

$$\hat{\alpha} \overset{a}{\sim} N\left(\alpha, \frac{1}{n}\sigma^2\right).$$

We may base our test statistic on this asymptotic distribution for $\hat{\alpha}$, for when $\alpha = 1$ (*i.e.* H_0 is true), it has mean 1, and standard deviation:

```
s = √( (1/n)  α / (-1 + α PolyGamma[1, α]) ) /. {α → 1, n → 799} // N
```

0.0440523

Because the alternative hypothesis H_1 is uninformative (two-sided), H_0 will be rejected if the observed value of $\hat{\alpha}$ (1.17382 was obtained in *Example 2*) is either much larger than unity, or much smaller than unity. The *p*-value (short for 'probability value'; see, for example, Mittelhammer (1996, pp. 535–538)) for the test is given by

$$P\bigl(\bigl|\hat{\alpha}-1\bigr|>1.17382-1\bigr)=1-P(0.82618<\hat{\alpha}<1.17382)$$

which equals:

$$g=\frac{1}{s\sqrt{2\pi}}\operatorname{Exp}\Bigl[-\frac{(\hat{\alpha}-1)^2}{2s^2}\Bigr];\qquad \texttt{domain[g]}=\{\hat{\alpha},\ -\infty,\ \infty\};$$

$$1-(\texttt{Prob[1.17382, g]}-\texttt{Prob[0.82618, g]})$$

0.0000795469

As the *p*-value is very small, this is strong evidence against H_0. ∎

⊕ **Example 4:** Constructing a Confidence Interval

In this example, we construct an approximate confidence interval for the mean $\mu=\alpha\beta$ of the Gamma(α, β) distribution using an asymptotic distribution for the MLE of the mean.

From the previous example, we know $\sqrt{n}\,(\hat{\theta}-\theta_0)\xrightarrow{d}N(0,i_0^{-1})$, where $\hat{\theta}$ is the MLE of $\theta_0=(\alpha,\beta)$. As μ is a function of the elements of θ_0, we may apply the Invariance Property (see §11.4 E) to find

$$\sqrt{n}\,(\hat{\mu}-\mu)\xrightarrow{d}N\Bigl(0,\ \frac{\partial\mu}{\partial\theta_0^{\mathrm{T}}}\times i_0^{-1}\times\frac{\partial\mu}{\partial\theta_0}\Bigr).$$

mathStatica derives the variance of the limit distribution as:

$$f=\frac{x^{\alpha-1}\,e^{-x/\beta}}{\Gamma[\alpha]\,\beta^{\alpha}};\qquad \texttt{domain[f]}=\{x,\ 0,\ \infty\}\ \&\&\ \{\alpha>0,\ \beta>0\};$$

```
Grad[α β, {α, β}].Inverse[FisherInformation[{α, β}, f]].
    Grad[α β, {α, β}] // Simplify
```

$\alpha\beta^2$

Consequently, we may write $\sqrt{n}\,(\hat{\mu}-\mu)\overset{a}{\sim}N(0,\alpha\beta^2)$. Unfortunately, a confidence interval for μ cannot be constructed from this asymptotic distribution, due to the presence of the unknown parameters α and β. However, if we replace α and β with, respectively, the estimates $\hat{\alpha}$ and $\hat{\beta}$, we find[6]

$$\hat{\mu}\overset{a}{\sim}N\Bigl(\mu,\ \tfrac{1}{n}\hat{\alpha}\hat{\beta}^2\Bigr).$$

From this asymptotic distribution, an approximate $100\,(1-\omega)\,\%$ confidence interval for μ can be constructed; it is given by

$$\hat{\mu}\pm z_{1-\omega/2}\sqrt{\hat{\alpha}\hat{\beta}^2/n}$$

§12.4 A MAXIMUM LIKELIHOOD ESTIMATION IN PRACTICE

where $z_{1-\omega/2}$ is the inverse cdf of the $N(0, 1)$ distribution evaluated at $1 - \omega/2$.

For the Nerve data of *Example 2*, with ML estimates of 1.17382 for α, and 0.186206 for β, an approximate 95% confidence interval for μ is:[7]

```
α̂ = 1.17382;      β̂ = 0.186206;      μ̂ = α̂ β̂;

z = √2 InverseErf[0, -1 + 2 (1 - 0.05/2)];

{μ̂ - z √(α̂ β̂² / 799), μ̂ + z √(α̂ β̂² / 799)}

{0.204584, 0.232561}
```

12.4 B Standard Errors and *t*-statistics

When reporting estimation results, it is important to mention, at the very least, the estimates, the standard errors of the estimators, and the *t*-statistics (*e.g.* see Table 1). For ML estimation, such details can be obtained from an asymptotic distribution for the estimator. It is insufficient to present just the parameter estimates. This, for example, occurred for the ARCH model estimated in §12.3, where standard errors and *t*-statistics were not presented (see Table 2). This is because FindMinimum / FindMaximum only returns point estimates of the parameters, and the optimised value of the observed log-likelihood. To report standard errors and *t*-statistics, further programming must be done.

For regular ML estimators such that

$$\sqrt{n}\left(\hat{\theta} - \theta_0\right) \xrightarrow{d} N(0, i_0^{-1})$$

with an asymptotic distribution:

$$\hat{\theta} \stackrel{a}{\sim} N\left(\theta_0, (n\, i_0)^{-1}\right)$$

we require a consistent estimator of the matrix $(n\, i_0)^{-1}$ in order to operationalise asymptotic inference, and to report estimation results. Table 3 lists three such estimators.

Fisher	$\left(n\, i_{\hat{\theta}}\right)^{-1}$
Hessian	$\left(-\dfrac{\partial^2}{\partial\theta\, \partial\theta^{\mathrm{T}}} \log L(\hat{\theta})\right)^{-1}$
Outer-product	$\left(\sum_{i=1}^{n} \left(\dfrac{\partial}{\partial\theta} \log f(X_i; \hat{\theta})\right)\left(\dfrac{\partial}{\partial\theta} \log f(X_i; \hat{\theta})\right)^{\mathrm{T}}\right)^{-1}$

Table 3: Three asymptotically equivalent estimators of $(n\, i_0)^{-1}$

Each estimator relies on the consistency of the MLE $\hat{\theta}$ for θ_0. All three are asymptotically equivalent in the sense that n times each estimator converges in probability to i_0^{-1}. The first estimator, labelled 'Fisher', was used in *Example 4*. The second, 'Hessian', is based on regularity condition 5a(ii) (see §11.5 A). This estimator is quite popular in practice, having the advantage over the Fisher estimator that it does *not* require solving an expectation. The 'Outer-product' estimator is based on the definition of Fisher Information (see §10.2 D, and condition 4a in §11.5 A). While it would appear more complicated than the others, it can come in handy if computation of the Hessian estimator becomes costly, for it requires only one round of differentiation.

If the MLE in a *non-identically distributed* sample (see §11.5 B) is such that,

$$\sqrt{n}\,(\hat{\theta} - \theta_0) \xrightarrow{d} N\!\left(0,\, (i_0^{(\infty)})^{-1}\right)$$

then to operationalise asymptotic inference, the Hessian and Outer-product estimators given in Table 3 may be used to estimate $(n\, i_0^{(\infty)})^{-1}$; however, the Fisher estimator is now $I_{\hat{\theta}}^{-1}$, where I_θ denotes the Sample Information on θ (see §10.2 E).

⊕ **Example 5:** Income and Education: An Exponential Regression Model

In *Example 15* of Chapter 11, we considered the simple Exponential regression model:

$$Y \,|\, (X = x) \sim \text{Exponential}(\exp(\alpha + \beta x)) \tag{12.5}$$

where regressor $X = x \in \mathbb{R}$, and parameter $\theta = (\alpha, \beta) \in \mathbb{R}^2$. Here is the pdf $f(y\,|\,X = x; \theta)$:

```
f  =  ─────────────── Exp[ - ─────────────── ];
      Exp[α + β x]            Exp[α + β x]

domain[f] = {y, 0, ∞} && {α ∈ Reals, β ∈ Reals, x ∈ Reals};
```

Greene (2000, Table A4.1) gives hypothetical data on the pair (Y_i, X_i) for $n = 20$ individuals, where Y denotes Income ($000s per annum) and X denotes years of Education. Here is the Income data:

```
Income = {20.5, 31.5, 47.7, 26.2, 44.0, 8.28,
          30.8, 17.2, 19.9, 9.96, 55.8, 25.2, 29.0,
          85.5, 15.1, 28.5, 21.4, 17.7, 6.42, 84.9};
```

… and here is the Education data:

```
Education = {12, 16, 18, 16, 12, 12, 16, 12,
             10, 12, 16, 20, 12, 16, 10, 18, 16, 20, 12, 16};
```

Figure 6 illustrates the data in the form of a scatter diagram.

§12.4 B   MAXIMUM LIKELIHOOD ESTIMATION IN PRACTICE

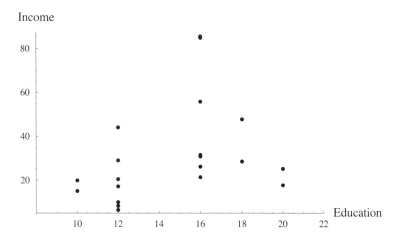

Fig. 6: The Income–Education data

Using ML methods, we fit the Exponential regression model (12.5) to this data. We begin by entering the observed log-likelihood:

```
obslogLθ = Log[∏_{i=1}^{n} (f /. {y → y_i, x → x_i})] /.
          {n → 20, y_i_ ⧴ Income[[i]], x_i_ ⧴ Education[[i]]}
```

$-42.9\, e^{-\alpha-20\beta} - 76.2\, e^{-\alpha-18\beta} - 336.1\, e^{-\alpha-16\beta} - 135.36\, e^{-\alpha-12\beta} - 35.\, e^{-\alpha-10\beta} - 20\alpha - 292\beta$

We obtain the ML estimates using `FindMaximum`'s Newton–Raphson algorithm:

```
solθ = FindMaximum[obslogLθ, {α, 0.1}, {β, 0.2},
                   Method → Newton]
```

$\{-88.1034, \{\alpha \to 1.88734, \beta \to 0.103961\}\}$

Thus, the observed log-likelihood is maximised at a value of -88.1034, with ML estimates of α and β reported as 1.88734 and 0.103961, respectively.

Next, we compute the Fisher, Hessian and Outer-product estimators given in Table 3. The Fisher estimator corresponds to the inverse of the (2×2) Sample Information matrix derived in *Example 15* of Chapter 11. It is given by:

```
Fisher = Inverse[
   ( n              ∑_{i=1}^{n} x_i  )
   ( ∑_{i=1}^{n} x_i   ∑_{i=1}^{n} x_i^2 )  /. {n → 20, x_i_ ⧴ Education[[i]]}] // N
```

$\begin{pmatrix} 1.20346 & -0.0790043 \\ -0.0790043 & 0.00541126 \end{pmatrix}$

The Hessian estimator is easily computed using **mathStatica**'s Hessian function:

$$\text{hessian} = \text{Inverse}[-\text{Hessian}[\text{obslogL}\theta, \{\alpha, \beta\}] \,/.\, \text{sol}\theta[\![2]\!]]$$

$$\begin{pmatrix} 1.54467 & -0.102375 \\ -0.102375 & 0.00701196 \end{pmatrix}$$

Calculating the Outer-product estimator is more involved, so we evaluate it in four stages. First, we calculate the score $\frac{\partial}{\partial \theta} \log f(x_i; \theta)$ using **mathStatica**'s Grad function:

$$\text{grad} = \text{Grad}[\text{Log}[f \,/.\, \{y \to y_i, x \to x_i\}], \{\alpha, \beta\}]$$

$$\{-1 + e^{-\alpha - \beta x_i} y_i, \; x_i (-1 + e^{-\alpha - \beta x_i} y_i)\}$$

Next, we form the outer product of this vector with itself—the distinctive operation that gives the estimator its name:

$$\text{op} = \text{Outer}[\text{Times}, \text{grad}, \text{grad}]$$

$$\begin{pmatrix} (-1 + e^{-\alpha - \beta x_i} y_i)^2 & x_i (-1 + e^{-\alpha - \beta x_i} y_i)^2 \\ x_i (-1 + e^{-\alpha - \beta x_i} y_i)^2 & x_i^2 (-1 + e^{-\alpha - \beta x_i} y_i)^2 \end{pmatrix}$$

We then Map the sample summation operator across each element of this matrix (this is achieved by using the level specification {2}):

$$\text{opS} = \text{Map}\left[\sum_{i=1}^{n} \# \,\&, \text{op}, \{2\}\right]$$

$$\begin{pmatrix} \sum_{i=1}^{n} (-1 + e^{-\alpha - \beta x_i} y_i)^2 & \sum_{i=1}^{n} x_i (-1 + e^{-\alpha - \beta x_i} y_i)^2 \\ \sum_{i=1}^{n} x_i (-1 + e^{-\alpha - \beta x_i} y_i)^2 & \sum_{i=1}^{n} x_i^2 (-1 + e^{-\alpha - \beta x_i} y_i)^2 \end{pmatrix}$$

Finally, we substitute the ML estimates and the data into opS, and then invert the resulting matrix:

$$\text{outer} = \text{Inverse}[\text{opS} \,/.\, \text{Flatten}[\{\text{sol}\theta[\![2]\!], \\ n \to 20, y_{i\_} :\to \text{Income}[\![i]\!], x_{i\_} :\to \text{Education}[\![i]\!]\}]]$$

$$\begin{pmatrix} 5.34805 & -0.342767 \\ -0.342767 & 0.0225022 \end{pmatrix}$$

In this particular case, the three estimators yield different estimates of the asymptotic variance-covariance matrix (this generally occurs). The estimation results for the trio of estimators are given in Table 4.

	Fisher		Hessian		Outer	
Estimate	SE	TStat	SE	TStat	SE	TStat
α 1.88734	1.09702	1.72042	1.24285	1.51856	2.31259	0.81612
β 0.10396	0.07356	1.41326	0.08374	1.24151	0.15001	0.69304

Table 4: Estimation results for the Income–Education data

The ML estimates appear in the first column (Estimate). The associated estimated asymptotic standard errors appear in the SE columns (these correspond to the square root of the elements of the leading diagonal of the estimated asymptotic variance-covariance matrices). The t-statistics are in the TStat columns (these correspond to the estimates divided by the estimated asymptotic standard errors; these are statistics for the tests of $H_0 : \alpha = 0$ and $H_0 : \beta = 0$). These results suggest that Education is not a significant explanator of Income, assuming model (12.5). ∎

12.5 Optimisation Algorithms

12.5 A Preliminaries

The numerical optimisation of a function (along with the related task of solving for the roots of an equation) is a problem that has attracted considerable interest in many areas of science and technology. Mathematical statistics is no exception, for as we have seen, optimisation is fundamental to estimation, whether it be for ML estimation or for other estimation methods such as the method of moments or the method of least squares. Optimisation algorithms abound, as even a cursory glance through Polak's (1971) classic reference work will reveal. Some of these have been coded into FindMinimum, but for each one that has been implemented in that function, there are dozens of others omitted. Of course, the fact that there exist so many different types of algorithms is testament to the fact that every problem is unique, and its solution cannot necessarily be found by applying one algorithm. The various attempts at estimating the ARCH model in §12.3 provide a good illustration of this.

We want to solve two estimation problems. The first is to maximise a real, single-valued observed log-likelihood function with respect to the parameter θ. The point estimate of θ_0 is to be returned, where θ_0 denotes (as always) the true parameter value. The second is to estimate the asymptotic standard errors of the parameter estimator and the asymptotic t-statistics. This can be achieved by returning, for example, the Hessian evaluated at the point estimate of θ_0 (*i.e.* the Hessian estimator given in Table 3 in §12.4). It is fair to say that obtaining ML estimates is the more important task; however, the two taken together permit inference using the asymptotic distribution.

The algorithms that we discuss in this section address the dual needs of the estimation problem; in particular, we illustrate the *Newton–Raphson* (NR) and the *Broydon–Fletcher–Goldfarb–Shanno* (BFGS) algorithms. The NR and BFGS algorithms are options in FindMinimum using Method→Newton and Method→QuasiNewton, respectively. However, in its Version 4 incarnation, FindMinimum returns only a point estimate of θ_0; other important pieces of information such as the final Hessian matrix are not recoverable from its output. This is clearly a weakness of FindMinimum which will hopefully be rectified in a later version of *Mathematica*.

Both the NR and BFGS algorithms, and every other algorithm implemented in FindMinimum, come under the broad category of *gradient methods*. Gradient methods form the backbone of the literature on optimisation and include a multitude of approaches including quadratic hill-climbing, conjugate gradients, and quasi-Newton methods. Put simply, gradient methods work by estimating the gradient and Hessian of the observed log-likelihood at a given point, and then jumping to a superior solution which estimates the optimum. This process is then repeated until convergence. Amongst the extensive literature on optimisation using gradient methods, we refer in particular to Polak (1971), Luenberger (1984), Gill *et al.* (1981) and Press *et al.* (1992). A discussion of gradient methods applied to optimisation in a statistical context appears in Judge *et al.* (1985).

Alternatives to optimisation based on gradient methods include direct search methods, simulated annealing methods, taboo search methods, and genetic algorithms. The first — direct search — involves the adoption of a search pattern through parameter space comparing values of the observed log-likelihood at each step. Because it ignores information (such as gradient and Hessian), direct search methods are generally regarded as inferior. However, the others — simulated annealing, taboo search, and genetic algorithms — fare better and have much to recommend them. Motivation for alternative methods comes primarily from the fact that a gradient method algorithm is unable to escape from regions in parameter space corresponding to local optima, for once at a local optimum a gradient algorithm will not widen its search to find the global optimum — this is termed the problem of multiple local optima.[8]

The method of simulated annealing (Kirkpatrick *et al.* (1983)) attempts to overcome this by allowing the algorithm to move to worse locations in parameter space, thereby skirting across local optima; the method performs a slow but thorough search. An attempt to improve upon the convergence speed of the annealing algorithm is Ingber's (1996) simulated quenching algorithm. Yet another approach is the taboo method (Glover *et al.* (1993)) which is a strategy that forces an algorithm (typically a gradient method) to move through regions of parameter space that have not previously been visited. Genetic algorithms (Davis (1991)) offer an entirely different approach again. Based on the evolutionary notion of natural selection, combinations of the best intermediate solutions are paired together repeatedly until a single dominant optimum emerges.

When applying a gradient method to an observed log-likelihood which has, or may have, multiple local optima, it is advisable to initiate the algorithm from different locations in parameter space. This approach is adequate for the examples we present here, but it can become untenable in higher-dimensional parameter spaces.

As outlined, the estimation problem has two components: estimating parameters and estimating the associated standard errors of the estimator. Fortunately, by focusing on the solution for the first component, we will, as a by-product, achieve the solution for the second. We begin by defining the *penalty function*, which is the negative of the observed log-likelihood function:

$$p(\theta) = -\log L(\theta). \qquad (12.6)$$

Minimising the penalty function for choices of θ yields the equivalent result to maximum likelihood. The reason for defining the penalty function is purely because the optimisation literature is couched in terms of minimisation, rather than maximisation. Finally, we assume the parameter θ, a $(k \times 1)$ vector, is of dimension $k \geq 2$ and such that $\theta \in \Theta = \mathbb{R}^k$. Accordingly, optimisation corresponds to unconstrained minimisation over choice variables defined everywhere in two- or higher-dimensional real space.

Before proceeding further, we make brief points about the $k = 1$ case. The case of numerical optimisation over the real line (*i.e.* corresponding to just one parameter, since $k = 1$) is of lesser importance in practice. If univariate optimisation is needed, line search algorithms such as Golden Search and methods due to Brent (1973) should be applied; see Luenberger (1984) for discussion of these and other possibilities. *Mathematica*'s FindMinimum function utilises versions of these algorithms, and our experience of its performance has, on the whole, been good. Determining in advance whether the derivative of the penalty function can be constructed (equivalent to the negative of the score) will usually cut down the number of iterations, and can save time. If so, a single starting point need only be supplied (*i.e.* it is unnecessary to compute the gradient and supply it to the function through the Gradient option). Univariate optimisation can, however, play an important role in multivariate optimisation by determining step-length optimally.

Finally, as we have seen, parametric constraints often arise in statistical models. In these cases, the parameter space $\Theta = \{\theta : \theta \in \Theta\}$ is a proper subset of k-dimensional real space or may be degenerate upon it (*i.e.* $\Theta \subset \mathbb{R}^k$). This means that maximum likelihood/minimum penalty estimation requires constrained optimisation methods. Our opinion on FindMinimum as a *constrained* optimiser — in its Version 4 incarnation — is that we cannot recommend its use. The approach that we advocate is to transform a constrained optimisation into an unconstrained optimisation, and use FindMinimum on the latter. This can be achieved by re-defining parameter θ to a new parameter $\lambda = g(\theta)$ in a manner such that $\lambda \in \mathbb{R}^q$, where $q \leq k$. Of course, the trick is to determine the appropriate functional form for the transformation g. Once we have determined g and optimised with respect to λ, recovery of estimation results (via the Invariance Property) pertinent to θ can be achieved by using replacement rules, as well as by exploiting *Mathematica*'s excellent differentiator.

12.5 B Gradient Method Algorithms

An algorithm (gradient method or otherwise) generates a finite-length sequence such as the following one:

$$\hat{\theta}_{(0)}, \hat{\theta}_{(1)}, \ldots, \hat{\theta}_{(r)} \tag{12.7}$$

where the bracketed subscript indicates the iteration number. Each $\hat{\theta}_{(j)} \in \mathbb{R}^k$, $j = 0, \ldots, r$, resides in the same space as the θ, and each can be regarded as an estimate of $\hat{\theta}$:

$$\hat{\theta} = \arg\max_{\theta \in \mathbb{R}^k} \log L(\theta) = \arg\min_{\theta \in \mathbb{R}^k} p(\theta).$$

The sequence (12.7) generally depends on three factors: (i) the point at which the algorithm starts $\hat{\theta}_{(0)}$, (ii) how the algorithm progresses through the sequence; that is, how

$\hat{\theta}_{(j+1)}$ is obtained from $\hat{\theta}_{(j)}$, and (iii) when the process stops. Of the three factors, our attention focuses mainly on the second — the iteration method.

○ *Initialisation*

Starting values are important in all types of optimisation methods — more so, perhaps, for gradient method algorithms because of the multiple local optima problem. One remedy is to start from different locations in the parameter space in order to trace out the surface of the observed log-likelihood, but this may not appeal to the purist. Alternative methods have already been discussed in §12.5 A, with simulated annealing methods probably worthy of first consideration.

○ *The Iteration Method*

Typically, the link between iterations takes the following form,

$$\hat{\theta}_{(j+1)} = \hat{\theta}_{(j)} + \mu_{(j)} d_{(j)} \tag{12.8}$$

where the step-length $\mu_{(j)} \in \mathbb{R}_+$ is a scalar, and the direction $d_{(j)} \in \mathbb{R}^k$ is a vector lying in the parameter space. In words, we update our estimate obtained at iteration j, namely $\hat{\theta}_{(j)}$, by moving in the direction $d_{(j)}$ by a step of length $\mu_{(j)}$.

The fundamental feature of algorithms coming under the umbrella of gradient methods is that they are *never worsening*. That is,

$$p(\hat{\theta}_{(0)}) \geq p(\hat{\theta}_{(1)}) \geq \cdots \geq p(\hat{\theta}_{(r-1)}) \geq p(\hat{\theta}_{(r)}).$$

Thus, each member in the sequence (12.7) traces out an increasingly better approximation for minimising the penalty function. Using these inequalities and the relationship (12.8), for any $j = 0, \ldots, r$, we must have

$$p(\hat{\theta}_{(j)} + \mu_{(j)} d_{(j)}) - p(\hat{\theta}_{(j)}) \leq 0. \tag{12.9}$$

The structure of a gradient method algorithm is determined by approximating the left-hand side of (12.9) by truncating its Taylor series expansion. To see this, replace $\mu_{(j)}$ in (12.9) with μ, and take a (truncated) Taylor series expansion of the first term about $\mu = 0$, to yield

$$\mu \, p_g(\hat{\theta}_{(j)}) \cdot d_{(j)}$$

where the $(k \times 1)$ vector p_g denotes the gradient of the penalty function. Of course, p_g is equivalent to the negative of the score, and like the score, it too disappears at $\hat{\theta}$; that is, $p_g(\hat{\theta}) = \vec{0}$. Replacing the left-hand side of (12.9) with the Taylor approximation, finds

$$p_g(\hat{\theta}_{(j)}) \cdot d_{(j)} \leq 0 \tag{12.10}$$

for μ is a positive scalar. Expression (12.10) enables us to construct a range of differing possibilities for the direction vector. For example, for a symmetric matrix $W_{(j)}$, we might select direction according to

$$d_{(j)} = -W_{(j)} \cdot p_g(\hat{\theta}_{(j)}) \tag{12.11}$$

because then the left-hand side of (12.10) is a weighted quadratic form in the elements of vector p_g, the weights being the elements of matrix $W_{(j)}$; that is,

$$p_g(\hat{\theta}_{(j)}) \cdot d_{(j)} = -p_g(\hat{\theta}_{(j)}) \cdot W_{(j)} \cdot p_g(\hat{\theta}_{(j)}). \tag{12.12}$$

This quadratic form will be non-positive provided that the matrix of weights $W_{(j)}$ is positive semi-definite (in practice, $W_{(j)}$ is taken positive definite to ensure strict improvement). Thus, the algorithm improves from one iteration to the next until a point $p_g(\hat{\theta}_{(r)}) = \vec{0}$ is reached within numerical tolerance.

Selecting different weight matrices defines various iterating procedures. In particular, four choices are NR = Newton–Raphson, Score = Method of Scoring, DFP = Davidon–Fletcher–Powell and BFGS = Broyden–Fletcher–Goldfarb–Shanno:

$$\text{NR:} \quad W_{(j)} = -H_{(j)}^{-1} \tag{12.13}$$

$$\text{Score:} \quad W_{(j)} = I_{(j)}^{-1} \tag{12.14}$$

$$\text{DFP:} \quad W_{(j+1)} = W_{(j)} + \frac{\Delta\hat{\theta} \times (\Delta\hat{\theta})^T}{\Delta\hat{\theta} \cdot \Delta p_g} - \frac{(W_{(j)} \cdot \Delta p_g) \times (W_{(j)} \cdot \Delta p_g)^T}{\Delta p_g \cdot W_{(j)} \cdot \Delta p_g} \tag{12.15}$$

$$\text{BFGS:} \quad W_{(j+1)} = (\text{as for DFP}) + (\Delta p_g \cdot W_{(j)} \cdot \Delta p_g)$$
$$\times \left(\frac{\Delta\hat{\theta}}{\Delta\hat{\theta} \cdot \Delta p_g} - \frac{W_{(j)} \cdot \Delta p_g}{\Delta p_g \cdot W_{(j)} \cdot \Delta p_g} \right) \times \left(\frac{\Delta\hat{\theta}}{\Delta\hat{\theta} \cdot \Delta p_g} - \frac{W_{(j)} \cdot \Delta p_g}{\Delta p_g \cdot W_{(j)} \cdot \Delta p_g} \right)^T \tag{12.16}$$

The notation used here is the following:

$H_{(j)}$ is the Hessian of the observed log-likelihood function evaluated at $\theta = \hat{\theta}_{(j)}$

$I_{(j)}$ is the Sample Information matrix evaluated at $\theta = \hat{\theta}_{(j)}$

$\Delta\hat{\theta} = \hat{\theta}_{(j)} - \hat{\theta}_{(j-1)}$ is the change in the estimate from the previous iteration, and

$\Delta p_g = p_g(\hat{\theta}_{(j)}) - p_g(\hat{\theta}_{(j-1)})$ is the change in the gradient.

The DFP and BFGS weighting matrices appear complicated, but as we shall see in the following section, implementing them with *Mathematica* is reasonably straightforward. Of the algorithms (12.13)–(12.16), FindMinimum includes the NR algorithm (Method → Newton) and the BFGS algorithm (Method → QuasiNewton).

To illustrate, we shall obtain the iterator for the Method of Scoring. Combine (12.8), (12.11) and (12.14), to yield

$$\hat{\theta}_{(j+1)} = \hat{\theta}_{(j)} - \mu_{(j)} I_{(j)}^{-1} \cdot p_g(\hat{\theta}_{(j)})$$

which, to be complete, requires us to supply a step-length $\mu_{(j)}$. We might, for instance, select step-length to optimally improve the penalty function when moving in direction $d_{(j)} = -I_{(j)}^{-1} \cdot p_g(\hat{\theta}_{(j)})$ from $\hat{\theta}_{(j)}$; this is achieved by solving

$$\mu_{(j)} = \arg\min_{\mu \in \mathbb{R}_+} p(\hat{\theta}_{(j)} - \mu I_{(j)}^{-1} \cdot p_g(\hat{\theta}_{(j)})).$$

Of course, this is a univariate optimisation problem that can be solved by numerical means using `FindMinimum`. Unfortunately, experience suggests that determining step-length in this manner can be computationally inefficient, and so a number of alternatives have been proposed. In particular, one due to Armijo is implemented in the examples given below.

A final point worth noting concerns estimating the asymptotic standard error of the estimator. As mentioned previously, this estimate is obtained as a by-product of the optimisation. This is because an estimate of the asymptotic variance-covariance matrix is given by the final weighting matrix $W_{(r)}$, since the estimates of the asymptotic standard error are the square root of the main diagonal of this matrix. The NR weight (12.13) corresponds to the Hessian estimator, and the Score weight (12.14) to the Fisher estimator (see Table 3); the DFP and BFGS weights are other (consistent) estimators. However, the default algorithm implemented in `FindMinimum` (the conjugate gradient algorithm) does not yield, as a by-product, the estimate of the asymptotic variance-covariance matrix.

○ *Stopping Rules*

Algorithms converge (asymptotically) to $\hat{\theta}$; nevertheless, from a practical view, the sequence (12.7) must be terminated in finite time, and the estimate $\hat{\theta}_{(r)}$ of $\hat{\theta}$ must be reported. This therefore requires that we define numerical convergence. How this is done may vary. Possibilities include the following:

(i) convergence defined according to epsilon change in parameter estimates:

$$\text{stop if } \|\hat{\theta}_{(r)} - \hat{\theta}_{(r-1)}\| < \epsilon_1$$

(ii) convergence defined according to epsilon change in the penalty function:

$$\text{stop if } |p(\hat{\theta}_{(r)}) - p(\hat{\theta}_{(r-1)})| < \epsilon_2$$

(iii) convergence defined according to the gradient being close to zero:

$$\text{stop if } \|p_g(\hat{\theta}_{(r)})\| < \epsilon_3$$

(iv) convergence defined according to the gradient element with the largest absolute value being close to zero:

$$\text{stop if } \max(|p_g(\hat{\theta}_{(r)})|) < \epsilon_4$$

where ϵ_1, ϵ_2, ϵ_3 and ϵ_4 are small positive numbers, $|\cdot|$ denotes the absolute value of the argument, and $\|\cdot\|$ denotes the Euclidean distance of the argument vector from the origin (the square root of the dot product of the argument vector with itself). The method we favour is (iv).

Of course, picking just one rule out of this list may be inappropriate as a stopping rule, in which case numerical convergence can be defined according to combinations of (i), (ii), (iii) or (iv) holding simultaneously. Finally, (i)–(iv) hold if $\hat{\theta}_{(r)}$ happens to locate either a local maximum or a saddle point of the penalty function, so it is usually necessary to check that the Hessian of the penalty function (equal to the negative of the Hessian of the observed log-likelihood) is positive definite at $\hat{\theta}_{(r)}$.

12.6 The BFGS Algorithm

In this section, we employ the Broyden–Fletcher–Goldfarb–Shanno (BFGS) algorithm to estimate a Poisson two-component-mix model proposed by Hasselblad (1969).

○ *Data, Statistical Model and Log-likelihood*

Our data — the Death Notice data — appears in Table 5. The data records the number of death notices for women aged 80 or over, each day, in the English newspaper, *The Times*, during the three-year period, 1910–1912.

Death Notices per day (X) :	0	1	2	3	4	5	6	7	8	9
Frequency (no. of days) :	162	267	271	185	111	61	27	8	3	1

Table 5: Death Notice data

The data is interpreted as follows: there were 162 days in which no death notices appeared, 267 days in which one notice appeared, ... and finally, just 1 day on which the newspaper listed nine death notices. We enter the data as follows:

count = {162, 267, 271, 185, 111, 61, 27, 8, 3, 1};

As the true distribution of X = 'the number of death notices published daily' is unknown, we shall begin by specifying the Poisson(γ) model for X,

$$P(X = x) = \frac{e^{-\gamma} \gamma^x}{x!}, \quad x \in \{0, 1, 2, ...\} \text{ and } \gamma \in \mathbb{R}_+$$

Then, the log-likelihood is:

Clear[G]; $\text{logL}\gamma = \text{Log}\left[\prod_{x=0}^{G}\left(\frac{e^{-\gamma}\gamma^x}{x!}\right)^{n_x}\right]$

$-\gamma \sum_{x=0}^{G} n_x + \text{Log}[\gamma] \sum_{x=0}^{G} x\, n_x - \sum_{x=0}^{G} \text{Log}[x!]\, n_x$

where, in order for `SuperLog` to perform its magic, we have introduced the Subscript n_x to index, element by element, the data in list `count` (so $n_0 = 162$, $n_1 = 267, ..., n_9 = 1$).[9] Define $G < \infty$ to be the largest number of death notices observed in the sample, so $G = 9$ for our data.[10] ML estimation in this model is straightforward because the log-likelihood is concave with respect to γ.[11] This ensures that the ML estimator is given by the solution to the first-order condition:

```
solγ = Solve[Grad[logLγ, γ] == 0, γ] // Flatten
```

$$\left\{ \gamma \to \frac{\sum_{x=0}^{G} x\, n_x}{\sum_{x=0}^{G} n_x} \right\}$$

For our data, the ML estimate is obtained by inputting the data into the ML estimator, `solγ`, using a replacement rule:

```
solγ = solγ /. {G → 9, nx_ :→ count[[x + 1]]} // N
```

$\{\gamma \to 2.15693\}$

We leave estimation of the standard error of the estimator as an exercise for the reader.[12]

When Hasselblad (1969) examined the Death Notice data, he suggested that the sampled population was in fact made up of two sub-populations distinguished according to season, since death rates in winter and summer months might differ. As the data does not discriminate between seasons, Hasselblad proceeded by specifying an unknown mixing parameter between the two sub-populations. We denote this parameter by ω (for details on component-mix models, see §3.4 A). He also specified Poisson distributions for the sub-populations. We denote their parameters by ϕ and ψ. Hasselblad's Poisson two-component mix model is

$$P(X = x) = \omega \frac{e^{-\phi} \phi^x}{x!} + (1 - \omega) \frac{e^{-\psi} \psi^x}{x!}, \quad x \in \{0, 1, 2, ...\}$$

where the mixing parameter ω is such that $0 < \omega < 1$, and the Poisson parameters satisfy $\phi > 0$ and $\psi > 0$. For Hasselblad's model, the observed log-likelihood can be entered as:

```
obsLogLλ - Log[∏_{x=0}^{G} (ω (e^-φ φ^x)/x! + (1 - ω) (e^-ψ ψ^x)/x!)^{nx}] /.
{G → 9, nx_ :→ count[[x + 1]], ω → 1/(1 + e^a), φ → e^b, ψ → e^c};
```

Note that we have implemented a re-parameterisation of $\theta = (\omega, \phi, \psi)$ to $\lambda = g(\theta) = (a, b, c) \in \mathbb{R}^3$ by using a replacement rule (see the second line of input).

Due to the non-linear nature of the first-order conditions, ML estimation of the unknown parameters requires iterative methods for which we choose the BFGS algorithm.[13] Using `FindMaximum`, initialised at $(a, b, c) = (0.0, 0.1, 0.2)$, finds:[14]

§12.6 MAXIMUM LIKELIHOOD ESTIMATION IN PRACTICE

```
FindMaximum[obslogLλ, {a, 0.0}, {b, 0.1}, {c, 0.2},
    Method → QuasiNewton]
```

{-1989.95, {a → 0.575902, b → 0.227997, c → 0.9796}}

Using the estimates 0.575902, 0.227997 and 0.9796, for a, b and c, respectively, we can obtain the ML estimates for ω, ϕ and ψ. Alas, with `FindMinimum`/`FindMaximum`, it is not possible to inspect the results from each iteration in the optimisation procedure; nor, more importantly, can we recover estimates of the asymptotic variance-covariance matrix of the ML estimator of λ. Without the asymptotic variance-covariance matrix, we cannot, for example, undertake the inference described in §12.4. Thus, `FindMinimum`/`FindMaximum` does not do all that we might hope for.

○ *BFGS Algorithm*

We now code the BFGS algorithm, and then apply it to estimate the parameters of Hasselblad's model. We begin by converting the re-parameterised observed log-likelihood into a penalty function:

```
p = -obslogLλ;
```

Our task requires the unconstrained minimisation of the penalty p with respect to λ. Our BFGS code requires that we define the penalty function pf and its gradient gradpf as *Mathematica* functions of the parameters, using an immediate evaluation:

```
pf[{a_, b_, c_}] = p;
gradpf[{a_, b_, c_}] = Grad[p, {a, b, c}];
```

To see that this has worked, evaluate the gradient of the penalty function at $a = b = c = 0$:

```
g = gradpf[{0, 0, 0}]
```

{0, -634, -634}

We now present some simple code for each part of the BFGS algorithm (12.16). The following module returns the updated approximation $W_{(j)}$ to the negative of the inverse Hessian matrix at each iteration:

```
BFGS[Δθ_, Δgrad_, W_] :=
  Module[{t1, t2, t3, t4, t5, t6, t7},
          t1 = Outer[Times, Δθ, Δθ];
          t2 = Δθ.Δgrad;
          t3 = W.Δgrad;
          t4 = Outer[Times, t3, t3];
          t5 = Δgrad.t3;
    (* For DFP ignore the remaining lines
            and return W+t1/t2-t4/t5 *)
          t6 = Δθ / t2 - t3 / t5;
          t7 = Outer[Times, t6, t6];
          W + t1 / t2 - t4 / t5 + t5 t7]
```

The BFGS updating expression can, of course, be coded as a one-line command. However, this would be inefficient as a number of terms are repeated; hence, the terms t1 to t7 in BFGS.

The next component that is needed is a line search method for determining step-length μ. There happen to be quite a few to choose from. For simplicity, we select a relatively easy version of Armijo's method as given in Polak (1971) (for a more detailed version, see Luenberger (1984)):

```
Armijo[f_, θ_, grad_, dir_] :=
  Module[{α = 0.5, β = 0.65, μ = 1., f0, gd},
    f0 = f[θ];   gd = grad.dir;
    While[ f[θ + μ dir] - f0 - μ α gd > 0, μ = β μ];  μ]
```

This module essentially determines a feasible step-length μ, but not necessarily an optimal one. The first argument, f, denotes the objective function (our penalty function). Because Armijo needs to evaluate f at many points, the Armijo function assumes that f is a *Mathematica* function like pf (not p). A more advanced method is Goldstein's (again, see Polak (1971) or Luenberger (1984)), where bounds are determined within which an optimising search can be performed using, for example, FindMinimum (but remember to transform to an unconstrained optimisation). The cost in undertaking this method is the additional time it takes to determine an optimal step-length.

To set BFGS on its way, there are two initialisation choices required—$\hat{\lambda}_{(0)}$ and $W_{(0)}$—which are the beginning parameter vector 'guess' and the beginning inverse Hessian matrix 'guess', respectively. The success of our search can depend crucially on these two factors. To illustrate, suppose we set $\hat{\lambda}_{(0)} = (0, 0, 0)$ and $W_{(0)} = I_3$. From (12.8), (12.11), and our earlier output, it follows that $\hat{\lambda}_{(1)} = \mu_{(0)} \times (0, 634, 634)$. Determining step-length, we find:

```
Armijo[pf, {0, 0, 0}, g, {0, 634, 634}]
```

- General::unfl : Underflow occurred in computation.
- General::unfl : Underflow occurred in computation.
- General::unfl : Underflow occurred in computation.
- General::stop : Further output of
    General::unfl will be suppressed during this calculation.

1.

... and the algorithm has immediately run into troubles. The cause of these difficulties is scaling (quantities such as Exp[-634] are involved in numeric computations). Fortunately, a heuristic that can help to overcome this type of ill-conditioning is to enforce scale dependence onto $W_{(0)}$. A simple one that can often work is:

```
W0[θ_, grad_] := √(θ.θ / grad.grad)  IdentityMatrix[Length[θ]]
```

W0 ensures that the Euclidean length of the initial direction vector from the origin matches that of the initial starting parameter; that is, $W_{(0)}$ is forced to be such that direction $d_{(0)} = -W_{(0)} \cdot g(\hat{\lambda}_{(0)})$ satisfies

$$\sqrt{d_{(0)} \cdot d_{(0)}} = \sqrt{\hat{\lambda}_{(0)} \cdot \hat{\lambda}_{(0)}} \ .$$

Of course, forcing $W_{(0)}$ to behave in this way always rules out selecting $\hat{\lambda}_{(0)}$ as a zero vector as the initial parameter guess. For further details on other generally better methods of scaling and pre-conditioning, see Luenberger (1984).

We now implement the BFGS algorithm using the parts constructed above. As the starting point, we shall select:

```
λ0 = {0.0, 0.1, 0.2};
```

The code here closely follows Polak's (1971) algorithm structure (given for DFP, but equally applicable to BFGS). If convergence to tolerance is achieved, the Do loop outputs the list 'results' which contains: (i) the number of iterations performed, (ii) the value of the objective function at the optimum, (iii) the optimal parameter values, and (iv) the final weight matrix W. If no output is produced, then convergence to tolerance has not been achieved within 30 iterations. Irrespective of whether convergence has been achieved or not, the final values of the parameters and the weight matrix are stored in memory and can be inspected. Finally, the coding that is given here is very much in 'bare bones' form; embellishments that the user might like (such as the output from each iteration) can be added as desired.

```
(* Start iteration (iter=0) *)
λ0 = {0.0, 0.1, 0.2};
g0 = gradpf[λ0];
W = W0[λ0, g0];

Do[ (* Subsequent iterations (maximum 30) *)
        d = -W.g0;
        λ1 = λ0 + Armijo[pf, λ0, g0, d] d;
        g1 = gradpf[λ1];
   If[ Max[Abs[g1]] < 10^-6,
       W = BFGS[λ1 - λ0, g1 - g0, W];
       Break[results = {iter, -pf[λ1], λ1, W}] ];
        Δλ = λ1 - λ0;
        Δg = g1 - g0;
  (* Reset λ0 and g0 for the next iteration *)
        λ0 = λ1; g0 = g1;
        W = BFGS[Δλ, Δg, W],     {iter, 30}]
```

$$\left\{ 26, -1989.95, \{0.575862, 0.228008, 0.979605\}, \begin{pmatrix} 0.775792 & -0.245487 & -0.0810482 \\ -0.245487 & 0.084435 & 0.0246111 \\ -0.0810482 & 0.0246111 & 0.0093631 \end{pmatrix} \right\}$$

The output states that the BFGS algorithm converged to tolerance after 26 iterations. The ML estimates are $\hat{a} = 0.575862$, $\hat{b} = 0.228008$ and $\hat{c} = 0.979605$; almost equivalent to the point estimates returned by `FindMaximum`. At the estimates, the observed log-likelihood is maximised at a value of -1989.95. The BFGS estimate of the asymptotic variance-covariance matrix is the (3×3) matrix in the output. Table 6 summarises the results.

	Estimate	SE	TStat
a	0.575862	0.880791	0.653801
b	0.228008	0.290577	0.784673
c	0.979605	0.0967631	10.1237

Table 6: ML estimation results for the unrestricted parameters

Our stopping rule focuses on the gradient, stopping if the element with the largest magnitude is smaller than 10^{-6}. Our choice of 10^{-6} corresponds to the default for `AccuracyGoal` in `FindMinimum`. It would not pay to go much smaller than this, and may even be wise to increase it with larger numbers of parameters.[15] Other stopping rules can be tried.[16] Finally, the outputted W is an estimate of the asymptotic variance-covariance matrix.

To finish, we present a summary of the ML estimates and their associated standard errors and t-statistics for the parameters of the original Poisson two-component-mix model. To do this, we use the Invariance Property, since the unrestricted parameters λ are linked to the restricted parameters θ by the re-parameterisation $\lambda = g(\theta)$. Here, then, are the ML estimates of the Poisson two-component-mix parameters $\theta = (\omega, \phi, \psi)$:

```
solλ = { a → results[[3, 1]],
         b → results[[3, 2]],
         c → results[[3, 3]] };

solθ = {ω →  1/(1 + e^a) , φ → e^b, ψ → e^c} /. solλ
```

$\{\omega \to 0.359885, \phi \to 1.2561, \psi \to 2.6634\}$

That is, the ML estimate of the mixing parameter is $\hat{\omega} = 0.359885$, and the ML estimates of the Poisson component parameters are $\hat{\phi} = 1.2561$ and $\hat{\psi} = 2.6634$. Here is the estimate of the asymptotic variance-covariance matrix (see (11.17)):

```
G = Grad[{ 1/(1 + e^a) , e^b, e^c}, {a, b, c}];

G.W.Transpose[G] /. solλ
```

$$\begin{pmatrix} 0.0411708 & 0.0710351 & 0.0497281 \\ 0.0710351 & 0.133219 & 0.0823362 \\ 0.0497281 & 0.0823362 & 0.0664192 \end{pmatrix}$$

We summarise the ML estimation results obtained using the BFGS algorithm in Table 7.

	Estimate	SE	TStat
ω	0.359885	0.202906	1.77366
ϕ	1.2561	0.364992	3.44143
ψ	2.6634	0.257719	10.3345

Table 7: ML estimation results for the Poisson two-component-mix model

Finally, it is interesting to contrast the fit of the Poisson model with that of the Poisson two-component-mix model. Here, as a function of $x \in \{0, 1, 2, ...\}$, is the fitted Poisson model:

$$\texttt{fitP} = \frac{e^{-\gamma} \gamma^x}{x!} \text{ /. sol}\gamma$$

$$\frac{0.115679 \; 2.15693^x}{x!}$$

... and here is the fitted Poisson two-component-mix model:

$$\texttt{fitPcm} = \left(\omega \frac{e^{-\phi} \phi^x}{x!} + (1-\omega) \frac{e^{-\psi} \psi^x}{x!} \right) \text{ /. sol}\theta$$

$$\frac{0.102482 \; 1.2561^x}{x!} + \frac{0.0446227 \; 2.6634^x}{x!}$$

Table 8 compares the fit obtained by each model to the data. Evidently, the Poisson two-component-mix model gives a closer fit to the data than the Poisson model in every category. This improvement has been achieved as a result of introducing two additional parameters, but it has come at the cost of requiring a more complicated estimation procedure.

	Count	Mixed	Poisson
0	162	161.227	126.784
1	267	271.343	273.466
2	271	262.073	294.924
3	185	191.102	212.044
4	111	114.193	114.341
5	61	57.549	49.325
6	27	24.860	17.732
7	8	9.336	5.464
8	3	3.089	1.473
9	1	0.911	0.353

Table 8: Fitted Poisson and Poisson two-component-mix models

12.7 The Newton–Raphson Algorithm

In this section, we employ the Newton–Raphson (NR) algorithm to estimate the parameters of an Ordered Probit model.

○ *Data, Statistical Model and Log-likelihood*

Random variables that cannot be observed are termed *latent*. A common source of such variables is individual sentiment because, in the absence of a rating scale common to all individuals, sentiment cannot be measured. Even without an absolute measurement of sentiment, it is often possible to obtain partial information by using categorisation; a sampling device that can achieve this is the ubiquitous 'opinion survey'. Responses to such surveys are typically ordered — e.g. choose one of 'disliked Brand X', 'indifferent to Brand X', or 'liked Brand X' — which reflects the ordinal nature of sentiment. Such latent, ordered random variables are typically modelled using cumulative response probabilities. Well-known models of this type include the *proportional-odds* model and the *proportional-hazards* model (e.g. see McCullagh and Nelder (1989)), and the *Ordered Probit* model due to McKelvey and Zavoina (1975) (see also Maddala (1983) and Becker and Kennedy (1992)). In this section, we develop a simple form of the ordered probit model (with cross-classification), estimating parameters using the Newton–Raphson (NR) algorithm.

During consultations with a general medical practitioner, patients were asked a large number of lifestyle questions. One of these was (the somewhat morbid), "Have you recently found that the idea of taking your own life kept coming into your mind?". Goldberg (1972) reports count data for 295 individuals answering this question in Table 9.

Illness class	Definitely not ($j=1$)	Do not think so ($j=2$)	Has crossed my mind ($j=3$)	Definitely has ($j=4$)
Normal ($i=1$)	90	5	3	1
Mild ($i=2$)	43	18	21	15
Severe ($i=3$)	34	8	21	36

Table 9: Psychiatric data — cross-classified by illness

The data is assumed to represent a categorisation of the 'propensity to suicidal thought', a latent, ordered random variable. Responses are indexed by j, running across the columns of the table. In addition, all individuals had been cross-classified into one of three psychiatric classes: normal ($i = 1$), mild psychiatric illness ($i = 2$), and severe psychiatric illness ($i = 3$). For example, of the 167 individuals responding "Definitely not", 90 were classified as normal, 43 as having mild psychiatric illness and 34 as suffering severe psychiatric illness. Enter the data:

```
freq = {{90, 5, 3, 1}, {43, 18, 21, 15}, {34, 8, 21, 36}};
```

Due to the cross-classification, the issue of interest is whether the propensity to suicidal thought can be ranked according to illness. Upon inspection, the data seems to suggest that the propensity to suicidal thought increases with mental illness. In order to quantify this view, we define three latent, ordered random variables,

Y_i^* = Propensity to suicidal thought of an individual classified with illness i

and we specify the following linear model for each,

$$Y_i^* = \beta_i + U_i, \quad i \in \{1, 2, 3\} \tag{12.17}$$

where U_i is an unknown disturbance term with zero mean. The (cross-classified) Ordered Probit model is characterised by assuming a trivariate Normal distribution (see §6.4 B) with independent components for the disturbances, namely,

$$\begin{pmatrix} U_1 \\ U_2 \\ U_3 \end{pmatrix} \sim N(\vec{0}, I_3) \tag{12.18}$$

which is scale invariant because observations are categorical. The class-specific parameter β_i enables us to quantify the differences between the psychiatric classes. In parametric terms, if propensity to suicidal thought can be ranked increasingly in respect of psychiatric illness, then we would expect $\beta_1 < \beta_2 < \beta_3$ — a testable hypothesis.[17]

Of course, it is *not* Y_i^* that is observed in Table 9; rather, observations have been recorded on another trio of random variables which we define as

Y_i = the response to the survey question of an individual classified with illness i.

To establish the link between response Y_i and propensity Y_i^*, we assume Y_i is a categorisation of Y_i^*, and that

$$P(Y_i = j) = P(\alpha_{j-1} < Y_i^* < \alpha_j) \tag{12.19}$$

for all combinations of indexes i and j. The parameters $\alpha_0, \ldots, \alpha_4$ are cut-off parameters which, because of the ordered nature of Y_i^*, satisfy the inequalities $\alpha_0 < \alpha_1 < \cdots < \alpha_4$. Given the Normality assumption (12.18), we immediately require $\alpha_0 = -\infty$ and $\alpha_4 = \infty$ to ensure that probabilities sum to unity. Substituting (12.17) into (12.19), yields

$$\begin{aligned} P(Y_i = j) &= P(\alpha_{j-1} - \beta_i < U_i < \alpha_j - \beta_i) \\ &= \Phi(\alpha_j - \beta_i) - \Phi(\alpha_{j-1} - \beta_i) \end{aligned} \tag{12.20}$$

where Φ denotes the cdf of a $N(0, 1)$ random variable (which is the marginal cdf of U_i):[18]

```
Clear[Φ];         Φ[x_] = 1/2 (1 + Erf[x/√2]);
```

Then, the observed log-likelihood is given by:

```
obslogLθ =
```

$$\text{Log}\left[\prod_{i=1}^{3} (\Phi[\alpha_1 - \beta_i])^{\text{freq}[\![i,1]\!]} (\Phi[\alpha_2 - \beta_i] - \Phi[\alpha_1 - \beta_i])^{\text{freq}[\![i,2]\!]}\right.$$

$$\left. (\Phi[\alpha_3 - \beta_i] - \Phi[\alpha_2 - \beta_i])^{\text{freq}[\![i,3]\!]} (1 - \Phi[\alpha_3 - \beta_i])^{\text{freq}[\![i,4]\!]}\right];$$

As it stands, the parameters of this model cannot be estimated uniquely. To see this, notice that in the absence of any restriction, it is trivially true that for any non-zero constant γ, the categorical probability in the ordered probit model satisfies

$$\Phi(\alpha_j - \beta_i) - \Phi(\alpha_{j-1} - \beta_i) = \Phi\big((\alpha_j + \gamma) - (\beta_i + \gamma)\big) - \Phi\big((\alpha_{j-1} + \gamma) - (\beta_i + \gamma)\big)$$

for *all possible* i and j. Thus, the probability determined from values assigned to the parameters $(\alpha_1, \alpha_2, \alpha_3, \beta_1, \beta_2, \beta_3)$ cannot be distinguished from the probability resulting from values assigned as per $(\alpha_1 + \gamma, \alpha_2 + \gamma, \alpha_3 + \gamma, \beta_1 + \gamma, \beta_2 + \gamma, \beta_3 + \gamma)$ for any arbitrary $\gamma \neq 0$. This phenomenon is known as a *parameter identification problem*. To overcome it, we must break the equivalence in probabilities for at least one combination of i and j. This can be achieved by fixing one of the parameters, thus effectively removing it from the parameter set. Any parameter will do, and any value can be chosen. In practical terms, it is better to remove one of the cut-off parameters $(\alpha_1, \alpha_2, \alpha_3)$, for this reduces by one the number of inequalities to which these parameters must adhere. Conventionally, the identifying restriction is taken to be:

```
α₁ = 0;
```

The parameter $\theta = (\alpha_2, \alpha_3, \beta_1, \beta_2, \beta_3)$ is defined over the space

$$\Theta = \{(\alpha_2, \alpha_3) : (\alpha_2, \alpha_3) \in \mathbb{R}_+^2, 0 < \alpha_2 < \alpha_3\} \times \{(\beta_1, \beta_2, \beta_3) : (\beta_1, \beta_2, \beta_3) \in \mathbb{R}^3\}$$

and therefore Θ is a proper subset of $\mathbb{R}^5$. For unconstrained optimisation, a transformation to new parameters $\lambda = g(\theta) \in \mathbb{R}^5$ is required. Clearly, the transformation need only act on the cut-off parameters, and one that satisfies our requirements is:

$$\text{prm} = \left\{\alpha_2 == \frac{\alpha_3}{1 + e^{a2}},\ \alpha_3 == e^{a3},\ \beta_1 == b1,\ \beta_2 == b2,\ \beta_3 == b3\right\};$$

where $\lambda = (a2, a3, b1, b2, b3) \in \mathbb{R}^5$. Notice that α_3 will be positive for all a3, and that it will always be larger than α_2 for all a2, so the constraints $0 < \alpha_2 < \alpha_3$ will always be satisfied.[19] From inspection of prm, it is apparent that we have not yet determined $g(\theta)$, for α_2 depends on α_3. However, by inputting:

```
θToλ = Solve[prm, {α₂, α₃, β₁, β₂, β₃}] // Flatten
```

$$\left\{\alpha_2 \to \frac{e^{a3}}{1 + e^{a2}},\ \beta_1 \to b1,\ \beta_2 \to b2,\ \beta_3 \to b3,\ \alpha_3 \to e^{a3}\right\}$$

we now have $g(\theta)$ in the form of a replacement rule. We now enter into *Mathematica* the observed log-likelihood function in terms of λ (obslogLλ):

§12.7 MAXIMUM LIKELIHOOD ESTIMATION IN PRACTICE

```
obslogLλ = obslogLθ /. θToλ;
```

Similar to §12.6, we can use `FindMaximum` to estimate the parameters using the NR algorithm:

```
FindMaximum[obslogLλ, {a2, 0}, {a3, 0},
    {b1, 0}, {b2, 0}, {b3, 0}, Method → Newton]

{-292.329, {a2 → 0.532781, a3 → -0.0507304,
    b1 → -1.34434, b2 → 0.0563239, b3 → 0.518914}}
```

But, as has previously been stated, the main drawback to using `FindMaximum` is that it does not supply the final Hessian matrix — we cannot construct an estimate of the asymptotic variance-covariance matrix of the ML estimator of λ from `FindMaximum`'s output.

○ *NR Algorithm*

We shall estimate λ and the asymptotic variance-covariance matrix using the NR algorithm. From (12.8), (12.11) and (12.13), the NR algorithm is based on the updating formulae:

$$\hat{\lambda}_{(k+1)} = \hat{\lambda}_{(k)} + \mu_{(k)} d_{(k)}$$
$$d_{(k)} = -W_{(k)} \cdot p_g(\hat{\lambda}_{(k)})$$
$$W_{(k)} = -H_{(k)}^{-1}$$

where k is the iteration index, p_g is the gradient of the penalty function, W is the inverse of the Hessian of the penalty function and H is the Hessian of the observed log-likelihood function. We obtain the penalty function, the gradient and the Hessian as follows:

```
p = -obslogLλ;
pf[{a2_, a3_, b1_, b2_, b3_}] = p;

g = Grad[p, {a2, a3, b1, b2, b3}];
gradpf[{a2_, a3_, b1_, b2_, b3_}] = g;

H = Hessian[obslogLλ, {a2, a3, b1, b2, b3}];
hessf[{a2_, a3_, b1_, b2_, b3_}] = H;
```

These are very complicated expressions, so unless your computer has loads of memory capacity, and you have loads of spare time, we strongly advise using the humble semicolon ';' (as we have done) to suppress output to the screen! Here, `gradpf` and `hessf` are functions with a `List` of `Symbol` arguments matching exactly the elements of λ. The reason for constructing these two functions is to avoid coding the NR algorithm with numerous replacement rules, since such rules can be computationally inefficient and more cumbersome to code. The vast bulk of computation time is spent on the Hessian matrix.

This is why NR algorithms are costly, for they evaluate the Hessian matrix at every iteration. It is possible to improve computational efficiency by compiling the Hessian.[20]

Another way to proceed is to input the mathematical formula for the Hessian matrix directly into *Mathematica*; Maddala (1983), for instance, gives such formulae. This method has its cost too, not least of which is that it runs counter to the approach taken throughout this volume, which is to ask *Mathematica* to do the work. Yet another approach is to numerically evaluate/estimate the first- and second-order derivatives, for clearly there will exist statistical models with parameter numbers of such magnitude that there will be insufficient memory available for *Mathematica* to derive the symbolic Hessian — after all, our example only has five parameters, and yet computing the symbolic Hessian already requires around 7 MB (on our reference machine) of free memory. In this regard, the standard add-on package NumericalMath`NLimit` may assist, for its ND command performs numerical approximations of derivatives.

In §12.3, we noted that the NR algorithm is useful as a 'finishing-off' algorithm which fine tunes our estimates. This is because NR uses the actual Hessian matrix, whereas quasi-Newton algorithms (like BFGS) only use estimates of the Hessian matrix. But, for this example, we will apply the NR algorithm from scratch. Fortunately for us, the log-likelihood of the Ordered Probit model can be shown to be globally concave in its parameters; see Pratt (1981). Thus, the Hessian matrix is negative definite for all θ, and therefore negative definite for all λ, as the two parameters are related one-to-one.

In principle, given concavity, the NR algorithm will reach the global maximum from wherever we choose to start in parameter space. Numerically, however, it is nearly always a different story! Sensible starting points nearly always need to be found when optimising, and the Ordered Probit model is no exception. For instance, if a starting value for a3 equal to 3.0 is chosen, then one computation that we are performing is the integral under a standard Normal distribution curve up to $\exp(3) \simeq 20$. In this case, it would not be surprising to see the algorithm crash, as we will run out of numerical precision; see Sofroniou (1996) for a discussion of numerical precision in *Mathematica*. Sensible starting values usually require some thought and are typically problem-specific — even when we are fortunate enough to have an apparently ideal globally concave log-likelihood, as we do here.

Our implementation of the NR algorithm follows. Like our BFGS algorithm, we have left it very much without any bells and whistles. Upon convergence to tolerance, the output is recorded in results which has four components: results[[1]] is the number of iterations taken to achieve convergence to tolerance; results[[2]] is the value of the maximised observed log-likelihood; results[[3]] is the ML point estimates; and results[[4]] is the negative of the inverse Hessian evaluated at the ML point estimates. The origin would seem to be a sensible starting value at which to initiate the algorithm.

```
Armijo[f_, θ_, grad_, dir_] :=
  Module[{α = 0.5, β = 0.65, μ = 1., f0, gd},
    f0 = f[θ];  gd = grad.dir;
    While[ f[θ + μ dir] - f0 - μ α gd > 0, μ = β μ];  μ]
```

§12.7 MAXIMUM LIKELIHOOD ESTIMATION IN PRACTICE

```
λ0 = {0., 0., 0., 0., 0.};
g0 = gradpf[λ0];

Do[        H0 = hessf[λ0];
           W0 = -Inverse[H0];
            d = -W0.g0;
           λ1 = λ0 + Armijo[pf, λ0, g0, d] d;
           g1 = gradpf[λ1];
If[Max[Abs[g1]] < 10⁻⁶, Break[results =
     {iter, -pf[λ1], λ1, -Inverse[hessf[λ1]]}]];
           λ0 = λ1;
           g0 = g1,
                          {iter, 1, 20}];
```

From its starting point, the NR algorithm takes just over 10 seconds to converge to tolerance on our reference machine. In total, it takes five iterations:

results[[1]]

5

The returned estimate of λ is:

results[[3]]

{0.532781, -0.0507304, -1.34434, 0.0563239, 0.518914}

at which the value of the observed log-likelihood is:

results[[2]]

-292.329

Table 10 gives estimation results for the parameters of our original Ordered Probit model (found using the Invariance Property). Because $\hat{\beta}_1 < \hat{\beta}_2 < \hat{\beta}_3$, our quantitative results lend support to the qualitative assessment made at the very beginning of this example — that propensity to suicidal thought increases with severity of psychiatric illness.[21]

	Estimate	SE	TStat
α_2	0.35157	0.059407	5.91804
α_3	0.95054	0.094983	10.00740
β_1	-1.34434	0.174219	-7.71641
β_2	0.05632	0.118849	0.47391
β_3	0.51891	0.122836	4.22446

Table 10: ML estimation results for the Ordered Probit model

12.8 Exercises

1. Generate 10 pseudo-random drawings from $X \sim N(0, 1)$ as follows:

   ```
   data = Table[√2 InverseErf[0, -1 + 2 Random[]], {10}]
   ```

   Use ML estimation to fit the $N(\mu, \sigma^2)$ distribution to the artificial data using each of the following:

   ```
   FindMaximum[obslogLθ, {μ, 0}, {σ, 1}]
   FindMaximum[obslogLθ, {μ, {-1, 1}}, {σ, {0.5, 2}}]
   FindMaximum[obslogLθ, {μ, 0, -3, 3}, {σ, 1, 0, 4}]
   FindMaximum[obslogLθ, {μ, 0}, {σ, 1}, Method → Newton]
   FindMaximum[obslogLθ, {μ, 0}, {σ, 1}, Method → QuasiNewton]
   ```

   where `obslogLθ` is the observed log-likelihood for $\theta = (\mu, \sigma)$. Contrast your answers against the estimates computed from the exact ML estimator

   $$\hat{\mu} = \frac{1}{n} \sum_{i=1}^{n} X_i \quad \text{and} \quad \hat{\sigma} = \sqrt{\frac{1}{n} \sum_{i=1}^{n} (X_i - \hat{\mu})^2}.$$

2. Let $X \sim \text{Waring}(a, b)$ with pmf

   $$P(X = x) = (b - a) \frac{\Gamma(x+a)\,\Gamma(b)}{\Gamma(a)\,\Gamma(x+b+1)}, \quad x \in \{0, 1, 2, \ldots\}$$

   where the parameters are such that $b > a > 0$. Use `FindMaximum` to obtain ML estimates of a and b for the Word Count data, which is loaded using:

   ```
   ReadList["WordCount.dat"]
   ```

   Hint: re-parameterise $\theta = (a, b)$ to $\lambda = (c, d) \in \mathbb{R}^2$, where $a = e^c$ and $b = e^c(1 + e^d)$. Estimate the variance-covariance matrix of the asymptotic distribution of the ML estimator using the Hessian estimator.

3. Let $X \sim \text{NegativeBinomial}(r, p)$.
   (i) Show that $\mu < \sigma^2$, where $\mu = E[X]$ and $\sigma^2 = \text{Var}(X)$.
   (ii) Let $(X_1, X_2, \ldots, X_n)$ denote a random sample of size n on X. Now it is generally accepted that $\overline{X}$, the sample mean, is the best estimator of μ. Using the log-likelihood concentrated with respect to the estimator $\overline{X}$ for μ, obtain the ML estimate of r for the data sets NB1 (enter as `ReadList["NB1.dat"]`) and NB2 (enter as `ReadList["NB2.dat"]`).
   (iii) Comment on the fit of each model. Can you find any reason why the ML estimate for the NB2 data seems so erratic?

4. Answer the following using the Nerve data given in §12.2. Let $X \sim \text{Gamma}(\alpha, \beta)$ with pdf $f(x; \theta)$, where $\theta = (\alpha, \beta)$. Example 2 derived the ML estimate as $\hat{\theta} = (\hat{\alpha}, \hat{\beta}) = (1.17382, 0.186206)$.
   (i) Derive Fisher's estimate of the asymptotic variance-covariance matrix of the ML estimator of θ (hint: see Example 3).
   (ii) Given $\hat{\theta}$, use the Invariance Property (§11.4 E) to derive ML estimates of:
      (a) $\lambda = (\mu, \nu)$, where $\mu = E[X]$ and $\nu = \text{Var}(X)$, and
      (b) the asymptotic variance-covariance matrix of the ML estimator of λ.

(iii) Re-parameterise the pdf of X to $f(x; \lambda)$.

    (a) Use FindMaximum's BFGS algorithm (Method → QuasiNewton) to obtain the ML estimate of λ.

    (b) Estimate the asymptotic variance-covariance matrix of the ML estimator of λ using the Fisher, Hessian and Outer-product estimators.

    (c) Compare your results for parts (a) and (b) to those obtained in (ii).

(iv) Using the Invariance Property (§11.4 E), report ML estimates of:

    (a) $\delta = (\mu, \sigma)$, where $\mu = E[X]$ and $\sigma = \alpha$, and

    (b) the asymptotic variance-covariance matrix of the ML estimator of δ.

(v) Re-parameterise the pdf of X to $f(x; \delta)$.

    (a) Use FindMaximum's BFGS algorithm to obtain the ML estimate of δ.

    (b) Estimate the asymptotic variance-covariance matrix of the ML estimator of δ using the Fisher, Hessian and Outer-product estimators.

    (c) Compare your results for parts (a) and (b) to those obtained in (iv).

5. The Gamma regression model specifies the conditional distribution of $Y \mid X = x$, with pdf

$$f(y \mid X = x; \theta) = \frac{1}{\Gamma(\sigma)} \left(\frac{\mu}{\sigma}\right)^{-\sigma} \exp\left(-\frac{y\sigma}{\mu}\right) y^{\sigma-1}$$

where $\mu = \exp(\alpha + \beta x)$ is the regression function, σ is a scaling factor and parameter $\theta = (\alpha, \beta, \sigma) \in \{\alpha \in \mathbb{R}, \beta \in \mathbb{R}, \sigma \in \mathbb{R}_+\}$. Use ML estimation to fit the Gamma regression model to Greene's data (see *Example 5*). By performing a suitable test on σ, determine whether the fitted model represents a significant improvement over the Exponential regression model for Greene's data.

6. Derive ML estimates of the ARCH model of §12.3 based on the BFGS algorithm of §12.6. Obtain an estimate of the variance-covariance matrix of the asymptotic distribution of the ML estimator. Report your ML estimates, associated asymptotic standard errors and t-statistics.

Appendix

A.1 Is That the Right Answer, Dr Faustus?

○ *Symbolic Accuracy*

Many people find the vagaries of integration to be a less than salubrious experience. Excellent statistical reference texts can make 'avoidance' a reasonable strategy, but one soon comes unstuck when one has to solve a non-textbook problem. With the advent of computer software like **mathStatica**, the Faustian joy of computerised problem solving is made ever more delectable. Indeed, over time, it seems likely that the art of manual integration will slowly wither away, much like long division has been put to rest by the pocket calculator. As we become increasingly reliant on the computer, we become more and more dependent on its accuracy. *Mathematica* and **mathStatica** are, of course, not always infallible, they are not panaceas for solving all problems, and it is possible (though rare) that they may get an integral or summation problem wrong. Lest this revelation send some readers running back to their reference texts, it should be stressed that those same reference texts suffer from exactly the same problem and for the same reason: mistakes usually occur because something has been 'typeset' incorrectly. In fact, after comparing **mathStatica**'s output with thousands of solutions in reference texts, it is apparent that even the most respected reference texts are peppered with surprisingly large numbers of errors. Usually, these are typographic errors which are all the more dangerous because they are hard to detect. A healthy scepticism for both the printed word and electronic output is certainly a valuable (though time-consuming) trait to develop.

One advantage of working with a computer is that it is usually possible to test almost any symbolic solution by using numerical methods. To illustrate, let us suppose that $X \sim$ Chi-squared(n) with pdf $f(x)$:

$$f = \frac{x^{n/2-1} \, e^{-x/2}}{2^{n/2} \, \Gamma[\frac{n}{2}]} \, ; \qquad \text{domain}[f] = \{x, \, 0, \, \infty\} \, \&\& \, \{n > 0\};$$

We wish to find the mean deviation $E\big[\,|X - \mu|\,\big]$, where μ denotes the mean:

$$\mu = \text{Expect}[x, \, f]$$

n

Since *Mathematica* does not handle absolute values well, we shall enter $|x-\mu|$ as the expression If[x < μ, μ - x, x - μ]. Then, the mean deviation is:

sol = Expect[If[x < μ, μ - x, x - μ], f]

$$\frac{4\,\text{Gamma}[1+\frac{n}{2},\frac{n}{2}] - 2\,n\,\text{Gamma}[\frac{n}{2},\frac{n}{2}]}{\Gamma[\frac{n}{2}]}$$

If, however, we refer to an excellent reference text like Johnson *et al.* (1994, p. 420), the mean deviation is listed as:

$$\text{JKBsol} = \frac{e^{-\frac{n}{2}}\,n^{n/2}}{2^{\frac{n}{2}-1}\,\Gamma[\frac{n}{2}]};$$

First, we check if the two solutions are the same, by choosing a value for *n*, say $n = 6$:

{sol, JKBsol} /. n → 6.

{2.6885, 1.34425}

Clearly, at least one of the solutions is wrong! Generally, the best way to check an answer is to use a completely different methodology to derive it again. Since our original attempt was *symbolic*, we now use *numerical* methods to calculate the answer. This can be done using functions such as NIntegrate and NSum. Here is the mean deviation as a numerical integral when $n = 6$:

NIntegrate[(Abs[x - μ] f) /. n → 6., {x, 0, ∞}]

- General::unfl : Underflow occurred in computation.
- General::unfl : Underflow occurred in computation.
- General::stop : Further output of
    General::unfl will be suppressed during this calculation.
- NIntegrate::ncvb :
    NIntegrate failed to converge to prescribed accuracy after 7
    recursive bisections in x near x = 5.918918918918919`.

2.68852

The warning messages can be ignored, since a rough approximation serves our purpose here. The numerical answer shows that **mathStatica**'s symbolic solution is correct; further experimentation reveals that the solution given in Johnson *et al.* is out by a factor of two. This highlights how important it is to check all output, from both reference books and computers.

Finally, since μ is used frequently throughout the text, it is good housekeeping to:

Clear[μ]

… prior to leaving this example. ■

○ *Numerical Accuracy*

> "A rapacious monster lurks within every computer,
> and it dines exclusively on accurate digits."
>
> McCullough (2000, p. 295)

Unfortunately, numerical accuracy is treated poorly in many common statistical packages, as McCullough (1998, 1999a, 1999b) has detailed.

> "Many textbooks convey the impression that all one has to do is use a computer to solve the problem, the implicit and unwarranted assumption being that the computer's solution is accurate and that one software package is as good as any other."
>
> McCullough and Vinod (1999, p. 635)

As a general 'philosophy', we try to avoid numerical difficulties altogether by treating problems symbolically (exactly), to the extent that this is possible. This means that we try to solve problems in the most general way possible, and that we also try to stop machine-precision numbers from sneaking into the calculation. For example, we can input one-and-a-half as $\frac{3}{2}$ (an exact symbolic entity), rather than as 1.5. In this way, *Mathematica* can solve many problems in an exact way, even though other packages would have to treat the same problem numerically. Of course, some problems can only be treated numerically. Fortunately, *Mathematica* provides two numerical environments for handling them:

(i) *Machine-precision numbers* (also known as floating-point): Almost all computers have optimised hardware for doing numerical calculations. These machine-precision calculations are very fast. However, using machine-precision forces all numbers to have a fixed precision, usually 16 digits of precision. This may not be enough to distinguish between two close numbers. For more detail, see Wolfram (1999, Section 3.1.6).

(ii) *Arbitrary-precision numbers:* These numbers can contain any number of digits, and *Mathematica* keeps track of the precision at all points of the calculation. Unfortunately, arbitrary-precision numerical calculations can be very slow, because they do not take advantage of a computer's hardware floating-point capabilities. For more detail, see Wolfram (1999, Section 3.1.5).

Therein lies the trade-off. If you use machine-precision numbers in *Mathematica*, the assumption is that you are primarily concerned with efficiency. If you use arbitrary-precision numbers, the assumption is that you are primarily concerned with accuracy. For more detail on numerical precision in *Mathematica*, see Sofroniou (1996). For a definitive discussion of *Mathematica*'s accuracy as a statistical package, see McCullough (2000). For *Mathematica*, the news is good:

> "By virtue of its variable precision arithmetic and symbolic power, *Mathematica*'s performance on these reliability tests far exceeds any finite-precision statistical package."
>
> McCullough (2000, p. 296)

⊕ **Example 1:** Machine-Precision and Arbitrary-Precision Numbers

Let $X \sim N(0, 1)$ with pdf $f(x)$:

$$f = \frac{e^{-\frac{x^2}{2}}}{\sqrt{2\pi}}; \quad \text{domain}[f] = \{x, -\infty, \infty\};$$

The cdf, $P(X \le x)$, as a symbolic entity, is:

$$F = \text{Prob}[x, f]$$

$$\frac{1}{2}\left(1 + \text{Erf}\left[\frac{x}{\sqrt{2}}\right]\right)$$

This is the exact solution. McCullough (2000, p. 290) considers the point $x = -7.6$, way out in the left tail of the distribution. We shall enter -7.6 using exact integers:

$$\text{sol} = F \,/.\, x \to -\frac{76}{10}$$

$$\frac{1}{2}\left(1 - \text{Erf}\left[\frac{19\sqrt{2}}{5}\right]\right)$$

... so this answer is exact too. Following McCullough, we now find the numerical value of `sol` using both machine-precision `N[sol]` and arbitrary-precision `N[sol,20]` numbers:

`N[sol]`

1.48215×10^{-14}

`N[sol, 20]`

$1.4806537490048047086 \times 10^{-14}$

Both solutions are correct up to three significant digits, 0.0000000000000148, but they differ thereafter. In particular, the machine precision number is incorrect at the fourth significant digit. By contrast, all twenty requested significant digits of the arbitrary-precision number 0.000000000000014806537490048047086 are correct, as we may verify with:

$$\text{NIntegrate}\left[f, \left\{x, -\infty, -\frac{76}{10}\right\}, \text{WorkingPrecision} \to 30, \text{PrecisionGoal} \to 20\right]$$

$1.4806537490048047086 1 \times 10^{-14}$

In the next input, we start off by using machine-precision, since −7.6 is entered with 2 digit precision, and we then ask *Mathematica* to render the result at 20-digit precision. Of course, this is meaningless—the extra added precision N[·,20] cannot eliminate the problem we have created:

N[F /. x → -7.6, 20]

1.48215×10^{-14}

If numerical accuracy is important, the moral is not to let machine-precision numbers sneak into one's workings. ∎

A.2 Working with Packages

Packages contain programming code that expand *Mathematica*'s toolset in specialised fields. One can distinguish *Mathematica* packages from *Mathematica* notebooks, because they each have different file extensions, as Table 1 summarises.

file extension	description
.m	*Mathematica* package
.nb	*Mathematica* notebook

Table 1: Packages and notebooks

The following suggestions will help avoid problems when using packages:

(i) Always load a package in its own Input cell, separate from other calculations.

(ii) Prior to loading a package, it is often best to first quit the kernel (type Quit in the front end, or use **Kernel Menu ▷ Quit Kernel**). This avoids so-called 'context' problems. In particular, **mathStatica** should *always* be started from a fresh kernel.

(iii) The Wolfram packages are organised into families. The easiest way to load a specific Wolfram package is to simply load its family. For instance, to use any of the Wolfram statistics functions, simply load the statistics context with:

    << Statistics`

Note that the ` used in <<Statistics` is not a ', nor a ', but a `.

(iv) **mathStatica** is also a package, and we can load it using:

    << mathStatica.m

or

    << mathStatica`

A.3   Working with =, →, == and :=

ClearAll[x, y, z, q]

○ *Comparing* Set (=) *With* Rule (→)

Consider an expression such as:

y = 3 x²

3 x²

We want to find the value of y when x = 3. Two standard approaches are: (i) Set (=), and (ii) Rule (→).

(i) Set (=): Here, we set x to be 3 :

x = 3; y

27

By entering x = 3 in *Mathematica*, we lose the generality of our analysis — x is now just the number 3 (and not a general variable x). Thus, we can no longer find, for example, the derivative D[y,x]; nor can we Plot[y,{x,1,2}]. In order to return y to its former pristine state, we first have to clear x of its set value:

Clear[x]; y

3 x²

To prevent these sorts of problems, we tend to avoid using approach (i).

(ii) Rule (→): Instead of *setting* x to be 3, we can simply *replace* x with 3 in just a single expression, by using a rule; see also Wolfram (1999, Section 2.4.1). For example, the following input reads, "Evaluate y when x takes the value of 3":

y /. x → 3

27

This time, we have not permanently changed y or x. Since everything is still general, we can still find, for example, the derivative of y with respect to x:

D[y, x]

6 x

○ **Comparing** `Set ( = )` **With** `Equal ( == )`

In some situations, both = and → are inappropriate. Suppose we want to solve the equation `z == Log[x]` in terms of x. If we input `Solve[z = Log[x], x]` (with one equal sign), we are actually asking *Mathematica* to `Solve[Log[x], x]`, which is not an equation. Consequently, the = sign should never be used with the `Solve` function. Instead, we use the == sign to represent a symbolic equation:

    `Solve[z == Log[x], x]`

    $\{\{x \to e^z\}\}$

If, by mistake, we enter `Solve[z = Log[x], x]`, then we must first `Clear[z]` before evaluating `Solve[z == Log[x], x]` again.

○ **Comparing** `Set ( = )` **With** `SetDelayed ( := )`

When defining functions, it is usually better to use `SetDelayed ( := )` than an *immediate* `Set ( = )`. When one uses `Set ( = )`, the right-hand side is immediately evaluated. For example:

    `F1[x_] = x + Random[]`

    `0.733279 + x`

So, if we call `F1` four times, the same pseudo-random number appears four times:

    `Table[F1[q], {4}]`

    `{0.733279 + q, 0.733279 + q, 0.733279 + q, 0.733279 + q}`

But, if we use `SetDelayed ( := )`, as follows:

    `F2[x_] := x + Random[]`

then each time we call the function, we get a different pseudo-random number:

    `Table[F2[q], {4}]`

    `{0.143576 + q, 0.77971 + q, 0.778795 + q, 0.618496 + q}`

While this distinction may appear subtle at first, it becomes important when one starts writing *Mathematica* functions. Fortunately, it is quite easy to grasp after a few examples.

In similar vein, one can use `RuleDelayed` :→ instead of an immediate `Rule` →.

A.4 Working with Lists

Mathematica uses curly braces {} to denote lists, *not* parentheses ( ). Here, we enter the list $X = \{x_1, \ldots, x_6\}$:

X = {x₁, x₂, x₃, x₄, x₅, x₆};

The fourth element, or part, of list X is:

X[[4]]

x_4

Sometimes, X⟦4⟧ is used rather than X[[4]]. The fancy double bracket ⟦ is obtained by entering ESC [[ESC . We now add 5 to each element of the list:

X + 5

$\{5 + x_1,\ 5 + x_2,\ 5 + x_3,\ 5 + x_4,\ 5 + x_5,\ 5 + x_6\}$

Other common manipulations include:

Plus @@ X

$x_1 + x_2 + x_3 + x_4 + x_5 + x_6$

Times @@ X

$x_1\, x_2\, x_3\, x_4\, x_5\, x_6$

Power @@ X

Here is a more sophisticated function that constructs an alternating sum:

Fold[(#2 - #1) &, 0, Reverse[X]]

$x_1 - x_2 + x_3 - x_4 + x_5 - x_6$

Next, we construct an Assumptions statement for the x_i, assuming they are all positive:

Thread[X > 0]

$\{x_1 > 0,\ x_2 > 0,\ x_3 > 0,\ x_4 > 0,\ x_5 > 0,\ x_6 > 0\}$

Here is a typical **mathStatica** 'domain' statement assuming $x_i \in (-\infty, 0)$:

```
Thread[{X, -∞, 0}]
```

$\{\{x_1, -\infty, 0\}, \{x_2, -\infty, 0\}, \{x_3, -\infty, 0\},$
$\{x_4, -\infty, 0\}, \{x_5, -\infty, 0\}, \{x_6, -\infty, 0\}\}$

Finally, here is some data:

```
data = Table[Random[], {6}]
```

$\{0.530808, 0.164839, 0.340276,$
$0.595038, 0.674885, 0.562323\}$

which we now attach to the elements of X using rules →, as follows:

```
Thread[X → data]
```

$\{x_1 \to 0.530808, x_2 \to 0.164839, x_3 \to 0.340276,$
$x_4 \to 0.595038, x_5 \to 0.674885, x_6 \to 0.562323\}$

These tricks of the trade can sometimes be very useful indeed.

A.5 Working with Subscripts

In mathematical statistics, it is both common and natural to use subscripted notation such as $y_1, ..., y_n$. This section first discusses "The Wonders of Subscripts" in *Mathematica*, and then provides "Two Cautionary Tips".

○ *The Wonders of Subscripts*

```
Clear[μ]
```

Subscript notation $\mu_1, \mu_2, ..., \mu_8$ offers many advantages over 'dead' notation such as $\mu 1, \mu 2, ..., \mu 8$. For instance, let:

```
r = Range[8]
```

$\{1, 2, 3, 4, 5, 6, 7, 8\}$

Then, to create the list $z = \{\mu_1, \mu_2, ..., \mu_7, \mu_8\}$, we enter:

```
z = Thread[μ_r]
```

$\{\mu_1, \mu_2, \mu_3, \mu_4, \mu_5, \mu_6, \mu_7, \mu_8\}$

We can now take advantage of *Mathematica*'s advanced pattern matching technology to convert from subscripts to, say, powers:

```
z /. μx_ → sˣ
```

$\{s, s^2, s^3, s^4, s^5, s^6, s^7, s^8\}$

and back again:

```
% /. sˣ_. → μx
```

$\{\mu_1, \mu_2, \mu_3, \mu_4, \mu_5, \mu_6, \mu_7, \mu_8\}$

Next, we convert the μ_i into functional notation $\mu[i]$:

```
z /. μx_ → μ[x]
```

$\{\mu[1], \mu[2], \mu[3], \mu[4], \mu[5], \mu[6], \mu[7], \mu[8]\}$

Now, suppose that μ_t ($t = 1, \ldots, 8$) denotes μ at time t. Then, we can go 'back' one period in time:

```
z /. μt_ → μt-1
```

$\{\mu_0, \mu_1, \mu_2, \mu_3, \mu_4, \mu_5, \mu_6, \mu_7\}$

Or, try something like:

$$z \;/.\; \mu_{t\_} \to \frac{\mu_t}{\mu_{9-t}^t}$$

$\left\{ \dfrac{\mu_1}{\mu_8}, \dfrac{\mu_2}{\mu_7^2}, \dfrac{\mu_3}{\mu_6^3}, \dfrac{\mu_4}{\mu_5^4}, \dfrac{\mu_5}{\mu_4^5}, \dfrac{\mu_6}{\mu_3^6}, \dfrac{\mu_7}{\mu_2^7}, \dfrac{\mu_8}{\mu_1^8} \right\}$

Because the index t is 'live', quite sophisticated pattern matching is possible. Here, for instance, we replace the even-numbered subscripted elements with $\mathbb{A}$:

```
z /. μt_ :> If[EvenQ[t], 𝔸t, μt]
```

$\{\mu_1, \mathbb{A}_2, \mu_3, \mathbb{A}_4, \mu_5, \mathbb{A}_6, \mu_7, \mathbb{A}_8\}$

Now suppose that a random sample of size $n = 8$, say:

```
data = {0, 1, 3, 0, 1, 2, 0, 2};
```

is collected from a random variable $X \sim \text{Poisson}(\lambda)$ with pmf $f(x)$:

```
f = e^-λ λˣ / x! ;   domain[f] = {x, 0, ∞} && {λ > 0} && {Discrete};
```

Then, using subscript notation, the *symbolic* likelihood can be entered as:

$$L = \prod_{i=1}^{n} (f \;/.\; x \to x_i)$$

$$\prod_{i=1}^{n} \frac{e^{-\lambda} \lambda^{x_i}}{x_i!}$$

while the *observed* likelihood is obtained via:

$$L \;/.\; \{n \to 8,\; x_{i\_} :> data[\![i]\!]\}$$

$$\frac{1}{24} e^{-8\lambda} \lambda^9$$

○ *Two Cautionary Tips*

Caution 1: While subscript notation has many advantages in *Mathematica*, its use also requires some care. This is because the internal representation in *Mathematica* of the subscript expression y_1 is quite different to the Symbol y. Technically, this is because Head[y] == Symbol, while Head[y_1] == Subscript. That is, *Mathematica* thinks of y as a Symbol, while it thinks of y_1 as Subscript[y,1]; see also Appendix A.8. Because of this difference, the following is important.

Suppose we set y = 3. To clear y, we would then enter Clear[y]:

y = 3; Clear[y]; y

y

For this to work, y must be a Symbol. It will not work for y_1, because the internal representation of y_1 is Subscript[y,1], which is not a Symbol. The same goes for $\bar{y}, \hat{y}, y^*$, and other notational variants of y. For instance:

y_1 = 3; Clear[y_1]; y_1

— Clear::ssym : y_1 is not a symbol or a string.

3

Instead, to clear y_1, one must use either y_1 = ., as in:

y_1 = 3; y_1 = .; y_1

y_1

or the more savage Clear[Subscript]:

y_1 = 3;   Clear[Subscript]; y_1

y_1

Note that Clear[Subscript] will clear *all* subscripted variables. This can be used as a nifty trick to clear all of $\{y_1, y_2, \ldots, y_n\}$ simultaneously!

Caution 2: In *Mathematica* Version 4.0, there are still a few functions that do not handle subscripted variables properly (though this seems to be mostly fixed as of Version 4.1). This problem can usually be overcome by wrapping `Evaluate` around the relevant expression. For instance, under Version 4.0, the following generates error messages:

```
f = Exp[x₁]; NIntegrate[f, {x₁, -∞, 2}]
```

- Function::flpar :
    Parameter specification {x₁} in Function[{x₁}, {f}]
        should be a symbol or a list of symbols.
- General::stop : Further output of Function::flpar will
        be suppressed during this calculation.
- NIntegrate::inum :
    Integrand {4.} is not numerical at {x₁} = {1.}.

```
NIntegrate[f, {x₁, -∞, 2}]
```

Wrapping `Evaluate` around f overcomes this 'bug' by forcing *Mathematica* to evaluate f prior to starting the numerical integration:

```
f = Exp[x₁]; NIntegrate[Evaluate[f], {x₁, -∞, 2}]
```

7.38906

Alternatively, the following also works fine:

```
NIntegrate[ Exp[x₁], {x₁, -∞, 2}]
```

7.38906

Similarly, the following produces copious error messages under Version 4.0:

```
f = x₁ + x₂; Plot3D[f, {x₁, 0, 1}, {x₂, 0, 1}]
```

but if we wrap `Evaluate` around f, the desired plot is generated:

```
f = x₁ + x₂; Plot3D[Evaluate[f], {x₁, 0, 1}, {x₂, 0, 1}]
```

As a different example, the following works fine:

```
D[x₁³ x, x₁]
```

3 x x₁²

but the next input does not work as we might expect, because x_1 = `Subscript[x, 1]` is interpreted by *Mathematica* as a function of x:

```
D[x₁³ x, x]
```

x₁³ + 3 x x₁² Subscript$^{(1,0)}$[x, 1]

A.6 Working with Matrices

This appendix gives a brief overview of matrices in *Mathematica*. A good starting point is also Wolfram (1999, Sections 3.7.1 – 3.7.11). Two standard *Mathematica* add-on packages may also be of interest, namely `LinearAlgebra`MatrixManipulation`` and `Statistics`DataManipulation``.

○ *Constructing Matrices*

In *Mathematica*, a matrix is represented by a list of lists. For example, the matrix

$$A = \begin{pmatrix} 1 & 2 & 3 \\ 4 & 5 & 6 \\ 7 & 8 & 9 \\ 10 & 11 & 12 \end{pmatrix}$$

can be entered into *Mathematica* as follows:

```
A = {{1, 2, 3}, {4, 5, 6}, {7, 8, 9}, {10, 11, 12}}
```

$$\begin{pmatrix} 1 & 2 & 3 \\ 4 & 5 & 6 \\ 7 & 8 & 9 \\ 10 & 11 & 12 \end{pmatrix}$$

If **mathStatica** is loaded, this output will appear on screen as a fancy formatted matrix. If **mathStatica** is not loaded, the output will appear as a `List` (just like the input). If you do not like the fancy matrix format, you can switch it off with the **mathStatica** function `FancyMatrix`—see Appendix A.8.

Keyboard entry: Table 2 describes how to enter fancy matrices directly from the keyboard. This entry mechanism is quite neat, and it is easily mastered.

short cut	description
CTRL ,	add a column
CTRL RET	add a row

Table 2: Creating fancy matrices using the keyboard

For example, to enter the matrix $\begin{pmatrix} 1 & 2 & 3 \\ 4 & 5 & 6 \end{pmatrix}$, type the following keystrokes in an Input cell:

(1 CTRL , 2 CTRL , 3 CTRL RET 4 TAB 5 TAB 6 →)

While this may appear as the fancy matrix $\begin{pmatrix} 1 & 2 & 3 \\ 4 & 5 & 6 \end{pmatrix}$, the internal representation in *Mathematica* is still `{{1, 2, 3}, {4, 5, 6}}`.

A number of *Mathematica* functions are helpful in constructing matrices, as the following examples illustrate. Here is an identity matrix:

IdentityMatrix[5]

$$\begin{pmatrix} 1 & 0 & 0 & 0 & 0 \\ 0 & 1 & 0 & 0 & 0 \\ 0 & 0 & 1 & 0 & 0 \\ 0 & 0 & 0 & 1 & 0 \\ 0 & 0 & 0 & 0 & 1 \end{pmatrix}$$

... a diagonal matrix:

DiagonalMatrix[{a, b, c, d}]

$$\begin{pmatrix} a & 0 & 0 & 0 \\ 0 & b & 0 & 0 \\ 0 & 0 & c & 0 \\ 0 & 0 & 0 & d \end{pmatrix}$$

... a more general matrix created with `Table`:

Table[a[i, j], {i, 2}, {j, 4}]

$$\begin{pmatrix} a[1,1] & a[1,2] & a[1,3] & a[1,4] \\ a[2,1] & a[2,2] & a[2,3] & a[2,4] \end{pmatrix}$$

... an example using subscript notation:

Table[$a_{i,j}$, {i, 2}, {j, 4}]

$$\begin{pmatrix} a_{1,1} & a_{1,2} & a_{1,3} & a_{1,4} \\ a_{2,1} & a_{2,2} & a_{2,3} & a_{2,4} \end{pmatrix}$$

... an upper-triangular matrix:

Table[If[i ≤ j, ☺, 0], {i, 5}, {j, 5}]

$$\begin{pmatrix} ☺ & ☺ & ☺ & ☺ & ☺ \\ 0 & ☺ & ☺ & ☺ & ☺ \\ 0 & 0 & ☺ & ☺ & ☺ \\ 0 & 0 & 0 & ☺ & ☺ \\ 0 & 0 & 0 & 0 & ☺ \end{pmatrix}$$

... and a Hilbert matrix:

Table[1 / (i + j - 1), {i, 3}, {j, 3}]

$$\begin{pmatrix} 1 & \frac{1}{2} & \frac{1}{3} \\ \frac{1}{2} & \frac{1}{3} & \frac{1}{4} \\ \frac{1}{3} & \frac{1}{4} & \frac{1}{5} \end{pmatrix}$$

Operating on Matrices

Consider the matrices:

$$M = \begin{pmatrix} a & b \\ c & d \end{pmatrix}; \qquad B = \begin{pmatrix} 1 & 2 \\ 3 & 4 \end{pmatrix};$$

For detail on getting pieces of matrices, see Wolfram (1999, Section 3.7.2). In particular, here is the first row of M:

M[[1]]

{a, b}

An easy way to grab, say, the second column of M is to select it with the mouse, copy, and paste it into a new Input cell. If desired, this can then be converted into InputForm (Cell Menu ▷ ConvertTo ▷ InputForm). Alternatively, we can obtain the second column with:

M[[All, 2]]

{a, c}

The dimension (2×2) of matrix M is obtained with:

Dimensions[M]

{2, 2}

The transpose of M is:

Transpose[M]

$$\begin{pmatrix} a & c \\ b & d \end{pmatrix}$$

The determinant of M is given by:

Det[M]

$-b c + a d$

The inverse of M is:

Inverse[M]

$$\begin{pmatrix} \frac{d}{-bc+ad} & -\frac{b}{-bc+ad} \\ -\frac{c}{-bc+ad} & \frac{a}{-bc+ad} \end{pmatrix}$$

The trace is the sum of the elements on the main diagonal:

Tr[M]

a + d

Here are the eigenvalues of M:

Eigenvalues[M]

$\{\frac{1}{2}(a + d - \sqrt{a^2 + 4bc - 2ad + d^2}),$
$\frac{1}{2}(a + d + \sqrt{a^2 + 4bc - 2ad + d^2})\}$

To illustrate matrix addition, consider $B + M$:

B + M

$\begin{pmatrix} 1+a & 2+b \\ 3+c & 4+d \end{pmatrix}$

To illustrate matrix multiplication, consider $B\,M$:

B.M

$\begin{pmatrix} a+2c & b+2d \\ 3a+4c & 3b+4d \end{pmatrix}$

... which is generally not equal to $M\,B$:

M.B

$\begin{pmatrix} a+3b & 2a+4b \\ c+3d & 2c+4d \end{pmatrix}$

Similarly, here is the product $B\,M\,B$:

B.M.B

$\begin{pmatrix} a+2c+3(b+2d) & 2(a+2c)+4(b+2d) \\ 3a+4c+3(3b+4d) & 2(3a+4c)+4(3b+4d) \end{pmatrix}$

... which is generally not equal to $B^T\,M\,B$:

Transpose[B].M.B

$\begin{pmatrix} a+3c+3(b+3d) & 2(a+3c)+4(b+3d) \\ 2a+4c+3(2b+4d) & 2(2a+4c)+4(2b+4d) \end{pmatrix}$

Powers of a matrix, such as $B^3 = B\,B\,B$, can either be entered as:

MatrixPower[B, 3]

$$\begin{pmatrix} 37 & 54 \\ 81 & 118 \end{pmatrix}$$

or as:

B.B.B

$$\begin{pmatrix} 37 & 54 \\ 81 & 118 \end{pmatrix}$$

but *not* as:

B^3

$$\begin{pmatrix} 1 & 8 \\ 27 & 64 \end{pmatrix}$$

Mathematica does not provide a function for doing Kronecker products, so here is one we put together for this Appendix:

**Kronecker[A_, B_] :=
    Partition[
        Flatten[
                Map[Transpose, Outer[Times, A, B]]
        ], Dimensions[A][[2]] Dimensions[B][[2]]]**

For example, here is the Kronecker product $B \otimes M$:

Kronecker[B, M]

$$\begin{pmatrix} a & b & 2a & 2b \\ c & d & 2c & 2d \\ 3a & 3b & 4a & 4b \\ 3c & 3d & 4c & 4d \end{pmatrix}$$

and here is the Kronecker product $M \otimes B$:

Kronecker[M, B]

$$\begin{pmatrix} a & 2a & b & 2b \\ 3a & 4a & 3b & 4b \\ c & 2c & d & 2d \\ 3c & 4c & 3d & 4d \end{pmatrix}$$

A.7 Working with Vectors

There are two completely different ways to enter a vector in *Mathematica*:

(i) *The* List *Approach:* This is the standard *Mathematica* method. It does not distinguish between column and row vectors. Thus, Transpose cannot be used on these vectors.

(ii) *The* Matrix *Approach:* Here, a vector is entered as a special case of a matrix. This does distinguish between column and row vectors, so Transpose can be used with these vectors. Entering the vector this way takes more effort, but it can be less confusing and more 'natural' than the List approach.

In this book, we use approach (i). Mixing the two approaches is not recommended, as this may cause error and confusion.

○ *Vectors as Lists*

The standard *Mathematica* way to represent a vector is as a List {...}, not a matrix {{...}}. Consider, for example:

vec = {15, -3, 5}

{15, -3, 5}

Mathematica thinks vec is a vector:

VectorQ[vec]

True

Is vec a column vector or a row vector? The answer is *neither*. Importantly, when the List approach is used, *Mathematica* makes no distinction between column and row vectors. Instead, *Mathematica* carries out whatever operation is possible. This can be confusing and disorienting. To illustrate, suppose we are interested in the (3×1) column vector $\vec{v}$ and the (1×3) row vector $\vec{u}$, given by

$$\vec{v} = \begin{pmatrix} a \\ b \\ c \end{pmatrix} \quad \text{and} \quad \vec{u} = (1 \ 2 \ 3).$$

Using the List approach, we enter both of them into *Mathematica* in the same way:

v = {a, b, c}
u = {1, 2, 3}

{a, b, c}
{1, 2, 3}

Although we can find the Transpose of a matrix, there is no such thing as a Transpose of a *Mathematica* Vector:

Transpose[v]

— Transpose::nmtx : The first two levels of the
one-dimensional list {a, b, c} cannot be transposed.

Transpose[{a, b, c}]

Once again, this arises because *Mathematica* does not distinguish between column vectors and row vectors. To stress the point, this means that the *Mathematica* input for $\vec{v}$ and $\vec{v}^T$ is exactly the same.

When the Dot operator is applied to two vectors, it returns a scalar. Thus, v.v is equivalent to $\vec{v}^T \vec{v}$ (1×1):

v.v

$a^2 + b^2 + c^2$

while u.u is equivalent to $\vec{u} \vec{u}^T$ (1×1):

u.u

14

In order to obtain $\vec{v} \vec{v}^T$ (3×3) and $\vec{u}^T \vec{u}$ (3×3), we have to derive the outer product using the rather cumbersome expression:

Outer[Times, v, v]

$$\begin{pmatrix} a^2 & ab & ac \\ ab & b^2 & bc \\ ac & bc & c^2 \end{pmatrix}$$

Outer[Times, u, u]

$$\begin{pmatrix} 1 & 2 & 3 \\ 2 & 4 & 6 \\ 3 & 6 & 9 \end{pmatrix}$$

Next, suppose:

$$M = \begin{pmatrix} 1 & 0 & 0 \\ 4 & 5 & 6 \\ 0 & 0 & 9 \end{pmatrix};$$

Then, $\vec{v}^T M \vec{v}$ (1×1) is evaluated with:

    v.M.v

    $5 b^2 + a (a + 4 b) + c (6 b + 9 c)$

and $\vec{u} M \vec{u}^T$ (1×1) is evaluated with:

    u.M.u

    146

Once again, we stress that we do not use `u.M.Transpose[u]` here, because one cannot find the `Transpose` of a *Mathematica* `Vector`.

    The **mathStatica** function `Grad[f, x̄]` calculates the gradient of scalar *f* with respect to $\vec{x} = \{x_1, \ldots, x_n\}$, namely

$$\left\{ \frac{\partial f}{\partial x_1}, \ldots, \frac{\partial f}{\partial x_n} \right\}.$$

Here, then, is the gradient of $f = a b^2$ with respect to $\vec{v}$:

    f = a b² ;   Grad[f, v]

    $\{b^2, 2 a b, 0\}$

The derivative of a vector with respect to a vector yields a matrix. If $\vec{f}$ is an *m*-dimensional vector, and $\vec{x}$ is an *n*-dimensional vector, then `Grad[f⃗, x̄]` calculates the $(m \times n)$ matrix:

$$\begin{pmatrix} \frac{\partial f_1}{\partial x_1} & \cdots & \frac{\partial f_1}{\partial x_n} \\ \vdots & \ddots & \vdots \\ \frac{\partial f_m}{\partial x_1} & \cdots & \frac{\partial f_m}{\partial x_n} \end{pmatrix}$$

This is also known as the Jacobian matrix. Here is an example:

    f = {a b², a, b, c², 1};   Grad[f, v]

$$\begin{pmatrix} b^2 & 2 a b & 0 \\ 1 & 0 & 0 \\ 0 & 1 & 0 \\ 0 & 0 & 2 c \\ 0 & 0 & 0 \end{pmatrix}$$

○ **Vectors as Matrices**

Column vectors $(m \times 1)$ and row vectors $(1 \times n)$ are, of course, just special cases of an $(m \times n)$ matrix. In this vein, one can force *Mathematica* to distinguish between a column vector and a row vector by entering them both as matrices {{...}}, rather than as a single List {...}. To illustrate, suppose we are interested again in the (3×1) column vector $\vec{v}$ and the (1×3) row vector $\vec{u}$, given by

$$\vec{v} = \begin{pmatrix} a \\ b \\ c \end{pmatrix} \quad \text{and} \quad \vec{u} = (1 \ 2 \ 3).$$

This time, we shall enter both $\vec{v}$ and $\vec{u}$ into *Mathematica* as if they were matrices. So, we enter the column vector $\vec{v}$ as:

V = {{a}, {b}, {c}}

$$\begin{pmatrix} a \\ b \\ c \end{pmatrix}$$

As far as *Mathematica* is concerned, this is *not* a Vector:

VectorQ[V]

False

Rather, *Mathematica* thinks it is a Matrix:

MatrixQ[V]

True

Similarly, we enter the row vector $\vec{u}$ as if it is the first row of a matrix:

U = {{1, 2, 3}}   (* not {1,2,3} *)

{{1, 2, 3}}

VectorQ[U]

False

MatrixQ[U]

True

Because V and U are *Mathematica* matrices, Transpose now works:

> **Transpose[V]**
>
> {{a, b, c}}
>
> **Transpose[U]**
>
> $\begin{pmatrix} 1 \\ 2 \\ 3 \end{pmatrix}$

We can now use standard notation to find $\vec{v}^T \vec{v}$ (1×1):

> **Transpose[V].V**
>
> {{a² + b² + c²}}

and $\vec{u}\,\vec{u}^T$ (1×1):

> **U.Transpose[U]**
>
> {{14}}

To obtain $\vec{v}\,\vec{v}^T$ (3×3) and $\vec{u}^T \vec{u}$ (3×3), we no longer have to use Outer products. Again, the answer is obtained using standard notation. Here is $\vec{v}\,\vec{v}^T$:

> **V.Transpose[V]**
>
> $\begin{pmatrix} a^2 & ab & ac \\ ab & b^2 & bc \\ ac & bc & c^2 \end{pmatrix}$

and $\vec{u}^T \vec{u}$:

> **Transpose[U].U**
>
> $\begin{pmatrix} 1 & 2 & 3 \\ 2 & 4 & 6 \\ 3 & 6 & 9 \end{pmatrix}$

Next, suppose:

$$M = \begin{pmatrix} 1 & 0 & 0 \\ 4 & 5 & 6 \\ 0 & 0 & 9 \end{pmatrix};$$

Then, $\vec{v}^T M \vec{v}$ (1×1) is evaluated with:

> **Transpose[V].M.V**
>
> {{5 b² + a (a + 4 b) + c (6 b + 9 c)}}

and $\vec{u} M \vec{u}^T$ (1×1) is evaluated with:

> **U.M.Transpose[U]**
>
> {{146}}

... not with U.M.U.

The Matrix approach to vectors has the advantage that it allows one to distinguish between column and row vectors, which seems more natural. However, on the downside, many *Mathematica* functions (including Grad) have been designed to operate on a single List (Vector), not on a matrix; these functions will often *not* work with vectors that have been entered using the Matrix approach.

A.8   Changes to Default Behaviour

mathStatica makes a number of changes to default *Mathematica* behaviour. These changes only take effect after you load **mathStatica**, and they only remain active while **mathStatica** is running. This section lists three 'visual' changes.

Case 1: Γ[x]

If **mathStatica** is not loaded, the expression Γ[x] has no meaning to *Mathematica*. If **mathStatica** is loaded, the expression Γ[x] is interpreted as the *Mathematica* function Gamma[x]:

> **Γ[x] == Gamma[x]**
>
> True

Case 2: Subscript and Related Notation in Input Cells

> **Quit**

If **mathStatica** is *not* loaded, it is best to avoid mixing x with its variants {x_1, $\hat{x}$, ...} in Input cells. To see why, let us suppose we set x = 3:

> **x = 3**
>
> 3

and then evaluate:

$$\{x_1,\ x^*,\ \bar{x},\ \vec{x},\ \tilde{x},\ \hat{x},\ \dot{x},\ \grave{x}\}$$

$$\{3_1,\ 3^*,\ \bar{3},\ \vec{3},\ \tilde{3},\ \hat{3},\ \dot{3},\ \grave{3}\}$$

This output is not the desired behaviour in standard notational systems.

```
Quit
```

However, if **mathStatica** is loaded, we can work with x and its variants $\{x_1, \hat{x}, \ldots\}$ at the same time without any 'problems':

```
<< mathStatica.m
x = 3
```

3

This time, *Mathematica* treats the variants $\{x_1, \hat{x}, \ldots\}$ in the way we want it to:

$$\{x_1,\ x^*,\ \bar{x},\ \vec{x},\ \tilde{x},\ \hat{x},\ \dot{x},\ \grave{x}\}$$

$$\{x_1,\ x^*,\ \bar{x},\ \vec{x},\ \tilde{x},\ \hat{x},\ \dot{x},\ \grave{x}\}$$

This change is implemented in **mathStatica** simply by adding the attribute HoldFirst to the following list of functions:

```
lis = {Subscript, SuperStar, OverBar, OverVector,
    OverTilde, OverHat, OverDot, Overscript,
    Superscript, Subsuperscript, Underscript,
    Underoverscript, SubPlus, SubMinus, SubStar,
    SuperPlus, SuperMinus, SuperDagger, UnderBar};
```

This idea was suggested by Carl Woll. In our experience, it works brilliantly, without any undesirable side effects, and without the need for the Notation package which can interfere with the subscript manipulations used by **mathStatica**. If, for some reason, you do not like this feature, you can return to *Mathematica*'s default behaviour by entering:

```
ClearAttributes[Evaluate[lis], HoldFirst]
```

Of course, if you do this, some Input cells in this book may no longer work as intended.

Case 3: Matrix Output

If **mathStatica** is not loaded, matrices appear as lists. For example:

```
Quit
```

```
m = Table[i - j, {i, 4}, {j, 5}]
```

{{0, -1, -2, -3, -4}, {1, 0, -1, -2, -3},
 {2, 1, 0, -1, -2}, {3, 2, 1, 0, -1}}

If, however, **mathStatica** is loaded, matrices automatically appear nicely formatted as matrices. For example:

```
Quit

<< mathStatica.m

m = Table[i - j, {i, 4}, {j, 5}]
```

$$\begin{pmatrix} 0 & -1 & -2 & -3 & -4 \\ 1 & 0 & -1 & -2 & -3 \\ 2 & 1 & 0 & -1 & -2 \\ 3 & 2 & 1 & 0 & -1 \end{pmatrix}$$

Standard matrix operations still operate flawlessly:

m[[1]]

{0, -1, -2, -3, -4}

m + 2

$$\begin{pmatrix} 2 & 1 & 0 & -1 & -2 \\ 3 & 2 & 1 & 0 & -1 \\ 4 & 3 & 2 & 1 & 0 \\ 5 & 4 & 3 & 2 & 1 \end{pmatrix}$$

Moreover, it is extremely easy to extract a column (or two): simply select the desired column with the mouse, copy, and paste it into a new Input cell. If desired, you can then convert into InputForm (Cell Menu ▷ ConvertTo ▷ InputForm).

This trick essentially eliminates the need to use the awkward MatrixForm command. If, for some reason, you do not like this fancy formatted output (*e.g.* if you work with very large matrices), you can return to *Mathematica*'s default behaviour by simply evaluating:

FancyMatrix[Off]

— FancyMatrix is now Off.

Then:

m

{{0, -1, 2, -3, -4}, {1, 0, -1, -2, -3},
 {2, 1, 0, -1, -2}, {3, 2, 1, 0, -1}}

You can switch it on again with FancyMatrix[On].

A.9 Building Your Own mathStatica Function

The building blocks of mathematical statistics include the expectations operator, variance, probability, transformations, and so on. A lot of effort and code has gone into creating these functions in **mathStatica**. The more adventurous reader can create powerful custom functions by combining these building blocks in different ways — much like a LEGO® set. To illustrate, suppose we want to write our own function to automate kurtosis calculations for an arbitrary univariate density function f. We recall that kurtosis is defined by

$$\beta_2 = \mu_4 / \mu_2^2$$

where $\mu_r = E[(X - \mu)^r]$. How many arguments should our Kurtosis function have? In other words, should it be Kurtosis[x, μ, f], or Kurtosis[x, f], or just Kurtosis[f]? If our function is smart, we will not need the 'x', since this information can be derived from domain[f]; nor do we need the 'μ', because this can also be calculated from density f. So, the neat solution is simply Kurtosis[f]. Then, we might proceed as follows:

```
Kurtosis[f_] := Module[{xx,mean,var,sol,b=domain[f]},
    xx   = If[ Head[b] === And, b[[1,1]], b[[1]]];
    mean =   Expect[xx, f];
    var  =      Var[xx, f];
    sol  = Expect[(xx - mean)^4, f] / var^2;
    Simplify[sol]        ]
```

In the above, the term xx picks out the random variable x from any given domain[f] statement. We also need to set the Attributes of our Kurtosis function:

SetAttributes[Kurtosis, HoldFirst]

What does this do? The HoldFirst expression forces the Kurtosis function to hold the f as an 'f', rather than immediately evaluating it as, say, $f = e^{-\lambda} \lambda^x / x!$. By holding the f, the function can then find out what domain[f] has been set to, as opposed to domain[$e^{-\lambda} \lambda^x / x!$]. Similarly, it can evaluate Expect[x, f] or Var[x, f]. More generally, if we wrote a function MyFunc[n_, f_], where f is the second argument (rather than the first), we would use SetAttributes[MyFunc, HoldRest], so that the f is still held. To illustrate our new function, suppose $X \sim \text{Poisson}(\lambda)$ with pmf $f(x)$:

$$f = \frac{e^{-\lambda} \lambda^x}{x!}; \quad \text{domain}[f] = \{x, 0, \infty\} \,\&\&\, \{\lambda > 0\} \,\&\&\, \{\text{Discrete}\};$$

Then, the kurtosis of the distribution is:

Kurtosis[f]

$3 + \dfrac{1}{\lambda}$

Notes

Chapter 1   Introduction

1. *Nota bene* Take note.

Chapter 2   Continuous Random Variables

1. **Warning:** On the one hand, σ is often used to denote the standard deviation. On the other hand, some distributions use the symbol σ to denote a parameter, even though this parameter is not equal to the standard deviation; examples include the Lognormal, Rayleigh and Maxwell–Boltzmann distributions.

2. The textbook reference solution, as listed in Johnson *et al.* (1994, equation (18.11)), is incorrect.

3. Black and Scholes first tried to publish their paper in 1970 at the *Journal of Political Economy* and the *Review of Economics and Statistics*. Both journals immediately rejected the paper without even sending it to referees!

4. The assumption that investors are risk-neutral is a simplification device: it can be shown that the solutions derived are valid in all worlds.

Chapter 3   Discrete Random Variables

1. The more surface area a face has, the greater the chance that it will contact the table-top. Hence, shaving a face increases the chance that it, and its opposing face, will occur. Now, because the die was a perfect cube to begin with, shaving the 1-face is no different from shaving the 6-face. The chance of a 1 or 6 is therefore uniformly increased. To see intuitively what happens to the probabilities, imagine throwing a die that has been shaved to extreme — the die would be a disk with two faces, 1 and 6, and almost no edge, so that the chance of outcomes 2, 3, 4 or 5 drop (uniformly) to zero.

2. The interpretation of the limiting distribution is this: after the process has been operating for a long duration ('burnt-in'), the (unconditional) probability p_k of the process being in a state k is independent of how the process first began. For given states k and j, p_k is defined as $\lim_{t \to \infty} P(X_t = k \mid X_0 = j)$ and is invariant to the value of j. Other terms for the limiting distribution include 'stationary distribution', 'steady-state distribution', and 'equilibrium distribution'. For further details on Markov chains see,

for example, Taylor and Karlin (1998), and for further details on asymptotic statistics, see Chapter 8.

3. In the derivations to follow, it is easier to think in terms of draws being made one-by-one without replacement. However, removing at once a single handful of m balls from the urn is probabilistically equivalent to m one-by-one draws, if not physically so.

4. To see that $f(x)$ has the same probability mass under domain[f]={x,0,n} as under domain[f]={x,0,Min[n,r]}, consider the two possibilities: If $n \le r$, everything is clearly fine. If $n > r$, the terms added correspond to every $x \in \{r+1, ..., n\}$. In this range, $x > r$, and hence $\binom{T-n}{r-x}$ is always 0, so that the probability mass $f(x) = 0$ for $x > r$. Thus, the probability mass is not affected by the inclusion of the extra terms.

5. It is *not* appropriate to treat the component-mix X as if it is a weighted average of random variables. For one thing, the domain of support of a weighted average of random variables is more complicated because the values of the weights influence the support. To see this, consider two Bernoulli variables. The domain of support of the component-mix is the union $\{0, 1\} \cup \{0, 1\} = \{0, 1\}$, whereas the domain of support of the weighted average is $\{0, \omega_1, \omega_2, 1\}$.

6. The assumption of Normality is not critical here. It is sufficient that Y_i has a finite variance. Then approximate Normality for $Y = \sum_{i=1}^{t} Y_i$ follows by a suitable version of the Central Limit Theorem; see, for example, Taylor and Karlin (1998, p. 75).

7. When working numerically, the trick here is to ensure that the variance of the Normal pdf σ^2 matches the variance of the parameter-mix model given by Expect[$t\omega^2, g$] = ω^2 / p. Then, taking say $\sigma^2 = 1$, we require $p = \omega^2$ for the variances to match. The values used in Fig. 11 ($\sigma = 1$, $\omega = \sqrt{0.1}$, $p = 0.1$) are consistent with this requirement.

8. Lookup tables are built by DiscreteRNG using *Mathematica*'s Which function. To illustrate, here is a lookup table for *Example 17*, where u = Random[]:

```
Which[ 0    < u < 0.1,    -1. ,
       0.1  < u < 0.5,    1.5 ,
       0.5  < u < 0.8,    Pi  ,
       True,              4.4  ]
```

Chapter 4   Distributions of Functions of Random Variables

1. Notes:

    (i) For a more detailed proof, see Walpole and Myers (1993, Theorem 7.3).

    (ii) Observe that $J = \dfrac{dx}{dy} = \dfrac{1}{dy/dx}$.

2. Let $X \sim$ Exponential($\frac{1}{a}$) with pdf $h(x)$:

$$h = a\, e^{-ax} \; ; \quad \text{domain}[h] = \{x, 0, \infty\} \;\&\&\; \{a > 0\} \;\&\&\; \{b > 0\};$$

Then, the pdf of $Y = b\, e^X$ $(b > 0)$ is:

```
Transform[y == b e^x, h]
TransformExtremum[y == b e^x, h]
```

$a\, b^a\, y^{-1-a}$

$\{y, b, \infty\}\ \&\&\ \{a > 0, b > 0\}$

3. The multivariate case follows analogously; see, for instance, Roussas (1997, p. 232) or Hogg and Craig (1995, Section 4.5).

Chapter 5   Systems of Distributions

1. The area defining $I(J)$ in Fig. 1 was derived symbolically using *Mathematica*. A comparison with Johnson *et al.* (1994) shows that their diagram is actually somewhat inaccurate, as is Ord's (1972) diagram. By contrast, Stuart and Ord's (1994) diagram seems fine.

2. For somewhat cleaner results, note that:
   (i) §7.2 B discusses unbiased estimators of central moments calculated from sample data;
   (ii) The 'quick and dirty' formulae used here for calculating moments from grouped data assume that the frequencies occur at the mid-point of each interval, rather than being spread over the interval. A technique known as Sheppard's correction can sometimes correct for this effect: see, for instance, Stuart and Ord (1994, Section 3.18).

3. The reader comparing results with Stuart and Ord (1994) should note that there is a typographic error in their solution to μ_3.

4. Two alternative methods for deriving Hermite polynomials (as used in statistics) are H1 and H2, where:

```
H1[j_] := 2^(-j/2) HermiteH[j, z/√2] // Expand
```

and:

```
Clear[g];   g'[z] = -z g[z];

H2[j_] := (-1)^j  D[g[z], {z, j}] / g[z]  // Expand
```

H1 makes use of the built-in HermiteH function, while H2 notes that if density $g(z)$ is $N(0, 1)$, then $g'(z) = -z\, g(z)$. While both H1 and H2 are more efficient than H, they are somewhat less elegant in the present context.

5. The original source of the data is Schwert (1990). Pagan and Ullah then adjusted this data for calendar effects by regressing out twelve monthly dummies.

Chapter 6   Multivariate Distributions

1. In order to ascribe a particular value to the conditioning variable, say $f(x_1 \mid X_2 = \frac{1}{2})$, proceed as follows:

    Conditional[x₁, f] /. x₂ → $\frac{1}{2}$

    — Here is the conditional pdf $f(x_1 \mid x_2)$:

    $\frac{1}{2} + x_1$

    Do *not* use Conditional[x₁, f /. x₂ → $\frac{1}{2}$]. In **mathStatica** functions, the syntax f /. x₂ → $\frac{1}{2}$ may only be used for replacing the values of parameters (not variables).

2. Some texts refer to this as the Farlie–Gumbel–Morgenstern class of distributions; see, for instance, Kotz *et al.* (2000, p. 51).

3. More generally, if $Z \sim N(0, 1)$, its cdf is $\Phi(z) = \frac{1}{2}\left(1 + \text{Erf}\left[\frac{z}{\sqrt{2}}\right]\right)$. Then, in a zero correlation *m*-variate setting with $\vec{Z} = (Z_1, \ldots, Z_m) \sim N(\vec{0}, I_m)$, the joint cdf will be:

    $$\left(\frac{1}{2}\right)^m \left(1 + \text{Erf}\left[\frac{z_1}{\sqrt{2}}\right]\right) \cdots \left(1 + \text{Erf}\left[\frac{z_m}{\sqrt{2}}\right]\right).$$

    This follows because $(Z_1, \ldots, Z_m)$ are mutually stochastically independent (Table 3 (i)).

4. *Mathematica*'s Multinormal statistics package contains a special CDF function for the multivariate Normal density. Under *Mathematica* Version 4.0.x, this function does not work if any $\rho_{ij} = 0$, irrespective of whether the 0 is a symbolic zero (0) or a numerical zero (0.). For instance, $P(X \le -2, Y \le 0, Z \le 2)$ fails to evaluate under zero correlation:

    CDF[dist3 /. ρ_ → 0, {-2, 0, 2}]

    — Solve::svars :
        Equations may not give solutions for all "solve" variables.

    — CDF::mnormfail : etc ...

    Fortunately, this problem has been fixed, as of *Mathematica* Version 4.1.

5. Under *Mathematica* Version 4.0, the CDF function in *Mathematica*'s Multinormal statistics package has two problems: it is very slow, and it consumes unnecessarily large amounts of memory. For example:

    G[1, -7, 3] // Timing

    {7.25 Second, 1.27981×10^{-12}}

Rolf Mertig has suggested (in email to the authors) a fix to this problem that does not alter the accuracy of the solution in any way. Simply enter:

```
Unprotect[MultinormalDistribution];

UpValues[MultinormalDistribution] =
    UpValues[MultinormalDistribution] /.
    HoldPattern[NIntegrate[a_, b__]] :>
    NIntegrate[Evaluate[a], b];
```

and then the CDF function is suddenly more than 40 times faster, and it no longer hogs memory:

```
G[1, -7, 3] // Timing
```

$\{0.11\text{ Second, }1.27981 \times 10^{-12}\}$

Under *Mathematica* Version 4.1, none of these problems occur, so there is no need to fix anything.

6. A random vector $\vec{X}$ is said to be spherically distributed if its pdf is equivalent to that of $\vec{Y} = H\vec{X}$, for all orthogonal matrices H. The zero correlation bivariate Normal is a member of the spherical class, because its pdf

$$\frac{1}{2\pi} \exp\left(-\frac{\vec{x}^T \vec{x}}{2}\right)$$

depends on $\vec{x}$ only through the value of the scalar $\vec{x}^T \vec{x}$, and so $(H\vec{x})^T(H\vec{x}) = \vec{x}^T(H^T H)\vec{x} = \vec{x}^T \vec{x}$, because $H^T H = I_2$. An interesting property of spherically distributed variables is that a transformation to polar co-ordinates yields mutually stochastically independent random variables. Thus, in the context of *Example 20* (Robin Hood) above, when $\rho = 0$, the angle Θ will be independent of the radius (distance) R (see density $g(r, \theta)$). For further details on the spherical family of distributions, see Muirhead (1982).

7. The multinomial coefficient

$$\binom{n}{x_1, x_2, \ldots, x_m} = \frac{n!}{x_1! x_2! \cdots x_m!}$$

is provided in *Mathematica* by the function `Multinomial[`$x_1, x_2, \ldots, x_m$`]`. It gives the number of ways to partition n objects into m sets of size x_i.

8. Alternatively, one can find the solution 'manually' as follows:

$$\begin{aligned}E[e^{t_1 Y_1 + t_2 Y_2 + (t_1 + t_2) Y_0}] &= E[e^{t_1 Y_1}] E[e^{t_2 Y_2}] F[e^{(t_1+t_2) Y_0}] \quad \text{by Table 3 (ii)} \\ &= \exp\left((e^{t_1} - 1)\lambda_1 + (e^{t_2} - 1)\lambda_2 + (e^{t_1 + t_2} - 1)\lambda_0\right).\end{aligned}$$

The same technique can be used to derive the pgf.

Chapter 7 Moments of Sampling Distributions

1. Chapter 2 introduced a suite of converter functions that allow one to express any population moment ($\acute{\mu}$, μ, or κ) in terms of any other population moment ($\acute{\mu}$, μ, or κ). These functional relationships also hold between the sample moments. Thus, by combining the moment converter functions with equation (7.2), we can convert any sample moment (raw, central or cumulant) into power sums. For instance, to convert the fourth central sample moment m_4 into power sums, we first convert from central m to raw $\acute{m}$ moments using CentralToRaw[4, m, $\acute{m}$] (note the optional notation arguments m and $\acute{m}$), and then use (7.2) to convert the latter into power sums. Here is m_4 in terms of power sums:

   CentralToRaw$\left[4\text{, m, }\acute{m}\right]$ /. $\acute{m}_{i\_}$:> $\frac{s_i}{n}$

   $$m_4 \to -\frac{3\, s_1^4}{n^4} + \frac{6\, s_1^2\, s_2}{n^3} - \frac{4\, s_1\, s_3}{n^2} + \frac{s_4}{n}$$

   This is identical to:

   SampleCentralToPowerSum[4]

   $$m_4 \to -\frac{3\, s_1^4}{n^4} + \frac{6\, s_1^2\, s_2}{n^3} - \frac{4\, s_1\, s_3}{n^2} + \frac{s_4}{n}$$

2. Kendall's comment on the term 'polykays' can be found in Stuart and Ord (1994, Section 12.22).

3. Just as we can think of moments as being 'about zero' (raw) or 'about the mean' (central), one can think of cumulants as also being 'about zero' or 'about the mean'. The *moment of moment* functions that are expressed in terms of cumulants, namely:

   RawMomentToCumulant
   CentralMomentToCumulant
   CumulantMomentToCumulant

   ... do their internal calculations *about the mean*. That is, they set $\mu_1 = \kappa_1 = 0$. As such, if p = PolyK[{1,2,3}][[2]], then RawMomentToCumulant[1, p] will return 0, not $\kappa_1\, \kappa_2\, \kappa_3$. To force **mathStatica** to do its ___ToCumulant calculations about zero rather than about the mean, add **z** to the end of the function name; e.g. use RawMomentToCumulant**z**. For example, given:

   p = PolyK[{1, 2, 3}][[2]];

   ... compare:

   RawMomentToCumulant[1, p]

   0

with:

```
RawMomentToCumulantZ[1, p]
```

$\kappa_1 \; \kappa_2 \; \kappa_3$

Working 'about zero' requires greater computational effort than working 'about the mean', so the various __CumulantZ functions are often significantly slower than their Z-less cousins.

4. PowerSumToAug, AugToPowerSum and MonomialToPowerSum are the only **mathStatica** functions that allow one to use shorthand notation such as $\{1^4\}$ to denote $\{1, 1, 1, 1\}$. This feature does not work with any other **mathStatica** function.

Chapter 8   Asymptotic Theory

1. The discussion of Calculus`Limit` has benefitted from detailed discussions with Dave Withoff of Wolfram Research.

2. Some texts (*e.g.* Billingsley (1995)) separate the definition into two parts: (i) terming (8.1) the weak convergence of $\{F_n\}_{n=1}^{\infty}$ to F, and (ii) defining convergence in distribution of $\{X_n\}_{n=1}^{\infty}$ to X only when the corresponding cdf's converge weakly.

3. van Beek improved upon the original version of the bounds referred to in the so-called Berry–Esseen Theorem; for details see, amongst others, Bhattacharya and Rao (1976).

4. Φ is the limiting distribution of W_* by the Lindeberg–Feller version of the Central Limit Theorem. This theorem is not discussed here, but details about it can be found in Billingsley (1995) and McCabe and Tremayne (1993), amongst others.

5. Under Version 4.0 of *Mathematica*, some platforms give the solution for μ_3^+ as

$$\frac{1}{(2+\theta)(4+\theta)\Gamma[\tfrac{\theta}{2}]} \left(e^{-\theta/2} \left(2^{4-\tfrac{\theta}{2}} \, \theta^{\tfrac{4+\theta}{2}} (2+\theta) - 2 e^{\theta/2} \left(32(4+3\theta)\Gamma\!\left[1+\tfrac{\theta}{2}\right] + 8(-4+\theta^2)\Gamma\!\left[3+\tfrac{\theta}{2}\right] - \theta^4(6+\theta)\Gamma\!\left[\tfrac{\theta}{2}\right] - 64\,\mathrm{Gamma}\!\left[3+\tfrac{\theta}{2}, \tfrac{\theta}{2}\right] \right) \right) \right)$$

Although this solution appears different to the one derived in the text, the two are nevertheless equivalent.

6. We emphasise that for any finite choice of n, this pseudo-random number generator is only approximately $N(0, 1)$.

7. For example, it makes no sense to consider the convergence in probability of $\{X_n\}_{n=1}^{\infty}$ to X, if all variables in the sequence are measured in terms of pounds of butter, when X is measured in terms of numbers of guns.

8. Letting MSE = $E[(\bar{X}_n - \theta)^2]$, write

$$\text{MSE} = E\left[\left(\frac{1}{n}\sum_{i=1}^{n}(X_i - \theta)\right)^2\right] = \frac{1}{n^2}\sum_{i=1}^{n}\sum_{j=1}^{n}E\left[(X_i - \theta)(X_j - \theta)\right].$$

Of the n^2 terms in the double-sum there are n when the indices are equal, yielding expectations in the form of $E[(X_i - \theta)^2]$; the remaining $n(n-1)$ terms are of the form $E[(X_i - \theta)(X_j - \theta)]$. Due to independence, the latter expectation can be decomposed into the product of expectations: $E[X_i - \theta]\,E[X_j - \theta]$. Thus,

$$\text{MSE} = \frac{1}{n^2}\sum_{i=1}^{n}E[(X_i - \theta)^2] + \frac{1}{n^2}\sum_{\substack{i=1 \\ i \neq j}}^{n}\sum_{j=1}^{n}E[X_i - \theta]\,E[X_j - \theta].$$

As each of the random variables in the random sample is assumed to be a copy of a random variable X, replace $E[(X_i - \theta)^2]$ with $E[(X - \theta)^2]$, as well as $E[X_i - \theta]$ and $E[X_j - \theta]$ with $E[X - \theta]$. Finally, then,

$$\text{MSE} = \frac{1}{n^2}\sum_{i=1}^{n}E[(X - \theta)^2] + \frac{1}{n^2}\sum_{\substack{i=1 \\ i \neq j}}^{n}\sum_{j=1}^{n}(E[X - \theta])^2$$

$$= \frac{1}{n}E[(X - \theta)^2] + \frac{n-1}{n}(E[X] - \theta)^2.$$

Chapter 9   Statistical Decision Theory

1. Sometimes, we do not know the functional form of $g(\hat{\theta}; \theta)$; if this is the case then an alternative expression for risk involves the multiple integral:

$$R_{\hat{\theta}}(\theta) = \int \cdots \int L(\hat{\theta}(x_1, \ldots, x_n), \theta)\,f(x_1, \ldots, x_n; \theta)\,dx_1 \cdots dx_n$$

where we let $\hat{\theta}(X_1, \ldots, X_n)$ express the estimator in terms of the variables in the random sample $X_1, \ldots, X_n$, the latter having joint density f (here assumed continuous). For the examples encountered in this chapter, we shall assume the functional form of $g(\hat{\theta}; \theta)$ is known.

2. The pdf of $X_{(r)}$ can be determined by considering the combinatorics underlying the rearrangement of the random sample. In all, there are n candidates from $(X_1, \ldots, X_n)$ for $X_{(r)}$, and $n-1$ remaining places that fall into two classes: $r-1$ places below x (x represents values assigned to $X_{(r)}$), and $n-r$ places above x. Those that fall below x do so with probability $F(x)$, and those that lie above x do so with probability $1 - F(x)$, while the successful candidate contributes the value of the pdf at x, $f(x)$.

3. Johnson et al. (1995, equation (24.14)) give an expression for the pdf of $X_{(r)}$ which differs substantially to the (correct) output produced by **mathStatica**. It is not difficult to show that the former is incorrect. Furthermore, it can be shown that equations

(24.15), (24.17) and (24.18) of Johnson *et al.* (1995) are incorrectly deflated by a factor of two.

4. *Mathematica* solves many integrals by using a large lookup table. If the expression we are trying to integrate is not in a standard form, *Mathematica* may not find the expression in its lookup table, and the integral will fail to evaluate.

Chapter 10 Unbiased Parameter Estimation

1. Many texts use the term Fisher Information when referring to either measure. Sample Information may be viewed as Fisher Information per observation on a size n random sample $\vec{X} = (X_1, \ldots, X_n)$.

2. *Example 10* is one such example. See Theorem 10.2.1 in Silvey (1995), or Gourieroux and Monfort (1995, pp. 81–82) for the conditions that a given statistical model must meet in order that the BUE of a parameter exists.

3. If the domain of support of X depends on unknown parameters (*e.g.* θ in $X \sim \text{Uniform}(0, \theta)$), added care needs to be taken when using (10.13). In this book, we shall not concern ourselves with cases of this type; instead, for further details, we refer the interested reader to Stuart and Ord (1991, pp. 638–641).

4. This definition suffices for our purposes. For the full version of the definition, see, for example, Hogg and Craig (1995, p. 330).

5. Here, $E[T] = \big(0 \times P(X_n \leq k)\big) + \big(1 \times P(X_n > k)\big) = P(X_n > k)$. Since X_n is a copy of X, it follows that T is unbiased for $g(\lambda)$.

Chapter 11 Principles of Maximum Likelihood Estimation

1. If θ is a vector of k elements, then the first-order condition requires the simultaneous solution of k equations, and the second-order condition requires establishing that the $(k \times k)$ Hessian matrix is negative definite.

2. It is conventional in the Normal statistical model to discuss estimation of the pair (μ, σ^2) rather than (μ, σ). However, because *Mathematica* treats σ^2 as a Power and not as a Symbol, activities such as differentiation and equation-solving involving σ^2 can not be undertaken. This can be partially overcome by entering SuperD[On] which invokes a **mathStatica** function that allows *Mathematica* to differentiate with respect to Power variables. Unfortunately, **mathStatica** does not contain a similar enhancement for equation-solving in terms of Power variables.

3. The following input generates an information message:

    NSum::nslim: Limit of summation n is not a number.

This has no bearing on the correctness of the output so this message may be safely ignored. We have deleted the message from the text.

4. Of course, biasedness is just one aspect of small sample performance. Chapter 9 considers other factors, such as performance under Mean Square Error.

5. The mgf of the Gamma$(n, \frac{1}{n\theta})$ distribution may be derived as:

```
g = y^a-1 e^-y/b  /. {a → n, b → 1};
    ─────────────              ──
      Γ[a] b^a                 nθ

domain[g] = {y, 0, ∞} && {n > 0, n ∈ Integers, θ > 0};

Expect[e^tY, g]
```

$$\left(1 - \frac{t}{n\theta}\right)^{-n}$$

Using simple algebra, this output may be re-written $(\theta/(\theta - \frac{t}{n}))^n$, which matches the mgf of $\overline{\log X}$.

6. Let $\{Y_n\}$ be a sequence of random variables indexed by n, and Y a random variable such that $Y_n \xrightarrow{d} Y$. Let g denote a continuous function (it must be independent of n) throughout the domain of support of $\{Y_n\}$. The Continuous Mapping Theorem states that $g(Y_n) \xrightarrow{d} g(Y)$; see, for example, McCabe and Tremayne (1993). In our case, we set $g(y) = y^{-1}$, and because convergence in distribution to a constant implies convergence in probability to the same constant (§8.5 A), the theorem may be applied.

7. Alternatively, the limiting distribution of $\sqrt{n}\left(\hat{\theta} - \theta\right)$ can be found by applying Skorohod's Theorem (also called the delta method). Briefly, let the sequence of random variables $\{Y_n\}$ be such that $\sqrt{n}\,(Y_n - c) \xrightarrow{d} Y$, where c is a constant and Y a random variable, and let a function g have a continuous first derivative with $G = \partial g(c)/\partial y$. Then $\sqrt{n}\,(g(Y_n) - g(c)) \xrightarrow{d} G Y$. In our case, we have $\sqrt{n}\,(\hat{\theta}^{-1} - \theta^{-1}) \xrightarrow{d} \theta^{-1} Z$. So $\{Y_n\} = \{\hat{\theta}^{-1}\}$, $c = \theta^{-1}$, $Y = \theta^{-1} Z$, where $Z \sim N(0, 1)$. Now set $g(y) = 1/y$, so $G = -1/c^2$. Applying the theorem yields:

$$\sqrt{n}\left(\hat{\theta} - \theta\right) \xrightarrow{d} (-\theta^2)\,\theta^{-1} Z \sim N(0, \theta^2).$$

8. The log-likelihood can be concentrated with respect to the MLE of α_0. Thus, if we let $((Y_1, X_1), \ldots, (Y_n, X_n))$ denote a random sample of size n on the pair (Y, X), the MLE of α_0 can, as a function of β, be shown to equal

$$\hat{\alpha} = \hat{\alpha}(\beta) = \log\left(\frac{1}{n}\sum_{i=1}^{n} Y_i\, e^{-\beta X_i}\right).$$

The concentrated log-likelihood function is given by $\log L(\hat{\alpha}(\beta), \beta)$, which requires numerical methods to be maximised with respect to β (numerical optimisation is discussed in Chapter 12).

Chapter 12   Maximum Likelihood Estimation in Practice

1. Of course, elementary calculus may be used to symbolically maximise the observed log-likelihood, but our purpose here is to demonstrate FindMaximum. Indeed, from *Example 5* of Chapter 11, the ML estimator of λ is given by the sample mean. For the Nerve data, the ML estimate of λ is:

   SampleMean[xdata]

   0.218573

2. For commentary on the comparison between ML and OLS estimators in the Normal linear regression model see, for example, Judge *et al.* (1985, Chapter 2).

3. Just for fun, another (equivalent) way to construct urules is:

   urules = MapThread[(u#1 → #2) &, {Range[n], uvec}];
   Short[urules]

   $\{u_1 \to 0., u_2 \to 0.13, \ll 236 \gg, u_{239} \to -0.11\}$

4. FindMaximum / FindMinimum may sometimes work with subscript parameters if Evaluate is wrapped around the expression to be optimised (*i.e.* FindMinimum[Evaluate[expr],{...}]); however, this device will not always work, and so it is best to avoid using subscript notation with FindMaximum / FindMinimum.

5. In practice, of course, a great deal of further experimentation with different starting values is usually necessary. For space reasons, we will not pursue our search any further here. However, we do encourage the reader to experiment further using their own choices in the above code.

6. In general, if $X_n \xrightarrow{d} X$ and $Y_n \xrightarrow{p} c$, where c is a constant, then $X_n Y_n \xrightarrow{d} c X$. We can use this result by defining

$$Y_n = \sqrt{\frac{\alpha \beta^2}{\hat{\alpha} \hat{\beta}^2}}.$$

Because of the consistency property of the MLE, we have $Y_n \xrightarrow{p} c = 1$. Thus,

$$\sqrt{\frac{\alpha \beta^2}{\hat{\alpha} \hat{\beta}^2}} \sqrt{n} \, (\hat{\mu} - \mu) \xrightarrow{d} 1 \times N(0, \alpha \beta^2) = N(0, \alpha \beta^2).$$

Therefore, at the estimates of $\hat{\alpha}$ and $\hat{\beta}$,

$$\sqrt{\frac{\alpha \beta^2}{\hat{\alpha} \hat{\beta}^2}} \sqrt{n} \, (\hat{\mu} - \mu) \stackrel{a}{\sim} N(0, \alpha \beta^2)$$

Thus,

$$\sqrt{n} \, (\hat{\mu} - \mu) \stackrel{a}{\sim} N\!\left(0, \hat{\alpha} \hat{\beta}^2\right).$$

7. The inverse cdf of the $N(0, 1)$ distribution, evaluated at $1 - \omega/2$, is derived as follows:

$$f = \frac{e^{-\frac{x^2}{2}}}{\sqrt{2\pi}}; \quad \text{domain}[f] = \{x, -\infty, \infty\};$$

$$\text{Solve}[y == \text{Prob}[x, f], x] /. y \to \left(1 - \frac{\omega}{2}\right)$$

$$\left\{\left\{x \to \sqrt{2}\ \text{InverseErf}\left[0, -1 + 2\left(1 - \frac{\omega}{2}\right)\right]\right\}\right\}$$

8. *Mathematica*'s on-line help for `FindMinimum` has an example of this problem in a one-dimensional case; see also Wolfram (1999, Section 3.9.8).

9. If we used `count[[x + 1]]` instead of n_x, the input would fail. Why? Because the product,

$$\prod_{x=0}^{G} \left(\frac{e^{-\gamma}\gamma^x}{x!}\right)^{\text{count}[[x+1]]}$$

is taken with x increasing from 0 to G, where G is a symbol (because it has not been assigned any numerical value). Since the numerical value of G is unknown, *Mathematica* can not evaluate the product. Thus, *Mathematica* must treat x as a symbol. This, in turn, causes `count[[x + 1]]` to fail.

10. In the previous input, it would not be advisable to replace G with 9, for then *Mathematica* would expand the product, and `SuperLog` would not take effect.

11. The log-likelihood is concave with respect to γ because:

$$\text{Hessian}[\text{logL}\gamma, \gamma]$$

$$-\frac{\sum_{x=0}^{G} x\, n_x}{\gamma^2}$$

... is strictly negative.

12. For example, an estimate of the standard error of the ML estimator of γ, using the Hessian estimator given in Table 3, is given by:

$$\sqrt{\frac{1}{-\text{Hessian}[\text{logL}\gamma, \gamma] /. \{G \to 9, n_{x\_} :\to \text{count}[[x + 1]]\}}} /. \text{sol}\gamma$$

0.0443622

13. It is a mistake to use the Newton–Raphson algorithm when the Hessian matrix is positive definite at points in the parameter space because, at these points, (12.12) must be positive-valued. This forces the penalty function/log-likelihood function to increase/decrease in value from one iteration to the next—the exact opposite of how a

gradient method algorithm is meant to work. The situation is not as clear if the Hessian matrix is indefinite at points in the parameter space, because (12.12) can still be negative-valued. Thus, the Newton–Raphson algorithm can work if the Hessian matrix happens to be indefinite, but it can also fail. On the other hand, the BFGS algorithm will work properly wherever it is located in parameter space, for (12.12) will always be negative.

In our example, it is easy to show that the Hessian matrix is *not* negative definite throughout the parameter space. For example, at $(a, b, c) = (0, 1, 2)$, the Hessian matrix is given by:

```
h = Hessian[obslogLλ, {a, b, c}] /. {a → 0, b → 1, c → 2} // N
```

$$\begin{pmatrix} -180.07 & -63.7694 & -374.955 \\ -63.7694 & -2321.03 & 75.9626 \\ -374.955 & 75.9626 & 489.334 \end{pmatrix}$$

The eigenvalues of this matrix are:

```
Eigenvalues[ h ]
```

$\{-2324.45, 660.295, -347.606\}$

Thus, h is indefinite since it has both positive and negative eigenvalues. Consequently, the Hessian matrix is not negative definite throughout the parameter space.

14. If Method→QuasiNewton or Method→Newton is specified, then it is unnecessary to supply the gradient through the option Gradient → Grad[obslogLλ,{a,b,c}], since these methods calculate the gradient themselves. If Method→QuasiNewton or Method→Newton is specified, but *Mathematica* cannot find symbolic derivatives of the objective function, then FindMaximum will not work.

15. To illustrate, let the scalar function $f(x)$ be such that the scalar x_0 minimises f; that is, $f'(x_0) = 0$. Now, for a point x close to x_0, and for f quadratic in a region about x_0, a Taylor series expansion about x_0 yields $f(x) = f(x_0) + f''(x_0)(x - x_0)^2 / 2$. Point x will be numerically distinct from x_0 provided at least that $(x - x_0)^2$ is greater than precision. Therefore, if $MachinePrecision is equal to 16, it would not be meaningful to set tolerance smaller than 10^{-8}.

16. It is inefficient to include a check of the positive-definiteness of $W_{(j)}$. This is because, provided $W_{(0)}$ is positive definite, BFGS will force all $W_{(j)}$ in the sequence to be positive definite.

17. Our analysis of this test is somewhat informal. We determine whether or not the ML point estimates satisfy the inequalities — that is, whether $\hat{\beta}_1 < \hat{\beta}_2 < \hat{\beta}_3$ holds — for our main focus of attention in this section is the computation of the ML parameter estimates using the NR algorithm.

18. The cdf of a N(0, 1) random variable is derived as follows:

$$f = \frac{e^{-\frac{x^2}{2}}}{\sqrt{2\pi}}; \quad \text{domain}[f] = \{x, -\infty, \infty\}; \quad \text{Prob}[x, f]$$

$$\frac{1}{2}\left(1 + \text{Erf}\left[\frac{x}{\sqrt{2}}\right]\right)$$

19. We refrain from using Subscript notation for parameters because FindMinimum / FindMaximum, which we apply later on, does not handle Subscript notation well.

20. The Hessian can be compiled as follows:

    hessfC = Compile[{a2, a3, b1, b2, b3}, Evaluate[H]];

    Mathematica requires large amounts of memory to successfully execute this command. In fact, around 43 MB of free RAM in the Kernel is needed for this one calculation; use MemoryInUse[] to check your own memory performance (Wolfram (1999, Section 2.13.4)). We can now compare the performance of the compiled function hessfC with the uncompiled function hessf. To illustrate, evaluate at the point $\lambda = (0, 0, 0, 0, 0)$:

    λval = {0., 0., 0., 0., 0.};

    Here is the compiled function:

    hessfC @@ λval // Timing

    $\{0.55 \text{ Second},$
    $\begin{pmatrix} -20.5475 & 10.608 & -0.999693 & -5.00375 & -3.83075 \\ 10.608 & -174.13 & 4.08597 & 31.0208 & 45.4937 \\ -0.999693 & 4.08597 & -65.9319 & -4.54747 \times 10^{-13} & -2.72848 \times 10^{-12} \\ -5.00375 & 31.0208 & 0. & -77.5857 & 2.27374 \times 10^{-13} \\ -3.83075 & 45.4937 & -7.7307 \times 10^{-12} & 1.3074 \times 10^{-12} & -78.8815 \end{pmatrix}\}$

    … while here is the uncompiled function:

    hessf[λval] // Timing

    $\{2.03 \text{ Second},$
    $\begin{pmatrix} -20.5475 & 10.608 & -0.999693 & -5.00375 & -3.83075 \\ 10.608 & -174.13 & 4.08597 & 31.0208 & 45.4937 \\ -0.999693 & 4.08597 & -65.9319 & -9.09495 \times 10^{-13} & 4.54747 \times 10^{-13} \\ -5.00375 & 31.0208 & 4.54747 \times 10^{-13} & -77.5857 & 2.27374 \times 10^{-13} \\ -3.83075 & 45.4937 & 1.36424 \times 10^{-12} & 7.67386 \times 10^{-13} & -78.8815 \end{pmatrix}\}$

    The compiled function is about four times faster.

21. The strength of support for this would appear to be overwhelming judging from an inspection of the estimated asymptotic standard errors of $\hat{\beta}_1$, $\hat{\beta}_2$ and $\hat{\beta}_3$. A rule of thumb that compares the extent of overlap of intervals constructed as

$$\text{estimate} \pm 2 \times \text{(estimated standard deviation)}$$

finds only a slight overlap between the intervals about the second and third estimates. Formal statistical evidence may be gathered by performing a hypothesis test of multiple inequality restrictions. For example, one testing scenario could be to specify the maintained hypothesis as $\beta_1 = \beta_2 = \beta_3$, and the alternative hypothesis as $\beta_1 < \beta_2 < \beta_3$.

References

Abbott, P. (1996), Tricks of the trade, *Mathematica Journal*, **6**(3), 22–23.

Amemiya, T. (1985), *Advanced Econometrics*, Harvard: Cambridge, MA.

Andrews, D. F. and Herzberg, A. M. (1985), *Data: A Collection of Problems from Many Fields for the Student and Research Worker*, Springer-Verlag: New York.

Azzalini, A. (1985), A class of distributions which includes the normal ones, *Scandinavian Journal of Statistics*, **12**, 171–178.

Balakrishnan, N. and Rao, C. R. (eds.) (1998a), *Order Statistics: Theory and Methods*, Handbook of Statistics, volume 16, Elsevier: Amsterdam.

Balakrishnan, N. and Rao, C. R. (eds.) (1998b), *Order Statistics: Applications*, Handbook of Statistics, volume 17, Elsevier: Amsterdam.

Balanda, K. P. and MacGillivray, H. L. (1988), Kurtosis: a critical review, *The American Statistician*, **42**, 111–119.

Bates, G. E. (1955), Joint distributions of time intervals for the occurrence of successive accidents in a generalized Pólya scheme, *Annals of Mathematical Statistics*, **26**, 705–720.

Becker, W. E. and Kennedy, P. E. (1992), A graphical exposition of the ordered probit, *Econometric Theory*, **8**, 127–131.

Bhattacharya, R. N. and Rao, R. R. (1976), *Normal Approximation and Asymptotic Expansions*, Wiley: New York.

Billingsley, P. (1995), *Probability and Measure*, 3rd edition, Wiley: New York.

Black, F. and Scholes, M. (1973), The pricing of options and corporate liabilities, *Journal of Political Economy*, **81**, 637–654.

Bowman, K. O. and Shenton, L. R. (1980), Evaluation of the parameters of S_U by rational functions, *Communications in Statistics – Simulation and Computation*, **B9**, 127–132.

Brent, R. P. (1973), *Algorithms for Minimization Without Derivatives*, Prentice-Hall: Englewood Cliffs, New Jersey.

Chow, Y. S. and Teicher, H. (1978), *Probability Theory: Independence, Interchangeability, Martingales*, Springer-Verlag: New York.

Clark, P. K. (1973), A subordinated stochastic process model with finite variance for speculative prices, *Econometrica*, **41**, 135–155.

Cook, M. B. (1951), Bivariate k-statistics and cumulants of their joint sampling distribution, *Biometrika*, **38**, 179–195.

Cox, D. R. and Hinkley, D. V. (1974), *Theoretical Statistics*, Chapman and Hall: London.

Cox, D. R. and Lewis, P. A. W. (1966), *The Statistical Analysis of Series of Events*, Chapman and Hall: London.

Cramer, J. S. (1986), *Econometric Applications of Maximum Likelihood Methods*, Cambridge University Press: Cambridge.

Csörgö, S. and Welsh, A. S. (1989), Testing for exponential and Marshall-Olkin distributions, *Journal of Statistical Planning and Inference*, **23**, 287–300.

Currie, I. D. (1995), Maximum likelihood estimation and *Mathematica*, *Applied Statistics*, **44**, 379–394.

David, F. N. and Barton, D. E. (1957), *Combinatorial Chance*, Charles Griffin: London.

David, F. N. and Kendall, M. G. (1949), Tables of symmetric functions – Part I, *Biometrika*, **36**, 431–449.

David, F. N., Kendall, M. G. and Barton, D. E. (1966), *Symmetric Function and Allied Tables*, Cambridge University Press: Cambridge.

David, H. A. (1981), *Order Statistics*, Wiley: New York.

Davis, L. (1991), *Handbook of Genetic Algorithms*, Van Nostrand Reinhold: New York.

Dhrymes, P. J. (1970), *Econometrics: Statistical Foundations and Applications*, Harper and Row: New York.

Dressel, P. L. (1940), Statistical seminvariants and their estimates with particular emphasis on their relation to algebraic invariants, *Annals of Mathematical Statistics*, **11**, 33–57.

Dwyer, P. S. (1937), Moments of any rational integral isobaric sample moment function, *Annals of Mathematical Statistics*, **8**, 21–65.

Dwyer, P. S. (1938), Combined expansions of products of symmetric power sums and of sums of symmetric power products with application to sampling, *Annals of Mathematical Statistics*, **9**, 1–47.

Dwyer, P. S. and Tracy, D. S. (1980), Expectation and estimation of product moments in sampling from a finite population, *Journal of the American Statistical Association*, **75**, 431–437.

Elderton, W. P. and Johnson, N. L. (1969), *Systems of Frequency Curves*, Cambridge University Press: Cambridge.

Ellsberg, D. (1961), Risk, ambiguity and the Savage axioms, *Quarterly Journal of Economics*, **75**, 643–649.

Engle, R. F. (1982), Autoregressive conditional heteroscedasticity with estimates of the variance of United Kingdom inflations, *Econometrica*, **50**, 987–1007.

Fama, E. (1965), The behaviour of stock market prices, *Journal of Business*, **38**, 34–105.

Feller, W. (1968), *An Introduction to Probability Theory and Its Applications*, volume 1, 3rd edition, Wiley: New York.

Feller, W. (1971), *An Introduction to Probability Theory and Its Applications*, volume 2, 2nd edition, Wiley: New York.

Fisher, R. A. (1928), Moments and product moments of sampling distributions, *Proceedings of the London Mathematical Society*, series 2, volume 30, 199–238 (reprinted in Fisher, R. A. (1950), *Contributions to Mathematical Statistics*, Wiley: New York).

Fraser, D. A. S. (1958), *Statistics: An Introduction*, Wiley: New York.

Gill, P. E., Murray, W. and Wright, M. H. (1981), *Practical Optimization*, Academic Press: New York.

Glover, F., Taillard, E. and de Werra, D. (1993), A user's guide to tabu search, *Annals of Operations Research*, **41**, 3–28.

Goldberg, D. P. (1972), *The Detection of Psychiatric Illness by Questionnaire; a Technique for the Identification and Assessment of Non-Psychotic Psychiatric Illness*, Oxford University Press: London.

Gourieroux, C. and Monfort, A. (1995), *Statistics and Econometric Models*, volume 1, Cambridge University Press: Cambridge.

Greene, W. H. (2000), *Econometric Analysis*, 4th edition, Prentice-Hall: Englewood Cliffs, New Jersey.

Gumbel, E. J. (1960), Bivariate exponential distributions, *Journal of the American Statistical Association*, **55**, 698–707.

Haight, F. A. (1967), *Handbook of the Poisson Distribution*, Wiley: New York.

Halmös, P. R. (1946), The theory of unbiased estimation, *Annals of Mathematical Statistics*, **17**, 34–43.

Hamdan, M. A. and Al-Bayyati, H. A. (1969), A note on the bivariate Poisson distribution, *The American Statistician*, **23**, 32–33.

Hasselblad, V. (1969), Estimation of finite mixtures of distributions from the exponential family, *Journal of the American Statistical Association*, **64**, 1459–1471.

Hoffmann-Jørgensen, J. (1993), Stable densities, *Theory of Probability and Its Applications*, **38**, 350–355.

Hogg, R. V. and Craig, A. T. (1995), *Introduction to Mathematical Statistics*, 5th edition, MacMillan: New York.

Ingber, L. (1996), Adaptive simulated annealing (ASA): lessons learned, *Journal of Control and Cybernetics*, **25**, 33–54.

Joe, H. (1997), *Multivariate Models and Dependence Concepts*, Chapman and Hall: London.

Johnson, N. L. (1949), Systems of frequency curves generated by methods of translation, *Biometrika*, **36**, 149–176.

Johnson, N. L., Kotz, S. and Balakrishnan, N. (1994), *Continuous Univariate Distributions*, volume 1, 2nd edition, Wiley: New York.

Johnson, N. L., Kotz, S. and Balakrishnan, N. (1995), *Continuous Univariate Distributions*, volume 2, 2nd edition, Wiley: New York.

Johnson, N. L., Kotz, S. and Balakrishnan, N. (1997), *Discrete Multivariate Distributions*, Wiley: New York.

Johnson, N. L., Kotz, S. and Kemp, A. W. (1993), *Univariate Discrete Distributions*, 2nd edition, Wiley: New York.

Judge, G. G., Griffiths, W. E., Carter Hill, R., Lütkepohl, H. and Lee, T.-C. (1985), *The Theory and Practice of Econometrics*, 2nd edition, Wiley: New York.

Kerawala, S. M. and Hanafi, A. R. (1941), Tables of monomial symmetric functions, *Proceedings of the National Academy of Sciences, India*, **11**, 51–63.

Kerawala, S. M. and Hanafi, A. R. (1942), Tables of monomial symmetric functions, *Proceedings of the National Academy of Sciences, India*, **12**, 81–96.

Kerawala, S. M. and Hanafi, A. R. (1948), Tables of monomial symmetric functions of weight 12 in terms of power sums, *Sankhya*, **8**, 345–59.

Kirkpatrick, S., Gelatt, C. D. and Vecchi, M. P. (1983), Optimization by simulated annealing, *Science*, **220**, 671–680.

Kotz, S., Balakrishnan, N. and Johnson, N. L. (2000), *Continuous Multivariate Distributions*, volume 1, 2nd edition, Wiley: New York.

Lancaster, T. (1992), *The Econometric Analysis of Transition Data*, Cambridge University Press: Cambridge.

Le Cam, L. (1986), The Central Limit Theorem around 1935, *Statistical Science*, **1**, 78–96.

Lehmann, E. (1983), *Theory of Point Estimation*, Wiley: New York.

Luenberger, D. G. (1984), *Linear and Non-Linear Programming*, 2nd edition, Addison-Wesley: Reading, MA.

Maddala, G. S. (1983), *Limited-Dependent and Qualitative Variables in Econometrics*, Cambridge University Press: Cambridge.

McCabe, B. and Tremayne, A. (1993), *Elements of Modern Asymptotic Theory with Statistical Applications*, Manchester University Press: Manchester.

McCullagh, P. and Nelder, J. A. (1989), *Generalized Linear Models*, 2nd edition, Chapman and Hall: London.

McCulloch, J. H. (1996), Financial applications of stable distributions, Chapter 13 in Maddala, G. S. and Rao, C. R. (eds.) (1996), *Handbook of Statistics*, volume 14, Elsevier: Amsterdam.

McCullough, B. D. (1998), Assessing the reliability of statistical software: Part I, *The American Statistician*, **52**, 358–366.

McCullough, B. D. (1999a), Assessing the reliability of statistical software: Part II, *The American Statistician*, **53**, 149–159.

McCullough, B. D. (1999b), The reliability of econometric software: E-Views, LIMDEP, SHAZAM and TSP, *Journal of Applied Econometrics*, **14**, 191–202.

McCullough, B. D. (2000), The accuracy of *Mathematica* 4 as a statistical package, *Computational Statistics*, **15**, 279–299.

McCullough, B. D. and Vinod, H. D. (1999), The numerical reliability of econometric software, *Journal of Economic Literature*, **37**, 633–665.

McKelvey, R. D. and Zavoina, W. (1975), A statistical model for the analysis of ordinal level dependent variables, *Journal of Mathematical Sociology*, **4**, 103–120.

Merton, R. C. (1990), *Continuous-Time Finance*, Blackwell: Cambridge, MA.

Mittelhammer, R. C. (1996), *Mathematical Statistics for Economics and Business*, Springer-Verlag: New York.

Morgenstern, D. (1956), Einfache Beispiele Zweidimensionaler Verteilungen, *Mitteilingsblatt für Mathematische Statistik*, **8**, 234–235.

Mosteller, F. (1987), *Fifty Challenging Problems in Probability with Solutions*, Dover: New York.

Muirhead, R. J. (1982), *Aspects of Multivariate Statistical Theory*, Wiley: New York.

Nelsen, R. B. (1999), *An Introduction to Copulas*, Springer-Verlag: New York.

Nerlove, M. (2002), *Likelihood Inference in Econometrics*, forthcoming.

Nolan, J. P. (2001), *Stable Distributions: Models for Heavy-Tailed Data*, Birkhäuser: Boston.

Ord, J. K. (1968), The discrete Student's t distribution, *Annals of Mathematical Statistics*, **39**, 1513–1516.

Ord, J. K. (1972), *Families of Frequency Distributions*, Griffin: London.

O'Toole, A. L. (1931), On symmetric functions and symmetric functions of symmetric functions, *Annals of Mathematical Statistics*, **2**, 101–149.

Pagan, A. and Ullah, A. (1999), *Nonparametric Econometrics*, Cambridge University Press: Cambridge.

Parzen, E. (1979), Nonparametric statistical data modeling, *Journal of the American Statistical Association*, **74**, 105–131 (including comments).

Pearson, K. (1902), On the probable error of frequency constants, *Biometrika*, **2**, 273–281.

Polak, E. (1971), *Computational Methods in Optimization*, Academic Press: New York.

Pratt, J. W. (1981), Concavity of the log-likelihood, *Journal of the American Statistical Association*, **76**, 103–106.

Press, W. H., Flannery, B. P., Teukolsky, S. A. and Vetterling, W. T. (1992), *Numerical Recipes in C: the Art of Scientific Computing*, 2nd edition, Cambridge University Press: Cambridge.

Rose, C. (1995), A statistical identity linking folded and censored distributions (with application to exchange rate target zones), *Journal of Economic Dynamics and Control*, **19**, 1391–1403.

Rose, C. and Smith, M. D. (1996a), On the multivariate normal distribution, *The Mathematica Journal*, **6**, 32–37.

Rose, C. and Smith, M. D. (1996b), Random[Title]: manipulating probability density functions, Chapter 16 in Varian, H. (ed.) (1996), *Computational Economics and Finance*, Springer-Verlag/TELOS: New York.

Rose, C. and Smith, M. D. (1997), Random number generation for discrete variables, *Mathematica in Education and Research*, **6**, 22–26.

Rose, C. and Smith, M. D. (2000), Symbolic maximum likelihood estimation with *Mathematica*, *The Statistician: Journal of the Royal Statistical Society*, Series D, **49**, 229–240.

Roussas, G. G. (1997), *A Course in Mathematical Statistics*, 2nd edition, Academic Press: San Diego.

Schell, E. D. (1960), Samuel Pepys, Isaac Newton, and probability, in Rubin, E. (ed.) (1960), Questions and answers, *The American Statistician*, **14**, 27–30.

Schervish, M. J. (1984), Multivariate normal probabilities with error bound, *Applied Statistics: Journal of the Royal Statistical Society*, Series C, **33**, 81–94 (see also: Corrections (1985), *Applied Statistics: Journal of the Royal Statistical Society*, Series C, **34**, 103–104).

Schwert, G. W. (1990), Indexes of United States stock prices from 1802 to 1987, *Journal of Business*, **63**, 399–426.

Sheather, S. J. and Jones, M. C. (1991), A reliable data-based bandwidth selection method for kernel density estimation, *Statistical Methodology: Journal of the Royal Statistical Society*, Series B, **53**, 683–690.

Silverman, B. W. (1986), *Density Estimation for Statistics and Data Analysis*, Chapman and Hall: London.

Silvey, S. D. (1975), *Statistical Inference*, Chapman and Hall: Cambridge.

Simonoff, J. S. (1996), *Smoothing Methods in Statistics*, Springer-Verlag: New York.

Sofroniou, M. (1996), Numerics in *Mathematica* 3.0, *The Mathematica Journal*, **6**, 64–73.

Spanos, A. (1999), *Probability Theory and Statistical Inference: Econometric Modeling with Observational Data*, Cambridge University Press: Cambridge.

StatLib, `http://lib.stat.cmu.edu/`

Stine, R. A. (1996), Data analysis using *Mathematica*, Chapter 14 in Varian, H. R. (ed.), *Computational Economics and Finance: Modeling and Analysis with Mathematica*, Springer-Verlag/TELOS: New York.

Stuart, A. and Ord, J. K. (1994), *Kendall's Advanced Theory of Statistics*, volume 1, 6th edition, Edward Arnold: London (also Wiley: New York).

Stuart, A. and Ord, J. K. (1991), *Kendall's Advanced Theory of Statistics*, volume 2, 5th edition, Edward Arnold: London.

'Student' (1908), The probable error of a mean, *Biometrika*, **6**, 1–25.

Sukhatme, P. V. (1938), On bipartitional functions, *Philosophical Transactions* A, **237**, 375–409.

Tadikamalla, P. R. and Johnson, N. L. (1982), Systems of frequency curves generated by transformations of logistic variables, *Biometrika*, **69**, 461–465.

Taylor, H. M. and Karlin, S. (1998), *An Introduction to Stochastic Modeling*, 3rd edition, Academic Press: San Diego.

Thiele, T. N. (1903), *Theory of Observations*, C. and E. Layton: London (reprinted in Thiele, T. N. (1931), *Annals of Mathematical Statistics*, **2**, 165–306).

Titterington, D. M., Smith, A. F. M. and Makov, U. E. (1985), *Statistical Analysis of Finite Mixture Distributions*, Wiley: Chichester.

Tjur, T. (1980), *Probability Based on Radon Measures*, Wiley: New York.

Tracy, D. S. and Gupta, B. C. (1974), Generalised h-statistics and other symmetric functions, *Annals of Statistics*, **2**, 837–844.

Tukey, J. W. (1956), Keeping moment-like computations simple, *Annals of Mathematical Statistics*, **25**, 37–54.

Uchaikin, V. V. and Zolotarev, V. M. (1999), *Chance and Stability: Stable Distributions and Their Applications*, VSP: Holland.

Varian, H. R. (1975), A Bayesian approach to real estate assessment, in Fienberg, S. and Zellner, A. (1975) (eds.), *Studies in Bayesian Econometrics and Statistics*, North-Holland: Amsterdam.

Varian, H. R. (ed.) (1996), *Computational Economics and Finance*, Springer-Verlag / TELOS: New York.

Walley, P. (1991), *Statistical Reasoning with Imprecise Probabilities*, Chapman and Hall: London.

Walley, P. (1996), Inferences from multinomial data: learning about a bag of marbles, *Statistical Methodology: Journal of the Royal Statistical Society, Series B*, **58**, 3–57 (with discussion).

Walpole, R. E. and Myers, R. H. (1993), *Probability and Statistics for Engineers and Scientists*, 5th edition, Macmillan: New York.

Wishart, J. (1952), Moment coefficients of the k-statistics in samples from a finite population, *Biometrika*, **39**, 1–13.

Wolfram, S. (1999), *The Mathematica Book*, 4th edition, Wolfram Media/Cambridge University Press: Cambridge.

Ziaud-Din, M. (1954), Expression of the k-statistics k_9 and k_{10} in terms of power sums and sample moments, *Annals of Mathematical Statistics*, **25**, 800–803.

Ziaud-Din, M. (1959), Expression of k-statistic k_{11} in terms of power sums and sample moments, *Annals of Mathematical Statistics*, **30**, 825–828.

Ziaud-Din, M. and Ahmad, M. (1960), On the expression of the k-statistic k_{12} in terms of power sums and sample moments, *Bulletin of the International Statistical Institute*, **38**, 635–640.

Zolotarev, V. M. (1995), On representation of densities of stable laws by special functions, *Theory of Probability and Its Applications*, **39**, 354–362.

Index

A

abbreviations 25
absolute values 41, 59, 284, 422
accuracy
    numerical 116, 230–231, 423–425
    symbolic 421–422
admissible estimator 302
Ali–Mikhail–Haq 212, 249
ancillary statistic 337
animations
    approximation error 286
    bivariate Exponential pdf (Gumbel Model II) 11
    bivariate Gamma pdf (McKay) 248
    bivariate Normal pdf 217
    bivariate Normal quantiles 219
    bivariate Normal–Uniform pdf 214
    bivariate Uniform pdf 213
    conditional mean and variance 215
    contours of bivariate Normal component-mix 249
    contours of the trivariate Normal pdf 227
    limit distribution of Binomial is Poisson 281
    Lorenz curve for a Pareto distribution 44
    non-parametric kernel density estimate 183
    Pearson system 150
    pmf of sum of two dice (fair vs shaved) 87
    Robin Hood 223
arbitrary-precision numbers 423–424
Arc–Sine distribution 6
ARCH model 384–392
assumptions technology 8–9
asymptotic distribution 282–286
    definition 282
    of MLE (invariance property) 369–371
    of MLE (maximum likelihood estimator) 367
    of MLE (multiple parameters) 371–374
    of MLE (with hypothesis testing) 393–394
    of sample mean 287
    of sample sum 287
asymptotic Fisher Information 375, 376
asymptotic theory 277–300
asymptotic unbiased estimator 366
asymptotic variance-covariance matrix 395–399, 404, 407, 410, 415, 418–419
augmented symmetric function 272–276
Azzalini's skew-Normal distribution 80, 225

B

bandwidth 181
Bates's distribution 139, 289–290
Bernoulli distribution 89–91
    cumulant generating function 271
    distribution of sample sum 141
    likelihood 352
    Logit model 90–91
    method of moments estimator 184
    pmf 89
    sample mean vs sample median 309–310
    sufficiency in Bernoulli trials 337
Berry–Esseen Theorem 453
best unbiased estimator (**BUE**) 325, 335–336, 362, 364
Beta distribution
    as defining Pearson *Type I(J)* 185
    as member of Pearson family 158
    cumulants 64
    fitted to censored marks 353–354
    MLE 363
    pdf 64
Beta–Binomial distribution 106
bias 306
Binomial distribution 91–95
    as limiting distribution of Ehrenfest urn 95
    as sum of n Bernoulli rv's 91, 141
    cdf 92
    kurtosis 93
    limit distribution 280, 281
    mgf 141, 281
    Normal approximation 93, 281, 299
    pmf 91
    Poisson approximation 95, 280, 300
    product cumulant 270
biology 107, 380
Birnbaum–Saunders distribution
    cdf, pdf, quantiles 38–39
    pseudo-random number generation 78
bivariate Cauchy distribution 237
bivariate Exponential distribution
    Gumbel Model I, 204
    Gumbel Model II, 11–13
bivariate Gamma (McKay) 248
bivariate Logistic distribution (Gumbel) 248, 249
bivariate Normal distribution 216–226
    cdf 216, 217, 229–231

bivariate Normal distribution (*cont.*)
   characteristic function 221
   component-mixture 249
   conditional distribution 220
   contour plot 218
   marginal distributions 220
   mgf 220
   orthant probability 231
   pdf 216, 217
   pseudo-random number generation 232–234
   quantiles 218–219
   truncated bivariate Normal 224–226
   variance-covariance matrix 220
   visualising random data 234
bivariate Normal–Uniform distribution 213–215
bivariate Poisson 243–248
   mgf 246
   moments 246–248
   pgf 244
   pmf 244–245
bivariate Student's t 237–238
bivariate Uniform (à la Morgenstern) 212–213
Black–Scholes option pricing 70–71, 447
Brownian motion 70

C

Cauchy distribution
   as a stable distribution 58
   as ratio of two Normals 134
   as transformation of Uniform 119
   characteristic function 143
   compared to Sinc^2 pdf 35–36
   distribution of sample mean 143
   mean 36
   pdf 35, 143
   product of two Cauchy rv's 148
cdf (cumulative distribution function)
   definitions
      - continuous multivariate 191
      - continuous univariate 31
      - discrete multivariate 194
      - discrete univariate 81
   limit distribution 279
   numerical cdf 39
   of Arc–Sine 7
   of Binomial 92
   of Birnbaum–Saunders 39
   of bivariate Exponential
      - Gumbel Model I, 204
      - Gumbel Model II, 12
   of bivariate Normal 216, 217, 229–231
   of bivariate Normal–Uniform 214
   of bivariate Uniform 213
   of half-Halo 75
   of Inverse Triangular 13
   of Levy 74
   of Maxwell–Boltzmann 32
   of Pareto distribution 38
   of Pascal 10
   of Reflected Gamma 33
   of stable distribution 59
   of trivariate Normal 229–231
   see also inverse cdf
censored data 354
censored distribution 68–69
   and option pricing 70–71
   and pseudo-random number generation 114
   censored Lognormal 71
   censored Normal 69
   censored Poisson 327
Central Limit Theorem 286–292, 365
   Generalised Central Limit Theorem 56
   Lindeberg–Feller 453
   Lindeberg–Lévy 287, 366, 368, 373
central moment 45, 200
characteristic function 50–60
   definition
      - multivariate 203
      - univariate 50
   inversion of cf
      - numerical 53, 55, 60
      - symbolic 53–60
   Inversion Theorem 53
   of bivariate Normal 221
   of Cauchy 58, 143
   of Levy 58
   of Lindley 51
   of Linnik 54
   of Normal 50, 57
   of Pareto 51
   of stable distribution 56–57
   relation to pgf 84
   transformations 131
   Uniqueness Theorem 52
Chebyshev's Inequality 295–296
Chi-squared distribution
   as square of a Normal rv 129, 131, 299
   asymptotic distribution of sample mean 283
   distribution of sample sum 142
   mean deviation 41, 421–422
   method of moments estimator 283
   mgf 131
   mode 36
   pdf 36, 41
   ratio of two Chi-squared rv's 135
   relation to Fisher F 135
   van Beek's bound 284–285
   see also noncentral Chi-squared
coefficient of variation 40
complete sufficient statistic 343, 346
component-mix distribution 102–104
   bivariate Normal component-mixture 249
   estimating a Poisson two-component-mix 405–411

conditional expectation $E[X \mid a<X \leq b]$ 66–67
    odd-valued Poisson rv 97–98
    truncated Normal 67
conditional expectation $E[X \mid Y=y]$ 197–199
    definitions: continuous 197, discrete 199
    deriving conditional mean and variance
      - continuous 198, 215
      - discrete 199
    Normal Linear Regression model 221–222
    Rao–Blackwell Theorem 342
    regression function 197, 221–222
conditional pdf $f(X \mid a<X \leq b)$ 65–67
conditional pdf $f(X \mid Y=y)$ 197
    of bivariate Exponential (Gumbel Model II) 12
    of bivariate Normal 220
    of bivariate Normal–Uniform 215
    Normal Linear Regression model 221–222
conditional pmf $f(X \mid Y=y)$ 199
conditional probability 65, 97
confidence interval 394–395
consistency 292–294, 367, 457
consistent estimator 294, 297
Continuous Mapping Theorem 366, 456
contour plot 188, 218, 227
convergence
    in distribution 278–282, 293
    in probability 292–298
    to a constant 294
copulae 211–215
correlation 201
    and independence 125, 211
    and positive definite matrix 228
    between k-statistics 268
    between order statistics 314
    definition 201
    trivariate example 202
    visualising correlation 212–213
    see also covariance
covariance 201
    between sample moments 266
    definition 201
    derived from central mgf 205
    in terms of raw moments 206
    of bivariate Exponential (Gumbel Model II) 12
    trivariate example 202
    see also correlation
Cramér–Rao lower bound 333–335
    for Extreme Value 336
    for Inverse Gaussian 334–335
    for Poisson 334
cumulant generating function
    definition 60, 203
    of Bernoulli 271
    of Beta 64
    of Poisson 96
cumulants 60
    in terms of moments 62, 206–207
    of Bernoulli 271
    of Beta 64
    of k-statistics 267–271
    of Poisson 96
    product cumulant 209–210, 269
    unbiased estimator of cumulants 256–260
cumulative distribution function (*see* cdf)

D

data
    censored 354
    population vs sample 151
    raw vs grouped 151
    —
    American NFL matches 260
    Australian age profile 239
    Bank of Melbourne share price 384
    censored student marks 354
    death notices 405
    grain 153
    income and education 396
    medical patients and dosage 90
    NB1, NB2 418
    nerve (biometric) 380, 418
    psychiatric (suicide) 412
    sickness 155
    snowfall 181
    student marks 151, 162, 170, 177, 354
    Swiss bank notes 19, 185
    US stock market returns 185
    word count 418
degenerate distribution 103, 238, 280
delta method 456
density estimation
    Gram–Charlier 175–180
    Johnson 164–174
    non-parametric kernel density 181–183
    Pearson 149–163
dice 84–87
differentiation with respect to powers 326
Discrete Uniform distribution 115
distributions
    asymptotic
    censored
    component-mix
    degenerate
    elliptical
    empirical
    limit distribution
    mixing
    parameter-mix
    piecewise
    spherical
    stable family
    stopped-sum
    truncated
    zero-inflated

INDEX

distributions – Continuous
   α-Laplace (*see* Linnik)
   Arc–Sine
   Azzalini's skew-Normal
   Bates
   Beta
   Birnbaum–Saunders
   Cauchy
   Chi-squared
   Double Exponential (*see* Laplace)
   Exponential
   Extreme Value
   Fisher F
   Gamma
   Gaussian (*see* Normal)
   half-Halo
   half-Normal
   Hyperbolic Secant
   Inverse Gamma
   Inverse Gaussian
   Inverse Triangular
   Irwin–Hall
   Johnson family
   Laplace
   Levy
   Lindley
   Linnik
   Logistic
   Lognormal
   Maxwell–Boltzmann
   noncentral Chi-squared
   noncentral F
   Normal
   Pareto
   Pearson family
   Power Function
   Random Walk
   Rayleigh
   Rectangular (*see* Uniform)
   Reflected Gamma
   semi-Circular (*see* half-Halo)
   Sinc^2
   stable
   Student's t
   Triangular
   Uniform
   Weibull
distributions – Discrete
   Bernoulli
   Beta–Binomial
   Binomial
   Discrete Uniform
   Geometric
   Holla
   Hypergeometric
   Logarithmic
   Negative Binomial
   Pascal
   Poisson
   Pólya–Aeppli
   Riemann Zeta
   Waiting-time Negative Binomial
   Waring
   Yule
   Zero-Inflated Poisson
   Zipf (*see* Riemann Zeta)
distributions – Multivariate
   bivariate Cauchy
   bivariate Exponential (Gumbel Model I and II)
   bivariate Gamma (McKay)
   bivariate Logistic (Gumbel)
   bivariate Normal
   bivariate Normal–Uniform (à la Morgenstern)
   bivariate Poisson
   bivariate Student's t
   bivariate Uniform (à la Morgenstern)
   Multinomial
   multivariate Cauchy
   multivariate Gamma (Cheriyan and Ramabhadran)
   multivariate Normal
   multivariate Student's t
   Trinomial
   trivariate Normal
   truncated bivariate Normal
domain of support 31, 81–85
   circular 191
   non-rectangular 124, 125, 190–191, 314
   rectangular 124, 190
   triangular 191, 314, 317
dominant estimator 302
Dr Faustus 421

E

economics and finance 43–45, 56, 70–72, 108–109, 117, 121, 384
Ehrenfest urn 94–95
ellipse 218, 236
ellipsoid 227
elliptical distributions 234
empirical pdf 73, 77, 154, 381, 383
empirical pmf 16, 110, 111, 112
engineering 122
entropy 15
Epanechnikov kernel 182
estimator
   admissible 302
   asymptotic unbiased 366
   BUE (best unbiased) 325, 335–336, 362, 364
   consistent 294, 297
   density (*see* density estimation)
   dominant 302
   estimator vs estimate 357
   Fisher estimator 395–396, 397, 404

h-statistic 253–256
Hessian estimator 395–396, 398, 404
inadmissible 302, 321–322
k-statistic 256–261
maximum likelihood estimator (*see* MLE)
method of moments 183–184, 283
minimax 305
minimum variance unbiased 341–346, 364
non-parametric kernel density 181–183
ordinary least squares 385
Outer-product 395–396, 398
sample central moment 360
sample maximum 320–321
sample mean (*see* sample mean)
sample median 309–310, 318–320
sample range 320–321
sample sum 277, 287
unbiased estimator of parameters 325–347
unbiased estimator of population moments 251–261
expectation operator
   basic properties 32
   definitions
     - continuous 32
     - discrete 83
     - multivariate 200
   when applied to sample moments 263
Exponential distribution
   bivariate 11–13, 204
   difference of two Exponentials 139–140
   distribution of sample sum 141–142
   likelihood 351
   MLE (numerical) 381
   MLE (symbolic) 358
   order statistics 313–314
   pdf 141, 313, 344, 358
   relation to Extreme Value 121
   relation to Pareto 121
   relation to Rayleigh 122
   relation to Uniform 121
   sufficient statistic 344
   sum of two Exponentials 136
Exponential regression 375–376, 396
Extreme Value distribution
   Cramér–Rao lower bound 336
   pdf 336, 377
   relation to Exponential 121

F

factorial moment 60, 206–207, 247
factorial moment generating function 60, 203, 247
factorisation criterion 339–341
families of distributions
   Gram–Charlier 175–180
   Johnson 164–174
   Pearson 149–163

stable family 56–61
fat tails 56, 108–109
   see also kurtosis
first-order condition 21, 36, 357–361, 363
Fisher estimator 395–396, 397, 404
Fisher F distribution 135
Fisher Information 326–332
   and MLE (regularity conditions) 367–368, 372–373
   asymptotic Fisher Information 375, 376
   first derivative form vs second derivative 329
   for censored Poisson 327–328
   for Gamma 331–332
   for Inverse Gaussian 18
   for Lindley 326
   for Normal 330–331
   for Riemann Zeta 329
   for Uniform 330
Frank 212
frequency polygon 73, 77, 151, 154, 380
   see also plotting techniques
Function Form 82
functions of random variables 117–148
fundamental expectation result 274

G

games
   archery (Robin Hood) 222–224
   cards, poker 101
   craps 87–89, 115
   dice (fair and unfair) 84–87
Gamma distribution
   as member of Pearson family 157, 185
   as sum of n Exponential rv's 141–142
   bivariate Gamma (McKay) 248
   Fisher Information 331–332
   hypothesis testing 392–394
   method of moments estimator 184
   mgf 142, 456
   MLE (numerical) 382–383
   multivariate (Cheriyan & Ramabhadran) 208
   pdf 73, 142
   pseudo-random number generation 73
   relation to Inverse Gamma 147
Gamma regression model 419
gas molecules 32
Gaussian kernel 19, 182
generating functions 46–56, 203–205
Geometric distribution
   definition 98
   distribution of difference of two rv's 148
   pmf 98
Gini coefficient 40, 43–45
gradient 357–361
Gram–Charlier expansions 175–180
graphical techniques (*see* plotting techniques)
Greek alphabet 28

H

h-statistic 253–256
half-Halo distribution 75, 80
half-Normal distribution 225
Helmert transformation 145
HELP 5
Hermite polynomial 175, 179, 449
Hessian estimator 395–396, 398, 404
Hessian matrix 358, 360
histogram 18, 155 (*see also* plotting techniques)
Holla's distribution 105, 112
Hyperbolic Secant distribution 80
Hypergeometric distribution 100–101

I

inadmissible estimator 302, 321–322
income distribution 43–44, 121
independence
    correlation and dependence 125, 211
    mutually stochastically independent 210
independent product space 124, 190
Invariance Property 360, 369–371, 401, 410, 417
inverse cdf
    numerical inversion 38–39, 75–77, 109
    symbolic inversion 37–38, 74–75
    of Birnbaum–Saunders 38–39
    of half-Halo 75
    of Levy 74
    of Pareto 38, 43
Inverse Gamma distribution
    as member of Pearson family 185
    pdf 365
    relation to Gamma 147, 365
    relation to Levy 58
Inverse Gaussian distribution
    Cramér–Rao lower bound 334–335
    Fisher Information 18
    pdf 18, 334
    relation to Random Walk distribution 147
Inverse Triangular distribution 13–14
Inversion Theorem 53
Irwin–Hall distribution 55, 139
isobaric 272

J

Jacobian of the transformation 118, 123, 130, 223
Johnson family 164–174
    as transformation of a Logistic rv 185
    as transformation of a Normal rv 164
    Types and chart 164
        - S_L (Lognormal) 165–167
        - S_U (Unbounded) 168–172
        - S_B (Bounded) 173–174

K

k-statistic 20, 256–261
kernel density (*see* non-parametric kernel density)
Khinchine's Theorem 298
Khinchine's Weak Law of Large Numbers 278, 296–298, 366
Kronecker product 437
kurtosis
    building your own function 446
    definition 40–41
    of Binomial 93
    of Poisson 446
    of Weibull 42
    Pearson family 149–150

L

Laplace distribution
    as Linnik 54
    as Reflected Gamma 33
    order statistics of 23, 315–317
    relation to Exponential 139–140
latent variable 353, 412
Lehmann–Scheffé Theorem 346
Levy distribution
    as a stable distribution 58
    as an Inverse Gamma 58
    cdf, pdf, pseudo-random number 74
likelihood
    function 21, 350–357
    observed 22, 351–357
    see also log-likelihood
limit distribution
    definition 279
    of Binomial 280, 281
    of sample mean (Normal) 279
limits in *Mathematica* 278
Lindley distribution
    characteristic function 51
    Fisher Information 326–327
    pdf 51, 327
linear regression function 221
linex (linear–exponential) loss 322
linguistics 107
Linnik distribution 54
List Form 82, 111
log-likelihood
    concentrated 361, 382–383, 418
    function 21, 357–376, 381
    observed log-likelihood
        - ARCH model (stock prices) 387
        - Exponential model (nerve data) 381
        - Exponential regression (income) 396
        - Gamma model (nerve data) 382–383
        - Logit model (dosage data) 90
        - Ordered Probit model (psychiatric data) 414–415

- Poisson two-component-mix model 405–406
   see also likelihood
Logarithmic distribution 115
Logistic distribution
   as base for a Johnson-style family 185
   bivariate 248, 249
   pdf 23, 318
   order statistics of 23
   relation to Uniform 147
   sample mean vs sample median 318–320
Logit model 90–91
Lognormal distribution
   and stock prices 71
   as member of Johnson family 165–167
   as transformation of Normal 120, 165
   censored below 71
   moments of sample sum 276
   pdf 71, 120
Lorenz curve 43–44
loss function 301–305
   asymmetric 303–304
   asymmetric quadratic 322, 323
   linex (linear–exponential) 322
   quadratic 306

M

machine-precision numbers 423–425
marginal distribution 195–196
   and copulae 211
   joint pdf as product of marginals 210, 211, 351, 355
   more examples 12, 126, 133–137, 146, 204, 214, 220, 224–225, 237–238, 244
Markov chain 94, 447–448
Markov's inequality 295–296
Mathematica
   assumptions technology 8–9
   bracket types 27
   changes to default behaviour 443–445
   differentiation with respect to powers 326
   Greek alphabet 28
   how to enter μ_r 30
   kernel (fresh and crispy) 5, 425
   limits 278
   lists 428–429
   matrices 433–437, 445
   notation (common) 27
   notation entry 28–30
   packages 425
   replacements 27
   subscripts 429–432
   timings 30
   upper and lower case conventions 24
   using Γ in Input cells 443
   vectors 438–443
   see also plotting techniques

mathStatica
   Basic vs Gold version 4
   Continuous distribution palette 5
   Discrete distribution palette 5
   HELP 5
   installation 3
   loading 5
   registration 3
   working with parameters 8
maximum likelihood estimation (*see* MLE)
Maxwell–Boltzmann distribution 32
mean 35–36, 45
   see also sample mean
mean deviation 40, 41, 299, 421–422
mean square error (*see* MSE)
median 37
   of Pareto distribution 37–38
   see also sample median
medical 90–91, 155, 380, 405, 412
method of moments estimator 183–184
   for Bernoulli 184
   for Chi-squared 283
   for Gamma 184
mgf (moment generating function)
   and cumulant generating function 60
   and independence 210
   central mgf 93, 203, 205, 247
   definition 46, 203
   Inversion Theorem 53
   Uniqueness Theorem 52
   of Binomial 93, 141, 281
   of bivariate Exponential (Gumbel Model I) 204
   of bivariate Exponential (Gumbel Model II) 12
   of bivariate Normal 220
   of bivariate Poisson 246
   of Chi-squared 131
   of Gamma 142, 456
   of Multinomial 239, 241–242, 242–243
   of multivariate Gamma 208
   of multivariate Normal 249
   of noncentral Chi-squared 144
   of Normal 47
   of Pareto 49
   of sample mean 141
   of sample sum 141
   of sample sum of squares 141
   of Uniform 48
MGF Method 52–56, 130–132, 141–147
MGF Theorem 52, 141
   more examples 281, 364–365
minimax estimator 305
minimum variance unbiased estimation (*see* MVUE)
mixing distributions 102–109
   component-mix 102–104, 249, 405–411
   parameter–mix 105–109

MLE (maximum likelihood estimation) 357–376
    asymptotic properties 365–366, 371–376
    general properties 362
    invariance property 369–371
    more than one parameter 371–374
    non-iid samples 374–376
    numerical MLE (*see* Chapter 12)
      - ARCH model (stock prices) 387
      - Exponential model (nerve data) 381
      - Exponential regression model (income) 396
      - Gamma model (nerve data) 382–383
      - Logit model (dosage data) 90
      - Normal model (random data) 418
      - Ordered Probit model (psychiatric data) 414–415
      - Poisson two-component-mix model 405–406
    regularity conditions
      - basic 367–369
      - more than one parameter 371–372
      - non-iid samples 374–375
    small sample properties 363–365
    symbolic MLE (*see* Chapter 11)
      - for Exponential 358
      - for Normal 359–360, 418
      - for Pareto 360–361
      - for Power Function 362–363
      - for Rayleigh 21
      - for Uniform 377
mode 36
moment conversion functions
    univariate 62–64
    multivariate 206–210
moment generating function (*see* mgf)
moments
    central moment 45, 200
    factorial moment 60, 206–207
    fitting moments (*see* Pearson, Johnson, method of moments)
    negative moment 80
    population moments vs sample moments 251
    product moment 200, 266
    raw moment 45, 200
moments of moments 261–271
    introduction 20
moments of sampling distributions 251–276
monomial symmetric function 273
Monte Carlo 290
    see also pseudo-random number generation
    see also simulation
Morgenstern 212
MSE (mean square error)
    as risk 306–311
    comparing h-statistics with polyaches 264–266
    of sample median and sample mean (Logistic) 318–320
    of sample range and sample maximum (Uniform) 320–321
    weak law of large numbers 296–297
multinomial coefficient 451
Multinomial distribution 238–243
multiple local optima 400
multivariate Cauchy distribution 236
multivariate Gamma distribution (Cheriyan and Ramabhadran) 208
multivariate Normal distribution 216–235
multivariate Student's t 236
mutually stochastically independent 210
MVUE (minimum variance unbiased estimation) 341–346, 364

N

Negative Binomial distribution 99, 105, 418
noncentral Chi-squared distribution
    as Chi-squared–Poisson mixture 105
    derivation 144
    exercises 299
noncentral F distribution 135
non-parametric kernel density 181–183
    with bi-weight, tri-weight kernel 182
    with Epanechnikov kernel 182
    with Gaussian kernel 19, 182
non-rectangular domain 124, 125, 190–191, 320–321
Normal distribution
    and Gram–Charlier expansions 175
    as a stable distribution 57
    as limit distribution of a Binomial 93, 281, 299
    as member of Johnson family 164–165, 167
    as member of Pearson family 150, 158
    asymptotic distribution of MLE of (μ, σ^2) 372–374
    basics 8
    bivariate Normal 216–226
    censored below 69
    central moments 265
    characteristic function 50, 57
    characteristic function of $X_1 X_2$ 132
    conditional expectation of sample median, given sample mean 342–343
    distribution:
      - of product of two Normals 132, 133
      - of ratio of two Normals 134
      - of X^2 129, 131
      - of sample mean 143, 294–295
      - of sample sum of squares 144
      - of sample sum of squares about the mean 145
    estimators for the Normal variance 307–308
    finance 56, 108–109
    Fisher Information 330–331
    limit distribution of sample mean 279

limit Normal distribution 362, 367
- examples 369, 392–395
mgf 47
mgf of X^2 131
MLE of (μ, σ^2) 359–360, 418
MVUE of (μ, σ^2) 346
Normal approximation to Binomial 93, 281, 299
pseudo-random number generation
- approximate 291–292
- exact 72–73, 418
QQ plot 291
raw moments 46
relation to Cauchy 134
relation to Chi-squared 129, 131
relation to Lognormal 120
risk of a Normally distributed estimator 303–304
sample mean as consistent estimator of population mean 294–295
standardising a Normal rv 120
sufficient statistics for (μ, σ^2) 340–341
trivariate Normal 226–228
truncated above 65–66, 67
working with σ vs σ^2 326, 377, 455
see also Invariance Property
Normal linear regression model 221–222, 385, 457

notation
Mathematica notation
- bracket types 27
- Greek alphabet 28
- how to enter μ_r 30
- notation (common) 27
- notation entry 28–30
- replacements 27
- subscripts 429–432
- upper and lower case conventions 24
- using Γ in Input cells 443
statistics notation
- abbreviations 25
- sets and operators 25
- statistics notation 26
- upper and lower case conventions 24

O

one-to-one transformation 118
optimisation
differentiation with respect to powers 326
first-order condition 21, 36, 357–361, 363
gradient 357–361
Hessian matrix 358, 360
multiple local optima 400
score 357–361
second-order condition 22, 36–37, 357–360
unconstrained vs constrained numerical optimisation 369, 379, 388–389, 401, 414

optimisation algorithms 399–405
Armijo 408
BFGS (Broyden–Fletcher–Goldfarb–Shanno) 399–400, 403, 405–411, 459
DFP (Davidon–Fletcher–Powell) 403
direct search 400
genetic 400
Golden Search 401
Goldstein 408
gradient method 400, 401–405
line search 401
Method → Newton 390–391, 397, 403, 415, 459
Method → QuasiNewton 403, 406–407, 419, 459
NR (Newton Raphson) 390–391, 397, 399–400, 403, 412–417, 458–459
numerical convergence 404–405
Score 403–404
simulated annealing 400
taboo search 400
option pricing 70–72
order statistics 311–322
distribution of:
- sample maximum 312, 321
- sample minimum 312
- sample median 318–320
- sample range 320–321
for Exponential 313–314
for Laplace 23, 315–317
for Logistic 23
for Uniform 312
joint order statistics 23, 314, 316, 320
Ordered Probit model 412–417
ordinary least squares 385
orthant probability 231
Outer-product estimator 395–396, 398

P

p-value 393–394
parameter identification problem 414
parameter-mix distribution 105–109
Pareto distribution
characteristic function 51
median 37–38
mgf 49
MLE 360–361
pdf 37, 49, 51, 360
quantiles 38
relation to Exponential 121
relation to Power Function 147
relation to Riemann Zeta 107
Pascal distribution 10, 99
pdf (probability density function)
definition 31, 187
see also Distributions
see also pmf (for discrete rv's)

peakedness 40–41, 108–109
Pearson family 149–163
    animated tour 150
    Pearson coefficients in terms of moments
        159–160
    Types and chart 150
        - *Type I*, 17, 156, 158, 185
        - *Type II*, 158
        - *Type III*, 154, 157, 185
        - *Type IV*, 151–153, 157
        - *Type V*, 158, 185
        - *Type VI*, 158
        - *Type VII*, 157
    unimodal 179
    using a cubic polynomial 161–163
penalty function 400, 407, 415
pgf (probability generating function)
    definitions 60, 84, 203
    deriving probabilities from pgf 85, 85–86,
        86, 104, 245
    of bivariate Poisson 244–245
    of Hypergeometric 100
    of Negative Binomial 99
    of Pascal 11
    of Zero-Inflated Poisson 104
physics 32, 94–95
piecewise distributions
    Bates's distribution 289–290
    Inverse Triangular 13
    Laplace 23, 315–317
    order statistics of 23
    Reflected Gamma 33
plotting techniques (some examples)
    arrows 37, 81, 280
    contour plots 188, 218, 227
    data
        - bivariate / trivariate 233–235
        - grouped data 18, 155
        - raw 151
            see also frequency polygon
        - scatter plot 397
        - time-series 384
        see also empirical pdf / pmf
    domain of support (bivariate) 125, 138, 140
    empirical pdf 73, 77, 154, 381, 383
    empirical pmf 16, 110, 111, 112
    filled plot 44, 68
    frequency polygon 73, 77, 151, 154, 380
    graphics array 32, 38, 68, 109, 118, 124, 168,
        174, 218
    histogram 18, 155
    Johnson system 170
    non-parametric kernel density 19, 182–183
    parametric plot 167
    pdf plots 6, 139, *etc.*
        - as parameters change 8, 14, 32, 145, 165,
            225, 313, 315
        - 3D 11, 188, 198, 213, 214, 217, 316
    Pearson system 17, 152
    pmf plots 10, 83, 98, 101, 103
        - as parameters change 87, 92, 96
        - 3D 190
    QQ plots 291
    scatter plot 397
    superimposing plots 34, 35, 37, 42, 54, 55,
        69, 91, 133, 219, 302, 306
    text labels 32, 37, 54, 145, 302, 306, 313
    wireframe 228
    see also animations
pmf (probability mass function)
    definitions 82, 189
    see also Distributions – Discrete
    see also pdf (for continuous rv's)
Poisson distribution 95–98
    as limit distribution of Binomial 95, 280, 300
    bivariate Poisson 243–248
    censoring 327–328
    Cramér–Rao lower bound 334
    cumulant generating function 96
    distribution of sample sum 137
    kurtosis 446
    odd-valued Poisson 97–98
    pmf 16, 95, 110, 334
    Poisson two-component-mix 102–103, 406
    pseudo-random number generation 16, 110
    sufficient statistic for λ 340
    zero-inflated Poisson 104
poker 101
Pólya–Aeppli distribution 105
polyache 255–256
polykay 257–259
Power Function distribution
    as a Beta rv 185, 363
    as defining Pearson *Type I(J)* 185
    MLE 362–363
    relation to Pareto 147
    sufficient statistic 363–364
power sum 252, 272–276
probability
    conditional 65, 97
    multivariate 191–194
    orthant probability 231
    probability content of a region 192–193,
        230–231
    throwing a die 84–87
    see also cdf
probability density function (*see* pdf)
probability generating function (*see* pgf)
probability mass function (*see* pmf)
probit model 412–413
product moment 200, 266
products / ratios of random variables 133–136
    see also:
        - deriving the pdf of the bivariate *t*
            237–238
        - product of two Uniforms 126–127
Proportional-hazards model 412
Proportional-odds model 412

pseudo-random number generation
    methods
        - inverse method (numerical) 75–77, 109–115
        - inverse method (symbolic) 74–75
        - *Mathematica*'s Statistics package 72–73
        - rejection method 77–79
    and censoring 114
    computational efficiency 113, 115
    List Form 111
    of Birnbaum–Saunders 78
    of Gamma 73
    of half-Halo 75–77
    of Holla 112
    of Levy 74
    of multivariate Normal 232–234
    of Normal 291–292, 418
    of Poisson 16, 110
    of Riemann Zeta 113
    visualising random data in 2D, 3D 233–235

Q

QQ plot 291
quantiles 37
    of Birnbaum–Saunders 38–39
    of bivariate Normal 218–219
    of bivariate Student's *t* 237
    of Pareto 38
    of trivariate Normal 227–228

R

random number (*see* pseudo-random number)
random variable
    continuous 31, 81, 187
    discrete 81–82, 189
    see also Distributions
Random Walk distribution 147
random walk with drift 355, 384–386
Rao–Blackwell Theorem 342
raw moment 45, 200
Rayleigh distribution
    MLE 21
    relation to Exponential 122
rectangular domain 124, 190
reference computer 30
Reflected Gamma distribution 33–34
registration 3
regression 384–392
regression function 197, 221–222
regularity conditions
    for Fisher Information 329–330
    for MLE
        - basic 367–369
        - more than one parameter 371–372
        - non-iid samples 374–375

relative mean deviation 299
re-parameterisation 369, 388–389, 401, 406, 410, 414
Riemann Zeta distribution
    area of application 107
    Fisher Information 329
    pmf 113, 329
    pseudo-random number generation 113
risk 301–305
Robin Hood 222–224

S

sample information 332, 338, 376
sample maximum 311, 312, 320–321, 377
sample mean
    as consistent estimator (Khinchine) 298
    as consistent estimator (Normal) 294–295
    as MLE (for Exponential parameter) 358
    as MLE (for Normal parameter) 359–360
    asymptotic distribution of sample mean 287
    definition 277
    distribution of sample mean
        - for Cauchy 143
        - for Normal 143
        - for Uniform 139, 288–292
    Khinchine's Theorem 298
    limit distribution of sample mean (Normal) 279
    mgf of 141
    variance of the sample mean 264
    vs sample median, for Bernoulli trials 309–310
    vs sample median, for Logistic trials 318–320
sample median
    conditional expectation of sample median, given sample mean 342–343
    vs sample mean, for Bernoulli trials 309–310
    vs sample mean, for Logistic trials 318–320
sample minimum 311, 312
sample moment 251
    sample central moment 251, 360
        - covariance between sample central moments 266
        - in terms of power sums 252
        - variance of 264
    sample raw moment 251
        - as unbiased estimators of population raw moments 253
        - in terms of power sums 252
sample range 320–321
sample sum
    asymptotic distribution of sample sum 287
    definition 277
    distribution of sample sum
        - for Bernoulli 141
        - for Chi-squared 142

sample sum (*cont.*)
    distribution of sample sum (*cont.*)
        - for Exponential 141–142
        - for Poisson 137
        - for Uniform 55, 139
    mgf of sample sum 141
    moments of sample sum 261–271, 276
sample sum of squares
    distribution of (Normal) 144
    mgf of 141
sampling with or without replacement 100
scedastic function 197
score 357–361
second-order condition 22, 36–37, 357–360
security (stock) price 70–72, 108–109, 384
Sheather–Jones optimal bandwidth 19, 182
signal-to-noise ratio 299
Silverman optimal bandwidth 182
simulation 87–89, 126–127, 298–299
    see also Monte Carlo
    see also pseudo-random number
Sinc^2 distribution 35–36
skewness
    definition 40
    of Weibull 42
    Pearson family 149–150
Skorohod's Theorem 456
small sample accuracy 289–292
smoothing methods 181–183
spherical distributions 234, 451
stable distributions 56–61
standard deviation 40, 45
standard error 395, 399
standardised random variable 40, 120, 281, 287
statistic 251
stopped-sum distribution 108
Student's t distribution
    as member of Pearson family 157
    as Normal–InverseGamma mixture 105
    bivariate Student's t 237–238
    derivation, pdf 134
sufficient statistic 337–341, 344, 362, 363–364
sums of random variables 136–147
    deriving pmf of bivariate Poisson 244–245
    sum of Bernoulli rv's 141
    sum of Chi-squared rv's 142
    sum of Exponentials 136, 141–142
    sum of Poisson rv's 137
    sum of Uniform rv's 54–55, 138–139
    see also sample sum
Swiss bank notes 19, 185
symmetric function 253, 272–276
systems of distributions (*see* families) 149–180

T

t distribution (*see* Student's t)
t-statistic 395, 399

theorems
    Berry–Esseen 453
    Central Limit Theorem 286–292
    Continuous Mapping Theorem 366, 456
    Inversion Theorem 53
    Khinchine 298
    Lehmann–Scheffé 346
    Lindeberg–Feller 453
    Lindeberg–Lévy 287
    MGF Theorem 52, 141
    Rao–Blackwell Theorem 342
    Skorohod's Theorem 456
    transformation theorems
        - univariate 118
        - multivariate 123
        - not one-to-one 127
    Uniqueness Theorem 52
timings 30
transformations 117–148
    MGF Method 52–56, 130–132, 141–147
    transformation method 118–130
        - univariate 118
        - multivariate 123
        - manual 130
        - Jacobian 118, 123, 130, 223
        - one-to-one transformation 118
        - not one-to-one 127
    Helmert transformation 145
    non-rectangular domain 124, 125
    transformation to polar co-ordinates 222–223
    see also:
        - products / ratios of random variables
        - sums of random variables
Triangular distribution
    as sum of two Uniform rv's 55, 138–139
Trinomial distribution 239
trivariate Normal 226–228
    cdf 229–231
    orthant probability 231
    pseudo-random number generation 232–234
    visualising random data 235
truncated distribution 65–67
    truncated (above) standard Normal 65–66, 67
    truncated bivariate Normal 224–226

U

unbiased estimators of parameters 325–347
    asymptotic unbiasedness 366
unbiased estimators of population moments 251–261
    introduction 20
    multivariate 259–261
    of central moments 253–254, 259–261
    of cumulants 256–258, 260
    of Normal population variance 307–308
    of population variance 253, 254
    of raw moments 253

Uniform distribution
    bivariate Uniform (à la Morgenstern) 212–213
    Fisher Information 330
    mgf 48
    MLE 377
    order statistics 312
    other transformations of a Uniform rv 122
    pdf 48, 122, 312, 320, 330
    product of two Uniform rv's 126–127
    relation to Bates 139, 289–290
    relation to Cauchy 119
    relation to Exponential 121
    relation to Irwin–Hall 55, 139
    relation to Logistic 147
    sample mean and Central Limit Theorem 288–292
    sample range vs sample maximum 320–321
    sum of Uniform rv's 54–55, 138–139
unimodal 36, 179, 182–183
Uniqueness Theorem 52

V

van Beek bound 283–285, 453
variance
    definition 40, 45
    of sample mean 264
    of 2nd sample central moment 264
variance-covariance matrix
    asymptotic variance-covariance matrix 395–399, 404, 407, 410, 415, 418–419
    definition 201

variance-covariance matrix (*cont.*)
    of bivariate Exponential
        - Gumbel Model I, 205
        - Gumbel Model II, 12
    of bivariate Normal 220
    of bivariate Normal–Uniform 215
    of bivariate Uniform 213
    of trivariate models 202, 211
    of truncated bivariate Normal 226
    of unbiased estimators 333–335

W

Waiting-time Negative Binomial distribution 99
Waring distribution 418
weak law of large numbers 296–298
Weibull distribution 42

X

xenium (*see* book cover)

Y

Yule distribution 107

Z

zero-inflated distributions 103–104
Zipf distribution (*see* Riemann Zeta) 107

Springer Texts in Statistics *(continued from page ii)*

Nguyen and Rogers: Fundamentals of Mathematical Statistics: Volume II: Statistical Inference
Noether: Introduction to Statistics: The Nonparametric Way
Nolan and Speed: Stat Labs: Mathematical Statistics Through Applications
Peters: Counting for Something: Statistical Principles and Personalities
Pfeiffer: Probability for Applications
Pitman: Probability
Rawlings, Pantula and Dickey: Applied Regression Analysis
Robert: The Bayesian Choice: From Decision-Theoretic Foundations to Computational Implementation, Second Edition
Robert and Casella: Monte Carlo Statistical Methods
Rose and Smith: Mathematical Statistics with *Mathematica*
Santner and Duffy: The Statistical Analysis of Discrete Data
Saville and Wood: Statistical Methods: The Geometric Approach
Sen and Srivastava: Regression Analysis: Theory, Methods, and Applications
Shao: Mathematical Statistics
Shorack: Probability for Statisticians
Shumway and Stoffer: Time Series Analysis and Its Applications
Terrell: Mathematical Statistics: A Unified Introduction
Whittle: Probability via Expectation, Fourth Edition
Zacks: Introduction to Reliability Analysis: Probability Models and Statistical Methods

MATHEMATICA TRIAL VERSION

To activate your Trial Version of *Mathematica* 4, go to register.wolfram.com and enter the License Number below.

License Number: L2986-3155

Version 4.1.0   SD30

Install Today!

CD may not function if installed more than 2 weeks after you receive it. Valid for 30 days after installation.

Use of this Trial Version is subject to Wolfram Research license agreement at register.wolfram.com/trial-license.html.

Wolfram Research, Inc., reserves the right not to activate this Trial Version.